WITHDRAWN

BIOCHEMISTRY AND MOLECULAR BIOLOGY OF FISHES, 1

PHYLOGENETIC AND BIOCHEMICAL PERSPECTIVES

Biochemistry and molecular biology of fishes

Volume 1

Series Editors

P.W. Hochachka

Department of Zoology
Faculty of Science
University of British Columbia
Vancouver, B.C.
Canada V6T 2A9

and

T.P. Mommsen

Department of Biochemistry and Microbiology
University of Victoria
Victoria, B.C.
Canada V8W 2Y2

Volumes in the Series

Phylogenetic and biochemical perspectives

(ISBN 0-444-891242)

ELSEVIER
AMSTERDAM - NEW YORK - OXFORD - TOKYO

Biochemistry and molecular biology of fishes, 1

Phylogenetic and biochemical perspectives

Edited by

P.W. Hochachka

Department of Zoology, Faculty of Science, University of British Columbia,
Vancouver, B.C., Canada V6T 2A9

and

T.P. Mommsen

Department of Biochemistry and Microbiology, University of Victoria, Victoria,
B.C., Canada V8W 2Y2

ELSEVIER
AMSTERDAM - NEW YORK - OXFORD - TOKYO

© 1991, ELSEVIER SCIENCE PUBLISHERS B.V. All rights reserved

All rights reserved. No part of this publication may be reproduced, stored in a retrieval system, or transmitted in any form or by any means, electronic, mechanical, photocopying, recording or otherwise, without the prior written permission of the Publisher, Elsevier Science Publishers B.V., P.O. Box 1527, 1000 BM Amsterdam, The Netherlands.

No responsibility is assumed by the Publisher for any injury and/or damage to persons or property as a matter of products liability, negligence or otherwise, or from any use or operation of any methods, products, instructions or ideas contained in the material herein. Because of the rapid advances int he medical sciences, the Publisher recomments that independent verification of diagnoses and drug dosages should be made.

Special regulations for readers in the USA. This publication has been registered with the Copyright Clearance Center, Inc. (CCC), Salem, Massachusetts. Information can be obtained from the CCC about conditions under which the photocopying of parts of this publication may be made in the USA. All other copyright questions, including photocopying outside of the USA, should be referred to the Publisher.

ISBN 0-444-891242 (volume)
ISSN 0-444-891854 (series)

This book is printed on acid-free paper.

Sole distributors for the USA and Canada;
ELSEVIER SCIENCE PUBLISHING COMPANY, INC.
655 AVENUE OF THE AMERICAS
NEW YORK, NY 10010
U.S.A.

Published by:
ELSEVIER SCIENCE PUBLISHERS B.V.
P.O. BOX 1527
1000 AE AMSTERDAM
THE NETHERLANDS

Library of Congress Cataloging-in-Publication Data

Phylogenetic and biochemical perspectives/edited by
  P.W. Hochachka and T.P. Mommsen.
    p. cm. -- (Biochemistry and molecular biology of fishes; 1)
  Includes bibliographical references and index.
  ISBN 0-444-89124-2 (U.S.: acid-free paper). -- ISBN 0-444-89185-4 (series)
  1. Biochemistry. 2. Molecular biology. 3. Fish--Physiology.
  I. Hochachka, Peter W. II. Mommsen, T.P. (Thomas P.) III. Series.
QP514.2.P479 1991
597'.019'2--dc20   91-16080
                                                                                  CIP

PRINTED IN THE NETHERLANDS

Preface

The idea of editing a series of volumes on The Biochemistry and Molecular Biology of Fishes was born out of the present day lack of a forum for state-of-the-art review articles in this rapidly expanding field of research. On the one hand, researchers and students in this area always find themselves combing the literature on general (rat-dominated) biochemistry before discovering short and usually incomplete and disappointing coverage of the situation in the piscine setting. On the other hand, the rapidly expanding volume and quality of the primary literature in fish biochemistry and molecular biology supply convincing evidence for a maturing field. This discipline is no longer the younger sibling of rat or human biochemistry but has recently led to a number of major conceptual breakthroughs; for this reason, and because its activity domain is sometimes nonoverlapping with 'mainstream' biochemistry, the field is certainly ripe and ready for a review series of its own.

Comparative biochemistry and molecular biology and comparative physiology as disciplines by definition use organisms as a special kind of experimental parameter for probing general mechanisms and principles of function. In theory this approach is relatively blind to phylogenetic boundaries, but in practise the realities of funding and availability of experimental material greatly narrow the field of play. As a result, two phylogenetic groups — the insects and the fishes — have over the last several decades provided the bulk of the experimental data base in these disciplines. Interestingly, although comparative biochemistry in many ways grew out of comparative physiology, the growth and development of these two activities in the insect field have to major extent proceeded along independent paths. By contrast, the comparative physiology and biochemistry of fishes have not been so independent of one another and the tendency has been for the former to envelope the latter. We believe that the current conceptual developments in the fields as well as the simple logistics of dealing with massive data bases make this the right time for the reality of independence to match the perception of independence, which we feel is another important rationale for this review series.

Our goal is to provide researchers and students with a pertinent information source from theoretical and experimental angles. To be useful to students, theoreticians, and experimentalists alike, contributing authors are urged to emphasize concepts as well as to relate experimental results to the biology of the animals, to point our controversial issues, and to delineate as much as is possible directions for future research.

Peter W. Hochachka
Thomas P. Mommsen
Vancouver and Victoria, B.C.
December, 1990

Contributors

Fred W. Allendorf, *Division of Life Sciences, University of Montana, Missoula, MT 59812, U.S.A. (Chapter 2)*

John D. Altringham, *Department of Pure and Applied Biology, The University, Leeds LS2 9JT, W. Yorkshire, U.K. (Chapter 10)*

Paul M. Anderson, *Department of Biochemistry and Molecular Biology, University of Minnesota-Duluth, Duluth, MN 55812-2487, U.S.A. (Chapter 3)*

Barbara A. Block, *Department of Organismal Biology and the Committee on Evolutionary Biology, The University of Chicago, 1025 East 57th Street, Chicago, IL 60637, U.S.A. (Chapter 11)*

James W. Campbell, *Department of Biochemistry and Cell Biology, Rice University, Houston, TX 77251, U.S.A. (Chapter 3)*

J.S. Clegg, *Bodega Marine Laboratory, University of California, Davis, P.O. Box 247, Bodega Bay, CA 94923, U.S.A. (Chapter 1)*

W. Drost-Hansen, *Department of Chemistry, University of Miami, Coral Gables, FL 33124, U.S.A. (Chapter 1)*

Moira M. Ferguson, *Department of Zoology, University of Guelph, Guelph, Ont., Canada N1G 2W1 (Chapter 2)*

Raymond P. Henry, *Department of Zoology and Wildlife Science, 101 Cary Hall, Auburn University, Auburn, AL 36849-4201, U.S.A. (Chapter 8)*

Peter W. Hochachka, *Department of Zoology, University of British Columbia, 6270 University Boulevard, Vancouver, B.C., Canada V6T 2A9 (Chapter 12)*

Ian A. Johnston, *Gatty Marine Laboratory, Department of Biology & Preclinical Medicine, University of St. Andrews, St. Andrews, Fife KY16 8LB, Scotland, U.K. (Chapter 10)*

Margaret J. McFall-Ngai, *Department of Biological Sciences, University of Southern California, Los Angeles, CA 90089-0371, U.S.A. (Chapter 4)*

Thomas P. Mommsen, *Department of Biochemistry and Microbiology, University of Victoria, P.O. Box 3055, Victoria, B.C., Canada V8W 3P6 (Chapter 6)*

Charles F. Phleger, *Department of Natural Science, San Diego State University, San Diego, CA 92182, U.S.A. (Chapter 9)*

Joseph F. Siebenaller, *Department of Zoology and Physiology, Louisiana State University, Baton Rouge, LA 70803–1725, U.S.A. (Chapter 13)*

Peter W. Sorensen, *Department of Fisheries and Wildlife, 200 Hodson Hall, 1980 Folwell Avenue, University of Minnesota, St. Paul, MN 55108, U.S.A. (Chapter 5)*

Norm E. Stacey, *Department of Zoology, University of Alberta, Edmonton, Alta., Canada T6G 2E1 (Chapter 5)*

Wesley Toller, *Department of Biological Sciences, University of Southern California, Los Angeles, CA 90089–0371 (Chapter 4)*

Patrick J. Walsh, *Division of Marine Biology and Fisheries, Rosenstiel School of Marine and Atmospheric Science, University of Miami, 4600 Rickenbacker Causeway, Miami, FL 33149–1098, U.S.A. (Chapters 6 and 8)*

Stephen H. Wright, *Department of Physiology, University of Arizona, Tucson, AZ 85724, U.S.A. (Chapter 7)*

Contents

Contributors
Preface   v
1. On the biochemistry and cell physiology of water
   J.S. Clegg and W. Drost-Hansen   1
2. Evolution of the fish genome
   M.M. Ferguson and F.W. Allendorf   25
3. Evolution of mitochondrial enzyme systems in fish: the mitochondrial synthesis of glutamine and citrulline
   J.W. Campbell and P.M. Anderson   43
4. Frontiers in the study of the biochemistry and molecular biology of vision and luminescence in fishes
   M.J. McFall-Ngai and W. Toller   77
5. Function and evolution of fish hormonal pheromones
   N.E. Stacey and P.W. Sorensen   109
6. Urea synthesis in fishes: evolutionary and biochemical perspectives
   T.P. Mommsen and P.J. Walsh   137
7. The interface of animal and aqueous environment: strategies and constraints on the maintenance of solute balance.
   S.H. Wright   165
8. Carbon dioxide and ammonia metabolism and exchange
   P.J. Walsh and R.P. Henry   181
9. Biochemical aspects of buoyancy in fishes
   C.F. Phleger   209
10. Movement in water: constraints and adaptations
    I.A. Johnston and J.D. Altringham   249
11. Endothermy in fish: thermogenesis, ecology and evolution
    B.A. Block   269
12. Temperature: the ectothermy option
    P.W. Hochachka   313
13. Pressure as an environmental variable: magnitude and mechanisms of perturbation
    J.F. Siebenaller   323
Species index   345
Subject index   349

Hochachka and Mommsen (eds.). *Biochemistry and molecular biology of fishes, vol. 1*
© 1991 Elsevier Science Publishers B.V. (Academic Publishing Division)

CHAPTER 1

On the biochemistry and cell physiology of water

J.S. CLEGG * AND W. DROST-HANSEN **

*Bodega Marine Laboratory, University of California, Davis, Bodega Bay, CA 94923, U.S.A., and **Department of Chemistry, University of Miami, Coral Gables, FL 33124, U.S.A.*

Abstract
I. Introduction
II. Some general comments on pure water
III. Intracellular architecture of eukaryotes
IV. Water near interfaces
   1. Vicinal water
   2. The NMR titration model
   3. The work of Hazlewood and Rorschach
   4. The association–induction hypothesis (AIH)
   5. The reference phase technique
   6. 'Pore water'
   7. The cluster model
   8. The properties of water in *Artemia* Cells
V. Intracellular microviscosity
VI. Cell volume, ion binding and cell water
VII. Some additional consequences of the properties of cell water
VIII. Concluding comments
Acknowledgements
IX. References

Abstract: This chapter focuses on the most abundant molecule in all living systems. Without it enzymes do not function normally, DNA collapses into a tangled mess, and no dielectric continuum (solvent) exists, within which so many physiological processes take place. After brief discussion of pure water we focus here upon the cellular and molecular aspects; however it is quite evident that water participates at all levels of biological organization, from molecular to biosphere. We concentrate on the *aqueous phase* properties of cells and consider only briefly the importance of the primary hydration of intracellular solutes, a well studied and non-controversial issue. Most thought about cell biology has been built on the assumption that the structure and properties of intracellular water are not significantly different from those of pure water (or that in ordinary dilute aqueous solutions). That widely held assumption appears to be questionable in view of a large body of evidence to the contrary. Attention is paid to the large amount of data showing that the water adjacent to surfaces exhibits interesting and unusual physical properties compared to the ordinary bulk liquid. We make the case that the internal environment of cells is characterized by an enormous surface to volume ratio and argue that, on this basis alone, we should expect the water in cells to deviate in its properties. A number of examples are summarized which indicate that this expectation has been documented experimentally, but has thus far not been accommodated within popular prevailing paradigms. We consider some examples of the importance of these interesting properties of intracellular water to cell structure and function.

I. Introduction

Most active animal cells contain $75 \pm 10\%$ water by weight and volume — a well known fact. Perhaps more revealing is a consideration of mole ratios: for every 25,000 water molecules there are only about 100 inorganic ions and even fewer low molecular weight organics (intermediary metabolites, nucleotides, etc.) about 75 lipids and phospholipid molecules, and only one or two protein molecules (we can neglect nucleic acids in this exercise). Of course, the quintessential feature of intracellular water is that it is *the* primary solvent — the 'mother liquor'[100] — within which so much of the metabolic repertoire of cells occurs. At the same time, water is a major substrate for a variety of metabolic reactions (as well as a product) and the vital role played by water in macromolecular conformation is a well established fact. Indeed, the participation of water in cells is so pervasive that an understanding of virtually every cellular structure and its function would seem to require a detailed description of the participation of water. Be that as it may, relatively little attention is paid to the properties of water in cells. As Albert Szent-Györgyi[100] said: 'Biology has forgotten water, or never discovered it.' He also wrote that water is the last thing a deep sea fish would ever discover, a saying in keeping with this series.

As knowledge about the ultrastructure of animal cells has advanced, it has become increasingly evident that the aqueous phases within cells exist in the presence of an exceptionally high surface to volume ratio. Indeed, we will document later how extensive is this intracellular macromolecular surface area, and will advance the proposition that, as a result, the physical properties of at least a lot of the water in cells are altered with respect to those of ordinary aqueous solutions. Further, we will consider the consequences this may have for several aspects of cell physiology. In so doing, we are mindful of the fact that it is generally assumed that intracellular water is not different from pure water and that most consensus views of cell physiology are built on this assumption. Questioning widely held traditional belief might provoke in some a less than enthusiastic response; however, as Rudolf Arnheim[4] noted: '... if you try to make sure not to step on anybody's toes, you will have no space left to walk.' We begin with some general comments on pure water.

II. Some general comments on pure water

It is commonly emphasized that water is a most unusual liquid; moreover, it is also the only inorganic liquid to occur naturally on earth. Approximately one gram mole of grams of water exist on earth, in rough numbers, 10^{47} water molecules, and it is the only chemical compound on earth to occur naturally in all of its three phases: solid, liquid and vapor. More importantly, the properties of water differ, sometimes dramatically, from the properties expected for a compound of such low molecular weight: its melting and boiling points are high as is the heat of vaporization, and its heat capacity and dielectric permittivity are remarkably high.

Many other properties differ from those one would expect based on a comparison with other low molecular weight substances. The density of solid water is less than that of the liquid (at the melting point) and liquid water possesses a maximum in density near 4°C.

The explanation for the unusual properties of water is found in the ability of the water molecule to form hydrogen bonds (H-bonds). These bonds are tetrahedrally disposed around the molecule, leading to a very 'open' structure. In ordinary ice the number of nearest neighbors (n_x) is 4.00 and in liquid water this 'openness' is essentially retained with values for n_x around 4.2 to 4.4 in the range of temperatures and pressures of physiological interest.

Not withstanding nearly a century of study on the structure of water, a final, 'einwandfrei' description of the molecular architecture of water still escapes us. A monograph by Eisenberg and Kauzmann[44] provides a highly readable introduction to the structure and properties of water. A monumental summary of the state of water (and solutions) research during the late seventies and early eighties is available in the form of a seven volume exposition entitled 'Water — a Comprehensive Treatise,' edited by Felix Franks[52]. A brief but excellent introduction to research on water has been provided by Stillinger[99], a leading researcher on water, and more recent information is available in the volume edited by Packer[87].

Most current theories of pure water depict the structure as a disordered array of H-bonded water molecules such that a continuous 'lattice' of chains of water molecules can be found throughout the liquid. In all probability only a few water molecules are 'free' in the sense of not being hydrogen bonded to at least one other molecule, while only a small fraction of water molecules have all four possible bonds intact at any moment. Even in the momentary absence of an H-bond between a water molecule and its nearest neighbors, strong interactions are still operating (dipole-dipole, quadruple, and higher interactions).

It has proven difficult experimentally to elucidate liquid structure in general and the structure of water in particular. The only 'direct' methods of study are by scattering techniques, such as X-ray diffraction or neutron scattering. Such experiments at least provide a 'radial distribution function' (rdf) of the relative distances from a critical molecule to its various neighbors; however, to complete the structural analysis a model must be formulated and its 'goodness of fit' checked by the extent of agreement between the rdf and the calculated molecular positions based on the model. All other experimental structure studies are indirect, usually employing some sort of spectroscopy (IR, Raman, NMR, dielectric, etc.). None of these approaches delineate simultaneously the location of the molecules (coordinates) and their dynamics.

There exist essentially two classes of models for liquid water: mixture models and continuum models. The former category traces its origin to the end of the last century when several investigators proposed that liquid water might be a mixture of relatively bulky structured entities similar to ice in a medium of essentially monomeric, closely packed (i.e., 'dense') water molecules. In the continuum models H-bonds are not broken but merely allowed to bend and/or stretch in response to random thermal motion.

Kauzmann has long been an advocate of the continuum models for liquid water and has recently published another, rather detailed model[64]. This random network model is interesting and important in that it provides an equation of state of liquid water and successfully accounts for a variety of abnormal properties of water based primarily on (a) spectroscopic data, (b) data on the properties of ordinary, crystalline ice $I(h)$, and (c) information about the water molecule in the vapor phase. The theory predicts correctly such features of liquid water as the negative expansion coefficient at low temperature but the point of change-over from positive to negative is in error by about 50°C. Qualitatively the new theory predicts correctly a minimum in the isothermal compressibility but the value for the temperature of the minimum is off by 100°C while the absolute value for the compressibility coefficient is off by a factor of 3. Henn and Kauzmann's theory predicts the entropy of liquid water (at 0°C) to within about 1%, but the agreement between the calculated and observed specific heat is not nearly as satisfactory and the theory fails to predict the (very shallow) minimum near 30°C. Notwithstanding the less than good agreement between observed and predicted values for some of the properties discussed, the model is impressive because of its conceptual simplicity. The model makes reference directly to the liquid state only through the use of spectroscopic data and otherwise is based on data for the solid state and from the single water molecules in the vapor. Thus the success of the model lends credence to 'continuum' models of water.

Over the past 15 years, a large number of papers have appeared dealing with computer simulation of water structure. Simulations are based on accepting a reasonable expression for the pair-wise interaction of water molecules, namely the pair potential (energy) function, $u(r)$. Two types of approaches are possible: a Monte Carlo technique and the molecular dynamics method. In the first approach the energy change is calculated which results from a small displacement of the molecules (from the positions chosen for the 'first instance') from the sum of $u(r)$ over all the pairs of molecules. Only those displacements which lead to lower total energy are accepted. Ultimately, the positions of maximum likelihood are obtained and expressed in terms of the rdf which in turn may be compared to the rdf obtained from scattering experiments. In the molecular dynamics method the molecules are supplied with energy and the resulting motions monitored. At fixed time intervals (say 10^{-12} s) the coordinates of the molecules are obtained and the path to equilibrium may thus be determined. As a result, the calculations provide both equilibrium thermodynamic properties as well as time-dependent aspects (for instance, the self-diffusion coefficient).

Much has been learned from simulation studies and the results are surely valuable hints as to what the structure of liquid water may be. Because of computational limitations, the typical sample of water molecules used in these analyses is about 500. If these molecules formed a small droplet the radius would be 5 water molecules and half of these would be from the 'outside layer.' Thus one must expect some dramatic surface effects (even if periodic or cyclic boundary conditions are employed in order to minimize this effect). Furthermore, as discussed below, many of these calculations are based on a pair-wise potential energy function, probably a serious limitation.

A particularly intriguing model of water has been proposed by Stanley and Teixeira[9,56,97,98] especially aimed at elucidating water structure at low temperature. The model is based on the idea of percolation theory, i.e., interconnectivity on a 'lattice.' The structure is that of an 'infinite' H-bonded network, continuously being restructured. At any moment some bonds are strained or broken; within the network are 'patches' of lower (local) density and lower entropy than the overall (global) values. The spatial positions of the various types of 'connectedness' are not randomly distributed but are correlated. In particular the structure contains tiny 'patches' of four-bonded molecules. Among other things, the model allows calculations of the isothermal compressibility, specific heat and thermal expansion coefficient. The size of the patches increases with decreasing temperature. A novel feature is that rather than a lower cut-off value in energy for hydrogen bond interactions, any attraction is considered a state of H-bonding. An analysis of the resulting H-bond distributions suggests the lifetimes of the H-bonding fit a power law, implying that the H-bonds do not have a characteristic lifetime; as would be expected intuitively, the lifetimes increase rapidly as the temperature is lowered, particularly below 0°C. Regrettably the model is based on computer simulations involving only 216 molecules (which still requires a great deal of computing time) and no attempt has been made so far to simulate the influence of a confining wall, thus still leaving open the question of the predicted nature of interfacial water.

Many other impressive computer simulations have indeed been made in efforts to model the structure of liquid water. However, these calculations usually are based on pair-additivity of the potentials for the H-bonded water molecules, so the possibility exists that subtle effects may escape the theoretician as no means are provided to build-in the possibility of extensive cooperativity — an aspect which Henry Frank[51] has so eloquently stressed. Very likely, this is the crux of the problem of interfacially modified water which will be referred to here as 'vicinal water.' If nothing else, the thermal anomalies in the properties of vicinal water strongly implicate cooperativity on a large scale — a *collective* behavior of water molecules which no existing potential function is yet able to reproduce. The cooperativity reflects non-pair-additivity and it does not seem plausible that 'effective' potential energy functions can be devised which will remedy the specific lack of a detailed understanding of many-body interactions in water. Brave attempts to allow for cooperativity have been made by Finney, Barnes and coworkers[50].

A particularly readable account of computer ('machine') simulations of water is provided by Barnes[5]. This article is also one of the first to report on progress in the problem of computer simulations for water at interfaces — i.e., spatially confined water. The approach included allowance for cooperativity among the water molecules: 'The cooperative nature of hydrogen-bonding should not be visualized as a mere strengthening of the bond networks, but as part and parcel of their existence.'

It appears that we are still quite far from reaching a consensus as to the structure of pure water. In view of this it is hardly surprising that we know even less about the structure of water near interfaces, a question of direct importance when considering the properties (and structure) of intracellular water. It seems appropriate to first examine the conditions under which intracellular water exists; that is, we should ask what the intracellular environment consists of, and what the water 'sees'.

III. Intracellular architecture of eukaryotes

Everyone knows that cells are not 'bags of enzymes'; yet, there does not seem to be a consensus on the extent to which the interiors of cells are 'organized'. A variety of organelles (membrane bounded and otherwise) exist in eukaryotic cytoplasm, as do the classical cytoskeletal elements. However, the extent of interconnection existing between these and other cytoplasmic structures (including the plasma membrane) and the nucleus is a matter yet to be determined. The traditional (textbook) view is that these structures are mostly 'suspended' in a concentrated solution — 'the cytosol' — consisting of inorganic ions, metabolites, adenine and other nucleotides, and macromolecules, chiefly proteins of which many are enzymes of intermediary metabolism. In this view, most cytoplasmic water would find itself as part of this highly concentrated 'cytosol' (a very ambiguous term[20]) often referred to as a 20% protein solution. A similar picture could be drawn for the crowded aqueous phase of the nucleus (the 'nucleoplasm') and the interiors of membrane-bound cytoplasmic organelles[20].

An alternative to this description has been advanced in which the vast majority of the macromolecules are organized into complexes, and even small solutes such as certain inorganic ions, nucleotides and metabolites are also envisaged to participate in this organization. This paradigm is perhaps best visualized in the form of the microtrabecular lattice (MTL) proposed by Keith Porter on the basis of high voltage electron microscopy[90]. Fig. 1, donated by Porter, illustrates this image of cytoplasmic organization. It is worth noting that, while some believe the MTL to be an artifact of the preparative procedures of microscopy (see 10,74,93) evidence from *non-microscopic* studies on intact (or almost intact) cells, in general, supports the existence of the MTL as a real structure. Because this evidence has been the subject of extensive, and we believe compelling recent reviews[6,15,20,72,106], that effort will not be repeated here. Although many details are lacking, we find this 'organized description' to be most consistent with existing evidence. Thus, if this paradigm is accepted, then the great majority of the intracellular water finds itself in a *dilute* solution, containing relatively few macromolecules and other solutes. One might expect, therefore, that this water would behave like that in an ordinary bulk aqueous phase. However, as pointed out, the aqueous volume in between ultrastructurally recognized structures is in very close proximity to macromolecular surfaces (Fig. 1). Estimates of the surface area of the MTL have been carried out by image analysis of HVEM photographs[57]. This area is so vast, about 100,000 μm^2 for cultured mammalian cells, that at least 50% of the total water finds itself within 50 Å from some surface[20,22]. Indeed, estimates from fluorescence recovery after photobleaching measurements on the diffusion of fluorescently labeled probes (Ficoll and dextrans) suggest that the fractional surface area may be even greater than indicated from HVEM analysis[8,79]. Thus, the influence of surfaces on the properties of water near them becomes an issue of some importance.

IV. Water near interfaces

1. Vicinal water

Vicinal water is defined as water (or aqueous solution) the structure of which is modified by proximity to an interface but excluding chemically 'bound' water directly on the surface (the water of primary hydration). As for the depth of the structurally modified layer, Adamson[1] has eloquently argued "that a 'quiescent' liquid/air surface is actually in a state of violent agitation in the molecular scale with individual molecules passing rapidly back and forth between it and the bulk regions on either side." The depth of this disturbed layer is probably of the order of 50 Å (about 15 water layers.) While structurally modified, these surface layers are not considered 'vicinal' (in the sense to be delineated below). However, in the presence of a monolayer (or even with just partial coverage) deep surface structure modifications occur, possibly extending over distances of many hundred molecular layers (see refs. 3, 67, 68). While the pure air/water interface does not show any evidence of vicinal water structuring, surface tension measurements made by the capillary rise method have shown unexpected thermal anomalies, for instance near 30°C. [The first set of reliable data showing such an anomaly goes back to a German investigator named Brunner in a paper from 1847 (see refs. 33, 34). The anomaly appears as an inflection point in the surface tension versus temperature; because the surface tension is a free energy, the temperature derivative is an entropy of surface formation.] One of us has shown[33,34] that for water in narrow glass capillaries there is an increase in the entropy of about a factor of 2 over a temperature interval of only 3 or 4°C near 30°C. This effect is considered to be a manifestation of a relatively long-range vicinal restructuring of the water, originating from the glass surface and this vicinal water structure undergoes some type of transition, probably a higher order phase transition, around 29–32°C.

The proximity to a solid surface appears to induce long-range structuring of the proximate water and the structures thus induced undergo no less than four thermal transitions between the freezing and boiling point of water. The characteristic, critical temperatures occur near 14–16°C, 29–32°C, 44–46°C and 59–62°C (i.e., essentially equidistantly spaced at 15°C intervals).

It appears that vicinal water is induced by proximity to most (or all) 'solid interfaces' regardless of the detailed specific chemical nature of the surface. This puzzling result is referred to as the 'paradoxical effect' (see refs. 40, 41, 48). The phrase 'solid surfaces' is used in this context in a most general sense, from mica or quartz plates, mineral grains and membranes to large macromolecules (above a certain 'critical' molecular weight) and possibly some types of aggregates, such as glycogen particles and micelles. The paradoxical effect is an important consideration as it is experimentally far more difficult to study the intracellular aqueous environment of intact cells than to study physico-chemically relatively well-defined model-systems. If we are able to demonstrate that vicinal water occurs near *any* aqueous phase/solid interface, then we may conclude that vicinal water must also be induced by some (or all) of the 'surfaces' within cells, such as membranes and those of the cytomatrix (Fig. 1).

We are not concerned in this paper with the more traditional aspects of the hydration of the cell constituents. While this aspect is, of course, of immense direct

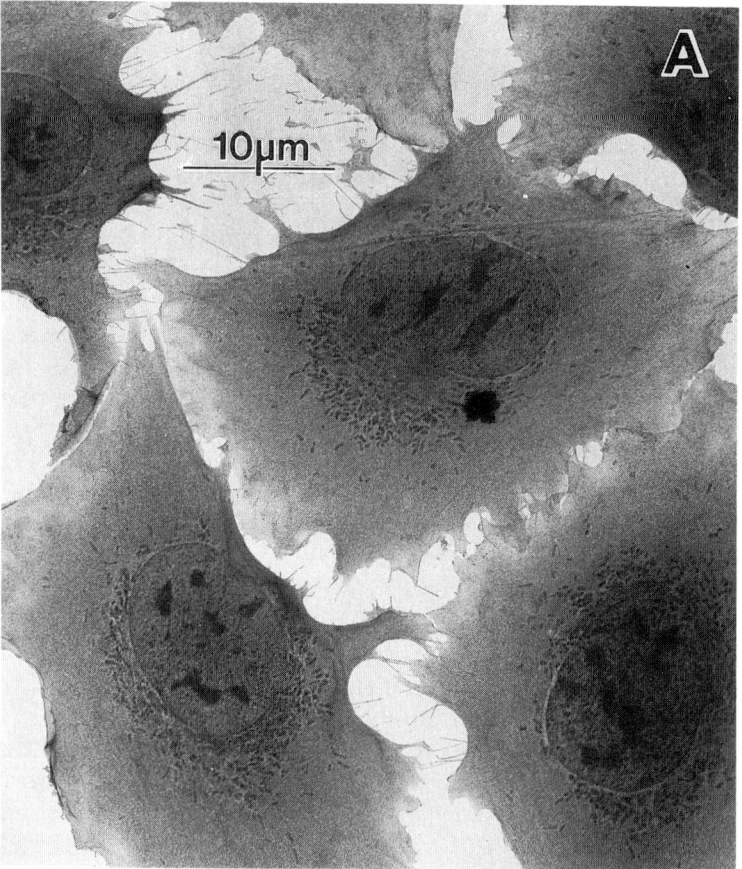

Fig. 1. High-voltage electron photomicrograph of cultured NRK cells (rat kidney) provided by K.R. Porter. A illustrates several individual cells, and B shows a small region of the cytoplasm magnified 90 times with respect to A. T (one of the numerous microtrabeculae), MT (microtubule), ER (endoplasmic reticulum), M (mitochondrion), P (polyribosome). The asterisks locate volumes between detectable cytoplasmic structure, which are presumably dilute aqueous solutions. Reference to details of preparative procedures can be found in the review by Porter[90]. The essential feature of this image of cells is that the cytoplasm is a 'unit structure', in which virtually all the formed elements are interconnected structurally and functionally.

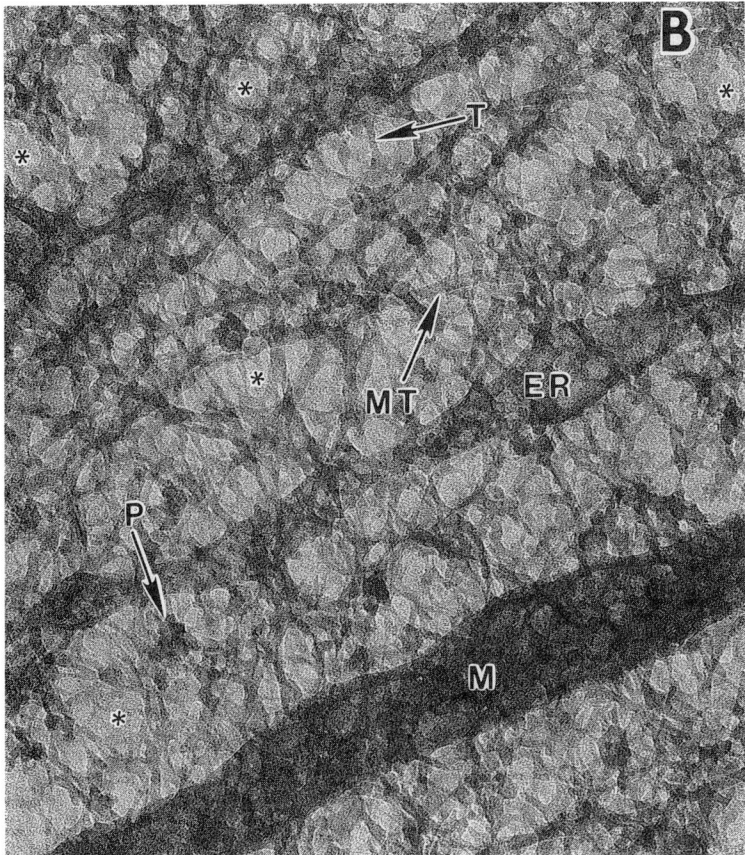

Fig. 1. (continued).

importance to macromolecular structure and function (see refs. 11, 54, 58, 86, 92) the main point of the present paper is that 'vicinal hydration' occurs at distances of at least 50 Å from the actual surface, and it is this aqueous phase upon which we focus our attention here.

Extensive discussions of the structure and properties of water at interfaces can be found in papers and reviews by various authors and the reader is referred to selected papers by Deryaguin and coworkers (see refs. 17, 27–30). Israelachvili et al.[69–71], Peshel and coworkers[89], Low[77] and Etzler[45], Etzler and Drost-Hansen[32–43;45–48]. For an objective and informative review of water at solid surfaces see also the classic paper by Clifford[18]. That paper examines in detail a vast literature on aqueous interfacial phenomena; a major conclusion is that surface induced changes in water structure may occur but that it is not likely that any such structures will extend more than 100Å from the surface. Clifford speculates that one possible mechanism for creation of interfacial water structures (different from the bulk structure) might be local geometric constraints preventing the full development of large structured elements of bulk water. A similar mechanism might explain the 'paradoxical effect': truly external geometric constraints might occur with any solid, non-smooth surface and thus account for vicinal water being induced by ionic, hydrophilic and hydrophobic surfaces, — i.e., regardless of the chemical nature of the surface. Given the enormous geometric complexity of a cell, the aqueous cytoplasm would be the obvious place to expect to find vicinal water.

For an excellent recent review of the types of forces which operate at interfaces in water, see the paper by Evans and Ninham[49]. These authors specifically delineate the role that physico-chemical insight and surface and colloid chemistry is likely to play in the description of biological systems. Their review progresses from general considerations of hydrophobic effects as the prime cause of aggregation in general to theories of micellar aggregation, in particular, and the question of vesicle stability. In their review Evans and Ninham provide a concise enumeration of the forces operating in colloidal systems; the authors note that classical theories were not able to deal successfully with some colloidal systems, such as clay swelling, interactions between air bubbles in salt solutions, or the characteristic spacing of lamellar lecithin multilayers. The difficulties are related to the relatively short range stabilizing forces, referred to as hydration forces or the structural component of the disjoining pressure. These authors also summarize the nature of hydration forces, hydrophobic forces and secondary hydration forces, and speculate on the likely role in cell biology of these forces. The monograph by Israelachvili[69] is highly recommended as an introduction to this topic.

Rand and Parsegian[91] have provided a thorough review of the hydration forces operating between phospholipid bilayers, much of this research coming from the laboratories of these authors. They point out the experimental and interpretive difficulties associated with such studies but, on balance, arrive at the conclusion that water is influenced over 10–30 Å from the bilayer surfaces.

Thus far we have concentrated on 'vicinal water', the particular view developed by one of us (W.D.-H.). As mentioned, the concept of vicinal water envisions modified water structures at interfaces, induced by mere proximity to the surface regardless of its specific chemical nature and extending over distances far larger than normally considered (say 30 to hundreds of molecular diameters). The most characteristic feature of this vicinal water is the occurrence of no less than four narrow temperature ranges in which the structural properties of the interfacial water changes more or less abruptly. The transitions near 30 and 45°C may notably affect the physiology of mammals, birds and many other species including many microorganisms[32,36]. The 15°C anomaly is also expected to affect the physiology of many marine organisms[36]. Some characteristic properties of vicinal water are listed in Table 1 (together with the corresponding bulk values). The table is taken from a recent paper[43], in which are listed the various publications from which the information has been compiled.

A number of other investigators have, over the years, presented their own descriptions of the nature of intracellular water and we take these up next. Although differing in detail, sometimes markedly, they all arrive at the conclusion

TABLE 1

A comparison of some properties of pure and vicinal water [a]

Property	Bulk	Vicinal
Density (g/cm^3)	1.00	0.97
Specific heat (cal/kg)	1.00	1.25 ± 0.05
Thermal expansion coefficient (°C^{-1})	$250 \cdot 10^{-6}$ (25°C)	$300-700 \cdot 10^{-6}$
(adiab.) Compressibility coefficient (Atm^{-1})	$45 \cdot 10^{-6}$	$60-100 \cdot 10^{-6}$
Excess sound absorption (α/v^2) cm$^{-1} \cdot$s^2	$7 \cdot 10^{-17}$	$\sim 35 \cdot 10^{-17}$
Heat conductivity (cal/sec)/cm^2/°C/cm)	0.0014	$\sim 0.01-0.05$
Viscosity (cP)	0.89	2–10
Energy of activation ionic conduction (kcal/mol)	~ 4	5–8
Dielectric relaxation frequency (Hz)	$19 \cdot 10^9$	$2 \cdot 10^9$

[a] See text for references to original citations.

that most (or all) of cell water differs importantly from that in dilute solution, and emphasize the considerable importance this could have to cell structure and function. Our coverage must be brief, so we refer the reader to books (6,15,85,87,106), all of which contain many additional references to the literature.

2. The NMR titration model

Ivan Cameron, Gary Fullerton and their colleagues in a series of studies applied nuclear magnetic resonance (NMR) spectroscopy to a variety of cells, tissues and macromolecules and have developed the idea of multiple water compartments in them (for recent coverage and reference to earlier work see refs. 14 and 110). Their approach involves the study of the motional behavior or water as a function of total water content of the system (achieved in cells by air-drying or by osmotic manipulation using impermeant solute addition, or medium dilution.) In some cases (*Xenopus* oocytes, mammalian lens) they conclude that *all* of the cell water is significantly perturbed (different) whereas in other systems (erythrocytes, sarcoma cells, sea urchin eggs) only about 50% seems to behave that way. Although one might question whether these severe manipulations of water content are excessively unphysiological and might produce 'artifacts', we believe they make a strong case for the existence of large fractions of cell water whose properties differ from those of dilute aqueous solutions. These differences are attributed to water-surface interactions within the cells under study, and the different hydration compartments are defined on the basis of distinct hydrogen bonding mechanisms.

3. The Work of Hazlewood and Rorschach

We should stress that the first work revealing altered motional properties of cell water by NMR was done concurrently on skeletal muscle by Cope[26] and Hazlewood et al.[61]. Hazlewood and his many colleagues, most notably Rorschach, have, since that time, produced a vast amount of evidence indicating that virtually none of the water in cells has motional properties similar to those of pure water. This group has studied a wide variety of cells and tissues, including malignant ones, and has utilized quasi-elastic neutron scattering (QNS) as well as NMR. Their publications are so numerous that only a few selected papers and reviews can be cited here[60-63,83,94,95]. Their work has also been covered in a number of recent books (see refs. 6, 85). This pioneering work has been challenged repeatedly but, in every case to our knowledge, the objections to their conclusions have been answered adequately and usually convincingly. It is our view that Hazlewood, Rorschach and coworkers have proven that the translational and rotational motions of the vast majority of animal cell water are indeed markedly slower than pure water, implying if not requiring that the structure is also different. Their work also stresses the enormous significance of these findings to cell biology, and laments the lack of attention paid to the subject.

4. The association–induction hypothesis

Hazlewood and colleagues have usually interpreted their results in the context of Gilbert Ling's association-induction hypothesis (AIH). First formulated in the late 1950s (see the massive book by Ling[76]) the major features of the AIH are as follows:

Virtually all of the water in cells is considered to exist as polarized multilayers arising from fixed charges on extended protein surfaces. The water multilayers exclude ions and other solutes to varying degree, and the contributions of membrane processes (active transport) are considered to be negligible. Ions are also associated with fixed charges on cellular macromolecules, notable proteins, the degree of binding for a given ion being influenced by a number of factors. 'Cardinal sites' exist on these particular proteins which, when filled with specific adsorbents such as ATP, initiate cooperative interactions within the protein–ion–water system. Hormones and regulatory cyclic nucleotides are included in the list of cardinal adsorbents. Binding at the cardinal site leads to cooperative alterations and the selective accumulation of K^+ over Na^+, and also generates the polarized multilayers of water; ATP splitting and the removal of ADP results in a movement of the system to a lower energy state in which the ion selectivity is lost, as is the polarization of water.

This hypothesis has been supported strongly by some, most notably by the extensive work of Hazlewood (as mentioned), and others (84,85). Nevertheless, the scientific community as a whole has been reluctant to accept the AIH. At least part of this reluctance seems to stem from less than compelling evidence for the actual existence of polarized multilayers of water in cells and for the extended proteins that generate them and selectively bind ions. While one might disagree with details of Ling's hypothesis it is not so easy to dismiss the objections he has raised against the validity of 'the traditional view.' Whether or not Ling is correct remains to be seen, in our opinion. Certainly, if he is correct, cell biology must undergo a revolution.

5. The reference phase technique

Horowitz, Paine and colleagues (see refs. 66, 88) have also provided experimental evidence that the traditional view of cellular water requires modification. They have developed an elegant method for examining the solvent properties of intracellular water *in situ* called the reference phase technique. A microdrop of gelatin sol (the reference phase) containing an appropriate radioactive solute is injected into the cytoplasm of amphibian oocytes and made to gel by temperature reduction within the physiological range for these cells. After diffusion equilibrium is achieved the cells are placed at $-160°$ to $-190°$, and the reference phase, nucleus, and selected areas of cytoplasm are microdissected at $-45°C$, and then analyzed for solute and water contents. The properties of the solute and water in the reference phase are considered to be the same as those of the surrounding Ringer's salt solution (hence, a 'reference'). Therefore, it can be predicted that if the traditional view is correct, then cytoplasmic concentrations must equal reference phase concentrations for the solute under study. However, that is not the result they observed for several ions and a number of non-metabolized compounds, indicating that cytoplasm exhibits solvent properties that differ distinctly from those of bulk water. They propose that the aqueous interior of cells is not a single homogeneous phase, and that total cell concentrations of solutes do not accurately reflect actual solute concentration distributions within different parts of cells; cytoplasmic concentrations can *not* be considered equal to chemical activities because of the occurrence of intracellular binding and the altered solvent properties of the cytoplasm.

These studies have important consequences to current views on cell physiology, the most obvious being solute transport. Since it is the electrochemical gradient across any cellular membrane that determines the direction and driving force for transport, current estimates of the energy requirements for, and even the direction of transport are likely to be less than accurate. Because Na^+ and K^+ concentrations in the nucleus differ appreciably from those of the cytoplasm and since these distributions appear to be determined by differences in binding and/or solvent properties the possibility arises that changes in binding and/or solvent properties of the cytoplasm could be intimately involved with the regulation of these cations in the nucleus. It should also be recognized that the data indicate that these cells do maintain certain solutes and ions against electrochemical gradients between cytoplasm and the cell exterior; therefore, the authors propose that active transport functions are performed by the plasma membrane in this system, and that altered solvent properties and ion binding are not the whole story.

6. 'Pore water'

In a remarkable series of recent studies (see refs. 107–109) Philippa Wiggins and her colleagues have developed the view that the structure of water confined to small volumes or pores (in contact with a bulk aqueous phase) can be altered by osmotic pressure gradients across the imaginary interface between the pore water and bulk phase. They find that the pore water exhibits different specificities for the critical inorganic ions, Na^+ and K^+, due to its altered structure. Likewise, different ions influence the structure (and properties) of pore water. One of several interesting features of this proposal is that altered water structures can arise in these small, confined volumes *without* the influence of water–surface interactions (of course, the presence of surfaces exists in these systems as well). These authors stress the implications of their results to the issues of osmotic balance, cell volume regulation and the properties of water in small membrane-bound organelles. Thus, although arrived at through a different experimental and conceptual approach, this research adds substantially to other evidence that at least much of intracellular water exhibits different density and solvent properties compared to the bulk liquid. Wiggins has contributed greatly to our understanding of intracellular water, and we urge the interested reader to study her original work — it is creative, and full of provocative ideas on the subject.

7. The cluster model

John Watterson has taken a more theoretical approach to the question of cell water structure, drawing from the literature on cell ultrastructure and the physics and biology of osmotic pressure (102–104). He begins with the idea that dynamic groupings of water molecules result from cooperative intermolecular binding and develops the notion that, at any instant, an unbroken interconnection exists, percolating from one side of the 'cluster' to the other. One of the most interesting features of his 'Cluster Model' is that water is envisaged as *the* coordinating vehicle in cells. In his words[105]:

'Subcellular movement takes place as though directed by an underlying co-ordination, implying a unifying principle which links metabolic chemical energy with macroscopic mechanical forces in a two- way partnership. This principle is clearly one of structure existing throughout subcellular space, and of all subcellular components, I think that water is the only one capable of fulfilling this role.'

Watterson's model is difficult to summarize briefly, and the interested reader is encouraged to examine his papers in detail to obtain the flavor of his argument as well as its substance. We believe Szent-Györgyi would enjoy Watterson's views since he once wrote[100] that 'Life is water dancing to the tune of solids.'

8. The properties of water in Artemia cells

Since the early 1970s an interdisciplinary, cooperative research program has been aimed at describing the physical properties of water in the cysts of *Artemia*, a useful model system for this purpose since they have the natural ability to reversibly lose (and regain) virtually all their cellular water. The cysts contain about 4000 cells with essentially no extracellular space, and their ultrastructure, biochemistry and development have been studied extensively (see several recent

TABLE 2

Some properties of pure water and the water in cysts of *Artemia* [a]

Parameter [b]	Pure water	Cyst water	Pw/Cw
NMR			
t_1	3 000	275	11
t_2	1 750	53	33
D	2.4	0.4	6
QNS			
D	2.5	0.75	3
γ	1	4	0.3
MD			
ϵ' (2 GHx)	78	40	2
ϵ' (35 GHz)	23	16	1.5
γ	8	10 –25	0.5
α	< 0.02	0.5	< 0.4
M–V	1.000	0.966	1.04

[a] Cysts were at their maximum water content (about 1.4 g/g) except for MD studies in which they contained 1.2 g/g.
[b] NMR = Nuclear magnetic resonance; QNS = quasi-elastic neutron scattering; M–V = mass–volume measurement; t_1 and t_2 are 'relaxation times,' in milliseconds; D, the self-diffusion coefficient of water, in cm^{-5} cm^2/s; γ, correlation times in 10^{-12} s; ϵ', the dielectric permittivity; α, the 'spread parameter' for dispersion over the frequency range 0.8–70 Ghz;, the density in g/cm^3, derived from mass–volume measurements (M–V). Sources for these data are given in Clegg[20,22].

books[12,81,101]). The rationale behind this research has been to probe the properties of cell water by applying as many techniques as possible to the same system in the hope that the weaknesses of each will, as it were, cancel out. It is not possible here to summarize all of the results; most of the work has been recently reviewed[20,21]. However, we have concluded from these studies that little (if any) of the water in these cells exhibits the kinetic and thermodynamic properties of the water in ordinary aqueous solutions. Table 2 tabulates some of these results for fully hydrated cysts, whose cells contain about 1.6 g H_2O/g dry mass (typical for yolky embryonic cells). While the 'unusual' nature of this system may raise doubts in the minds of some about the general applicability of the results, we have argued otherwise[21]. Finally, we note that concurrent studies on the metabolism of these cells have generated an hypothesis describing potential relationships between enzyme organization and surface-associated water, the 'vicinal water network model'[19].

V. Intracellular microviscosity

We consider this topic under a separate heading because it illustrates how seldom the properties of cell water are taken into account when interpreting experimental data. Many estimates of intracellular microviscosity have been published, the values ranging from 3–15 cP (centipoise). These data, obtained most recently by electron spin resonance (see ref. 82) and fluorescence methods (see refs. 31, 79) have been analyzed in detail, particularly with regard to potential artifacts due to binding of the probe utilized. To our knowledge the enhanced microviscosity of cell water (compared to the pure liquid of 1 cP) has not been attributed to the possible difference in structure between cell water and pure water. In fact, two recent and highly sophisticated studies[31,78] do not even mention the possibility that an altered structure of intracellular water might account for part or all of the observed high values for intracellular microviscosity. It seems reasonable to propose that the reduction in cell water rotation and translation, and other evidence for enhanced hydrogen-bonding in intracellular water (see previous sections) might be a major cause for the increased microviscosity values obtained in intact cells.

VI. Cell volume, ion binding and cell water

It is a rather remarkable fact that cells somehow 'know' what their volume should be under ordinary physiological circumstances. Equally remarkable, it seems that we do not know precisely how they measure this parameter. A vast literature has been produced on the matter of cell volume and its regulation (for recent literature see the book edited by Kleinzeller[72], and the reviews of Chamberlin and Strange[16], and Hoffmann and Simonsen[65]). In virtually all these studies (but see Macknight[80] for a very thoughtful and objective exception) it is assumed that the properties of intracellular water are identical to those of the external medium and that cell volume is achieved and maintained by balancing the external and internal osmotic pressure through accumulation or release of inorganic ions (notably Na^+, K^+, and Cl^-) and/or compatible solutes such as primary amines and polyhydroxyalcohols (see Somero[96]). There is certainly good evidence for that contention. However, we note that according to this view the *properties* of cell water are not even considered in explanations for cell volume and its regulation. In this regard it is important to note that an increasing body of evidence indicates that a substantial fraction of the inorganic ions in cells may not contribute to the colligative properties of cytoplasm — that is, they are 'bound'. This evidence comes from a number of different authors (many of whom have no 'ax to grind') and a wide array of studies, ranging from NMR and electron microprobe studies, to fluorescence ratio imaging, ion efflux and electron microscopy (see books[76,85], and articles[14,25,59,63] for literature coverage). Although doubt exists concerning the precise fraction of free versus bound ions, it appears reasonable to suspect that on the order of one-half the total inorganic ion complement is 'bound' in the sense that this fraction is not freely diffusing in cells. This has considerable significance to the issue of cell volume and the properties of cell water. Thus, in the case of mammalian cells, Cl^-, K^+, and Na^+ are considered to be the overwhelming contributors to intracellular osmotic pressure. Indeed, the sum of these three ions, roughly 250 mM, is commonly thought to balance most of the external osmotic pressure (about 300 mOsm/kg water). But if about 50% of these ions do not contribute to intracellular osmotic pressure then we must ask how the cell can be at osmotic equilibrium? Negendank[85] has considered this question at some length, and has argued that the answer must involve the properties of cell water, a contention with which we agree[23]. The implication of this analysis is that solutes dissolved in cell water may exert higher osmotic pressure than those same solutes in the external aqueous phase. An inescapable consequence of this explanation is that the structure of cell water must also be different. Given the evidence on ion binding, and the absence of significant mechanical resistance to cell volume changes near 'osmotic equilibrium' in animal cells, there seems to be no other reasonable explanation (see, however, the interesting analysis by Lechene[75], and by Bereiter-Hahn and Strohmeier[7]). It will be interesting to see what the future holds for these proposals, and for the traditional explanation of cell volume and its regulation.

VII. Some additional consequences of the properties of cell water

The preceding section certainly indicates that our thinking about at least one aspect of cell physiology would have to be altered appreciably if, as we suggest, the properties of cell water are not the same as those in the external medium. But there are other consequences worth noting:

1. Much of what is known about macromolecular function in cells is based on data obtained *in vitro*, almost always in highly dilute aqueous solutions. That approach is very convenient, but if intracellular water differs from that in test tubes, as we believe it does, then information obtained *in vitro* may not allow us to construct (or better 'reconstruct') an accurate description of these molecules and their activities when they operate *within* cells. We agree with Albrecht-Buehler[2] that reductionistic approaches are powerful, but they can never alone provide us with a reliable description of cells. His review of this issue is powerful and highly recommended.

2. It is widely accepted that direct interactions between macromolecules and their surrounding water of hydration play critical roles in their structure and function. There is no debate about this issue, and it seems very likely that water plays subtle but important roles in metabolism through water–enzyme interactions. However, to understand those roles we must know the details of the aqueous microenvironment in which this activity occurs.

3. Available evidence suggests, to us at least, that the solvent properties of at least a large fraction of the total cell water, notably in cytoplasm, differ from those of ordinary aqueous solutions. On this basis, some contribution to the uneven distribution of certain solutes across the plasma membrane, as well across membranes *within* cells (organelles), could arise from such solvent differences. In addition, small metabolites might 'partition' between various intracellular aqueous phases[55]. Even protein distribution within cells could be influenced in this fashion. A speculative 'model' on the organization of enzymes in the aqueous cytoplasm includes the possibility that a loose association of enzymes with the cytomatrix may

be driven by water interactions involving their respective surfaces, similar to those involved in association through hydrophobic interactions[19].

4. Assembly–disassembly processes are influenced by the properties of the aqueous phase within which they occur. Such mechanisms could be critical to enzyme–enzyme associations and the dynamic turnover of the cytomatrix, and possibly other cell structures. Watterson's[105] analysis, speculative though it may be, is worth careful study in this regard.

5. Many molecular interactions in cells involve electrostatic interactions which are, of course, very sensitive to the dielectric properties of the aqueous phase in which they occur. Thus, the possibility that the dielectric permittivity of cell water is reduced (Table 2, 24) relative to dilute solutions, may be of some importance.

6. A reasonably good correlation exists between modifications in the cytomatrix and changes in the amount and properties of cell water, both of which commonly, although not always, accompany cell transformation by viruses or carcinogens. While that may be fortuitous, it is notable that the usual observation is a reduction in cytomatrix surface area and an increase in the amount of cell water that has 'bulk-like' properties (see refs. 22, 60). That is consistent with the proposed relationship between the cytomatrix and its effects on the properties of the surrounding aqueous environment, and vice versa. It has also not escaped our attention[19,20,25] that many of the metabolic changes accompanying the transformation process are associated with 'soluble' enzymes which, in the view of some of us, are not really 'soluble' at all but are instead part of the water-cytomatrix system.

7. Without the concept of vicinal water, and its characteristic thermal anomalies at several different temperatures of physiological interest, it is difficult to see how a large body of usual thermal responses of organisms can be explained completely. On the other hand, accepting the thermal transitions in the vicinal water structures allow for relatively facile explanations of (sometimes dramatic) complex thermal responses of organisms including some very abrupt thermal death limits, selection of body temperatures and multiple temperature growth optima.

VIII. Concluding comments

Because the traditional (consensus) view on intracellular water considers it to be identical to that in dilute aqueous solutions we have chosen to concentrate here on the body of evidence that is not in accord with that opinion. As George Bernard Shaw put it: 'An idea that no one believes cannot be proved too often.'

Our position has been that the scientific community, most of whose research is not directly concerned with the matter, should be informed that alternative positions, based on substantial evidence, have been taken by a number of investigators. Perhaps more importantly, we believe that the consequences of altered cell water to virtually every aspect of cell structure and function could be extremely important, and would have great impact on the way we view cells and their activities. If nothing else, we hope that more attention will be given to the matter: there is sufficient reason to be concerned about paradigms of cell biology built on the assumption that the intracellular aqueous phase(s) is no different than, essentially, pure water. At the same time we recognize that the body of existing evidence, only some of which has been dealt with here, has apparently not been sufficiently compelling to change matters, or simply overlooked. Thus, we also recognize that the assumption of ordinary cell water has, thus far, worked surprisingly well in the development of contemporary thought in cell biology. Is this assumption a reasonable one? Can we consider paradigms built upon it as adequate first-approximations, as almost right? Perhaps so, but we are reminded of the saying by Mark Twain that 'the difference between the right... and the almost right... is the difference between lightning and lightning bug.'

Acknowledgments. We thank Professor Keith Porter for provision of HVEM photomicrographs and Diane Cosgrove for her usual skill and patience with manuscript preparation. The expense of publication came from NSF Grant DCB 8820347 (to J.S.C.). Most importantly, we dedicate this paper to Professor Carlton Hazlewood who, for over 25 years, has championed the importance of intracellular water and dedicated his research to its study.

IX. References

1. Adamson, A.W. *Physical Chemistry of Surfaces.* New York: Wiley, 1982.
2. Albrecht-Buehler, G. In defense of 'nonmolecular' cell biology. *Int. Rev. Cytol.* 120: 191–241, 1990.
3. Alpers, W. and H. Hühnerfuss. Molecular aspects of the system water/monomolecular surface film and the occurrence of a new anomalous dispersion regime at 1.43 GHz. *J. Phys. Chem.* 87: 5251–5258, 1983.
4. Arnheim, R. *Observations on Psychology, the Arts, and the Rest.* Berkeley: University of California Press, 1989, p. 369.
5. Barnes, P. Machine Simulation of Water. In: *Progress in Liquid Physics* edited by C.A. Croxton. New York: Wiley, 1978, pp. 391–428.
6. Benga, G. *Water Transport in Biological Membranes* Vol. 1. Boca Raton: CRC Press, 1989, p. 1–223.
7. Bereiter-Hahn, J. and R. Strohmeier. Hydrostatic pressure in metazoan cells in culture. In: *Cytomechanics,* edited by J. Bereiter-Hahn, O.R. Anderson and W.E. Reif. Berlin: Springer-Verlag, 1987, p. 261–272.
8. Blum, J.J., G. Lawler, M. Reed, and I. Shin. Effect of cytoskeletal geometry on intracellular diffusion. *Biophys. J.* 56: 995–1005, 1989.
9. Blumberg, R.L., H.E. Stanley, A. Geiger, and P. Mausbach, Connectivity of hydrogen bonds in liquid water. *J. Chem. Phys.* 80: 5230–5241, 1984.
10. Bridgman, P.C. and T.S. Reese. The structure of cytoplasm in directly frozen cells. I. Filamentous meshworks and the cytoplasmic ground substance. *J. Cell Biol.* 99: 1655–1660, 1984.
11. Brooks, C.L. and M. Karplus. Solvent effects on protein motion and protein effects on solvent motion. *J. Mol. Biol.* 208: 159–181, 1989.
12. Brown, R.A., P. Sorgeloos, and C.N.A. Trotman. *The Biology of Artemia.* Boca Raton: CRC Press, 1990, in press.
13. Cameron, I.L., W.E. Hardman, K.E. Hunter, C. Haskin, N.K.R. Smith, and G.D. Fullerton. Evidence that a major portion of cellular potassium is bound — a review paper. *Scan. Microsc.* 4: 89–102, 1990.
14. Cameron, I.L., P. Merta, and G.N. Fullerton. Osmotic and motional properties of intracellular water as influenced by osmotic swelling and shrinking of *Xenopus* oocytes. *J. Cell. Physiol.* 142: 592–602, 1990.
15. Cañedo, L.E., L.E. Todd, L. Packer, and J. Jaz. *Cell Function and Disease.* New York: Plenum, 1988, pp. 1–524.

16. Chamberlin, M.E. and K. Strange. Anisosmotic cell volume regulation: a comparative view. *Am. J. Physiol.* 257:C159-C173, 1989.
17. Churev, N.V. and B.V. Deryaguin. Inclusion of structural forces in the theory of colloids and films. *J. Coll. & Interface Sci.* 103: 542–553, 1985.
18. Clifford, J. Properties of water in capillaries and thin films. In: *Water — a Comprehensive Treatise*, Vol. 5, edited by F. Franks. New York: Plenum, 1975, pp. 75–132.
19. Clegg, J.S. Metabolism and the intracellular environment: the vicinal-water network model. In: *Cell-Associated Water*, edited by W. Drost-Hansen and J.S. Clegg. New York: Academic Press, 1979, p. 363–413.
20. Clegg, J.S. Properties and metabolism of the aqueous cytoplasm and its boundaries. *Am. J. Physiol.* 246:R133-R151, 1984.
21. Clegg, J.S. On the physical properties and potential roles of intracellular water. In: *Regulation of Cell Metabolism*, edited by G.R. Welch and J.S. Clegg. New York: Plenum, 1986, pp. 41–55.
22. Clegg, J.S. On the internal environment of animal cells. In: *Microcompartmentation*. Boca Raton: CRC Press, 1988, pp. 1–16.
23. Clegg, J.S. L-929 cells under hyperosmotic conditions: water, $Na+$ and $K+$. *Cell Biophys.* 13: 119–132, 1988.
24. Clegg, J.S., S. Szwarnowski, V.E.R. McClean, R.J. Sheppard, and E.H. Grant. Interrelationships between water and cell metabolism in *Artemia* cysts. X. Microwave dielectric studies. *Biochim. Biophys. Acta* 121: 458–468, 1982.
25. Clegg, J.S. and M.B. Barrios. The 'cytosol': a neglected and poorly understood compartment of eukaryotic cells. In: *Cell Function and Disease*, edited by L.E. Cañedo, L.E. Todd, L. Packer and J. Jaz. New York: Plenum, 1988, pp. 159–170.
26. Cope, F. Nuclear magnetic resonance evidence using D_2O for structured water in muscle and brain. *Biophys. J.* 9: 303–319, 1969.
27. Deryaguin, B.V. Recent research into the properties of water in thin films and micro-capillaries. In: *The State and Movement of Water in Living Organisms*, XIX Sympos. Soc. Exp. Biol. Cambridge: Cambridge Univ. Press, 1964, pp. 55–60.
28. Deryaguin, B.V. and N.V. Churev. Structural component of disjoining pressure. *J. Coll. Interface Sci.*, 1974, 49: 249–255.
29. Deryaguin, B.V., V.V. Karasev and E.N. Khromova. Thermal expansion of ordinary and heavy water in the pores of titanium dioxide. *J. Coll. Interface Sci.* 78: 274, 1980.
30. Deryaguin, B.V., V.V. Karasev and E.N. Khromova. Thermal expansion of water in fine pores. *J. Coll. Interface Sci.*, 109: 586–587, 1986.
31. Dix, J.A. and A.S. Verkman. Mapping of fluorescence anisotropy in living cells by ratio imaging. Application to cytoplasmic viscosity. *Biophys. J.* 57: 231–240, 1990.
32. Drost-Hansen, W. Temperature anomalies and biological temperature optima in the process of evolution. *Naturwiss*. 43: 512, 1956.
33. Drost-Hansen, W. Aqueous interfaces — methods of study and structural properties. Part II, In: *Chemistry and Physics of Interfaces*, edited by S. Ross. Washington: American Chemical Society, 1965.
34. Drost-Hansen, W. The effects on biological systems of higher-order phase transitions in water. *Ann. N.Y. Acad. Sci.* 125 (Art. 2): 471–501, 1965.
35. Drost-Hansen, W. Allowable thermal pollution limits – a physico-chemical approach. *Chesapeake Sci.*, 10: 281–288, 1969.
36. Drost-Hansen, W. Structure and properties of water at biological interfaces. In: *Chemistry of the Cell Interface*, Vol. B., edited by H.D. Brown. New York: Academic Press, 1971, pp. 1–184.
37. Drost-Hansen, W. Effects of pressure on the structure of water in various aqueous systems. *S.E.B. Symp. Proc.* 26: 61–101, 1972.
38. Drost-Hansen, W. Molecular aspects of aqueous interfacial structure. *J. Geophys. Res.* 77: 5132–5146, 1972.
39. Drost-Hansen, W. Water at biological interfaces — structural and functional aspects. *Phys. Chem. Liq.* 7: 243–346, 1978.
40. Drost-Hansen, W. Role of vicinal water in cellular evolution. In: *Water and Ions in Biological Systems*, edited by A. Pullman, V. Vasilescu and L. Packer. New York, Plenum Press, pp. 523–534, 1985.
41. Drost-Hansen, W. Anomalous volume properties of vicinal water and some recent thermodynamic (DSC) measurements relevant to cell physiology. In: *Water and Ions in Biological Systems*, edited by A. Pullman, V. Vasilescu and L. Packer. New York: Plenum, pp. 289–294, 1985.

42. Drost-Hansen, W. and M. Lavergne. Discontinuities in slope of the temperature dependence of the thermal expansion of water. *Naturwiss.* 43: 511, 1956.
43. Drost-Hansen, W. and J.L. Singleton. Our aqueous heritage. I. Evidence for vicinal water in cells. II. Role of vicinal water in cells. In: *Foundations of Medical Cell Biology*, edited by E.E. Bittar. Greenwich, CN: JAI Press, in press.
44. Eisenberg, D. and W. Kauzmann. *The Structure and Properties of Water*. Oxford: Clarendon, 1969.
45. Etzler, F.M. A statistical thermodynamic model of water near solid interfaces. *J. Coll. Interface Sci.* 92: 43–56, 1983.
46. Etzler, F.M. Enhancement of hydrogen bonding in vicinal water: heat capacity of water and deuterium oxide in silica pores. *Langmuir* 4: 878–883, 1988.
47. Etzler, F.M. and W. Drost-Hansen. A role for vicinal water in growth, metabolism and cellular organization. *Adv. Chem. Am. Chem. Soc.* 188: 485–497, 1980.
48. Etzler, F.M. and W. Drost-Hansen. Recent thermodynamic data on vicinal water and a model for their interpretation. *Croatica Chemica Acta* 56: 563– 592, 1983.
49. Evans, D.F. and B.W. Ninham. Molecular forces in the self-organization of amphiphiles. *J. Phys. Chem.* 90: 226–234, 1986.
50. Finney, J.L. Towards a molecular picture of liquid water. In: *Biophysics of Water*, edited by F. Franks and S. Mathias. New York: Wiley, 1982, pp. 73–95.
51. Frank, H.S. Structural models. In: *Water — a Comprehensive Treatise, Vol. 1*, edited by F. Franks. New York: Plenum, 1972, pp. 515–543.
52. Franks, F. (editor). *Water — a Comprehensive Treatise, Vols. 1–7*. New York: Plenum Press, 1972–1982.
53. Reference deleted.
54. Frauenfelder, H. and E. Gratton. Protein dynamics and hydration. *Methods Enzymol.* 127: 207–216, 1986.
55. Garlid, K.D. Aqueous phase structure in cells and organelles. In: *Cell-Associated Water*, edited by W. Drost-Hansen and J.S. Clegg. New York: Academic Press, 1979, pp. 293–362.
56. Geiger, A. and H.E. Stanley. Low-density 'patches' in the hydrogen-bond network of liquid water. Evidence from molecular-dynamics computer calculations. *Phys. Rev. Lett.* 49: 1749–1752, 1982.
57. Gershon, N.D, K.R. Porter, and B.L. Trus. The cytoplasmic matrix: its volume, surface area and the diffusion of molecules through it. *Proc. Natl. Acad. Sci. U.S.A.* 82: 5030–5035.
58. Goldanskii, V.I. and Y.F. Krupyanskii. Protein and protein-bound water dynamics studied by Rayleigh scattering of Mössbaner radiation (RSMR). *Quart. Rev. Biophys.* 22: 39–92, 1989.
59. Harootunian, A.T., J.P.Y. Kao, B.K. Eckert, and R.Y. Tsien. Fluorescence ratio imaging of cytosolic free Na^+ in individual fibroblasts and lymphocytes. *J. Biol. Chem.* 264: 19458–19467, 1989.
60. Hazlewood, C.F. A view of the significance and understanding of the physical properties of intracellular water. In: *Cell-Associated Water*, edited by W. Drost-Hansen and J.S. Clegg. New York: Academic Press, 1979, pp. 165–260.
61. Hazlewood, C.F., B.L. Nichols, and N.F. Chamberlain. Evidence for the existence of a minimum of two phases of ordered water in skeletal muscle. *Nature*, Lond. 222: 747–750, 1969.
62. Hazlewood, C.F. and M. Kellermayer. Ion and water retention by permeabilized cells. *Scan. Microsc.* 2: 267–273, 1988.
63. Hazlewood, C.F. and M. Kellermayer. The state of potassium in skeletal muscle and in non-muscle cells. *Scan. Microsc.*, 3: 1241–1245, 1989.
64. Henn, A.R. and W. Kauzmann. Equation of state of a random network, continuum model of liquid water. *J. Phys. Chem.* 93: 3770–3783, 1989.
65. Hoffmann, E.K. and L.O. Simonsen. Membrane mechanisms in volume and pH regulation in vertebrate cells. *Physiol. Rev.* 69: 315–382, 1989.
66. Horowitz, S.B. and D.S. Miller. Solvent properties of ground substance studied by cryomicrodissection and reference phase techniques. *J. Cell Biol.* 99: 172s–179s, 1984.
67. Hühnerfuss, H. Molecular aspects of organic surface films on marine water and the modification of water waves. *La Chimica L'Industria* 65: 97–101, 1983.
68. Hühnerfuss, H. and W. Wolfgang. The thermal anomaly of relaxation effects in monomolecular surface films. *J. Coll. Interface Sci.* 107: 476–480, 1983.
69. Israelachvili, J. *Intermolecular and Surface Forces*. New York: Academic Press, 1985.
70. Israelachvili, J. and G.E. Adams. Measurement of forces between two mica surfaces in aqueous electrolyte solutions in the range 0–100 nm. *J. Chem. Soc.* 74: 975–1001, 1977.
71. Israelachvili, J. and B.W. Ninham. Intermolecular forces – the long and the short of it. *J. Coll. Interface Sci.*, 57: 14–25, 1977.

72. Jones, D.P. *Microcompartmentation*. Boca Raton: CRC Press, 1988, pp. 1–261.
73. Kleinzeller, A. *Cell Volume Control: Fundamental and Comparative Aspects in Animal Cells*. New York: Academic Press, 1987, pp. 1–281.
74. Kondo, H. Reexamination of the reality or artifact of the microtrabeculae. *J. Ultrastruct. Res.* 87: 124–130, 1984.
75. Lechene, C. Cellular volume and cytoplasmic gel. *Biol. Cell.* 55: 177–180, 1985.
76. Ling, G.N. *In Search of the Physical Basis of Life*. New York: Plenum Press, 1984, pp. 1–791.
77. Low, P.F. Nature and properties of water in montmarillonite-water systems. *J. Soil. Sci.* 43: 174–191, 1979.
78. Luby-Phelps, K., D.L. Taylor, and F. Lanni. Probing the structure of cytoplasm. *J. Cell Biol.* 102: 2015–2022, 1986.
79. Luby-Phelps, K., F. Lanni, and D.L. Taylor. The submicroscopic properties of cytoplasm as a determinant of cellular function. *Ann. Rev. Biophys. Chem.* 17: 369–396, 1988.
80. MacNight, A.D.C. Volume maintenance in isomatic conditions. *Curr. Top. Memb. Transp.* 30: 3–43, 1987.
81. Macrae, T.H., J.C. Bagshaw, and A.H. Warner. *Biochemistry and Cell Biology of Artemia*. Boca Raton: CRC Press, 1989, pp. 1–264.
82. Mastro, A.M. and D.J. Hurley. Diffusion of a small molecule in the aqueous compartment of mammalian cells. In: *The Organization of Cell Metabolism*, edited by G.R. Welch and J.S. Clegg. New York: Plenum Press, 1986, pp. 57–74.
83. Misra, L.K., E.E. Kim, C.F. Hazlewood, and L.W. Dennis. Magnetic resonance relaxation times and imaging in the pathophysiology of muscle. In: *Cell Function and Disease*, edited by L.E. Cañedo, L.E. Todd, L. Packer, and J. Jaz. New York: Plenum Press, 1988, pp. 421–444.
84. Negendank, W. A cooperative transition theory applied to the kinetics of ion exchanges in cells. *Cell Biophys.* 13: 93–117, 1988.
85. Negendank, W. and L. Edelmann. *The State of Water in the Cell*. Chicago: Scanning Microscopy, 1988, pp. 1–114.
86. Ooi, T. and M. Oobatake. Intermolecular interactions between protein and other molecules including hydration effects. *J. Biochem.* 104: 440–444, 1988.
87. Packer, L. Biomembranes. Part O. Protons and Water. *Methods Enzymol.* 127: 1–830, 1986.
88. Paine, P.L. Diffusive and nondiffusive proteins *in vivo*. *J. Cell Biol.* 99: 188s–195s.
89. Peschel, G. and P. Belouschek. The problem of water structure in biological systems. In: *Cell-Associated Water*, edited by W. Drost-Hansen and J.S. Clegg. New York: Academic Press, 1979, pp. 3–52.
90. Porter, K.R. Structural organization of the cytomatrix. In: *The Organization of Cell Metabolism*, edited by G.R. Welch and J.S. Clegg. New York: Plenum Press, 1986, pp. 9–26.
91. Rand, R.P. and V.A. Parsegian. Hydration forces between phospholipid bilayers. *Biochim. Biophys. Acta* 988: 351–376, 1989.
92. Richards, W.G., P.M. King, and C.A. Reynolds. Solvation effects. *Protein Engineering* 2: 319–327, 1989.
93. Ris, H. The cytoplasmic filament system in critical point dried whole mounts and plastic embedded sections. *J. Cell Biol.* 100: 1474–1479, 1985.
94. Rorschach, H.E. and C.F. Hazlewood. Protein dynamics and the NMR relaxation time T1 of water in biological systems. *J. Magn. Reson.* 70: 79–88, 1986.
95. Rorschach, H.E.D.W. Beardon, C.F. Hazlewood, D.B. Heidorn, and R.M. Nicklow. Quasi-elastic scattering studies of water diffusion. *Scan. Microsc.* 1: 2043–2049, 1987.
96. Somero, G. Protons, osmolytes, and fitness of internal milieu for protein function. *Am. J. Physiol.* 251:R197–R213, 1986.
97. Stanley, H.E. and J. Teixeira. Interpretation of the unusual behaviour of H_2O at low temperatures: tests of a percolation model. *J. Chem. Phys.* 73: 3404–3422, 1980.
98. Stanley, H.E. and J. Teixeira. Are concepts of percolation and gelation of possible relevance to the behaviour of water at very low temperatures? *Ferroelectrics* 30: 213–226, 1980.
99. Stillinger, F.H. Water revisited. *Science* 209: 451–457, 1980.
100. Szent-Györgyi, A. Biology and the pathology of water. *Persp. Biol. Med.* 13: 239, 1971.
101. Warner, A.H., T.H. MacRae, and T.C. Bagshaw. *Cellular and Molecular Biology of Artemia Development*. New York: Plenum Press, 1989, pp. 1–431.
102. Watterson, J.G. Does solvent structure underlie osmotic mechanisms? *Phys. Chem. Liq.* 16: 313–316, 1987.
103. Watterson, J.G. Solvent cluster size and colligative properties. *Phys. Chem. Liq.* 16: 317–320, 1987.
104. Watterson, J.G. A role for water in cell structure. *Biochem. J.* 248: 615–617, 1987.

105. Watterson, J.G. The role of water in protein function. *Progr. Mol. Subcell. Biol.*, in press.
106. Welch, G.R. and J.S. Clegg. *The Organization of Cell Metabolism*. New York: Plenum Press, 1986, pp. 1–389.
107. Wiggins, P.M. Water structure in polymer membranes. *Prog. Polym. Sci.* 13: 1–35, 1988.
108. Wiggins, P.M. and B.A.E. MacClement. Two states of water found in hydrophobic clefts: possible contribution to mechanisms of cation pumps and enzymes. *Int. Rev. Cytol.* 108: 249–303, 1987.
109. Wiggins, P.M. and R.T. Van Ryn. Changes in ion selectivity with changes in density of water in gels and cells. *Biophys. J.*, 58: 585–596, 1990.
110. Zimmerman, S., I.L. Cameron, and A.M. Zimmerman. The relaxation state of water in sea urchin eggs. In: *The Cell Biology of Fertilization*, edited by H. Schatten and G. Schatten. New York: Academic Press, 1989, pp. 319–340.

CHAPTER 2

Evolution of the fish genome

MOIRA M. FERGUSON * AND FRED W. ALLENDORF **

*Department of Zoology, University of Guelph, Guelph, Ontario, Canada N1G 2W1, and **Division of Biological Sciences, University of Montana, Missoula, MT, U.S.A. 59812

I. Introduction
II. Physical organization
   1. The genome
   2. Karyotypic evolution – Chromosomal rearrangements
   3. Linkage
   4. Mobile genetic elements
III. Gene duplication
   1. Polyploidy
   2. Regional duplication
IV. Functional organization
   1. Gene structure
   2. Function and rates of evolution
V. Is the genome organized?
Acknowledgements
VI. References

I. Introduction

The functional relationship between the genome and the biochemical and molecular products necessary for metabolic processes is well established. Early studies of fishes concentrated on examining the genome at a cellular level by characterizing such things as amount of DNA and chromosomal morphology. The tools to study biochemical processes within the context of genome organization and function were not readily available. The development of molecular and biochemical techniques led to examinations at the protein and nucleotide levels and revolutionized our view of the genome. The new data and unexpected complexities suggested the need to understand genome structure and organization in conjunction with biochemical studies. For instance, the discovery of 'extra' gene copies whose products were free to evolve different functions attracted the attention of biochemists. It then became necessary for biochemists to know the 'genetics' of the biological products under study if the evolution of the different molecules in relation to environmental variation was to be understood.

   Our purpose is to overview the evolution of the fish genome with selected fish examples and provide an important framework to begin a series on biochemistry and molecular biology of fishes. We begin with the basics and discuss the physical organization of the genome describing the location of the genetic material and its structure and evolution. We then describe gene duplications and their anticipated significance to the biochemical and molecular evolution of fishes. The next section takes a different approach by examining the genome from the viewpoint of function rather than structure. Finally, we attempt to determine if the genome as we see it makes sense by asking 'is the genome organized?'. Even though molecular approaches to understanding genome evolution are still in their infancy, the analyses of biochemical and molecular processes are strengthened with knowledge of genes and evolution.

II. Physical organization

1. The genome

The genetic material of fishes (like all vertebrates) is located within the nucleus organized into chromosomes and in the cytoplasm as a closed circular molecule in the mitochondria. The vast majority of genes, even those that are required for mitochondrial function, are encoded in the nucleus. Nuclear DNA and mitochondrial DNA (mtDNA) differ in two important aspects, variability in gene content and transmission genetics. These differences are, in fact, what makes study of them interesting and powerful in regards to understanding the evolution of fishes. The vertebrate mtDNA molecule is conserved in size and content; it is about 15,000–18,000 bases long and codes for 2 ribosomal RNA genes, 22 transfer RNA genes, and 13 protein genes that code for subunits of enzymes functioning in electron transport or ATP synthesis[40].

The nuclear genome, however, is considerably more complex and shows extensive size variability (nuclear DNA content)[30]. However, there is no simple relationship between amount of DNA per nucleus and complexity of the adult organism (the so-called 'C-value paradox'). For instance, a lungfish species (*Protopterus aethiopicus*) has over 40 times more DNA per haploid genome than humans[12].

Gold and Amemiya[27] measured the nuclear DNA content of 200 individuals representing 20 species of North American cyprinid fishes and observed significant heterogeneity both between individuals within populations of species (maximum of 6%) and among species. Greater divergence in genome size was detected in a speciose genus suggesting that genome size change is concentrated in speciation episodes. Gold and Amemiya[27] proposed that the lack of adverse phenotypic effects of this variation supports the 'selfish DNA' hypothesis in that much of the variation reflects gains or losses of phenotypically inconsequential DNA[17].

Nuclear DNA and mtDNA also differ in their transmission to progeny. In contrast, to the conventional meiotic segregation of nuclear DNA (some exceptions will be discussed later), offspring usually inherit the mtDNA molecule of the mother[40].

2. Karyotypic evolution — Chromosomal rearrangements

The number and form of chromosomes of an organism is referred to as its karyotype (Fig. 1). Autosomes are the more numerous chromosomes in the

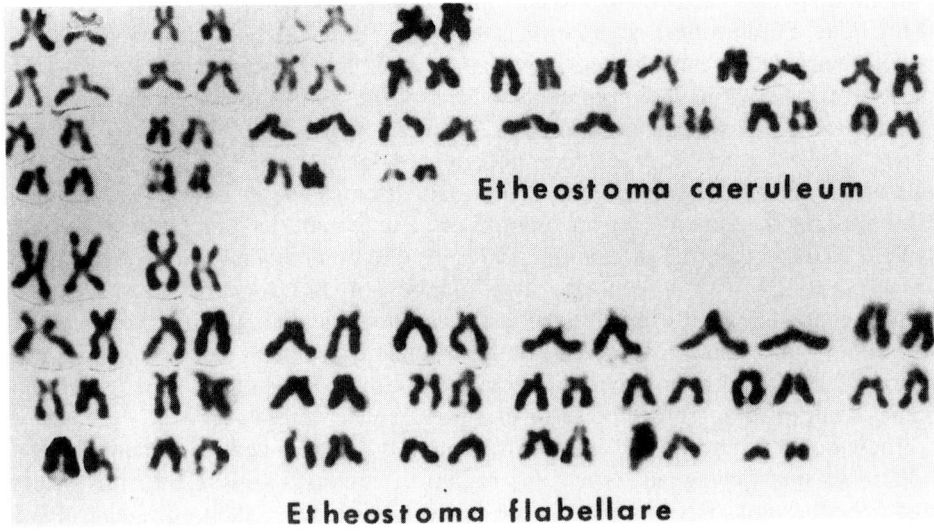

Fig. 1. Karyotypes of two percid species. The top row in each karyotype of etheostamine fish indicates metacentric and submetacentric chromosomes[14].

karyotype and do not differ between the sexes. Particular chromosomes are designated as sex chromosomes if they can be identified as carrying the sex determining genes; sometimes these differ morphologically between the sexes. Many fish species or populations within species do not have distinguishable sex chromosomes[26,48].

Thorgaard[60] was one of the first to report a heteromorphic sex chromosome pair in a male teleost. Male, but not female rainbow trout (*Oncorhynchus mykiss*), have one pair of morphologically distinct chromosomes. Interestingly some rainbow trout males from isolated populations or near the limits of the species range show no evidence of sex chromosome heteromorphism, the presumed primitive condition[60,62]. Apparently, rainbow trout are still in the process of evolving differentiated sex chromosomes because heteromorphic sex chromosomes are believed to have evolved from morphologically identical homologs once some mechanism to inhibit crossing over has evolved[44,55].

Chromosomes are further distinguished by the location of the centromere and relative size of arms (Fig. 1). Acrocentric chromosomes are rod-shaped with a terminal centromere (single arm), metacentrics are V-shaped with a centromere in the middle (two symmetrical arms), and telocentrics are an intermediate condition with one arm shorter than the other. The primitive vertebrate genome, comprising about 20% of the average mammalian cellular DNA content, is thought to have evolved into 48 acrocentric or telocentric chromosomes through a series of genome duplications[45]. The common karyotype in fishes consisting of 48 acrocentric chromosomes[56] supports this proposal. This karyotype has been conserved among members of single taxonomic groups (i.e., sunfishes) as well as phylogenetically unrelated species such as *Gambusia affinis* (Cyprinodontiformes), *Mugil cephalus* (Mugilidae, Perciformes), and *Pagellus erythrinus* (Sparidae, Perciformes)[56].

Fish karyotypes can also be remarkably variable[26,56]. Karyotypic changes are most often associated with chromosome breaks followed by joining of broken ends. A common type of chromosomal rearrangement in fishes is a change in chromosome number. Such Robertsonian rearrangements[67] involve the fusion of two one-armed chromosomes into one two-armed chromosome or fission (the reverse). Chromosome arm number or the amount of genetic material is not affected by this type of rearrangement. For example, chromosome numbers in rainbow trout range from 58 to 64 with a constant arm number of 104 (ref. 62). Robertsonian rearrangements can also involve the sex chromosomes; Thorgaard[61] observed that differences in chromosome counts between male and female sockeye salmon (*O. nerka*, 57 and 58 chromosomes, respectively) could be attributed to the Y chromosome fusing with an autosome to form a metacentric chromosome.

Individual chromosomes can be differentiated by staining with compounds that adhere to specific regions. Nucleolus organizer regions (NORs), the sites of 18S and 28S ribosomal RNA genes, can be detected with silver staining as achromatic stalks. The size of these regions varies substantially between and within species. For example, Foresti et al.[23] showed that one species of Gymnotiformes shows marked intraspecific variation; the NOR region in these fish varies six-fold. These authors suggest that an increase in the number of ribosomal genes in the NORs has occurred during the evolution of this group. Other methods for the identification of specific chromosomal regions include the staining of heterochromatin in the centromere or telomere[47].

3. Linkage

The location of different genes on the same chromosome has important implications to their inheritance. If they are in close proximity they will not be inherited independently because chromosomes are inherited as blocks unless recombination takes place. Most evidence of linkage in fishes has come from classical methods of constructing informative genetic crosses and tabulating numbers of parental and recombinant genotypes among locus pairs in backcross or F_2 progeny. This stands in contrast to the methods of somatic cell hybridization or by DNA–DNA *in situ* hybridization most commonly used in mammals.

Given the large scale karyotypic divergence that has occurred throughout the evolution of vertebrates, one would expect that few genetic linkages or chromosomal syntenies would be conserved across these groups. However, several apparent cases of gene arrangement conservation have been detected (Fig. 2)[42]. This means that some genetic linkages or chromosomal syntenies may have been conserved through the more than 400 million years since the divergence of fish and tetrapod lineages. Even though the proportion of apparent syntenic group conservations decreases as gene maps of more distantly related vertebrates are compared, striking examples exist. For example, Pep-B and Ldh-B are syntenic in man, many other mammals, salmonid fishes, and ranid frogs. The most parsimonious explana-

Poeciliid fish	Salmonid fish	Frogs (Rana)	Mouse	Mouse lemur	Man
II	13	HK2	1	LDHA	2
⌈GPI2	⌈MDH	MANA	⌈IDH1		⌈ACP1
IDH-M	IDH	PEPB	2	HEXA	IDH-S
LDHA	GPI	LDHB	⌈ACP2	PKM2	MDH1
LDHC	LDHA	MPI	SORD	MPI	10
PK2	LDHC		9	NP	⌈GOT-S
ENO2	GOT	GPI	⌈PK3	CKBB	HK
MPI	MPI	GOT2(M?)	MPI	SORD	11
III	3	PEPD	7		⌈ACP2
⌈GUK2	⌈GPI	GOT1(S?)	⌈PEPD	PEPB	HBB
GAPD1	PEPD	TPI	GPI	LDHB	LDHA
ME	SORD		IDH-M	CS	12
IV	7	IDH1	HBB	TPI	⌈PEPB
⌈PK1	⌈LDHB	HB	LDHA	GAPD	LDHB
GPI1	PEPB		8	ENO2	CS
IDH1	1		⌈GOT-M		ENO2
GOT-M	⌈GOT		19	GPI	GAPD
PEPD	MDH		⌈GOT-S	PEPD	TPI
V			6		14
⌈MDH2			⌈LDHB		⌈NP
GLYDH			GAPD		CKBB
VI			TPI		15
⌈NP2			14		⌈IDH-M
GLNS			⌈NP		MANA
TF			11		PKM2
UMPK			⌈HBA		HEXA
GUK3			10		SORD
AMY			⌈CS		MPI
VII			HK		16
⌈GALT2			APK		⌈GOT-M
IDH2					HBA
U3					19
⌈GOT3					⌈GPI
MDH1					PEPD
U4					MANB
⌈PEPA					CKMM
TPI1					
GAPD3					

Fig. 2. A summary of chromosomal synthenies in various vertebrates. Each group of loci represents a probable linkage group[42].

tion is that the different vertebrate classes received this linkage group from a common ancestor rather than being recently and independently derived in many lineages.

The extreme conservation of genetic linkages throughout the evolution of vertebrates can be explained by two classes of hypotheses[42]. Gene arrangements could be preserved by natural selection because of some functional significance. Alternatively, the origin of vertebrates via gene duplication has resulted in specific meiotic behaviour of chromosomes leading to the retention of gene arrangements[41]. Evidence for both explanations exists[42]. However, identification of the precise mechanisms awaits more research.

A different approach to construct gene maps is to determine the gene-centromere distances in gynogenetic diploids[64]. Gynogenetic diploids are usually produced by suppressing the extrusion of the second polar body at meiosis II by heat or pressure shock so that the egg is functionally diploid[63]. Irradiated sperm is used to activate development but does not contribute genetic material to the embryo. Gene centromere distances are estimated by counting the relative number of cross-overs between the locus and the centromere; the number of recombinational events is directly proportional to the distance. Extensive gene-centromere data in salmonid fishes suggest that the distances between two loci and the centromere differ between species[3,53,64,68]. These differences can be explained by chromosomal rearrangements or differential recombination rates between taxa. Such interspecific variation stands in contrast with the previously reported conservation of linkage distances reported in salmonid fishes.

The inheritance of genetic elements is further complicated if they are located on the sex chromosomes. The first indication of sex linkage is if the phenotypic ratios of a trait differ between males and females. For instance, differences in the frequency of variant allele at a Hexoseaminidase (Hex) locus between male and female Arlee rainbow trout led Gellman et al.[24] to test for sex linkage. Male rainbow trout are heterogametic (having two kinds of sex chromosomes)[60], and therefore linkage could only be detected when the male parent is segregating at Hex. A significant association between Hex and sex in families where the male parent was segregating for Hex was observed confirming that Hex is syntenic with a locus (Sex) that determines sex[24]. A different pattern is seen in other species, such as the southern platyfish (*Xiphophorus maculatus*), where females are heterogametic[33]. Many of the factors controlling pigment patterns are sex-linked in this species.

Even though the above approaches have made important advances in determining the physical relationship among loci and centromeres, assignment of a particular locus or linkage group to a chromosome is rarely possible in fishes. This is because methods for identifying chromosomes through staining and banding are relatively undeveloped. Methods such as Q-banding with a fluorescent dye or C-banding[47] rarely provide the degree of resolution to morphologically distinguish among different chromosomes. One exception to this would be where genetic loci are linked to sex determining loci and heteromorphic sex chromosomes exist. Significant progress awaits the utilization of the predominant gene mapping methods in mammals or probing chromosomes with radiolabelled or chromophore-labelled sequences.

4. Mobile genetic elements

Genes can move from one chromosome to another in the absence of major chromosomal rearrangements. These mobile elements or transposons have the ability to excise themselves from the chromosome and insert themselves at a new site. These elements have genes that code for the enzyme responsible for the excision and insertion. Studies with *Drosophila melanogaster* clearly illustrate the potential evolutionary importance of mobile genetic elements. *Drosophila* have a P-element transposon family that is 2,900 bases long and has the same base sequence on both ends but in reverse direction (inverted repeat). The P-elements change chromosomal position when individuals with P-elements are crossed to those without them. The effects of transposition are categorized under the term 'hybrid dysgenesis' and include inactivation or activation of genes at the locations where the P-elements are inserted and gonadal sterility[9]. Thus P-elements could result in reproductive isolation between different populations because hybrids may be sterile[57].

We know of no examples of transposable elements in fishes. However there is evidence that genes have been inserted into the genomes of fishes in the same manner as transposable elements. The nucleotide sequence of a protamine gene in rainbow trout shows many characteristics that suggest it was derived from a processed gene introduced into the genome by reverse transcription of a mature mRNA[31]. This could have been accomplished by horizontal transmission by infection with retroviruses. The nucleotide sequences flanking the trout protamine gene show significant homology to the avian sarcoma virus.

III. Gene duplication

The amount of nuclear DNA per cell has increased during evolution such that many genes are present in several copies[45]. Two general modes of duplication that have occurred in fishes are duplication of the entire genome (polyploidy) and regional duplication of specific regions. Regional duplication can result in distinct but similar genes that produce proteins of different forms or a set of genes that code for an identical protein product.

1. Polyploidy

Polyploidy has been reported in six orders of fishes: Acipenseriformes, Salmoniformes, Cypriniformes, Siluriformes, Atheriniformes, and Perciformes[5,11,19]. Most of the evidence for genome duplication in these groups comes from analyses of DNA content, karyotype, and more recently, duplicate gene expression of protein coding loci (Fig. 3). For example, cypriniform tetraploids show a nearly perfect doubling of their chromosome complement compared to their diploid counterparts.

Additional evidence for the polyploid constitution of proposed tetraploids has come from analysis of multilocus enzyme systems. For instance, it has been proposed that ancestors of vertebrates expressed one supernatant form of malate dehydrogenase (MDH)[21]. The supernatant form underwent an ancient duplication resulting in the typical pattern of two supernatant loci in diploid cypriniforms[49]. Tetraploid catestomids presumably went through additional polyploidizations because they express four supernatant loci. Polyploidy is a major evolutionary force because an additional locus is freed to acquire previously forbidden mutations and diverge to ultimately serve new functions[45].

Polyploid fishes have arisen from chromosome replication without cell division within the same taxon (autopolyploid) or genome doubling by combining the

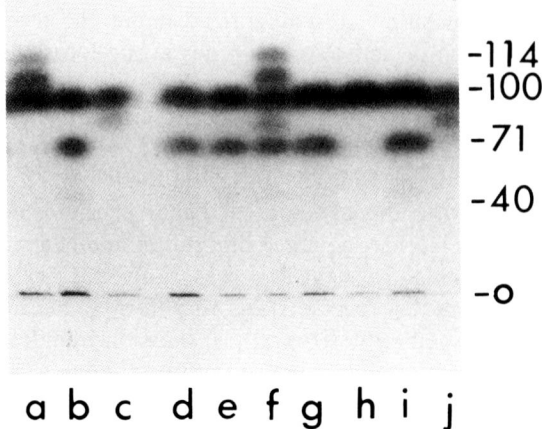

Fig. 3. A duplicated isocitrate dehydrogenase locus (Idh-3,4) in rainbow trout (*Oncorhynchus mykiss*). Each fish has four gene copies. Four different electromorphs, representing the protein products of different alleles, have been observed. These are designated by their relative electrophoretic mobilities such that the (100) allele is the most frequent. The origin is indicated by (o) and all migration is anodal. The presumed genotypes at Idh-3,4 are $114/100^3$ (a); $100^3/40$ (b,d,e,g,i); $100^3/71$ (c,j); $114/100^2/40$ (f); and 100^4 (h)[15].

chromosomes of different taxa involving hybridization (allopolyploid). Thus recent autotetraploids will have four chromosomes of similar structure while allopolyploids are likely to have two sets of chromosomes, each with different structure as each pair will have come from a different taxonomic origin.

Duplicate genes can be inherited in a variety of ways depending upon the type of polyploid and the time since the event. The chromosomes of recent tetraploids will form quadrivalents (all four chromosomes pair together) and undergo tetrasomic segregation. However, the evolution of structural differences between the two chromosome pairs will lead to the divergence of the duplicated locus in two loci. The chromosomes will then pair as bivalents (in two pairs) resulting in disomic segregation. The evolution of disomic segregation is expected to occur sooner in allotetraploids than autotetraploids because the chromosomes of allotetraploids will already be composed of two sets because of their hybrid origin. This prediction has been confirmed with extensive electrophoretic comparisons of the autotetraploid salmonids[5] and the allotetraploid catostomids[19]. Residual tetrasomic segregation has only been detected in the salmonids. The extensive disomic segregation in the salmonids is presumably due to the large amount of time since the duplication event (approximately 25–100 million years)[45].

A second feature in the evolution of fish polyploids is that a substantial proportion of the duplicate genes have become silenced through the fixation of alleles that produce no detectable enzyme product (null alleles) (Fig. 4). For example, electrophoretic studies of salmonids have shown that about two-thirds of examined loci do not show duplicate gene expression[5]. The expression of one of the duplicate genes is sometimes lost in all tissues or both loci may still retain

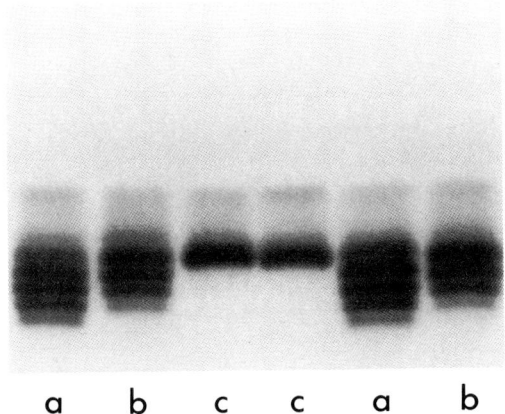

Fig. 4. A null allele (n) at a lactate dehydrogenase locus (Ldh-1) in brown trout (*Salmo trutta*). Fish with two copies of the null allele have 56% of the LDH enzyme activity in white muscle relative to homozygotes for the active allele (100). The remaining activity results from enzyme produced by other LDH loci. The presumed genotypes at Ldh-1 are a = 100/100, b = 100/n, c = n/n (ref. 4).

expression in different tissues (Fig. 5). Interestingly, the observed rate of gene silencing in salmonids is slower than that predicted by theoretical models[35] possibly because of selection against null alleles[18,58] or the retention of tetrasomic segregation[5,34].

Unusual inheritance occurs in some polyploid fishes. Hybrids between poeciliid bisexual species reproduce by gynogenesis and hybridogenesis[50]. In gynogenesis, a triploid hybrid female produces eggs with the same number of chromosomes as her somatic cells. Chromosomes from a coexisting sexual male are used to initiate embryogenesis but are not incorporated into the embryo. Thus the female's complement of are inherited clonally to her all-female offspring. In hybridogenesis,

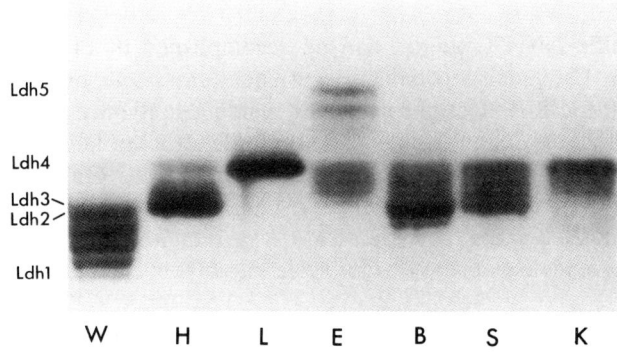

Fig. 5. Tissue-specific expression of five lactate dehydrogenase loci (Ldh) in brown trout (*Salmo trutta*). The location of the homotetramers produced by the common allele at each of the five loci is shown. W = white skeletal muscle, H = heart, L = liver, E = eye, B = brain, S = stomach, K = kidney[4].

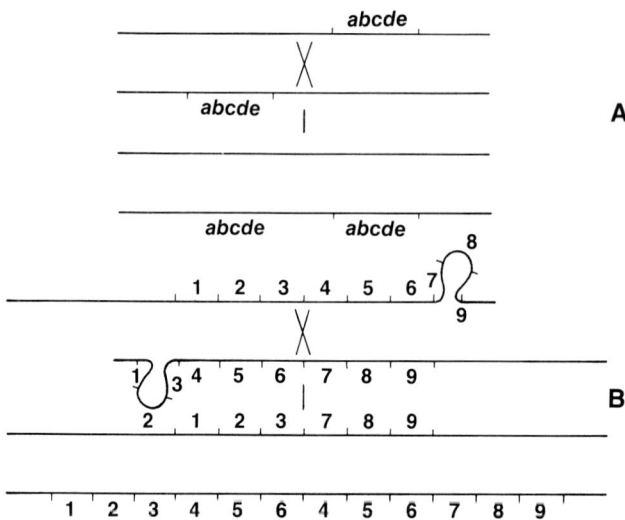

Fig. 6. Unequal crossing over of chromosomes. A illustrates the origin of a tandemly arranged DNA sequence, abcde. In B, two chromosomes each carry nine tandemly arranged copies of the same sequence. They pair out of register and crossing over gives rise to one chromosome with six copies and one with 12 copies[38].

a diploid hybrid female is produced. During meiosis, all eggs receive her maternally derived chromosomes; the paternally derived complement is differentially discarded in the polar body. The hybridogenetic females mate with males of a sexual species thus reestablishing the diploid condition. Both these modes of inheritance have been established with analyses of enzyme coding loci[66]. Moreover, restriction fragment length polymorphisms of mtDNA were used to establish that paternal leakage of mtDNA is extremely low or absent in the hybridogens[6].

2. Regional duplication

Gene duplication of a specific DNA sequence can be accomplished by errors in mitotic or meiotic processes. One such mechanism is unequal cross-over between homologous sequences resulting in the duplicate genes residing tandemly on the same chromosome (Fig. 6). Even though this may yield multiple structural cistrons, the regulatory gene controlling the activity is often not duplicated. Once the process begins, more copies can arise because there will be a greater tendency for the chromosomes to pair out of register. The duplicates can remain linked or be located on different chromosomes because of larger scale karyotypic changes such as translocations.

The products of regional duplication remain structurally similar and functionally constant if there is a need for large amounts of product in a relatively short time. For example, the tandemly linked and clustered antifreeze protein genes of winter flounder (*Pseudopleuronectes americanus*) show high sequence homology[51,52]. The

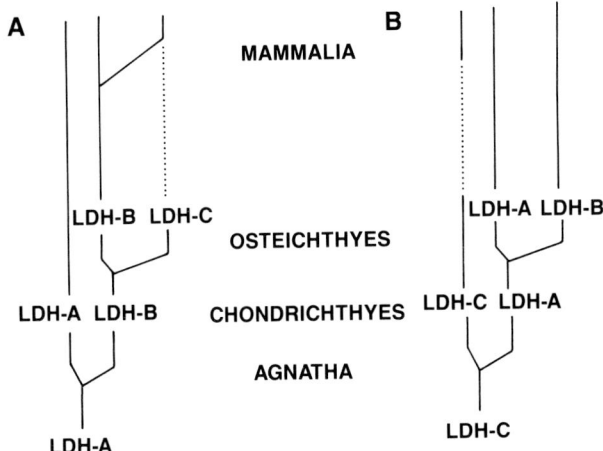

Fig. 7. Evolution of LDH isozymes. A illustrates the pattern of evolution suggested by Markert et al.[39] where Ldh-A is duplicated to form Ldh-A and Ldh-B. B illustrates the alternative scheme proposed by Crawford et al.[13] where Ldh-C gives rise to Ldh-C and Ldh-A[13].

production of antifreeze is strongly correlated with season such that the proteins are very abundant during the winter and decrease by at least 3 orders of magnitude in summer[22]. Interestingly, variation in the number and organization of antifreeze genes is correlated with serum antifreeze levels in pleuronectids and presumably the need for such a product[52]. For instance, the yellowtail flounder (*Limanda ferruginea*) has fewer copies than winter flounder and lower serum antifreeze levels. The increase in gene dosage of winter flounder might have helped this species colonize the ice-laden, shallow water of the east coast of North America.

The need for several similar but not identical products has led to structural and functional divergence among the products of regional duplications. For example, duplication of a single primitive lactate dehydrogenase locus (Ldh) in ancient vertebrates has resulted in at least three LDH isozymes based on similarities in amino acid compositions. LDH-A (muscle form) is best suited for pyruvate reduction in anaerobic tissues whereas LDH-B (heart form) is superior for lactate oxidation in aerobic tissues[39]. The distribution and biochemical properties of the third isozyme, LDH-C, depends upon the taxon. Markert et al.[39] proposed that a single Ldh-A type locus was duplicated to form Ldh-A and Ldh-B. Ldh-B was then duplicated to form Ldh-B and Ldh-C (Fig. 7). However, recent sequencing of a Ldh-B gene in *Fundulus heteroclitus* has questioned this proposed pathway; Ldh-C has been proposed as ancestral to the other genes (Fig. 7)[13]. The alternative evolutionary schemes can be resolved by DNA sequence analyses of representative vertebrates. A more extreme example of how regional duplication can result in products with divergent functions is the evolution of growth hormone, prolactin, and placental lactogen through repeated duplication from a common ancestor[2,43].

All the descendants of regional duplications do not produce functional products. These pseudogenes have sufficient sequence homology to functional genes to

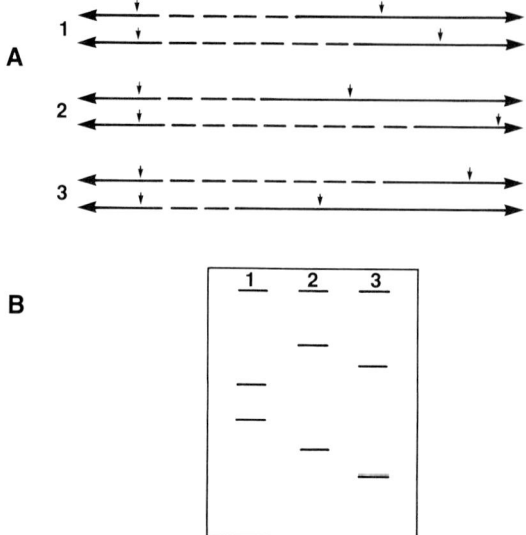

Fig. 8. Variation in the number of tandem repeats of a DNA sequence. Individuals 1, 2 and 3 have different numbers of repeats on their homologous chromosomes (A). For example, individual 1 has four copies on one chromosome and six copies on the homologue. The DNA is cut with a restriction enzyme (at arrows), electrophoresed in an agarose gel, and transferred to a filter. A DNA sequence from the repeating elements is radiolabelled as used as a hybridization probe. The DNA bands are visualized with autoradiography of the hybridized filter. Many allelic bands are revealed (B), reflecting different numbers of tandem repeats among chromosomes[29].

show that they have arisen by the duplication of functional genes. However, they contain mutations that make it certain that they are not translated into functional polypeptides. They can be linked to the functional locus or be located on other chromosomal segments. An immunoglobulin heavy chain variable region pseudogene, recently characterized in goldfish (*Carassius auratus*), illustrates the molecular mechanism of why such genes cannot produce a functional product[66]. This pseudogene DNA sequence contains a stop codon in the first coding region resulting in premature termination of mRNA synthesis and lacks the site necessary for the binding of the RNA polymerase enzyme (synthesizes mRNA). The structural similarity of this pseudogene with a second linked gene (86% nucleotide identity in the coding region) strongly suggests that they are the products of a gene duplication event.

Two other types of repeated DNA are present in the genome of vertebrates. Middle-repetitive dispersed DNA varies from a few hundred to a few thousand nucleotides and is usually present in tens or hundreds of copies per genome. Often a single copy may be present at each site or clusters may exist. Highly repetitive DNA is composed of short sequences, each present in very large numbers, often in tandemly arranged blocks of varying number of copies (Fig. 8). A particular sequence may be present on all chromosomes. Even though the evolutionary significance of these sequences is unknown, the large amount of variation in the

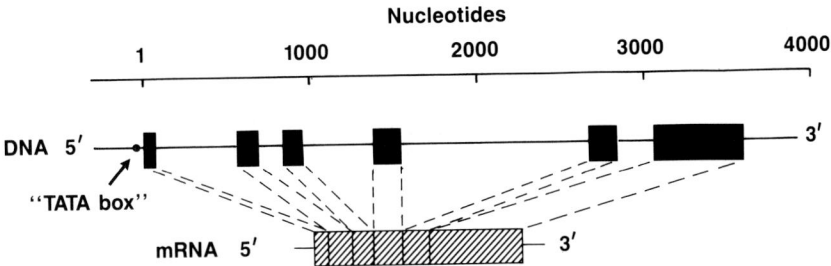

Fig. 9. A growth hormone gene in rainbow trout (*Oncorhynchus mykiss*) and its messenger RNA. The six coding sequences (exons) are represented by solid boxes and are interspersed with non-coding sequences (introns). The mature mRNA is composed of the exons only because the introns are excised[2].

number of tandem repeats at each locus among individuals can be used to estimate relatedness[32]. Fields *et al.*[20] found a high level of polymorphisms for such 'fingerprints' by analysis of DNA from a single-pair matings of rainbow trout. Each of the offspring showed a unique electrophoretic banding pattern (separated by the number of copies of each repeat) with each band being traced to one or both parents. The use of fingerprints to ascertain relatedness must proceed with caution, however. The allelic distributions in both the individuals of interest and the base population are usually unknown leading to biased estimates of relatedness[37].

IV. Functional organization

1. Gene structure

The development of techniques for the direct sequencing of the DNA molecule has revolutionized our understanding of the functional organization of the gene. Surprisingly, the sequence of a gene that codes for a particular macromolecule is not necessarily continuous but is split into exons by genomic regions that are not directly involved in the final product (intravening sequences or introns) (Fig. 9). Both the exons and introns are transcribed into mRNA but the section corresponding to the introns is excised before translation. In addition to the non-coding regions interspersed between the exons, eukaryotic genes have stretches of non-coding DNA both upstream and downstream from the gene proper. The upstream flanking region contains sequences (promoters) that act as signals for recognition and attachment of the transcribing enzyme. It is possible to identify promoters because their sequences in eukaryotes are generally conserved (e.g., the Goldberg-Hogness or TATA box).

A growth hormone (GH) gene of rainbow trout illustrates many of the above generalized features of a eukaryotic gene. The functional arrangement of the trout gene has recently been determined by comparing the genomic DNA nucleotide sequence (introns + exons) to that produced from making DNA (cDNA) of the mature mRNA transcript with a reverse transcriptase enzyme (exons only)[1,2] (Fig. 9). The organisation of the trout GH gene was determined by aligning the trout GH cDNA sequence with the corresponding regions of the genomic sequence. Introns were sequences of genomic DNA that had no homology with cDNA sequences; they had been excised before the production of the mature mRNA transcript. This trout gene is comprised of six exons, in contrast to the five exons in mammals. The additional intron in the trout gene interrupts translated regions that are analogous to the last exon of its mammalian counterpart. The promoter or 'TATA box' is located 22 nucleotides upstream from the beginning of the first exon. Two other important features are the putative cap site (+1) and the poly (A) signal; these are both important in the processing of the mRNA transcript.

The role of introns is unknown. One hypothesis is that introns are degenerate transposable elements which have lost their capacity for transposition. Alternatively, Gilbert[25] has proposed that modern genes evolved by bringing together a number of small and functional 'minigenes' (ancestors of exons) and that introns represent the neighbouring segments of DNA around these units.

All vertebrate genes do not contain introns. Sequencing studies have shown that rainbow trout protamine genes[16] and vertebrate mtDNAs have no interrupted genes[40]. The mtDNA molecule does, however, contain other types of non-coding regions; a 'control' region contains sequences that initiate replication and transcription. The control region in vertebrates contains a displacement loop (D-loop) which functions in replication.

2. Function and rates of evolution

Comparison of homologous amino acid and nuclear DNA sequences among organisms has shown clearly that different regions of the 'gene' evolve at different rates. Most substitutions, deletions, and insertions occur in dispensable sequences (intergenic sequences and introns)[36]. Presumably mutations within coding regions (exons) are not tolerated because of functional constraints. Generally regions that code for important macromolecules or subsections of macromolecules that are critical for function are more strictly conserved than less critical regions such as introns. For example, the greatest similarity among the amino-acid sequences of 14 LDH proteins from a wide variety of organisms (bacteria to human) is in the active site; the region that interacts with the substrate or coenzyme[13]. Similarly, Sekine *et al.*[54] suggested that the complete match within one region of the GH molecule of salmon, rat, cow, and human suggests that this region is critical for biological activity.

Molecular approaches to the study of fish sequences are so much in their infancy, that it would be premature to assume that the non-conserved regions of genes are functionally unimportant. For instance, Crawford *et al.*[13] have suggested that the non-conserved amino acids in LDH may be important in determining the biochemical parameters and physiological role of each isozyme. In addition, one cannot always assume that the structure of introns is under reduced selective constraint because an intron of the immunoglobulin heavy chain variable region gene shows unexpected conservation[66].

Differential evolution among different parts of the DNA molecule also occurs in mtDNA. The non-coding control region, especially the D-loop, changes very rapidly relative to rRNA genes[40]. Furthermore, marked changes in sequence length (overall size) in fishes have only been observed in the D-loop[7,8,10]. The extensive length polymorphism and heteroplasmy (multiple forms within the same individual) in the D-loop region in the mtDNA of white sturgeon (*Acipenser transmontanus*) is due to variable numbers of a perfect 82-bp direct repeat[10]. Interestingly, most base substitutions within coding regions do not result in the replacement of the particular amino acid residue. Sequence comparisons of mtDNA cloned from six Pacific salmon species detected variation at 338 silent and 12 non-silent positions within protein coding genes[59].

V. Is the genome organized?

Our ability to answer this question is biased by our perception of the way things should work and our limited knowledge of the way they do work. We might expect that functionally related genes would be physically contiguous so that the production of their products can be coordinated. Furthermore, we might propose that genes which control the timing and location of expression of other genes (regulatory genes) would be linked to the genes that they regulate (structural genes).

There is some evidence that genetic elements are organized as we might propose. For example, Graham et al.[28] observed that the gene order in homeobox complexes (responsible for body segmentation in *Drosophila*) has been conserved. The ordering of homeobox domains along an anterior-posterior axis corresponding to the timing of expression suggests that this arrangement is of functional significance.

We also know that many eukaryotic regulatory genes are located adjacent to the structural gene (*cis*-acting). For instance, the temporal regulatory gene that regulates the expression of phosphoglucomutase-1 enzyme in the liver of rainbow trout is apparently *cis*-acting (Allendorf, unpublished data). Examples such as this are not in the majority, however. For instance, functionally associated genes, such as those that code for enzymes within the major biochemical cycles, are not necessarily physically located near each other. Moreover, many regulatory genes are not linked to their structural genes.

The existence of the large amount of non-coding DNA interdispersed between and within coding regions is also hard to reconcile with an efficiently organized genome. This does not seem an especially efficient way of functioning; energy must be expended to remove these sequences before a mature mRNA molecule is produced. Perhaps, we do not understand the role of intervening sequences and more study will uncover the appropriate information. However, an equally appealing view is that such DNA is parasitic[17,46]. A resolution of this issue is not clear at present.

Acknowledgements. The authors thank the Natural Sciences and Engineering Research Council of Canada and the National Science Foundation, U.S.A. (BSR-8906451) for support. We also thank R.G. Danzmann for commenting on the manuscript and D. Morizot for providing a preprint.

VI. References

1. Agellon, L.B. and T.T. Chen. Rainbow trout growth hormone: molecular cloning and expression in *Escherichia coli*. *DNA* 5: 463–471, 1986.
2. Agellon, L.B., S.L. Davies, T.T. Chen and D.A. Powers. Structure of a fish (rainbow trout) growth hormone gene and its evolutionary implications. *Proc. Natl. Acad. Sci. U.S.A.* 85: 5136–5140, 1988.
3. Allendorf, F.W., J.E. Seeb, K.L. Knudsen, G.H. Thorgaard, and R.F. Leary. Gene-centromere mapping of 25 loci in rainbow trout. *J. Hered.* 77: 307–312, 1986.
4. Allendorf, F.W., G. Stahl and N. Ryman. Silencing of duplicate genes: a null allele polymorphism for lactate dehydrogenase in brown trout (*Salmo trutta*). *Mol. Biol. Evol.* 1: 238–248, 1984.
5. Allendorf, F.W. and G.H. Thorgaard. Tetraploidy and the evolution of salmonid fishes. In: *Evolutionary Genetics of Fishes*, edited by B. Turner. New York: Plenum Press, 1984, pp. 1–53.
6. Avise, J.C. and R.C. Vrijenhoek. Mode of inheritance and variation of mitochondrial DNA in hybridogenetic fishes of the genus *Poeciliopsis*. *Mol. Biol. Evol.* 4: 514–525, 1987.
7. Bentzen, P., W.C. Leggett and G.G. Brown. Length and restriction site heteroplasmy in the mitochondrial DNA of American shad (*Alosa sapidisssima*). *Genetics* 118: 509–518, 1988.
8. Bermingham, E., T. Lamb and J.C. Avise. Size polymorphism and heteroplasmy in the mitochondrial DNA of lower vertebrates. *J. Hered.* 77: 249–252, 1986.
9. Bregliano, J.C., G. Picard, A. Bucheton, A. Pelisson, J.M. Lavige and P. L'Heritier. Hybrid dysgenesis in *Drosophila melanogaster*. *Science* 207: 606–611, 1980.
10. Buroker, N.E., J.R. Brown, T.A. Gilbert, P.J. O'Hara, A.T. Beckenbach, W.K. Thomas and M.J. Smith. Length heteroplasmy of sturgeon mitochondrial DNA: an illegitimate elongation model. *Genetics* 124: 157–163, 1990.
11. Buth, D.G. Duplicate isozyme loci in fishes: origins, distribution, phyletic consequences, and locus nomenclature. In: *Isozymes, Current Topics in Biological and Medical Research, Vol. 10*, edited by M.C. Rattazzi, J.G. Scandalios and G.W. Whitt. New York: Liss, 1983, pp. 381–400.
12. Cavalier-Smith, T. *The Evolution of Genome Size*. New York: John Wiley, 1985.
13. Crawford, D.L., H.R. Constantino and D.A. Powers. Lactate dehydrogenase-B cDNA from the teleost *Fundulus heteroclitus*: evolutionary implications. *Mol. Biol. Evol.* 6: 369–383, 1989.
14. Danzmann, R.G. The karyology of eight species of fish belonging to the family Percidae. *Can. J. Zool.* 57: 2055–2060, 1979.
15. Danzmann, R.G., M.M. Ferguson and F.W. Allendorf. Allelic differences in initial expression of paternal alleles at an isocitrate dehydrogenase locus in rainbow trout (*Salmo gairdneri*). *Dev. Genet.* 5: 117–127, 1985.
16. Dixon, G.H., J.M. Aiken, J.M. Jankowski, D.I. McKenzie, R. Moir and J.C. States. Organization and evolution of the protamine genes of the salmonid fishes. In: *Chromosomal Proteins and Gene Expression*, edited by G.R. Reeck, G.A. Goodwin and P. Puigdomenech. New York: Plenum Press, 1986, pp. 287–314.
17. Doolittle, W.F. and F. Sapienza. Selfish genes, the phenotype paradigm and genome evolution. *Nature* 284: 617–618, 1980.
18. Ferguson, M.M., K.L. Knudsen, R.G. Danzmann and F.W. Allendorf. Developmental rate and viability of rainbow trout with a null allele at a lactate dehydrogenase locus. *Biochem. Genet.* 26: 177–189, 1988.
19. Ferris, S.D. Tetraploidy and the evolution of the catostomid fishes. In: *Evolutionary Genetics of Fishes*, edited by B.J. Turner. New York: Plenum Press, 1984, 55–93.
20. Fields, R.D., K.R. Johnson and G.H. Thorgaard. DNA fingerprints in rainbow trout detected by hybridization with DNA of bacteriophage M13. *Trans. Am. Fish. Soc.* 118: 78–81, 1989.
21. Fisher, S.E., J.B. Shaklee, S.D. Ferris and G.S. Whitt. Evolution of five multilocus isozyme systems in the chordates. *Genetica* 52/53: 73–85, 1980.
22. Fletcher, G.L. Circannual cycles of blood plasma freezing point and Na^+ and Cl^- concentrations in Newfoundland winter flounder (*Pseudopleuronectes americanus*): correlation with water temperature and photoperiod. *Can. J. Zool.* 55: 789–795, 1977.

23. Foresti, F., L.F. Almeida Toledo and S.A. Toledo F. Polymorphic nature of nucleolus organizer regions in fishes. Cytogenet. Cell Genet. 31: 137–144, 1981.
24. Gellman, W.A., F.W. Allendorf and G.H. Thorgaard. Hexosaminidase is sex linked in rainbow trout. Isozyme Bull. 20: 14, 1987.
25. Gilbert, W. Genes-in-pieces revisited. Science 228: 823–894, 1985.
26. Gold, J.R. Cytogenetics. In: Fish Physiology, Vol. VIII, edited by W.S. Hoar and D.J. Randall and J.R. Brett, New York: Academic Press, 1979, pp. 353–405.
27. Gold, J.R. and C.T. Amemiya. Genome size in North American minnows (Cyprinidae). II. Variation among 20 species. Genome 29: 481–489, 1987.
28. Graham, A., N. Papalopulu and R. Krumlauf. The murine and *Drosophila* homeobox gene complexes have common features of organization and expression. Cell 57: 367–378, 1989.
29. Hallerman, E.M. and J.S. Beckmann. DNA-level polymorphism as a tool in fisheries science. Can. J. Fish. Aquat. Sci. 45: 1075–1087, 1988.
30. Hinegardner, R. Evolution of genome size. In: *Molecular Evolution*, edited by F.J. Ayala, Sunderland, MA: Sinauer, 1976, pp. 179–199.
31. Jankowski, J.M., J.C. States and G.H. Dixon. Evidence of sequences resembling avian retrovirus long terminal repeats flanking the trout protamine gene. J. Mol. Evol. 23: 1–10, 1986.
32. Jeffreys, A.J., V. Wilson and S.L. Thein. Hypervariable 'minisatellite' regions in human DNA. Nature 314: 67–73, 1985.
33. Kallman, K.D. A new look a sex determination in poeciliid fishes. In: *Evolutionary Genetics of Fishes*, edited by B.J. Turner, New York: Plenum Press, 1984, pp. 95–171.
34. Li, W.-H. Rate of gene silencing at duplicate loci: a theoretical study and interpretation of data from tetraploid fishes. Genetics 95: 237–258, 1980.
35. Li, W.-H. Evolutionary change at duplicate genes. In: *Isozymes, Current Topics in Biological and Medical Research, Vol. 6*, edited by M.C. Rattazzi, J.G. Scandalios and G.W. Whitt. New York: Alan R. Liss, 1982, pp. 55–92.
36. Li, W.-H., C.C. Cheng and C.I. Wu. Evolution of DNA sequences. In: *Molecular Evolutionary Genetics*, edited by R.J. MacIntyre, New York: Plenum Press, 1985, pp. 1–94.
37. Lynch, M. Estimation of relatedness by DNA fingerprinting. Mol. Biol. Evol. 5: 584–599, 1988.
38. Maynard Smith, J. *Evolutionary Genetics*. New York: Oxford University Press, 1989.
39. Markert, C.L., J.B. Shaklee and G.S. Whitt. Evolution of a gene. Science 189: 102–114, 1975.
40. Moritz, C., T.E. Dowling and W.M. Brown. Evolution of animal mitochondrial DNA: relevance for population biology and systematics. Ann. Rev. Ecol. Syst. 18: 269–292, 1987.
41. Morizot, D.C. Comparative gene mapping evidence for chromosome duplications in chordate evolution. Isozyme Bull. 19: 9–10, 1986.
42. Morizot, D.C. Use of fish gene maps to predict ancestral vertebrate genome organization. In: *Sixth International Conference, Isozymes*, edited by C.L. Markert, New York: Alan R. Liss, in press.
43. Niall, H.D., M.L. Hogan, R. Sauer, I.Y. Roseblum and F.C. Greenwood. Sequences of pituitary and placental lactogenic and growth hormones: evolution from a primordial peptide by gene reduplication. Proc. Natl. Acad. Sci. U.S.A. 68: 866–869, 1971.
44. Ohno, S. *Sex Chromosomes and Sex Linked Genes*. New York: Springer-Verlag, 1967.
45. Ohno, S. *Evolution by Gene Duplication*. New York: Springer-Verlag. 1970.
46. Orgel, L.E. and F.H.C. Crick. Selfish DNA: the ultimate parasite. Nature 284: 645–646, 1980.
47. Phillips, R.B. and K.D. Zajicek. Q band chromosomal banding polymorphisms in lake trout (*Salvelinus namaycush*). Genetics 101: 227–234, 1982.
48. Price, D.J. Genetics of sex determination in fishes: a brief review. In: *Fish Reproduction: Strategies and Tactics*, edited by G.W. Potts and R.J. Wootton. London: Academic Press, 1984, pp. 77–89.
49. Rainboth, W.J. and G.S. Whitt. Analysis of evolutionary relationships among shiners of the subgenus *Luxilus* (Teleostei, Cypriniformes, Notropis) with the lactate dehydrogenase and malate dehydrogenase isozymes systems. Comp. Biochem. Physiol. 49B: 241–252, 1974.
50. Schultz, R.J. Evolution and ecology of unisexual fishes. In: *Evolutionary Biology*, edited by M.K. Hecht, W.C. Steere, and B. Wallace. New York: Plenum Press, 1977, 277–331.
51. Scott, G.K., C.L. Hew and P.L. Davies. Antifreeze protein genes are tandemly linked and clustered in the genome of the winter flounder. Proc. Natl. Acad. Sci. U.S.A. 82: 2613–2617, 1985.
52. Scott, G.K., P.L. Davies, M.H. Kao and G.L. Fletcher. Differential amplification of antifreeze protein genes in the Pleuronectinae. J. Mol. Evol., 27: 29–35, 1988.
53. Seeb, J.E. and L. Wishard Seeb. Gene mapping of isozyme loci in chum salmon. J. Heredity 77: 399–402, 1986.
54. Sekine, S., A. Saito, H. Itoh, H. Kawauchi and S. Itoh. Molecular cloning and sequence analysis of chum salmon gonadotrophin. Proc. Natl. Acad. Sci. U.S.A. 86: 8645–8649, 1989.

55. Singh, L., I.F. Purdom and K.W. Jones. Satellite DNA and evolution of sex chromosomes. *Chromosoma* 59: 43–62, 1976.
56. Sola, L., S. Cataudella and E. Capanna. New developments in vertebrate cytotaxonomy. III. karyology of bony fishes: a review. *Genetica* 54: 285–328, 1981.
57. Syvanen, M. The evolutionary implications of mobile genetic elements. *Ann. Rev. Genet.* 18: 271–294, 1984.
58. Takahata, N. and T. Maruyama. Polymorphism and loss of duplicate gene expression: a theoretical study with application to tetraploid fish. *Proc. Natl. Acad. Sci. U.S.A.* 76: 4521–4525, 1979.
59. Thomas, W.K. and A.T. Beckenbach. Variation in salmonid mitochondrial DNA: evolutionary constraints and mechanisms of substitution. *J. Mol. Evol.* 29: 233–245, 1989.
60. Thorgaard, G.H. Heteromorphic sex chromosomes in male rainbow trout. *Science* 196:900–902, 1977.
61. Thorgaard, G.H. Sex chromosomes in the sockeye salmon: a Y-autosome fusion. *Can. J. Genet. Cytol.* 20: 349–354, 1978.
62. Thorgaard, G.H. Chromosomal differences among rainbow trout populations. *Copeia* 1983: 650–662, 1983.
63. Thorgaard, G.H. and S.K. Allen. Chromosome manipulation and markers in fishery management. In: *Population Genetics and Fishery Management*, edited by N. Ryman and F. Utter, Seattle: University of Washington Press, 1987, pp. 319–332.
64. Thorgaard, G.H., F.W. Allendorf and K.L. Knudsen. Gene-centromere mapping in the rainbow trout: high interference over long map distances. *Genetics* 103: 771–783, 1983.
65. Vrijenhoek, R.C. The evolution of clonal diversity in *Poeciliopsis*. In: *Evolutionary Genetics of Fishes*, edited by B.J. Turner. New York: Plenum Press, 1984, pp. 399–429.
66. Wilson, M.R., D. Middleton and G.W. Warr. Immunogloblin heavy chain variable region gene evolution: structure and family relationships of two genes and a pseudogene in a teleost fish. *Proc. Natl. Acad. Sci. U.S.A.* 85: 1566–1570, 1988.
67. White, M.J.D. *Animal Cytology and Evolution*. London: University Press Cambridge, 1973.
68. Wright, J.E., K.R. Johnson and B. May. Synthetic linkage map of salmonids. In: *Genetic Maps 1987*: a compilation of linkage and restriction maps of genetically studied organisms, edited by S.J. O'Brien. New York: Cold Spring Harbor Laboratories, 1987, pp. 405–413.

CHAPTER 3

Evolution of mitochondrial enzyme systems in fish: the mitochondrial synthesis of glutamine and citrulline

JAMES W. CAMPBELL * AND PAUL M. ANDERSON **

*Department of Biochemistry and Cell Biology, Rice University, Houston, TX 77251, U.S.A., and **Department of Biochemistry and Molecular Biology, University of Minnesota-Duluth, Duluth, MN 55812, U.S.A.*

I. Introduction
II. Mitochondrial glutamine synthesis
  1. Occurrence and distribution of glutamine synthetase in fish
  2. General properties of elasmobranch glutamine synthetase
  3. Subcellular localization and metabolic function of hepatic glutamine synthetase
  4. Isozymes of glutamine synthetase in cartilaginous fish
III. Mitochondrial citrulline synthesis
  1. Occurrence and distribution of urea cycle enzymes in fish
  2. General properties of CPSs and associated enzymes
  3. Compartmentation of CPS-III and associated enzymes in elasmobranch tissues and functional significance
  4. Evolution of CPS and associated enzymes
  5. Evolutionary significance of the hepatic system for citrulline synthesis in elasmobranchs
  6. On the handling of mitochondrial ammonia in fish and other ammonoteles
  7. The special localization of arginase in elasmobranch and teleost mitochondria
Acknowledgements
IV. References

I. Introduction

Most studies in molecular evolution have focused on how gene mutations, and the consequent amino acid substitutions in the enzymes they encode, may alter the metabolism of organisms. It is this alteration in phenotype that affects interaction with the environment and ultimately, adaptation to specific niches[78,98]. Gene duplication is the coinage of molecular adaptation because it allows for the development of new enzymes via mutation of one of the duplicated genes while the constitutive function of the original gene is retained[123]. This process first gives rise to allozymes that may differ by only one to a few amino acids. These substitutions may affect not only catalytic efficiency[132] and thermal stability[181], but also gross regulatory parameters[100,102] of the allozymes. Continued divergence of duplicated genes during the early evolution of metabolism is thought to have resulted in the emergence of enzymes with modified substrate specificities for function in specific metabolic sequences[87,190] and eventually, to homologous enzymes functioning in different pathways[108,129].

In eukaryotic cells, the subcellular compartmentation of enzymes is also critical to their metabolic function. For example, most enzymes that generate hydrogen peroxide are localized within peroxisomes so this product can be detoxified by catalase[106]. Association of related enzymes within organelles such as occurs in mammalian liver mitochondria between the enzymes of transdeamination and the urea cycle[60,150] also allows for channeling of substrates and an increase in overall metabolic efficiency. Most organellar proteins are nuclearly encoded and translated in the cytosol. The information for targeting to specific organelles is contained in the primary structure of their mRNA translation products[186]. Thus, just as some mutations affect catalytic, stability, or regulatory properties of enzymes, others must also affect their subcellular localization. It might be expected that such mutations would result in fairly marked changes in metabolism. It has, for example, been postulated that the emergence of uricotely as a different type of excretory nitrogen metabolism during the evolution of the tetrapod vertebrates was dependent upon the translocation of glutamine synthetase into liver mitochondria. The localization of glutamine synthetase in this compartment allowed it to function more efficiently as the primary detoxifying system for ammonia during amino acid gluconeogenesis[29].

Differences in the subcellular localization of enzymes may be either tissue- or species-specific. The latter are of prime concern from the standpoint of the evolution of metabolism. One of the first examples of species-specific compartmentation was that of phospho*enol*pyruvate (PEP)-carboxykinase in mammals and birds[64,120]. This enzyme catalyzes a key reaction in gluconeogenesis and occurs in both the cytosolic and mitochondrial compartments in liver of birds and mammals as well as in other vertebrates, including fish[169]. Other examples of species-specific compartmentation in fish include allantoicase, an enzyme of the uric acid degradative pathway, which occurs in the cytosolic and peroxisomal compartments in liver tissue[76]. Changes in the subcellular distribution of an enzyme may require subsequent changes in its catalytic and regulatory properties as well as those of associated enzymes and transport systems for greater efficiency of function. The development of a uricotelic type hepatic metabolism among higher vertebrates required changes not only in the mitochondrial localization of glutamine synthetase, but also in the regulatory properties of enzymes of the purine biosynthetic pathway[25,29]. The exclusive localization of PEP-carboxykinase in the mitochondrial compartment requires changes in the source of cytosolic reducing equivalents for the conversion of triose phosphates to glucose[167,192] and therefore changes in transport systems and in the localization of related enzymes such as alanine aminotransferase[49]. In most cases, translocation from one compartment to another increases metabolic efficiency by bringing related reactions together. In other cases, it is necessary because of changes in membrane permeability. Ornithine transaminase is usually localized in mitochondria. However, in insects, whose mitochondria have become impermeable to Krebs cycle intermediates in order to sustain high rates of oxidative metabolism, it is localized in the cytosol where it functions in the formation of proline which is permeable to insect mitochondria[144].

More than 90% of mitochondrial proteins are nuclearly encoded and synthe-

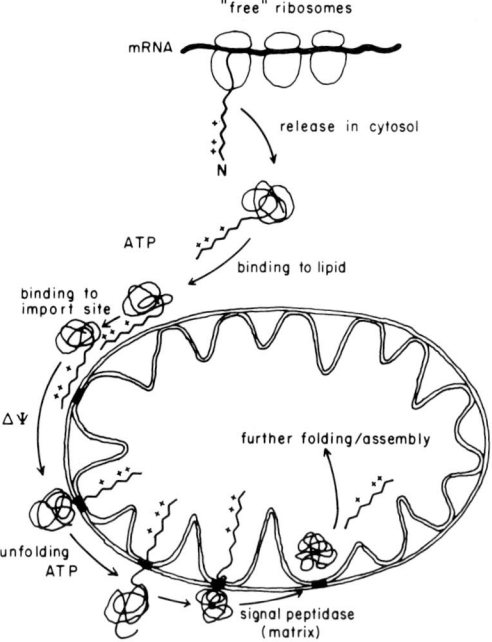

Fig. 1. Generalized mechanism for the synthesis and import of peptides targeted to mitochondria by an N-terminal, transient presequence.

sized in the cytosol[9]. A generalized outline of their import into mitochondria is given in Fig. 1. Although internal targeting signals are utilized by some mitochondrial peptides, especially those of the inner mitochondrial membrane, most of those targeted to the matrix utilize an N-terminal presequence consisting of from 12 to more than 70 amino acids[73,183]. There is no consensus sequence for targeting to mitochondria. The common shared properties of mitochondrial targeting signals are their basic, hydroxyl, and hydrophobic amino acid contents (and general absence of acidic amino acids) and their ability to form amphiphilic helices[187]. Following release from free ribosomes, the prepeptide interacts in an ATP-dependent manner with an unfoldase, also called a chaperonin, to assume an import-competent configuration[50]. These chaperonins either are, or are closely related to, heat shock proteins. There is an initial interaction of the signal sequence with cardiolipin and other acidic phospholipids of the mitochondrial membranes[56,170]. There is also interaction with a protein receptor molecule on the outer mitochondrial membrane[184] although the inner membrane contains all the necessary import machinery[83]. Translocation into the matrix requires an energized inner membrane ($\Delta\Psi$) and ATP. During this process, the targeting signal is removed by a signal peptidase[127] and further hydrolyzed[146]. Final assembly of the imported peptide to the mature form requires an additional chaperonin within the mitochondrion[37].

Compartmental isozymes, that is, enzymes localized to two different subcellular compartments either in different cells in a given tissue or in different tissues, very much illustrate the role of gene duplication in the evolution of metabolism. Aspartate aminotransferase[40], PEP-carboxykinase[42,79], and malate dehydrogenase (MDH)[88] are examples of compartmental isozymes occurring in mitochondrial and cytosolic compartments. The cytosolic and mitochondrial forms of each of these pairs of isozymes are coded for by separate genes that may be differentially regulated. They are, however, thought to be homologous, having arisen from an ancestral gene via duplication. Following duplication of the ancestral gene, it was necessary for one of the resulting genes to either acquire the necessary coding information for a mitochondrial targeting signal or for this information to be deleted. The peptide coded for by the gene lacking the targeting information remains in the cytosolic compartment. In any event, these compartmental isozymes have subsequently undergone different rates of evolution so they are now distinct with respect to their molecular, immunochemical, and kinetic properties. Cytosolic and mitochondrial MDHs show only 23% sequence similarity. In fact, the mitochondrial MDH isozyme shows more similarity to the *Escherichia coli* enzyme than to its cytosolic counterpart[88]. The isozyme within a given compartment nevertheless shows considerable sequence similarity to its counterpart in other species.

Chondrichthian fish (elasmobranchs and holocephalans) represent the most primitive of the jawed vertebrates[151] and are generally felt to have diverged from the main line of descent leading to the tetrapods[152]. A major physiological adaptation of these fish is their utilization of urea as an osmolyte. This required the development of membranes that are relatively impermeable to urea and a renal system for the active reabsorption of urea[131]. It also required development of the capacity for urea biosynthesis. The urea-biosynthetic pathway present in liver of chondrichthian fish is distinct from that in other ureotelic vertebrates in that a mitochondrial isozyme of glutamine synthetase is the primary ammonia-utilizing enzyme, and not carbamoyl-phosphate synthetase (CPS), the first enzyme in the classical urea cycle. Ornithine transcarbamylase and a CPS are also localized in the matrix of these mitochondria. The latter is a glutamine-dependent enzyme so citrulline is nevertheless the eventual end-product of mitochondrial ammonia metabolism. This unique mitochondrial citrulline-forming system also appears to be utilized by other fishes to detoxify ammonia on an opportunistic basis, especially under conditions of limited water availability. The co-localization of glutamine synthetase and CPS in hepatic mitochondria of primitive fish may therefore represent evolutionary precedents for their independent function as primary ammonia-detoxifying enzymes in land vertebrates. Here, we discuss the possible evolution of these and associated enzymes of the urea pathway and their targeting to different subcellular compartments.

II. Mitochondrial glutamine synthesis

1. Occurrence and distribution of glutamine synthetase in fish

Webb and Brown[194,195] determined the tissue distribution of glutamine synthetase in a variety of fish, both ureosmotic and non-ureosmotic. The results of this survey

TABLE 1

Glutamine synthetase in fish liver and brain tissues

Species	Activity (μmol/h/g tissue)	
	Liver	Brain
Oncorhynchus tshawytscha (Chinook salmon)	bld	2558
Platichthys stellatus (starry flounder)	bld	2425
Sebastes caurinus (copper rockfish)	bld	1214
Ictalurus punctatus (channel catfish)	19 * (C)	361 *
Lepomis macrochirus (bluegill)	bld	680 *
Pomoxis nigromaculatus (black crappie)	bld	476 *
Opsanus beta (Gulf toadfish)	323 (M)	–
O. tau (oyster toadfish)	90 (C)	–
Heteropneustes fossilis (mudskipper)	61 (M)	495 (C)
Squalus acanthias (spiny dogfish)	1212, 773 * (M)	1049, 488 * (C)
Raja binoculata (big skate)	1707	1207
Dasyatis sabina (Gulf stingray)	421 * (M)	184 * (C)
Hydrolagus colliei (ratfish)	456, 28 * (M)	1326, 76 * (C)

Values for activity are not corrected for assay temperature (26–37°). Values marked by * indicate the synthetic reaction was used for assay; other values are based on the transferase assay[172]. (C) indicates that the enzyme is localized in the cytosolic compartment and (M), in the mitochondrial compartment. Data are compiled from refs. 33, 36, 112, 148, 194, 198, and 204. bld = below level of detection.

confirmed the presence of high levels of the activity in brain tissue[204] and showed that liver tissue of ureosmotic fish also has high levels of activity, higher, in fact, than that in brain. As shown in Table 1, some non-ureosmotic teleosts also have low levels of glutamine synthetase activity in liver. Again, the higher levels appear to be in those species in which the enzyme is localized in the mitochondrial compartment. Glutamine synthetase also occurs in kidney, as well as other fish tissues (see Table 3). The enzyme is localized in the mitochondrial compartment in both elasmobranch[95] and teleost kidney[36].

2. General properties of elasmobranch glutamine synthetase

Glutamine synthetase catalyzes the ATP-dependent synthesis of glutamine as shown in reaction 1:

(1) Glutamate + ATP + NH_3 $\xrightarrow{Mg^{2+}}$ glutamine + ADP + P_i

Mn^{2+} or Co^{2+} may substitute for Mg^{2+}. Hydroxylamine may also substitute for ammonia resulting in γ-glutamylhydroxamate formation via reaction 2:

(2) Glutamate + ATP + NH_2OH $\xrightarrow{Mg^{2+}}$ γ-glutamylhydroxamate + ADP + P_i

The first step in glutamine formation is relatively non-stereospecific in that either L- or D-glutamate will serve as a substrate, at least for the sheep brain enzyme[172].

TABLE 2

Properties of vertebrate hepatic glutamine synthetases

Enzyme M_r	360,000	390,000	400,000
Subunit M_r	44,000	42,000	46,000
K_m ATP (Mg^{2+})	2 mM	0.9 mM	0.7 mM
K_m L-glutamate (Mg^{2+})	5 mM	6 mM	11 mM
K_m ammonia (Mg^{2+})	0.3 mM	0.5 mM	0.015 mM

General variation in M_r determination is $\pm 10\%$ (see e.g., Table 3). Data are compiled from refs. 51, 154, 173, and 188.

Because the resulting hydroxamate can be quantitated with acidic ferric chloride, reaction 2 can be utilized for the assay of glutamine synthetase activity[128,189]. A more sensitive transferase assay is based on reaction 3[203]:

$$(3) \text{Glutamine} + NH_2OH \xrightarrow{Mg^{2+}, ADP} \gamma\text{-glutamylhydroxamate} + NH_3$$

Prokaryotic glutamine synthetases are typically dodecameric in their quaternary structure and are regulated by adenylation--deadenylation. The eukaryotic enzymes are octameric in their quaternary structure and are regulated by small metabolites[43,97] (chapts. in 135). The only fish glutamine synthetase that has been purified and characterized is from liver of the spiny dogfish, *Squalus acanthias*[154]. As shown in Table 2, the properties of the dogfish enzyme are very similar to those of other vertebrate glutamine synthetases. It does, however, have an unusually high affinity for ammonia.

3. Subcellular localization and metabolic function of hepatic glutamine synthetase

In mammalian liver, glutamine synthetase is in the cytosolic compartment[205] of perivenule hepatocytes, the only cells in which it is expressed in mammals[160]. It appears to function in these cells as a 'fail-safe' mechanism for ammonia detoxification when the capacity of the urea pathway is exceeded[75]. It presumably also provides glutamine for biosynthetic reactions in the cytosol. In uricotelic liver, glutamine synthetase is localized in the mitochondrial matrix where it functions as the primary ammonia-detoxifying enzyme in a manner analogous to that of CPS in ureotelic liver[25]. Glutamine formed in uricotelic liver mitochondria exits to the cytosol where it serves as a precursor for uric acid synthesis. The mitochondrial formation of glutamine is therefore analogous to the mitochondrial formation of citrulline in ureotelic liver. Since citrulline and glutamine are neutral compounds, both ammonia-detoxifying mechanisms are presumed to have evolved in response to the need to convert ammonia formed intramitochondrially during amino acid catabolism to a form that would not bind protons during transit of the inner membrane and therefore, not uncouple oxidative phosphorylation[25].

In cartilaginous fish and some teleosts, glutamine synthetase is localized in mitochondria in liver and kidney and in the cytosol in brain and other tissues (Tables 1 and 3). The renal enzyme may function as part of a substrate cycle for ammonia excretion during acidosis[95]. While the mitochondrial localization of the enzyme in liver of cartilaginous fish is similar to that in uricotelic liver, this probably evolved more as a mechanism for glutamine-dependent urea synthesis for osmotic purposes[6] than as a primary ammonia-detoxifying system since most fish efficiently handle ammonia by direct excretion[131]. A significant metabolic consequence of the glutamine-dependent urea synthesis system in ureosmotic fish is that glutamine is utilized directly for carbamoyl phosphate (and, ultimately, urea) synthesis and may not be available as a substrate in the cytosol for pyrimidine nucleotide synthesis (or to other pathways requiring glutamine as a substrate in reactions catalyzed by amidotransferases). The absence of aspartate transcarbamylase activity in liver of spiny dogfish[5] is consistent with this view as is the absence of glutamine from blood serum in elasmobranchs[104]. Since glutamine is a significant energy source for tumor growth[43,113,136], the unavailability of circulating glutamine in elasmobranchs may contribute to their unusually low incidence of neoplasms[15,103].

The freshwater, air-breathing teleost *Heteropneustes fossilis*, which is able to switch from an ammonotelic- to a ureotelic-type metabolism during hyperammonemia or dehydration stress, shows a subcellular distribution of glutamine synthetase in liver similar to that in ureosmotic fish[36]. However, it has not been established if urea synthesis in this species involves glutamine formation as the initial step in ammonia assimilation. A relationship between the mitochondrial localization of glutamine synthetase in liver and glutamine-dependent urea synthesis is not universal in fish, at least in teleost species, since glutamine synthetase is present in the cytosol of liver in some that appear to have a glutamine-dependent CPS[30,112].

4. Isozymes of glutamine synthetase in cartilaginous fish

Most octameric eukaryotic glutamine synthetases have a subunit mass of around 45 kDa whereas the subunit mass of the dodecameric prokaryotic enzymes is around 50 kDa. The yeast (*Saccharomyces cerevisiae*) enzyme, which is cytosolic, appears to represent a transition form between the dodecameric prokaryotic glutamine synthetases and the eukaryotic octameric enzymes in that it is composed of 10 to 12 subunits of mass 42 kDa[111] but shows more sequence similarity with the eukaryotic enzyme than with the prokaryotic enzyme[94]. In plants, glutamine synthetase occurs as cytosolic and chloroplastal isozymes that appear to have arisen via gene duplication and are now coded for by a multigene family[44]. There also appear to be two glutamine synthetase isozymes in insects that differ in sequence and are coded for by separate genes[24]. The enzyme has been reported to be mitochondrial in insect muscle[55] but whether or not the two insect glutamine synthetases are compartmental isozymes has not been established.

As shown in Fig. 2, there appears to be a single gene for glutamine synthetase in elasmobranchs since only single hybridization signals are obtained with restriction

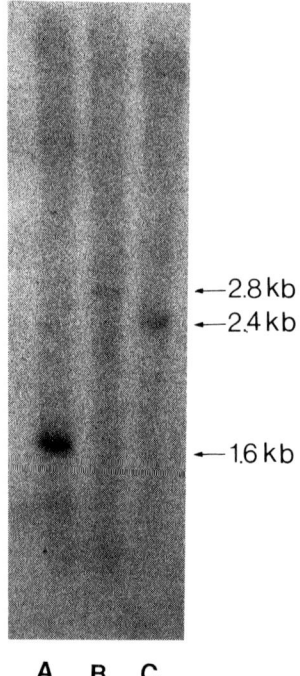

Fig. 2. Southern blot analysis of stingray (*Dasyatis sabina*) DNA using bovine retinal glutamine synthetase cDNA clone pGS1 as a probe[160]. Lane A: *Eco*RI; lane B: *Hind*III; lane C: *Bam*HI. Numbers indicate size of DNA in kilobases. (from ref. 147)

fragments of elasmobranch DNA utilizing a bovine retinal cDNA probe. The elasmobranch enzyme is also immunochemically cross-reactive with antisera to the chicken mitochondrial enzyme which further indicates a great deal of sequence conservation in vertebrate glutamine synthetases[162]. As shown in Fig. 3, glutamine synthetase mRNA in the stingray (*Dasyatis sabina*) is around 3 kB in size which is within the size range for most other vertebrate glutamine synthetase mRNAs[160]. Dogfish (*S. acanthias*) mRNA, on the other hand, is much larger at 4.3 kb. Despite the difference in the size of their respective mRNAs, elasmobranch glutamine synthetase subunits fall in the size range $M_r = 42,000$ to $47,000$ which is typical of other eukaryotic enzymes (Table 2)[154,161]. This is also true for the ratfish, *Hydrolagus colliei*[148].

As can be seen in Table 3, there is a tissue-specific expression of glutamine synthetase isozymes in both elasmobranchs and holocephalans. The larger isozyme is generally present in tissues in which the enzyme is localized in mitochondria which includes both liver[34] and kidney[95]. The smaller isozyme is expressed in brain[161] and spleen[5] where it is cytosolic. The smaller isozymes in gill, retina, heart, and rectal gland therefore also are assumed to be cytosolic in their localization.

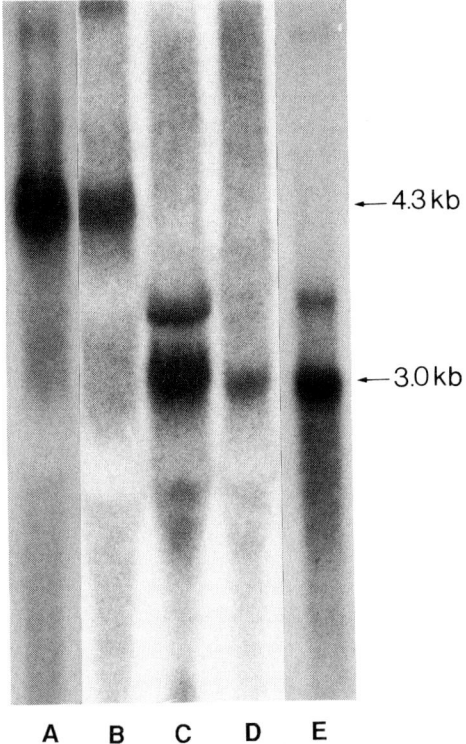

Fig. 3. Northern blot analysis of elasmobranch glutamine synthetase RNA using pGS1 as a probe (see Fig. 2). Lane A: dogfish (*Squalus acanthias*) kidney; lane B: dogfish brain; lane C: stingray (*D. sabina*) kidney; lane D: stingray brain; lane E: stingray liver. Numbers indicate size of RNA in kilobases. (from ref. 147)

The data in Table 3 also indicate that the targeting of pre-glutamine synthetase to mitochondria in elasmobranch liver involves a classical N-terminal, transient presequence. However, it has not been possible to demonstrate a time-dependent conversion of the larger precursor to the smaller mature form in either intact cells or mitochondrial lysates[147]. In liver, where the enzyme is mitochondrial, the mRNA translation product is 1.5 to 3.0 kDa larger that the mature form. In brain, the size of the translation product is the same as that of the mature peptide[161]. As shown in Fig. 4, lysed stingray liver mitochondria convert the 48 kDa stingray precursor to the 45 kDa mature form. It should also be noted that these mitochondria also convert the 50 kDa dogfish precursor to the 45 kDa mature stingray subunit, not to the 47 kDa mature dogfish subunit. This suggests there may be species-specific cleavage sites on elasmobranch pre-glutamine synthetase which would account for the species-specific isozymes in liver tissue. It has also been possible to demonstrate the synthesis of glutamine synthetase in tissue slices of stingray kidney. This synthesis is abolished by the cytosolic protein synthesis inhibitor, cycloheximide, but not by the mitochondrial inhibitor, chloramphenicol.

TABLE 3

M_r of glutamine synthetase subunit in cartilaginous fish

Source	M_r	
	mRNA translation product	Mature peptide
Squalus acanthias		
liver (mitochondrial)	50,000	47,000
kidney (mitochondrial)	–	47,000
brain (cytosolic)	45,000	45,000
gill	–	45,000
heart	–	45,000
rectal gland	–	45,000
retina	–	45,000
spleen	–	45,000
Dasyatis sabina		
liver (mitochondrial)	48,000	45,000
kidney (mitochondrial)	–	45,000
brain (cytosolic)	42,000	42,000
Hydrolagus colliei		
liver (mitochondrial)	–	45,000
brain (cytosolic)	–	43,500

Data compiled from refs. 5, 148, and 161.

Nevertheless, only the mature 45 kDa subunit can be observed under the experimental conditions used[147].

The tissue-specific expression of compartmental isozymes coded for by separate genes, which is most common, is due to tissue-specific promoters, enhancers, and other gene controlling elements so that one or the other gene is expressed[116]. However, in the case where both isozymes are coded for by the same gene, two mechanisms for expressing two different peptides have been described (Fig. 5). One involves the use of alternate transcription start sites so that two different mRNAs are transcribed[177]. The other involves the use of alternate translation start sites so that two peptides are translated[171]. Which of the latter two mechanisms are utilized for the formation of tissue-specific isozymes of glutamine synthetase in elasmobranchs is not known.

III. Mitochondrial citrulline synthesis

1. Occurrence and distribution of urea cycle enzymes in fish

Teleost fishes are ammonotelic, excreting the majority of their nitrogenous end-products of metabolism via the gills as ammonia[58]. Nevertheless, as shown in Tables 4 and 5, all five of the urea cycle enzyme activities have been reported in some species and significant quantities of urea are present in tissues and excreted by some species. In most species, the levels of enzyme activity are low. However, they can be relatively high as first demonstrated by Read[142] for the oyster toadfish

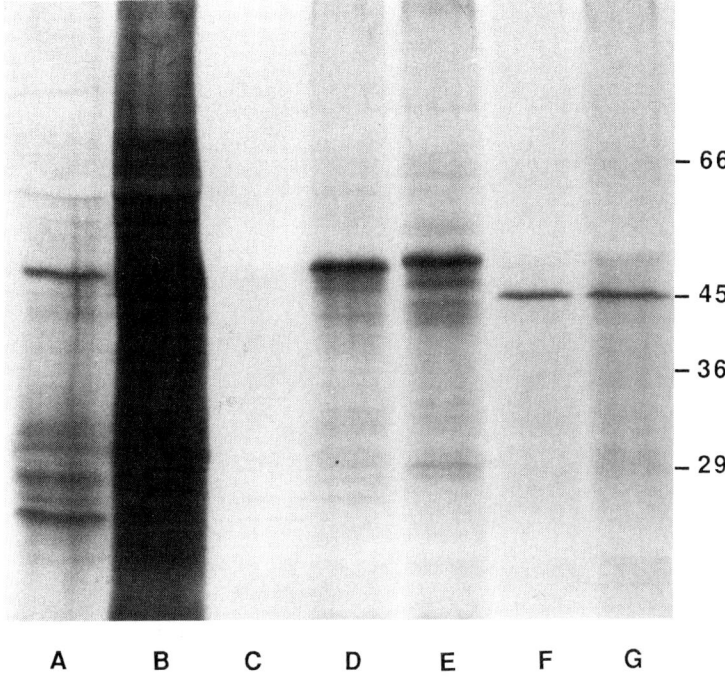

Fig. 4. Processing of dogfish (*S. acanthias*) and stingray (*D. sabina*) pre-glutamine synthetase by lysed stingray hepatic mitochondria. Dogfish and stingray poly(A$^+$) mRNAs were translated in a standard rabbit reticulocyte mixture containing [^{35}S]L-methionine[161]. The labeled translation product was then incubated with lysed stingray mitochondria. Following incubation, glutamine synthetase was immunoprecipitated from the mixture, separated by SDS-PAGE, and the gel radioautographed. Lane A: no RNA added; lane B: total stingray translation products; lane C: immunoprecipitated with pre-immune serum; lane D: stingray translation product; lane E: dogfish translation product; lane F: stingray translation product incubated with lysed stingray liver mitochondria; lane G: dogfish translation product incubated with lysed stingray liver mitochondria (from ref. 147).

(*Opsanus tau*). The expression of significant levels of urea cycle enzymes in teleosts appears to be correlated with unique environmental or evolutionary adaptations. For example, active ureogenesis and increased urea excretion occurs (1) in marine toadfishes[191] and in air-breathing teleosts[137,149] subjected to temporary dehydration, and (2) in a teleost that has adapted to an extremely alkaline (pH 9.6–10.0) condition[138,201].

Marine elasmobranchs (as well as holocephalans and the coelacanth) synthesize and retain high concentrations (0.3–0.4 M) of urea in their tissues for osmoregulation[21,65,131,166] (see Table 5). Most of the small amount of urea excreted is through the gills rather than via the kidney such as occurs in mammals. Urea is actively absorbed by the elasmobranch kidney. In contrast to teleosts where ammonia is the predominant end-product excreted by the gills, in elasmobranchs, 75% of the total nitrogen excreted by the gill is as urea[131]. Urea synthesis in elasmobranchs involves a pathway analogous to the classical urea cycle in mammals[21,65,131] (see Table 4), except for the unique exceptions discussed below.

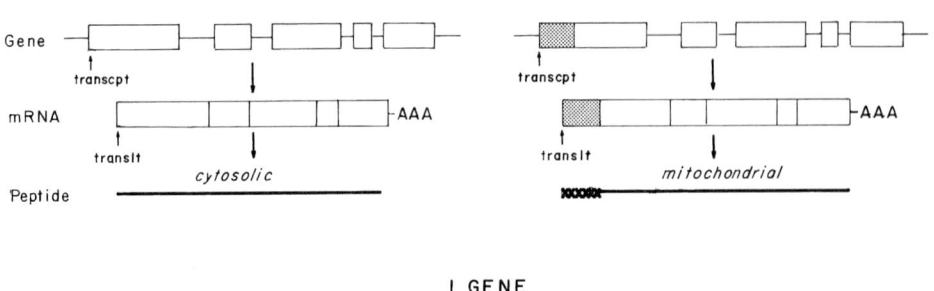

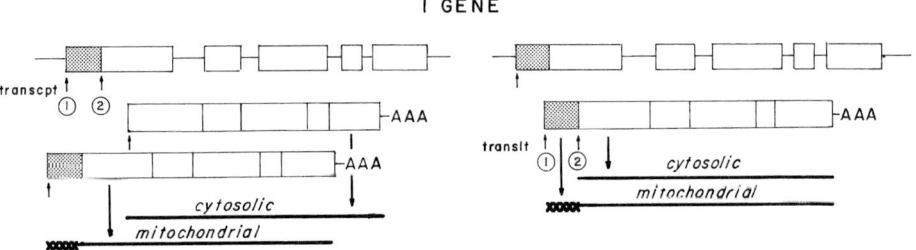

Fig. 5. Genetic mechanisms for formation of compartmental isozymes targeted by presence or absence of N-terminal, transient presequence. transcpt = transcriptional start site; translt = translational start site.

2. General properties of CPSs and associated enzymes

CPSs catalyze the formation of carbamoyl phosphate, which is the initial precursor in two major metabolic pathways, the urea cycle (and/or arginine biosynthetic pathway) and the pyrimidine biosynthetic pathway[110,139]. Two different types of synthetases are present in mammals. CPS-I is localized in the mitochondrial matrix of liver where it provides carbamoyl phosphate for urea synthesis[85,105,109,133]. The nitrogen-donating substrate for CPS-I is ammonia and the enzyme is subject to regulation by N-acetyl-L-glutamate (NAcGlu), which functions as a positive allosteric effector that is required for activity. CPS-I is comprised of a single subunit with mass of 160 kDa. The other synthetase, CPS-II, is present in the cytosol of liver and other tissues and provides carbamoyl phosphate for pyrimidine biosynthesis[59,89,93]. CPS-II is part of a multifunctional complex which also contains aspartate transcarbamylase and dihydroorotase. This complex has a mass of around 240 kDa. CPS-II (1) utilizes glutamine as the preferred nitrogen-donating substrate (ammonia will serve as a substrate with a V_{max} often higher than that obtained with glutamine, but the K_m is much higher than the K_m for glutamine), (2) does not require and is not affected by NAcGlu, and (3) is subject to regulation by the positive allosteric affector, 5-phosphoribosyl-1-pyrophosphate and the negative allosteric effector, UTP (end-product inhibition).

In mammals, the first two steps (reactions 4 and 5) of the urea cycle catalyzed by CPS-I and ornithine transcarbamylase, respectively, occur in mitochondria.

TABLE 4

Activities of carbamoyl-phosphate synthetase and other urea-cycle enzymes in liver of various species of fish

Species (ref.)	μmol/g liver (wet wt.)/h					
	CPS (NH$_3$)	CPS (Glutamine)	OTC	ASS	ASL	ARG
Teleosts (freshwater)						
Esox lucius (pike[82])	0.06	–	72	0.8	2.6	691
Salmo gairdneri (rainbow trout[82])	0.13	–	1.4	1.7	1.1	1,326
Salmo gairdneri (rainbow trout[39])	2.6	–	0.5	–	–	5,778
Cyprinus carpio (carp[82])	0.42	–	1	0.5	0.4	58
Carassius auratus (goldfish[82])	0.09	–	–	2.8	bld	71
Scardinus erithrophthalmus (rudd[82])	0.22	–	–	3.6	0.2	170
Rutilus rutilus (roach[82])	0.11	–	1	0.3	0.8	329
Tinca vulgaris (tench[82])	0.19	–	2.4	3.6	1.3	366
Siluris glanis (catfish[82])	bld	–	–	9.4	0.3	3,542
Perca fluviatilis (perch[82])	0.04	–	4	bld	1.4	990
Anguilla anguilla (eel[82])	0.16	–	6	1.6	1.8	4,565
Ictalurus punctatus (channel catfish[197])	2.97	–	221	bld	–	5,472
Heteropneustes fossilis (air-breather[149])	4.56	–	252	29	27	7,700
Oreochromis alcalicus grahami (tilapia, alkaline adapted[138])	–	1.24	439	–	–	703
Oreochromis nilotica (tilapia[138])	–	0.07	bld	–	–	795
Micropterus salmoides (largemouth bass[30])	–	0.6	126	–	–	2,340
Lepomis macrochirus (bluegill[30])	0.07	0.08	–	–	–	–
Pomoxis nigromaculatus (crappie[30])	0.10	0.08	–	–	–	–
Ictalurus natalis (bullhead[30])	0.06	0.04	–	–	–	–
Teleosts (marine)						
Opsanus tau (oyster toadfish[142])	9.8	–	10,210	14.6	184	31,800
Opsanus tau (oyster toadfish[112])	0.91	3.7	2,418	–	–	25,026
Opsanus beta (Gulf toadfish[112])	0.84	7.8	3,127	4.8	19.8	1,889
Porichthys notatus (plainfin midshipman[2])	0.08	0.13	–	–	–	–
Lepidopsetta bilineata (rock sole[2])	0.17	0.16	–	–	–	–
Platichthys stellatus (starry flounder[2])	0.19	0.20	–	–	–	–
Gadus macrocephalus (Pacific cod[2])	0.05	0.06	–	–	–	–
Sebastes caurinus (copper rockfish[2])	0.13	0.15	–	–	–	–
Porichthys notatus (plainfin midshipman[2])	0.08	0.13	–	–	–	–
Clupea harengus (herring[82])	0.23	–	1.7	bld	1.2	664
Gadus callarias (cod[82])	0.13	–	1	4	1	245
Merlangus merlangus (whiting[82])	–	–	–	0.5	1.2	644
Mullus barbatus (red mullet[82])	0.24	–	179	3.4	1.7	714
Solea vulgaris (common sole[82])	–	–	29.5	bld	1.1	702
Salmo salar (Atlantic salmon[82])	0.03	–	4	0.6	0.7	1,117
Chondrostei and Holostei						
Scaphirhynchus platorhynchus (sturgeon[45])	0.03 *	–	–	–	–	–
Lepisosteus platostomus (gar[45])	0.03 *	–	–	–	–	–
Amia calva (bowfin[45])	0.03 *	–	–	–	–	–
Polyodon spathula (paddlefish[45])	0.01 *	–	–	–	–	–

(table continued on page 56.)

TABLE 4 (continued)

Species (ref.)	μmol/g liver (wet wt.)/h					
	CPS (NH$_3$)	CPS (Glutamine)	OTC	ASS	ASL	ARG
Elasmobranchs						
Sphyrna tiburo (hammerhead shark[19])	11.5	–	–	–	–	–
Sphyrna zygaena (hammerhead shark[19])	bld	–	–	–	–	–
Dasyatis sabina (stingray[19])	23.8	–	–	–	–	–
Scoliodon terrae.novae (bull shark[19])	9.2	–	–	–	–	–
Sphyrna tudes (hammerhead shark[19])	53.7	–	–	–	–	–
Scyliorhinus canicula (cat shark[193])	26	26	–	–	–	–
Raja circularis (skate[193])	0	13	–	–	–	–
Dasyatis americana (freshwater stingray[66])	6.5	–	14,360	21.6	–	34,800
Urolophus jamaicensis (round ray[66])	4.5	–	8,450	16.5	–	13,920
Potamotrygon sp. (freshwater stingray[66])	0.36	–	1,600	9.4	–	4,310
Squalus acanthias (dogfish shark[33])	–	18	198	36	42	648
Squalus acanthias (dogfish shark[2])	0.5	5	–	–	–	–
Raja binoculata (big skate[2])	0.9	6.7	–	–	–	–
Taeniura lymma (blue-spotted stingray[2])	1.2	8.3	–	–	–	–
Holocephalan						
Hydrolagus colliei (ratfish[2])	1	4.0	–	–	–	–
Hydrolagus colliei (ratfish[140])	0.82 *	–	161 *	0.2 *	3.8 *	39 *
Coelacanth						
Latimeria chalumnae[67]	4.75	–	8,240	10	–	30,000

Standard errors are not listed. Values for CPS are listed as ammonia- or glutamine-dependent activity. – = not assayed; bld = below level of detection. OTC = ornithine transcarbamylase; ASS = argininosuccinate synthetase; ASL = argininosuccinate lyase; ARG = arginase. * Activities reported as μmol/mg protein/h.

Citrulline formed by these

(4) $NH_3 + 2ATP + HCO_3^- \xrightarrow{\text{NAcGlu, Mg}^{2+}} 2ADP + P_i + NH_2COPO_4^{2-}$

(5) $NH_2COPO_4^{2-} + \text{L-ornithine} \xrightarrow{\text{Mg}^{2+}} \text{L-citrulline} + P_i$

reactions is then translocated to the cytosol, where the remaining three reactions of the urea cycle (reactions 6–8), catalyzed by argininosuccinate synthetase, argininosuccinate lyase, and arginase, respectively, occur.

(6) L-citrulline + L-aspartate + $ATP_3 \to$ L-argininosuccinate + AMP + PP_i

(7) L-argininosuccinate + $H_2O \to$ fumarate + L-arginine

(8) L-arginine + $H_2O \to$ urea + L-ornithine

CPS activities in fish have not been well characterized and, until recently, any activity was assumed to be due to CPS-I. Thus, ammonia was utilized as the only

TABLE 5

Concentrations and rates of excretion of urea and ammonia in various species of fish

Species (ref.)	Ammonia (mM or μmol/g)	Urea (mM or μmol/g)		Rate of excretion (μmol/h/kg fish)	
				Ammonia	Urea
Teleosts (freshwater)					
Cyprinus carpio (carp[182])	–	(blood)	1.4	–	49
Salmo gairdneri (rainbow trout[125])	–	–		–	34
Salmo gairdneri (rainbow trout[48])	–	(embryo)	6.8	–	–
Carassius auratus (goldfish[125])	–	–		–	40
Ictalurus punctatus (catfish[199])	0.9	(serum)	0.9	–	–
	33.8	(liver)	24		
Oncorhynchus nerka (sockeye salmon[18])	–	–		609	80
Poecilia reticulata (guppy[48])	–	(embryo)	46	–	–
Oreochromis alcalicus grahami (tilapia alkaline adapted[138])	0.8	(plasma)	10.5	–	–
Oreochromis nilotica (tilapia[138])	1.0	(plasma)	2.3	–	–
Teleosts (marine)					
Leptocottus armatus (sculpin[202])	–	–		122	17
Platichthys stellatus (starry flounder[202])	–	–		253	13
Engraulis mordax (anchovy[107])	–	–		96	12
Engraulis ringens (anchovy[107])	–	–		357	38
Trachurus symmetricus (jack mackerel[107])	–	–		89	9
Teleosts (freshwater, amphibious)					
Periophthalmus sobrinus (mudskipper[69])	–	–		240	360
Sicyases sanguineus (Chilean clingfish[70])	0.2	(blood)	3	188	270
Periophthalmus cantonensis (mudskipper[71])	–	–		221	260
Periophthalmus cantonensis[114]	7	(plasma)	5	1,560	70
Boleophthalmus pectinirostris (mudskipper[114])	1.8	(plasma)	0.6	390	15
Blennius pholis (L.) (blenny[46])	–	–		93	11
Anabas scandens (climbing gourami[137])	1.6	(blood)	1.1	921	300
Channa gachua (snakehead[137])	1.6	(blood)	1.4	594	350
Mystus vittatus (bagrid catfish[137])	0.2	(blood)	0.1	883	280
Elasmobranchs					
Raja erinacea (little skate[68])	–	–		–	239
Potamotrygon sp. (freshwater stingray[66])	–	–		980	bld
Potamotrygon hystrix (freshwater stingray[14])	–	(serum)	0.8	–	–
		(serum)	6.0 *		
Hemiscyllium plagiosum (Bennett) (lip-shark[200])	–	–		–	668
Squalus acanthias (dogfish shark[12])	–	(plasma)	376	–	–
		(muscle)	319		
Holocephali					
Callorhynchus milii (plownose chimaeras[12])	–	(plasma)	225	–	–
		(muscle)	278		
Hydrolagus colliei (ratfish[141])	–	(serum)	245	–	–

Standard errors are not listed. See refs. 80 and 163–165 for earlier literature summaries of urea concentrations in fish tissues. * Adapted to 50% seawater. bld = below level of detection.

nitrogen-donating substrate and NAcGlu was present in the assay mixtures (and a dependency on this effector was not determined)[21]. Since a third type of CPS, CPS-III, is also active under these conditions, it is quite possible and also quite

TABLE 6

Distinguishing properties of carbamoyl-phosphate synthetases in vertebrates

Property	CPS-I	CPS-II	CPS-III
Nitrogen-donating substrate	NH_3	Glutamine (NH_3 also, high K_m)	Glutamine (NH_3 also, low V_{max})
Allosteric effectors			
Required (positive)	NAcGlu	–	NAcGlu
Positive	–	PRPP	–
Negative	–	UTP	–
Subunit M_r	160,000	240,000 (multifunctional: 3 enzyme activities)	160,000
Metabolic function	Urea synthesis (nitrogen catabolism)	Pyrimidine nucleotide synthesis	Urea (osmoregulation) and nitrogen catabolism)
Distribution	Amphibian, lungfish, and mammalian liver	Likely in most animal tissues	All cartilagenous and some teleost fishes
Subcellular localization	Mitochondria	Cytosol	Mitochondria

Information compiled from refs. 1–3, 6, 33, 34, 59, 105, 112, 174, and 175. PRPP = 5-phosphoribosyl-1-pyrophosphate.

likely that the activity measured was that of CPS-III. Trammel and Campbell[174,175] first reported this third type of CPS in several invertebrate species. Like CPS-II, CPS-III utilizes glutamine as the preferred nitrogen-donating substrate but, like CPS-I, requires NAcGlu. Also like CPS-I, CPS-III is localized in the matrix of mitochondria. Unlike CPS-II, CPS-III is not inhibited by pyrimidine nucleotides[174]. Anderson[1,2] subsequently reported the presence of CPS-III activity at low levels in liver of the freshwater teleost largemouth bass (*Micropterus salmoides*) and the marine teleost plainfin midshipman (*Porichthys notatus*) and at much higher levels in liver of marine elasmobranchs and the ratfish (*H. colliei*). CPS-III activity has also been reported recently in several other teleost species[112,138]. A summary of the properties of the various CPSs is given in Table 6.

CPS-III from both the spiny dogfish (*S. acanthias*) and largemouth bass (*M. salmoides*) has been purified and characterized[3,34]. The general properties of both enzymes are similar to CPS-I from mammals, except that glutamine serves as the nitrogen-donating substrate. Ammonia can also serve as a substrate, but the K_m is quite high (8 mM for bass CPS-III vs. 0.08 mM for glutamine) and the V_{max} is 25% of that obtained with glutamine. The dogfish enzyme is subject to significant inhibition by physiological concentrations of urea. Citrulline synthesis by isolated, actively respiring mitochondria is likewise inhibited by urea but respiration under these conditions is not, however, affected[4]. CPS-III from bass, which would not be expected to have a function related to osmoregulation by urea, is also inhibited by urea but less so than the dogfish enzyme.

The apparent K_m for ammonia of CPS-III is not affected by NAcGlu and vice versa. However, the apparent K_ms for NAcGlu and glutamine decrease significantly as the concentration of the other increases. This unique synergistic relationship may have physiological significance, since both compounds are derived directly from glutamate within the mitochondrial matrix. The glutamine-dependent activity of fish CPS-III, but not the ammonia-dependent activity, is inhibited by several glutamine analogs that are general active site-directed, irreversible inhibitors of amidotransferases. The presence of MgATP as well as NAcGlu is required for reaction with these analogs. After their reaction with the enzyme, a significant level of ammonia-dependent activity appears in the absence of NAcGlu, and the K_m for NAcGlu in the ammonia-dependent reaction is greatly lowered, providing additional evidence for a significant interaction between the glutamine and NAcGlu sites.

Ornithine transcarbamylase activity has been reported in a number of fish species, but has been purified and characterized only from liver of *S. acanthias*[206]. The dogfish enzyme has similar structural (trimer of identical subunits) and catalytic properties to those of other ornithine transcarbamylases. The only unusual feature of the purified dogfish enzyme is that the specific activity is considerably lower than that of the purified mammalian and bacterial enzymes. However, the low specific activity of the dogfish enzyme is similar to that of purified avian ornithine transcarbamylase[179].

3. Compartmentation of CPS-III and associated enzymes in Elasmobranch tissues and functional significance

The high levels of glutamine synthetase in elasmobranchs were the first indication that the first step in ammonia fixation in elasmobranchs might be the formation of glutamine, which is then utilized for carbamoyl phosphate synthesis[2,194,195]. Anderson and Casey[6], using isolated liver mitochondria from the spiny dogfish capable of citrulline synthesis with succinate as an energy source, showed that glutamine is, in fact, an obligatory intermediate in citrulline synthesis from glutamate and ammonia.

CPS-III, glutamine synthetase, ornithine transcarbamylase, and arginase are all localized within the mitochondrial matrix of dogfish hepatic mitochondria[33,35]. This is in contrast to mammals and amphibians where hepatic glutamine synthetase and arginase are exclusively or predominantly cytosolic enzymes[28]. It is, however, analogous to the situation in birds in which the two enzymes are co-localized in liver mitochondria. Thus, the initial step in ammonia metabolism, glutamine synthesis, in liver is identical in elasmobranchs, birds and reptiles. In elasmobranchs, the glutamine formed is utilized directly for carbamoyl phosphate formation whereas in birds and reptiles, it exits to the cytosol where it is utilized for uric acid formation[29]. The comparative relationships between the various vertebrate classes are illustrated in Fig. 6. In species capable of urea synthesis, a mitochondrial localization of hepatic arginase appears to be associated with the presence of CPS-III whereas a predominantly cytosolic localization is associated with CPS-I[30,112]. The urea cycle, which is monophyletic in vertebrates, thus appears to have undergone two key changes during vertebrate evolution: a switch from CPS-III to

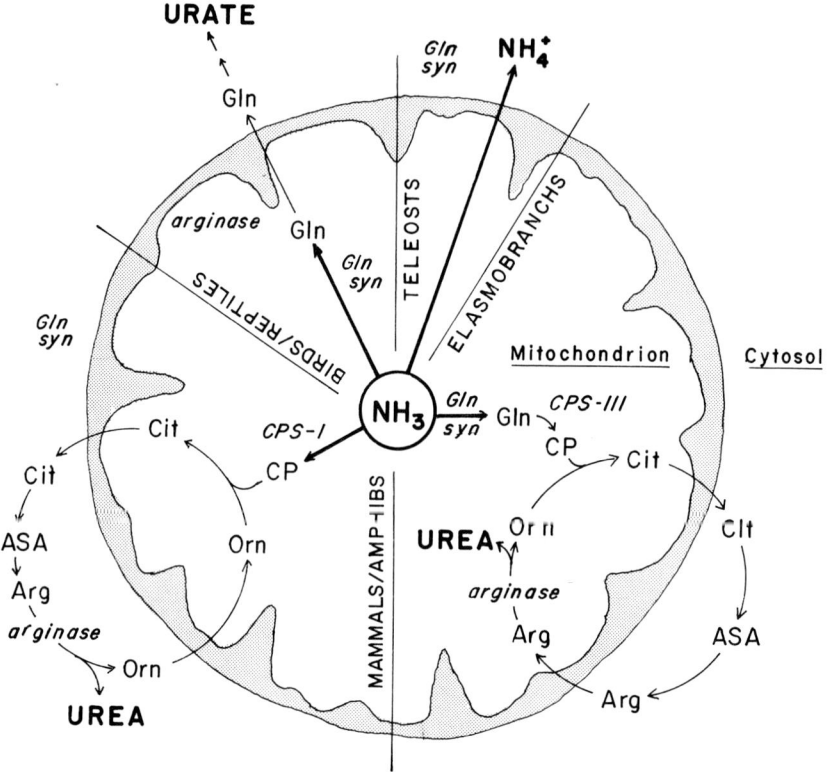

Fig. 6. Mitochondrial handling of ammonia in liver of vertebrates. *Gln syn* = glutamine synthetase; *CP* = carbamoyl phosphate; *ASA* = argininosuccinate; others are defined in text or are standard abbreviations.

CPS-I and loss of targeting of all except a small percentage of arginase to hepatic mitochondria.

The apparent exclusive use of glutamine formed intramitochondrially in liver of cartilaginous and other fish for urea synthesis raises the question of the source of glutamine and carbamoyl phosphate for cytosolic pyrimidine biosynthesis, a point only recently addressed[5]. Aspartate transcarbamylase activity cannot be detected in liver of the spiny dogfish which suggests that a pyrimidine biosynthetic pathway may be absent. However, all enzymes of the pyrimidine pathway, including CPS-II activity, are present in spleen and other extrahepatic tissues although in very low levels. The first three enzymes of the pyrimidine pathway, CPS-II, aspartate transcarbamylase, and dihydroorotase, are all localized in the cytosol, but do not appear to be associated as a multifunctional complex or a single polypeptide chain. Glutamine synthetase is also localized in the cytosol in spleen. Elasmobranchs thus appear to express two different glutamine-dependent CPS activities, but in different tissues that rely on two isozymes of glutamine synthetase[161] for their substrate[5].

The smallmouth bass, *M. salmoides*, also has CPS-II activity, but in contrast with the dogfish, it is present in the same tissue as CPS-III. Aspartate transcarbamylase activity is present in liver cytosol of the smallmouth bass along with CPS-II and dihydroorotase[1,30]. The three pyrimidine pathway enzymes appear to be associated as a multifunctional complex as they are in mammals[89]. The CPS-III activity present in the smallmouth bass is localized in liver mitochondria along with ornithine transcarbamylase and arginase. Glutamine synthetase, on the other hand, is localized in the cytosol. Isolated, actively respiring mitochondria from bass liver are not able to synthesize citrulline using either glutamine or glutamate plus ammonia as substrates. This suggests that the low level of CPS-III activity present in mitochondria may be physiologically insignificant. CPS-II is detectable in several marine teleosts that do not express CPS-III activity[2] and also in freshwater teleosts[30]. Although CPS-III activity is present in the largemouth bass, a member of the sunfish family, it is not detectable in several closely related species of the sunfish family (e.g., crappies and bluegills). A cDNA probe for dogfish CPS-III is now available which may be used to determine if the CPS-III gene has been deleted or simply 'silenced' at either the transcriptional or translational level in those species in which activity cannot be detected (Salo, Lusty and Anderson, unpublished). The molecular evolution of CPSs in general is discussed in more detail below.

A physiological consequence of the co-localization of glutamine synthetase and arginase in the mitochondrial matrix in elasmobranch liver as well as in liver of some teleosts that express CPS-III is that urea is formed within the mitochondria (Fig. 6). Urea is permeable to mitochondrial membranes[4,10] but the formation of urea inside could result in higher concentrations than that found in the extramitochondrial compartments. This could possibly be of regulatory importance if the observed inhibition of CPS-III by urea is of physiological significance.

Isolated mitochondria from dogfish liver catalyze stoichiometric citrulline synthesis from ornithine, bicarbonate, and glutamate at very low (< 0.1 mM) ammonia concentrations[6,35]. This high efficiency ammonia utilization is due not only to the co-localization of CPS-III and glutamine synthetase within the mitochondria but also to the latter enzyme's unusually low K_m for ammonia (15 μM or some 10-fold lower than that of other vertebrate glutamine synthetases; see Table 2)[154]. The apparent K_m of dogfish CPS-III for glutamine (0.16 mM) is lower than the K_m of CPS-I for ammonia. This and the fact that glutamine synthetase is present in excess of CPS-III[154] provide for a more efficient ammonia-utilizing system than the CPS-I system. The increased efficiency of such a system may more than compensate for the addition of an extra ATP-dependent step, that of glutamine synthetase. Leech et al.[104] have reported that considerable quantities of ammonia (but little or no glutamine in contrast to mammals) are released continuously from muscle into the circulatory system of spiny dogfish, both before and during starvation, and the prebranchial plasma concentrations of ammonia are relatively high. Since ammonia is efficiently removed by the gill, the glutamine synthetase–CPS-III system may represent an adaptive mechanism for utilization of low ammonia concentrations for the synthesis of urea for osmotic purposes. The functional role of CPS-III in teleosts, especially when glutamine synthetase is localized in the cytosol[30,112], is not known. Embryos from guppies and trout have

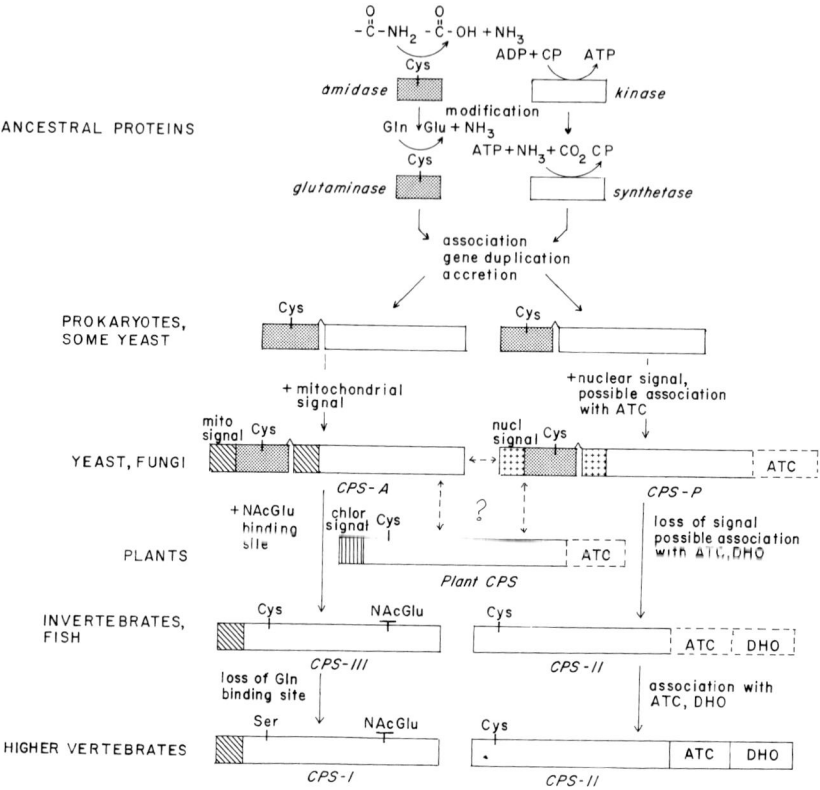

Fig. 7. Hypothetical scheme for evolution of the carbamoyl-phosphate synthetases. CP = carbamoyl phosphate; NAcGlu = N-acetyl-L-glutamate; ATC = aspartate transcarbamylase; and DHO = dihydroorotase. (Based, in part, on refs. 59, 121, 122, and 168.)

significant urea biosynthetic capability and the levels of urea during embryonic development are affected by the osmolarity of the medium[48] so perhaps urea functions as a critical osmolyte during this period.

4. Evolution of CPS and associated enzymes

A hypothetical scheme for the evolution of CPSs is shown in Fig. 7. In *E. coli* and other prokaryotes, carbamoyl phosphate is synthesized by a single glutamine-dependent CPS composed of a small and large subunit. The small subunit contains the glutamine-hydrolyzing site and the larger subunit, the carbamoyl phosphate synthetic site[74,110,178]. This enzyme provides carbamoyl phosphate for both arginine and pyrimidine biosynthesis. The prokaryote enzyme is felt to have originally evolved from a separate amidase and most likely a kinase. Carbamoyl phosphate is formed from cyanate and phosphate under 'prebiotic' conditions[90] so it is likely that it was first used by a kinase to form ATP. During the evolution of the prokaryote synthetase, there was both retention of the original sequences and acquisition of new coding information[121,122]. In the yeast *S. cerevisiae*, there are two CPSs: the arginine pathway-specific enzyme (CPS-A) and the pyrimidine pathway-specific enzyme (CPS-P). The latter is localized to the nucleus in both yeast[118] and the fungus *Neurospora crassa*[47] and is part of a multifunctional peptide also containing aspartate transcarbamylase activity[168]. In *S. cerevisiae*, CPS-A as well as ornithine transcarbamylase are in the cytosolic compartment whereas both enzymes are mitochondrial in other yeast species[180] and *N. crassa*[196]. The large subunit of CPS-A is translated as a larger molecular weight precursor presumably containing a mitochondrial targeting signal[119] so the mitochondrial targeting signal for CPS as well as for ornithine transcarbamylase must have been acquired early on in the evolution of eukaryotic cells. This is also true for the nuclear targeting signal although this signal appears to have been lost or modified soon after its acquisition. As indicated in Fig. 7, presumably both subunits possess targeting signals. There is even less consensus with respect to nuclear targeting signals than with mitochondrial signals although most contain basic amino acids, some of which are critical[53]. It is therefore possible to envision the evolution of a mitochondrial targeting signal from a nuclear signal or vice versa. A CPS-II-like enzyme, the only CPS in plants, is localized mainly in chloroplasts[126,155,156]. Chloroplastal targeting signals are similar to mitochondrial signals and will, in fact, target passenger peptides to mitochondria *in vitro* at low efficiency[187]. The chloroplastal targeting signal of the alga, *Chlamydomonas reinhardtii*, is in some ways more closely related to mitochondrial signals than to the chloroplastal signals of higher plants[62] so it is equally possible to envision the evolution of one of these targeting signals from the other.

In yeast, both CPS-P and aspartate transcarbamylase are coded for by a single gene and, in higher eukaryotes, a single gene also encodes these two enzymes plus dihydroorotase[168]. Aspartate transcarbamylase activity has been detected in several invertebrates[13,99,174] but whether it is associated with CPS-II is not known. CPS-II and aspartate transcarbamylase appear to be separate in the protozoans *Crithidia fasciculata*[7] and *Leishmania donovanii*[117]. In shark spleen, the two enzymes as well as dihydroorotase also appear to be separate[5].

Glutamine is the main substrate for both yeast and fungal CPS-A and CPS-P. A main difference between the glutamine-dependent CPSs, CPS-P and CPS-II, and ammonia-dependent CPS-I is the presence of a cysteine at the glutamine-binding site of the former[121,122,168]. In CPS-I, this is changed to a serine. In addition to a high degree of sequence similarity between the CPSs for which such data are available, *E. coli* CPS, CPS-A, CPS-P and CPS-I, -II, and -III are all immunochemically cross-reactive[52,119] suggesting that they are indeed homologous proteins that could have evolved via a scheme similar to that in Fig. 7.

The evolution of ornithine transcarbamylase appears to have been much the same as that of CPS-I. The catabolic prokaryotic enzyme is composed of from three to nine subunits but all anabolic ornithine transcarbamylases, that is, those involved in arginine or urea synthesis, are trimeric and show considerable sequence similarity, both among themselves and with other carbamoyl transferases[176]. Ornithine transcarbamylase is localized in mitochondria in invertebrates[174], elasmobranchs[33], and holocephalans[148]. As shown in Fig. 8, the enzyme is translated as

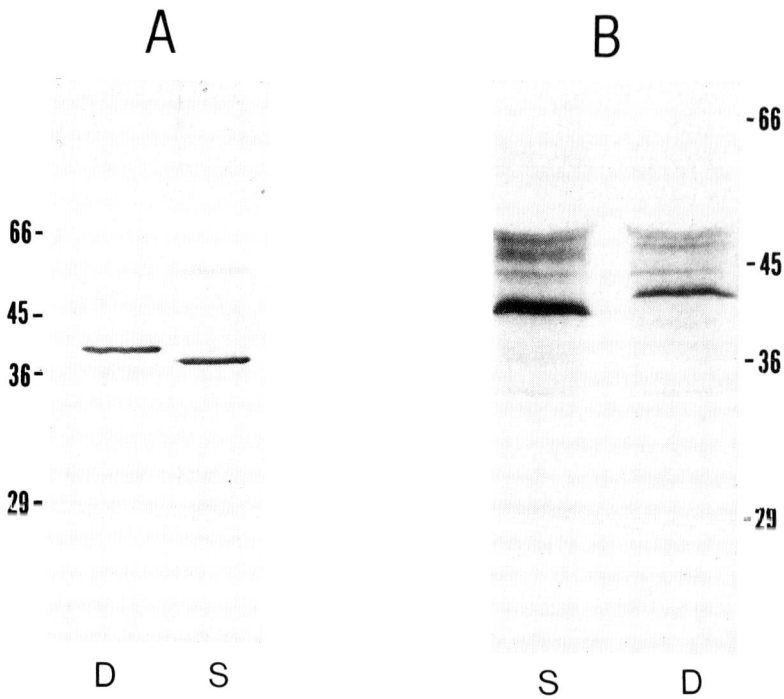

Fig. 8. Elasmobranch ornithine transcarbamylase. A: liver ornithine trancarbamylase mRNA translation products from stingray (*D. sabina*; D) and dogfish (*S. acanthias*; S). B: Western blots of mature peptides isolated from liver mitochondria. Anti-beef liver ornithine transcarbamylase antiserum was used for immunoprecipitation and immunoblotting (see ref. 27).

higher molecular weight precursors in both the stingray (43 kDa) and dogfish (40 kDa) which are presumably processed to the mature forms (39 and 37 kDa, respectively) during import. Again, as for glutamine synthetase, there thus appear to be species-specific isozymes of ornithine transcarbamylase in elasmobranchs based on the size differences of both the precursors and mature enzymes.

5. Evolutionary significance of the hepatic system for citrulline synthesis in elasmobranchs

The Chondrichthyes share common ancestors with the line of primitive vertebrates leading to the tetrapods. It is generally felt that this line gave rise to the ancestors of crossopterygians and tetrapods, on the one hand, and to the Dipnoi, on the other[152]. Although it was once felt that tetrapods were direct descendents of the crossopterygians, studies on the coelacanth (*Latimeria chalumnae*) have led to questions about this relationship[61]. Nevertheless, *L. chalumnae* also utilizes urea as an osmolyte[20] and the possible presence of CPS-III in its liver mitochondria[112] suggests that the primitive elasmobranch system for mitochondrial citrulline synthesis could have persisted through to the common ancestors of crossopterygians and early tetrapods. The presence of CPS-I in dipnoans[86], as well as in amphibians, suggests CPS-III may have been replaced by CPS-I in the ancestors of early tetrapods to serve a primary function in hepatic ammonia detoxification. The localization of glutamine synthetase and CPS-III within liver mitochondria of cartilaginous fish therefore may have preceded both a CPS and glutamine synthetase functioning independently as primary ammonia detoxifying enzymes in higher vertebrates. Glutamine synthetase is generally low in amphibians[29], except in certain tree frogs that are capable of excreting a high percentage of their excretory nitrogen as urate when water-deprived[157]. The enzyme is cytosolic in liver of these frogs[28] indicating the mitochondrial targeting mechanism utilized in elasmobranchs was lost during the evolution of the early tetrapods. The utilization of this cytosolic glutamine synthetase by the tree frogs for at least partial detoxification of hepatic ammonia under extreme environmental conditions may well represent the first tetrapod 'experiment' in utilizing this enzyme as a primary ammonia-detoxifying enzyme. When water is available to the frogs, urea is the main end-product so CPS-I in the mitochondrial matrix normally functions for hepatic ammonia detoxification as it does in other adult amphibia[28]. The presence of both a mitochondrial glutamine synthetase as well as CPS-I in tortoises indicates that during the evolution of the stem reptiles (cotylosaurians) from their amphibian-like, amniote ancestors, the genetic events resulting in re-targeting of glutamine synthetase to mitochondria occurred[27]. Since neither avian nor crocodilian glutamine synthetase is translated as a larger precursor[158,159], this targeting does not involve a transient N-terminal presequence as it did in elasmobranchs. Nevertheless, translocation into the mitochondrial matrix allowed for the more efficient hepatic ammonia detoxification mechanism under environmental conditions where detoxification via CPS-I might be unfavorable. During divergence of the stem reptiles, CPS-I was retained exclusively in the therapsid reptilian line leading to the mammals, most likely due to the role of urea in the formation of a hypertonic urine in the mammalian kidney[96]. While some early sauropsids may have retained the dual capacity for hepatic ammonia detoxification inherited from their cotylosaurian ancestors, the archosaurs – dinosaurs and their kin – relied exclusively on glutamine synthetase since the uricotelic hepatic ammonia detoxification system is now the only system present in birds and crocodilians, sole survivors of the archosaurian line[29,159]. The ability to excrete urate as a semi-solid allowed the archosaurian cloaca to function efficiently as a water-conserving mechanism which some feel is responsible for the rapid radiation of the archosaurs during the increasingly arid Triassic[29].

6. On the handling of mitochondrial ammonia in fish and other ammonoteles

The primary mitochondrial product of the ureotelic ammonia detoxification system – citrulline – and of the uricotelic system – glutamine – are neutral molecules and their exit from mitochondria therefore does not interfere with the normal proton gradient required for chemiosmotic ATP synthesis[25]. This raises the question of how ammonia exits liver mitochondria in fish and other ammonotelic species that do not possess one of these systems. The rate of glutamate oxidation by elasmobranch liver mitochondria is comparable to that of fatty acids, ketone bodies, and tricarboxylic acid cycle intermediates and is tightly coupled to ATP formation[115]. The rate of glutamate deamination by catfish liver mitochondria can account for 160% of the total ammonia excreted[26] so the rate of intramitochondrial ammonia formation in fish can be quite high. Because of the relative alkalinity of the mitochondrial matrix[11], much of the ammonia formed via glutamate deamination is present as NH_3 which is generally considered to be the permeant form. A high rate of NH_3 efflux across the inner mitochondrial membrane should therefore uncouple oxidative phosphorylation just as does OH^- efflux[25,153]. While conversion of ammonia to a neutral compound, citrulline or glutamine, prevents this in higher vertebrates, how this is done in fish and other ammonoteles is not known. Because of the special role of glutamine as a precursor of NH_4^+ excreted by the mammalian kidney, kidney mitochondria have been the main focus for research on ammonia translocation across mitochondrial membranes. Three mechanisms for its efflux from these mitochondria have been proposed[101] and in all, NH_3 is considered to be the permeant species. In two of the mechanisms, H^+ for NH_4^+ formation following efflux arises from the electrogenic proton pump. In the third mechanism, H^+ effluxes as an undissociated carboxylate of glutamate. The latter requires a stoichiometric appearance of ammonia and glutamate in the mitochondrial medium. In contrast to all three proposed mechanisms, ammonia efflux from catfish liver mitochondria is independent of proton flux and the efflux of ammonia and glutamate are not stoichiometric[26]. It thus appears that ammonia formed during amino acid catabolism in this and possibly other fish species exits hepatic mitochondria as NH_4^+ (Fig. 6). The development of a mitochondrial translocase or transport system for NH_4^+ in these mitochondria could therefore represent just as specific an adaptation for hepatic ammonia 'detoxification' by ammonoteles as does citrulline or glutamine formation by ureoteles and uricoteles[25].

7. The special localization of arginase in elasmobranch and teleost mitochondria

Both the structure and subcellular localization of arginase are quite flexible. The subunit mass is usually between 30 and 40 kDa. The enzyme occurs as a monomer in earthworms[84,143], a dimer in certain molluscs[32], either a trimer[23,130,185] or tetramer[77] in rat liver, and as hexamers and octamers in other species[17,145]. There are two genes for arginase in mammals that code for two compartmental isozymes, a predominantly cytosolic form in liver and a mitochondrial form in kidney. These two genes do not cross-hybridize nor are their products immunochemically cross-reactive[54]. The same also holds true for two cytosolic isozymes in *N. crassa*[17]. Arginase is equally eclectic in its subcellular localization. While it is predominantly cytosolic in mammalian liver, it appears to be particulate in liver cells in culture[57]. Arginase is mitochondrial in birds and has been reported to be localized in the intermembrane space in both kidney and liver mitochondria[72]. It has also been reported to be localized in the mitochondrial matrix in kidney mitochondria[91]. The subcellular localization of arginase, like its oligomeric structure, therefore appears to be quite labile.

Despite the lability of compartmentation of arginase, its subcellular localization probably reflects the efficiency of its metabolic function. Ornithine supplied to mammalian liver mitochondria is channeled directly to ornithine transcarbamylase in the matrix[41], due most likely to the association of this enzyme with the inner mitochondrial membrane[134]. Although most of the arginase in mammalian liver is cytosolic, approximately 10% is tightly bound to the outer mitochondrial membrane[38]. Neither the liver arginase gene nor its mRNA contains information for a mitochondrial targeting signal[92,124] so the nature of this binding is not known. The insertion of several other peptides into the outer membrane does not require a targeting signal but instead may involve a special conformation of the inserted peptide or a ubiquitin-dependent mechanism[207]. The conformation and/or oligomeric state of a fraction of the arginase synthesized in liver may therefore result in its binding to, or insertion in, the outer membrane. This association nevertheless allows for the very efficient conversion of arginine to ornithine and for transfer of the latter to ornithine transcarbamylase in the mitochondrial matrix[63]. The uptake of exogenous ornithine by liver mitochondria is via an energy-independent H^+ antiport system[8,81] but whether this is true for ornithine formed by the outer membrane-associated arginase is not known.

In the dogfish, S. acanthias, arginase is also mitochondrial in liver[33] but occurs in the matrix and is not membrane-associated[35]. Arginine is just as effective a substrate for citrulline formation by these mitochondria as is ornithine indicating it is efficiently converted to ornithine by this matrix arginase. Arginase is a Mn^{2+}-enzyme and most of the Mn^{2+} present in liver cells is in the particulate fraction[16]. It would therefore appear that arginase could function more efficiently within the mitochondrial compartment which contains relatively high Mn^{2+} concentrations. Its localization in the matrix of dogfish liver mitochondria may therefore represent this metabolic adaptation. Based on the Mn^{2+} content of the cytosol in mammalian liver, it has been suggested that cytosolic arginase is not fully activated, meaning only a fraction of the enzyme in this compartment is enzymatically active[16]. Because of the slow turnover of urea in elasmobranchs[131], whether there is need for a high rate of urea synthesis is questionable. However, elasmobranchs and other carnivorous fish must obtain much of their energy from amino acids and therefore require highly efficient amino acid catabolizing systems[25]. The combined actions of arginase and ornithine transaminase constitute such a system since arginine is converted to glutamate which can be efficiently oxidized by the fish mitochondrial system. Arginase is also localized in the mitochondrial matrix of the teleost fish *Genypterus maculatus*[31] where it is felt to function in this arginine-catabolizing capacity. The co-localization of arginase and ornithine transaminase in the cytosol of insect fat body, where the two enzymes also function in the formation of an oxidizable substrate[144], lends support to the idea that the special compartmentation of arginase in fish liver mitochondria may have resulted from an increased efficiency of arginine catabolism.

Acknowledgements. The work from the authors' laboratories was supported by National Science Foundation Grants DMB82-14901 (J.W.C.), DCB87-18042 (J.W.C.), and DCB 86-08090 (P.M.A.).

IV. References

1. Anderson, P.M. A glutamine- and *N*-acetyl-L-glutamate-dependent carbamoyl phosphate synthetase activity in the teleost *Micropterus salmoides. Comp. Biochem. Physiol.* 54B: 261–263, 1976.
2. Anderson, P.M. Glutamine- and *N*-acetylglutamate-dependent carbamoyl phosphate synthetase in elasmobranchs. *Science* 208: 291–293, 1980.
3. Anderson, P.M. Purification and properties of the glutamine- and *N*-acetyl-L-glutamate-dependent carbamoyl phosphate synthetase from liver of *Squalus acanthias. J. Biol. Chem.* 256: 12228–12238, 1981.
4. Anderson, P.M. Effects of urea, trimethylamine oxide, and osmolality on respiration and citrulline synthesis by isolated hepatic mitochondria from *Squalus acanthias. Comp. Biochem. Physiol.* 85B: 783–788, 1986.
5. Anderson, P.M. Glutamine-dependent carbamoyl-phosphate synthetase and other enzyme activities related to the pyrimidine pathway in spleen of *Squalus acanthias* (spiny dogfish). *Biochem. J.* 261: 523–529, 1989.
6. Anderson, P.M. and C.A. Casey. Glutamine-dependent synthesis of citrulline by isolated hepatic mitochondria from *Squalus acanthias. J. Biol. Chem.* 259: 456–462, 1984.
7. Aoki, T. and H. Oya. Kinetic properties of carbamoyl-phosphate synthetase II (glutamine-hydrolyzing) in the parasitic protozoan *Crithidia fasciculata* and separation of the enzyme from aspartate carbamoyltransferase. *Comp. Biochem. Physiol.* 87B: 143–150, 1987.
8. Aronson, D.L. and J.J. Diwan. Uptake of ornithine by rat liver mitochondria. *Biochemistry* 20: 7064–7068, 1981.
9. Attardi, G. and G. Schatz. Biogenesis of mitochondria. *Annu. Rev. Cell Biol.* 4: 289–333, 1988.
10. Ballantyne, J.S. and T.W. Moon. Solute effects on mitochondria from an elasmobranch (*Raja erinacea*) and a teleost (*Pseudopleuronectes americanus*). *J. Exp. Zool.* 239: 319–328, 1986.
11. Beavers, A.D. and M.F. Powers. On the regulation of the mitochondrial inner membrane anion channel by magnesium and protons. *J. Biol. Chem.* 264: 17148–17155, 1989.
12. Bedford, J.J. The composition of the fluid compartments of two chondrichthyians, *Callorhyncus millii* and *Squalus acanthias. Comp. Biochem. Physiol.* 76A: 75–80, 1983.
13. Bishop, S.H. and J.W. Campbell. Arginine and urea biosynthesis in the earthworm *Lumbricus terrestris. Comp. Biochem. Physiol.* 15: 51–71, 1965.
14. Bittner, A. and S. Lang. Some aspects of the osmoregulation of Amazonia freshwater stingrays (*Potamotrygon hystrix*) – I. Serum osmolality, sodium and chloride content, water content, hematocrit and urea level. *Comp. Biochem. Physiol.* 67A: 9–13, 1980.
15. Bodine, A.B., C.A. Luer and S. Gangjee. A comparative study of monoxygenase activity in elasmobranchs and mammals: activation of the model pro-carcinogen aflatoxin B by liver preparations of calf, nurse shark and clearnose skate. *Comp. Biochem. Physiol.* 82C: 255–257, 1985.
16. Bond, J.S., M.L. Failla and D.F. Unger. Elevated manganese concentration and arginase activity in livers of streptozotocin-induced diabetic rats. *J. Biol. Chem.* 258: 8004–8009, 1983.
17. Borkovich, K.A. and R.L. Weiss. Purification and characterization of arginase from *Neurospora crassa. J. Biol. Chem.* 262: 7081–7086, 1987.
18. Brett, J.R. and C.A. Zala. Daily pattern of nitrogen excretion and oxygen consumption of sockeye salmon (*Oncorhynchus nerka*) under controlled conditions. *J. Fish. Res. Bd. Can.* 32: 2479–2486, 1975.
19. Brown, G.W., Jr. Urea synthesis in elasmobranchs. In: *Taxonomic Biochemistry and Serology*, edited by C.A. Leone. New York: Ronald Press, 1964, pp. 407–416.
20. Brown, G.W., Jr. and S.G. Brown. Urea and its formation in coelacanth liver. *Science* 155: 570–573, 1967.
21. Brown, G.W., Jr. and S.G. Brown. On urea formation in primitive fishes. In: *Evolutionary Biology of Primitive Fishes*, edited by R.E. Foreman, A. Gorbman, J.M. Dodd, and R. Olsson. New York: Plenum Press, 1985, pp.321–337.
22. Brown, G.W., Jr. and P.P. Cohen. Comparative biochemistry of urea synthesis. 3. Activities of urea cycle enzymes in various higher and lower vertebrates. *Biochem. J.* 75: 82–91, 1960.

23. Brusdeilins, M., R. Kühner and K. Schumacher. Purification, affinity to anti-human arginase immunoglobulin-Sepharose 4B and subunit molecular weights of mammalian arginase. *Biochim. Biophys. Acta* 840: 79–90, 1985.
24. Caggese, C., R. Caizzi, M.P. Bozzetti, P. Barsanti and F. Ritossa. Genetic determinants of glutamine synthetase in *Drosophila melanogaster*: a gene for glutamine synthetase I resides in the 21B3–6 region. *Biochem. Genet.* 26: 571–584, 1988.
25. Campbell, J.W. Excretory nitrogen metabolism. In: *Environmental and Metabolic Animal Physiology. Comparative Animal Physiology*, 4th edn., edited by C.L. Prosser. New York: Wiley–Liss, 1991, pp. 277–324.
26. Campbell, J.W., P.L. Aster and J.E. Vorhaben. Mitochondrial ammoniagenesis in liver of the channel catfish *Ictalurus punctatus*. *Am. J. Physiol.* 244: R709–R717, 1983.
27. Campbell, J.W., D.D. Smith, Jr. and J.E. Vorhaben. Avian and mammalian mitochondrial ammonia-detoxifying systems in tortoise liver. *Science* 228: 349–351, 1985.
28. Campbell, J.W., J.E. Vorhaben and D.D. Smith, Jr. Hepatic ammonia metabolism in a uricotelic treefrog *Phyllomedusa sauvagei*. *Am. J. Physiol.* 246: R805-R810, 1984.
29. Campbell, J.W., J.E. Vorhaben and D.D. Smith, Jr. Uricoteley: its nature and origin during the evolution of tetrapod vertebrates. *J. Exp. Zool.* 243: 349–363, 1987.
30. Cao, X., J.R. Kemp and P.M. Anderson. Subcellular localization of two glutamine-dependent carbamoyl-phosphate synthetases and related enzymes in liver of *Micropterus salmoides* (largemouth bass) and properties of isolated liver mitochondria: comparative relationships with elasmobranchs. *J. Exp. Zool.* 258: 24–33, 1991.
31. Carvajal, N., E. Kessi and L. Ainol. Subcellular localization and kinetic properties of arginase from the liver of *Genypterus maculatus*. *Comp. Biochem. Physiol.* 88B: 229–231, 1987.
32. Carvajal, N., E. Kessi, J. Bidart and A. Rojas. Properties of arginase from the foot muscle of *Chiton latus*. *Comp. Biochem. Physiol.* 90B: 385–388, 1988.
33. Casey, C.A. and P.M. Anderson. Subcellular localization of glutamine synthetase and urea cycle enzymes in liver of spiny dogfish (*Squalus acanthias*). *J. Biol. Chem.* 257: 8449–8453, 1982.
34. Casey, C.A. and P.M. Anderson. Glutamine and N-acetyl-L-glutamate-dependent carbamoyl phosphate synthetase from *Micropterus salmoides*. Purification, properties, and inhibition by glutamine analogs. *J. Biol. Chem.* 258: 8723–8732, 1983.
35. Casey, C.A. and P.M. Anderson. Submitochondrial localization of arginase and other enzymes associated with urea synthesis and nitrogen metabolism in liver of *Squalus acanthias*. *Comp. Biochem. Physiol.* 82B: 307–315, 1985.
36. Chakravorty, J., N. Saha and B.K. Ratha. A unique pattern of tissue distribution and sub-cellular localization of glutamine synthetase in a freshwater air-breathing teleost, *Heteropneustes fossilis*. *Biochem. Int.* 19: 519–527, 1989.
37. Cheng, M.Y., F.-U. Hartl, J. Martin, R.A. Pollock, F. Kalousek, W. Neupert, E.M. Hallberg, R.L. Hallberg and A.L. Horwich. Mitochondrial heat-shock protein hsp 60 is essential for assembly of proteins imported into yeast mitochondria. *Nature* 337: 620–625, 1989.
38. Cheung, C.-W. and L. Raijman. Arginine, mitochondrial arginase and the control of carbamyl phosphate synthesis. *Arch. Biochem. Biophys.* 209: 643–649, 1981.
39. Chiu, Y.N., R.E. Austic and G.L. Rumsey. Urea cycle activity and arginine formation in rainbow trout (*Salmo gairdneri*). *J. Nutr.* 116: 1640–1650, 1986.
40. Christen, P., U. Graf-Hausner, F. Bossa and S. Doonan. Comparison of covalent structures of the isoenzymes of aspartate aminotransferase. In: *Transaminases*, edited by P. Christen, and D.E. Metzler. New York: Wiley, 1985, pp. 173–185.
41. Cohen, N.S., C.-W. Cheung and L. Raijman. Channeling of extramitochondrial ornithine to matrix ornithine transcarbamylase. *J. Biol. Chem.* 262: 203–208, 1987.
42. Cook, J.S., S.L. Weldon, J.P. Garcia-Ruiz, H. Hod and R.W. Hanson. Nucleotide sequence of the mRNA encoding the cytosolic form of phosphoenolpyruvate carboxykinase (GTP) from the chicken. *Proc. Nat. Acad. Sci. U.S.A.* 83: 7583–7587, 1986.
43. Cooper, J.L. Glutamine synthetase. In: *Glutamine and Glutamate in Mammals Vol. 1,*, edited by E. Kvamme. Boca Raton: CRC Press, 1988, pp. 7–31.
44. Coruzzi, G.M., J.W. Edwards, S.V. Tingey, F.-Y. Tsai and E.L. Walker. Glutamine synthetase – molecular evolution of an eclectic multi-gene family. In: *Molecular Basis of Plant Development (UCLA Symp. Mol. Cell Biol. Vol. 92)*, edited by R. Goldberg. New York: Alan R. Liss, 1989, pp. 223–232.
45. Cvancara, V.A. Liver carbamyl phosphate synthetase in the primitive freshwater bony fishes (*Chondrostei, Holostei*). *Comp. Biochem. Physiol.* 49B: 785–787, 1974.
46. Davenport, J. and M.D.J. Sayer. Ammonia and urea excretion in the amphibious teleost *Blennius pholis* (L.) in sea-water and in air. *Comp. Biochem. Physiol.* 84A: 189–194, 1986.

47. Davis, R.H. Metabolite distribution in cells. *Science* 178: 835–840, 1972.
48. Dépeche, J., R. Gilles, S. Daufresne and H. Chiapello. Urea content and urea production via the ornithine-urea cycle pathway during the ontogenic development of two teleost fishes. *Comp. Biochem. Physiol.* 63A: 51–56, 1979.
49. DeRosa, G. and R.W. Swick. Metabolic implications of the distribution of the alanine aminotransferase isoenzymes. *J. Biol. Chem.* 250: 7961–7967, 1975.
50. Deshaies, R.J., B.D. Koch, M. Werner-Washburne, E.A. Craig and R. Schekman. A subfamily of stress proteins facilitates translocation of secretory and mitochondrial precursor polypeptides. *Nature* 332: 800–805, 1988.
51. Deuel, T.F., M. Louie and A. Lerner. Glutamine synthetase from rat liver. *J. Biol. Chem.* 253: 6111–6118, 1978.
52. Devaney, M.A. and S.G. Powers-Lee. Immunological cross-reactivity between carbamyl phosphate synthetases I, II, and III. *J. Biol. Chem.* 259: 703–706, 1984.
53. Dingwall, C. and R. Laskey. Protein import into the cell nucleus. *Annu. Rev. Cell Biol.* 2: 367–390, 1986.
54. Dizikes, G.S., E.P. Spector and S.D. Cederbaum. Cloning of rat liver arginase cDNA and elucidation of regulation of arginase gene expression in H4 rat hepatoma cells. *Som. Cell. Mol. Genet.* 12: 375–384, 1986.
55. Dowton, N., I.R. Kennedy and M.C. Wang. Localization of glutamine synthetase in fleshfly flight muscle. *Insect Biochem.* 18: 717–727, 1988.
56. Eilers, M., T. Endo and G. Schatz. Adriamycin, a drug interacting with acid phospholipids, blocks import of precursor proteins by isolated yeast mitochondria. *J. Biol. Chem.* 264: 2945–2950, 1989.
57. Eliasson, E. Arginase in Chang's liver cells. *Acta Chem. Scand.* B28: 233–238, 1974.
58. Evans, D.H. and K.J. More. Modes of ammonia transport across the gill epithelium of the dogfish pup (*Squalus acanthias*). *J. Exp. Biol.* 138: 375–397, 1988.
59. Evans, D.R. CAD, a chimeric protein that initiates de novo pyrimidine biosynthesis in higher eukaryotes. In: *Multidomain Proteins – Structure and Evolution*, edited by D.G. Hardie and J.R. Coggins. New York: Elsevier, 1986, pp. 283–331.
60. Fahien, L.A., E.H. Kmiotek, G. Woldegiorgis, M. Evenso, E. Shrago and M. Marshall. Regulation of aminotransferase-glutamate dehydrogenase interactions by carbamyl phosphate synthetase-I, Mg^{2+} plus leucine versus citrate and malate. *J. Biol. Chem.* 260: 6069–6079, 1985.
61. Forey, P.L. Golden jubilee for the coelacanth *Latimeria chalumnae*. *Nature* 336: 727–732, 1988.
62. Franzén, L.-G., J.-D. Rochaix and G. Von Heijne. Chloroplast transit peptides from the green alga *Chlamydomonas reinhardtii* share features with both mitochondrial and higher plant chloroplast presequences. *FEBS Lett.* 260: 165–168, 1990.
63. Freedland, R.A., G.L. Ciozier, B.L. Hicks and A.J. Meijer. Arginine uptake by isolated rat liver mitochondria. *Biochim. Biophys. Acta* 802: 407–412, 1984.
64. Gevers, W. The regulation of phosphoenolpyruvate synthesis in pigeon liver. *Biochem. J.* 103: 141–152, 1967.
65. Goldstein, L. and R.P. Forster. Nitrogen metabolism in fishes. In: *Comparative Biochemistry of Nitrogen Metabolism, Vol. 2. The Vertebrates*, edited by J.W. Campbell. London: Academic Press, 1970, pp. 495–518.
66. Goldstein, L. and R.P. Forster. Urea biosynthesis and excretion in freshwater and marine elasmobranchs. *Comp. Biochem. Physiol.* 39B: 415–421, 1971.
67. Goldstein, L., S. Harley-Dewitt and R.P. Forster. Activities of ornithine-urea cycle enzymes and of trimethylamine oxidase in the coelacanth *Latimeria chalumnae*. *Comp. Biochem. Physiol.* 44B: 357–362, 1973.
68. Goldstein, L., R.P. Forster, D. Funkhouser and G. Fouty. Urea metabolism and osmoregulation in the skate, *Raja erinacea*. *Bull. Mt. Desert Is. Biol. Lab.* 8: 29–30, 1968.
69. Gordon, M.S., I. Boòtius, D.H. Evans, R. McCarthy, and L.C. Oglesby. Aspects of the physiology of terrestrial life in amphibious fishes. I. The mudskipper, *Periophthalmus sobrinus*. *J. Exp. Biol.* 50: 141–149, 1969.
70. Gordon, M.S., S. Fischer and S. Tarifeno. Aspects of the physiology of terrestrial life in amphibious fishes. II. The Chilean clingfish, *Sicyases sanguineus*. *J. Exp. Biol.* 53: 559–572, 1970.
71. Gordon, M.S., W.N. Wilson and Y.Y. Alice. Aspects of the physiology of terrestrial life in amphibious fishes. III. The Chinese mudskipper, *Periophthalmus cantonensis*. *J. Exp. Biol.* 72: 57–75, 1978.
72. Grazi, E., E. Magri and G. Balboni. On the control of arginine metabolism in chick kidney and liver. *Eur. J. Biochem.* 60: 431–436, 1975.
73. Grivell, L.A. Protein import into mitochondria. *Int. Rev. Cytol.* 111: 107–141, 1988.

74. Guillou, F., S.D. Rubino, R.S. Markovitz, D.M. Kenny and C.J. Lusty. *Escherichia coli* carbamoyl-phosphate synthetase: domains of glutaminase and synthetase subunit interaction. *Proc. Natl. Acad. Sci. U.S.A.* 86: 8304–8308, 1989.
75. Häussinger, D. and W. Gerok. Metabolism of amino acids and ammonia. In: *Regulation of Hepatic Metabolism*, edited by R.G. Thurman, F.C. Kauffman, and K. Jungermann. London: Plenum Press, 1986, pp. 253–295.
76. Hayashi, S., S. Fujiwara and T. Noguchi. Degradation of uric acid in fish liver peroxisomes. *J. Biol. Chem.* 264: 3211–3215, 1989.
77. Hirsch-Kolb, H., H.J. Kolb and D.M. Greenberg. Nuclear magnetic resonance studies of manganese binding of rat liver arginase. *J. Biol. Chem.* 246: 395–401, 1971.
78. Hilbish, T.J. and R.K. Koehn. Dominance in physiological phenotypes and fitness at an enzyme level. *Science* 229: 52–54, 1985.
79. Hod, Y., M.F. Utter and R.W. Hanson. The mitochondrial and cytosolic forms of avian phosphoenolpyruvate carboxykinase (GTP) are encoded by different messenger RNAs. *J. Biol. Chem.* 257: 13787–13794, 1982.
80. Holmes, W.N. and E.M. Donaldson. The body compartments and the distribution of electrolytes. In: *Fish Physiology*, edited by W.S. Hoar and D.J. Randall. New York: Academic Press, 1969, Vol. 1, pp. 1–89.
81. Hommes, F.A., A.G. Eller, B.A. Evans and A.L. Carter. Reconstitution of ornithine transport in liposomes with Lubrol extracts of mitochondria. *FEBS Lett.* 170: 131–134, 1984.
82. Huggins, A.K., G. Skutsch and E. Baldwin. Ornithine-urea cycle enzymes in teleostean fish. *Comp. Biochem. Physiol.* 28: 587–602, 1969.
83. Hwant, T., T. Jascur, D. Vestweber, L. Pon and G. Schatz. Disrupted yeast mitochondria can import precursor proteins directly through their inner membrane. *J. Cell Biol.* 109: 487–493, 1989.
84. Iino, T. and T. Shimadate. Purification and properties of gut arginase from earthworm *Pheretima communissima*. *Comp. Biochem. Physiol.* 83B: 79–84, 1986.
85. Jackson, M.J. Mammalian urea cycle enzymes. *Annu. Rev. Genet.* 20: 431–464, 1986.
86. Janssens, P.A. and P.P. Cohen. Ornithine-urea cycle enzyme in the African lungfish *Protopterus aethiopicus*. *Science* 152: 358–359, 1966.
87. Jensen, R.A. Enzyme recruitment in evolution of new function. *Annu. Rev. Microbiol.* 30: 409–425, 1976.
88. Joh, T., H. Takeshima, T. Tsuzuki, C. Setoyama, K. Shimada, S. Tanase, S. Kuramitsu, H. Kagamiyama and Y. Morino. Cloning and sequence analysis of cDNA encoding mammalian cytosolic malate dehydrogenase. *J. Biol. Chem.* 262: 15127–15131, 1987.
89. Jones, M.E. Pyrimidine nucleotide biosynthesis in animals: genes, enzymes and regulation of UMP biosynthesis. *Annu. Rev. Biochem.* 49: 253–279, 1980.
90. Jones, M.E., L. Spector and F. Lipmann. Carbamyl phosphate, the carbamyl donor in enzymatic citrulline synthesis. *J. Am. Chem. Soc.* 77: 819–820, 1955.
91. Kadowaki, H., H.W. Israel and M.C. Nesheim. Intracellular localization of arginase in chick kidney. *Biochim. Biophys. Acta* 437: 158–165, 1976.
92. Kawamoto, S., Y. Amaya, K. Murakami, F. Tokunaga, S. Iwanaga, K. Kobayashi, T. Saheki, S. Kimura and M. Mori. Complete nucleotide sequence of cDNA and deduced amino acid sequence of rat liver arginase. *J. Biol. Chem.* 262: 6280–6283, 1987.
93. Keppler, D. and A. Holstege. Pyrimidine nucleotide metabolism and its compartmentation. In: *Metabolic Compartmentation*, edited by H. Sies. New York: Academic Press, 1982, pp. 259–286.
94. Kim, K.H. and S.G. Rhee. Sequence of peptides from *Saccharomyces cerevisiae* glutamine synthetase. *J. Biol. Chem.* 263: 833–838, 1988.
95. King, P.A. and L. Goldstein. Renal ammoniagenesis and acid excretion in the dogfish, *Squalus acanthias*. *Am. J. Physiol.* 245: R581–R589, 1983.
96. King, P.A. and L. Goldstein. Renal excretion of nitrogenous compounds in vertebrates. *Renal Physiol., Basel* 8: 261–278, 1985.
97. Knight, T.J. and P.J. Langston-Unkefer. Adenine nucleotides as allosteric effectors of pea seed glutamine synthetase. *J. Biol. Chem.* 263: 11084–11089, 1988.
98. Koehn, R.K., A.J. Zera and J.G. Hall. Enzyme polymorphism and natural selection. In: *Evolution of Genes and Proteins*, edited by M. Nei and R.K. Koehn. Sunderland: Sinauer, 1983, pp. 115–136.
99. Koueta, N., M. Mathieu and E. Boucaud-Camou. Aspartate transcarbamylase activity in *Sepia officinalis* L. (Mollusca, Cephalopoda). *Comp. Biochem. Physiol.* 87B: 351–356, 1987.
100. Kuo, L.C., I. Zambidis and C. Caron. Triggering of allostery in an enzyme by a point mutation: ornithine transcarbamylase. *Science* 245: 522–524, 1989.

101. Lanoue, K.F. and A.C. Schoolwerth. Metabolite transport in mitochondria. *Annu. Rev. Biochem.* 48: 871–922, 1979.
102. Lau, F.T.-K. and A.R. Fersht. Conversion of allosteric inhibition to activation in phosphofructokinase by protein engineering. *Nature* 326: 811–812, 1987.
103. Lee, A. and R. Langer. Shark cartilage contains inhibitors of tumor angiogenesis. *Science* 221: 1185–1187, 1983.
104. Leech, A.R., L. Goldstein, C. Cha and J.M. Goldstein. Alanine biosynthesis during starvation in skeletal muscle of the spiny dogfish, *Squalus acanthias*. *J. Exp. Zool.* 207: 73–80, 1979.
105. Marshall, M. Carbamyl phosphate synthetase I from frog liver. In: *The Urea Cycle*, edited by S. Grisolia, R. Báguena and F. Mayor. New York: Wiley, 1976, pp. 133–156.
106. Masters, C. and R. Holmes. Peroxisomes: new aspects of cell physiology and biochemistry. *Physiol. Rev.* 57, 816–882, 1977.
107. McCarthy, J.J. and T.E. Whitledge. Nitrogen excretion by anchovy (*Engraulis mordax*) and jack mackerel (*Trachurus symmetricus*). *Fish. Bull.* 70: 395–401, 1972.
108. Mehta, P.K., T.I. Hale and P. Christen. Evolutionary relationships among aminotransferases. Tyrosine aminotransferase, histidinol-phosphate aminotransferase and aspartate aminotransferase are homologous proteins. *Eur. J. Biochem.* 186: 249–253, 1989.
109. Meijer, A.J. and H.E.S.J. Hensgens. Ureogenesis. In: *Metabolic Compartmentation*, edited by H. Sies. New York: Academic Press, 1982, pp. 259–286.
110. Meister, A. Mechanism and regulation of the glutamine-dependent carbamyl phosphate synthetase of *Escherichia coli*. In: *Advances in Enzymology and Related Areas of Molecular Biology*, edited by A. Meister. New York: Wiley, 1989, pp. 315–374.
111. Mitchell, A.P. and B. Magasanik. Purification and properties of glutamine synthetase from *Saccharomyces cerevisiae*. *J. Biol. Chem.* 258: 119–124, 1983.
112. Mommsen, T.P. and P.J. Walsh. Evolution of urea synthesis in vertebrates: the piscine connection. *Science* 243: 72–75, 1989.
113. Moreadith, R.W. and A.L. Lehninger. The pathways of glutamate and glutamine oxidation by tumor cell mitochondria. Role of mitochondrial $NAD(P)^+$-dependent malic enzyme. *J. Biol. Chem.* 259: 6215–6221, 1984.
114. Morii, H., K. Nishikata and O. Tamura. Ammonia and urea excretion from mudskipper fishes *Periophthalmus cantonensis* and *Boleophthalmus pectinirostris* transferred from land to water. *Comp. Biochem. Physiol.* 63A: 23–28, 1979.
115. Moyes, C.D., T.W. Moon and J.S. Ballantyne. Oxidation of amino acids, Krebs cycle intermediates, fatty acids and ketone bodies by *Raja erinacea* liver mitochondria. *J. Exp. Zool.* 237: 119–128, 1986.
116. Müller, M.M., T. Gerster and W. Schaffner. Enhancer sequences and the regulation of gene transcription. *Eur. J. Biochem.* 176: 485–495, 1988.
117. Murkherjee, T., M. Ray and A. Bhaduri. Aspartate transcarbamylase from *Leishmania donovani*. *J. Biol. Chem.* 263: 708–713, 1988.
118. Nagy, M., J. Laporte, B. Penerne and G. Hervé. Nuclear localization of aspartate transcarbamylase in *Saccharomyces cerevisiae*. *J. Cell Biol.* 92: 790–794, 1982.
119. Ness, S.A. and R.L. Weiss. Carbamoyl phosphate synthetases from *Neurospora crassa*. *J. Biol. Chem.* 260: 14355–14362, 1985.
120. Nordlie, R.C. and H.A. Lardy. Mammalian liver phosphoenolpyruvate carboxykinase activities. *J. Biol. Chem.* 238: 2259–2263, 1963.
121. Nyunoya, H., K.E. Broglie and C.J. Lusty. The rat liver carbamyl phosphate synthetase I gene evolved by fusion of an ancestral glutaminase gene and a synthetase gene. *Proc. Natl. Acad. Sci. U.S.A.* 82: 2244–2246, 1985.
122. Nyunoya, H., K.E. Broglie, E.E. Widgren and C.J. Lusty. Characterization and derivation of the gene coding for mitochondrial carbamyl phosphate synthetase I of rat. *J. Biol. Chem.* 260: 9346–9356, 1985.
123. Ohno, S. *Evolution by Gene Duplication*. Heidelberg: Springer-Verlag, 1970.
124. Ohtake, A., M. Takiguchi, Y. Shigeto, Y. Amaya, S. Kawamoto and M. Mori. Structural organization of the gene for rat liver-type arginase. *J. Biol. Chem.* 263: 2245–2249, 1988.
125. Olson, K.R. and P.O. Fromm. Excretion of urea by two teleosts exposed to different concentrations of ambient ammonia. *Comp. Biochem. Physiol.* 40A: 999–1008, 1971.
126. O'Neal, T.D. and A.W. Naylor. Some regulatory properties of pea leaf carbamoyl phosphate synthetase. *Plant Physiol.* 57: 23–28, 1976.
127. Ou, W.-J., A. Ito, H. Okazaki and T. Omura. Purification and characterization of a processing protease from rat liver mitochondria. *EMBO J.* 8: 2605–2612, 1989.

128. Pamiljans, V., P.R. Krisnaswamy, G. Dumville and A. Meister. Studies on the mechanism of glutamine synthetase: isolation and properties of the enzyme from sheep brain. *Biochemistry* 1: 153–158, 1962.
129. Parsot, C. A common origin for enzymes involved in the terminal step of the threonine and tryptophan biosynthetic pathways. *Proc. Natl. Acad. Sci. U.S.A.* 84: 5207–5210, 1987.
130. Penninckx, M., J.-P. Simon and J.-M. Wiame. Interaction between arginase and L-ornithine carbamoyltransferase in *Saccharomyces cerevisiae*. *Eur. J. Biochem.* 49: 429–442, 1974.
131. Perlman, D.F. and L. Goldstein. Nitrogen metabolism. In: *Physiology of Elasmobranch Fishes*, edited by T.J. Schuttleworth. Berlin: Springer-Verlag, 1988, pp. 253–275.
132. Place, A.R. and D.A. Powers. Genetic variation and relative catalytic efficiencies: lactate dehydrogenase B allozymes of *Fundulus heteroclitus*. *Proc. Natl. Acad. Sci. U.S.A.* 76: 2354–2358, 1979.
133. Powers, S.G. and A. Meister. Urea synthesis and ammonia metabolism. In: *The Liver: Biology and Pathobiology*, edited by I. Arias, H. Popper, D. Schacter and D.A. Shafritz. New York: Raven Press, 1982, pp. 251–263.
134. Powers-Lee, S.G., R.A. Matisco and M. Bendayan. The interaction of rat liver carbamoyl phosphate synthetase and ornithine transcarbamoylase with inner mitochondrial membrane. *J. Biol. Chem.* 262: 15683–15688, 1987.
135. Prusiner, S. and E.R. Stadtman (editors). *The Enzymes of Glutamine Metabolism*. New York: Academic Press, 1973.
136. Quesada, A.R., M.A. Medina, J. Márquez, F.M. Sánchez-jiménez and I. Nunez De Castro. Contribution by host tissues to circulating glutamine in mice inoculated with Ehrlich ascites tumor cells. *Cancer Res.* 48: 1551–1553, 1988.
137. Ramaswamy, M. and T.G. Reddy. Ammonia and urea excretion in three species of air-breathing fish subject to aerial exposure. *Proc. Indian Acad. Sci. (Animal Sci.)* 92: 293–297, 1983.
138. Randall, D.J., C.M. Wood, S.F. Perry, H. Bergman, G.M.O. Maloiy, T.P. Mommsen and P.A. Wright. Urea excretion as a strategy for survival in a fish living in a very alkaline environment. *Nature* 337: 165–166, 1989.
139. Ratner, S. Enzymes of arginine and urea synthesis. *Adv. Enzymol.* 39: 1–90, 1973.
140. Read, L.J. Enzymes of the ornithine-urea cycle in the chimaera *Hydrolagus colliei*. *Nature* 215: 1412–1413, 1967.
141. Read, L.J. Chemical constituents of body fluid and urine of the holocephalan *Hydrolagus colliei*. *Comp. Biochem. Physiol.* 39A: 185–192, 1971.
142. Read, L.J. The presence of high ornithine-urea cycle enzyme activity in the teleost *Opsanus tau*. *Comp. Biochem. Physiol.* 39B: 409–413, 1971.
143. Reddy, S.R.R. and J.W. Campbell. A low molecular weight arginase in the earthworm. *Biochim. Biophys. Acta* 159: 557–560, 1968.
144. Reddy, S.R.R. and J.W. Campbell. Arginine metabolism in insects: role of arginase in proline formation during silkmoth development. *Biochem. J.* 115: 495–503, 1969.
145. Reddy, S.R.R. and J.W. Campbell. Molecular weights of arginase from different species. *Comp. Biochem. Physiol.* 32: 499–509, 1970.
146. Ren, W.-P., H. Ono and S. Tuboi. evidence for intra-mitochondrial degradation of the extrapeptide of ornithine aminotransferase. *Biochem. Biophys. Res. Commun.* 163: 215–219, 1989.
147. Ritter, N.M. *Elasmobranch Glutamine Synthetase: Synthesis and Subcellular Location of Tissue Specific Isozymes*. Ph.D. Thesis, Rice University, 1988.
148. Ritter, N.M., D.D. Smith, Jr. and J.W. Campbell. Glutamine synthetase in liver and brain tissues of the holocephalan, *Hydrolagus colliei*. *J. Exp. Zool.* 243: 181–188, 1987.
149. Saha, N. and B.K. Ratha. Active ureogenesis in a freshwater air-breathing teleost, *Heteropneustes fossilis*. *J. Exp. Zool.* 241: 137–141, 1987.
150. Salerno, C., P. Fasella and L.A. Fahien. Interaction of aminotransferase with other metabolically linked enzymes. In: *Transaminases*, edited by P. Christen and D.E. Metzler. New York: Wiley, 1985, pp. 192–208.
151. Schaeffer, B. and M. Williams. Relationship of fossil and living elasmobranchs. *Am. Zool.* 17: 293–302, 1977.
152. Schultze, H.-P. Dipnoans as sarcopterygians. *J. Morph. (Suppl.)* 1: 39–74, 1986.
153. Selwyn, M.J. Holes in mitochondrial inner membrane. *Nature* 330: 424–425, 1987.
154. Shankar, R.A. and P.M. Anderson. purification and properties of glutamine synthetase from liver of *Squalus acanthias*. *Arch. Biochem. Biophys.* 239: 248–259, 1985.
155. Shibata, H., H. Ochiai, Y. Sawa and S. Miyoshi. Localization of carbamoylphosphate synthetase and aspartate carbamoyltransferase in chloroplasts. *Plant Physiol.* 80: 126–129, 1986.

156. Shibata, H., Y. Sawa, H. Ochiai, T. Kawashima and K. Yamane. a possible regulation of carbamoylphosphate synthetase and aspartate carbamoyltransferase in chloroplasts. *Plant Sci.* 51: 129–133, 1987.
157. Shoemaker, V.H., D. Balding, R. Ruibal and L.L. McClanahan. Uricotelism and low evaporative water loss in a South American frog. *Science* 175: 1018–1020, 1972.
158. Smith, D.D., Jr. and J.W. Campbell. Subcellular location of chicken brain glutamine synthetase and comparison with chicken liver mitochondrial glutamine synthetase. *J. Biol. Chem.* 258: 12265–12268, 1983.
159. Smith, D.D., Jr. and J.W. Campbell. Glutamine synthetase in liver of the American alligator, *Alligator mississippiensis. Comp. Biochem. Physiol.* 86B: 755–762, 1987.
160. Smith, D.D., Jr. and J.W. Campbell. Distribution of glutamine synthetase and carbamoyl-phosphate synthetase-I in vertebrate liver. *Proc. Natl. Acad. Sci. U.S.A.* 85: 160–164, 1988.
161. Smith, D. D, Jr., N.M. Ritter and J.W. Campbell. Glutamine synthetase isozymes in elasmobranch brain and liver tissues. *J. Biol. Chem.* 262: 198–202, 1987.
162. Smith, D.D., Jr., J.E. Vorhaben and J.W. Campbell. Preparation and cross-reactivity of anti-avian glutamine synthetase antibody. *J. Exp. Zool.* 226: 29–35, 1983.
163. Smith, H.W. The composition of the body fluids of elasmobranchs. *J. Biol. Chem.* 81: 407–419, 1929.
164. Smith, H.W. The absorption and excretion of water and salts by elasmobranch fishes. I. Fresh water elasmobranchs. *Am. J. Physiol.* 98: 279–295, 1931.
165. Smith, H.W. The absorption and excretion of water and salts by elasmobranch fishes. II. Marine elasmobranchs. *Am. J. Physiol.* 98: 296–310, 1931.
166. Smith, H.W. The retention and physiological role of urea in Elasmobranchii. *Biol. Rev.* 11: 49–82, 1936.
167. Söling, H.-D. and J. Kleinke. Species dependent regulation of hepatic gluconeogenesis in higher animals. In: *Gluconeogenesis: Its Regulation in Mammalian Species*, edited by R.W. Hanson and M.A. Mehlman. New York: Wiley, 1976, p. 369–462.
168. Souciet, J.-L., S. Potier, J.-C. Hebert and F. Lacroute. Nucleotide sequence of pyrimidine specific carbamoyl phosphate synthetase, a part of the yeast multifunctional protein encoded by the URA2 gene. *Mol. Gen. Genet.* 207: 314–319, 1987.
169. Suarez, R.K. and T.P. Mommsen. Gluconeogenesis in teleost fishes. *Can. J. Zool.* 65: 1869–1882, 1987.
170. Suenaga, M., S. Lee, N.G. Park, H. Aoyagi, T. Kato, A. Umeda and K. Amako. Basic amphipathic helical peptides induce destabilization and fusion of acid and neutral liposomes. *Biochim. Biophys. Acta* 981: 143–150, 1989.
171. Suzuki, T., M. Sato, T. Yoshida and S. Tuboi. Rat liver mitochondrial and cytosolic fumarases with identical amino acid sequences are encoded from a single gene. *J. Biol. Chem.* 264: 2581–2586, 1989.
172. Tate, S.S. and A. Meister. Glutamine synthetases of mammalian liver and brain. In: *The Enzymes of Glutamine Metabolism*, edited by S. Prusiner and E.R. Stadtman. New York: Academic Press, 1973, pp. 77–127.
173. Tate, S.S., F.-Y. Leu and A. Meister. Rat liver glutamine synthetase. *J. Biol. Chem.* 247: 5312–5321, 1972.
174. Tramell, P.R. and J.W. Campbell. Carbamyl phosphate synthesis in a land snail, *Strophocheilus oblongus. J. Biol. Chem.* 245: 6634–6641, 1970.
175. Tramell, P.R. and J.W. Campbell. Carbamyl phosphate synthesis in invertebrates. *Comp. Biochem. Physiol.* 40B: 395–406, 1971.
176. Tricot, C., J.-L. De Coen, P. Momin, P. Falmagne and V. Stalon. Evolutionary relationships among bacterial carbamoyltransferases. *J. Gen. Microbiol.* 135: 2453–2464, 1989.
177. Tropschug, M., D.W. Nicholson, F.-U. Hartl, H. Köhler, N. Pfanner, E. Wachter and W. Neupert. Cyclosporin A-binding protein (cyclophilin) of *Neurospora crassa. J. Biol. Chem.* 263: 14433–14440, 1988.
178. Trotta, P.P., L.M. Pinkus, R.H. Haschemyer and A. Meister. Reversible dissociation of the monomer of glutamine-dependent carbamyl phosphate synthetase into catalytically active heavy and light subunits. *J. Biol. Chem.* 249: 492–499, 1974.
179. Tsuji, S. Chicken ornithine transcarbamylase: purification and some properties. *J. Biochem. (Tokyo)* 94: 1307–1315, 1983.
180. Urrestarazu, L.A., S. Vissers and J.-M. Wiame. Change in location of ornithine carbamoyltransferase and carbamoylphosphate synthetase among yeasts in relation to the arginase/ornithine

carbamoyltransferase regulatory complex and the energy status of the cells. *Eur. J. Biochem.* 79: 473–481, 1977.
181. Van Beneden, R.J. and D.A. Powers. The isozymes of glucose-phosphate isomerase (GPI-A$_2$ and GPI-B$_2$) from the teleost fish *Fundulus heteroclitus* (L.). *J. Biol. Chem.* 260: 14596–14603, 1985.
182. Vellas, F. and A. Serfaty. Urea excretion in the carp (*Cyprinus carpio* L.). *Arch. Sci. Physiol.* 21: 185–192, 1967.
183. Verner, K. and G. Schatz. Protein translocation across membranes. *Science* 241: 1307–1313, 1988.
184. Vestweber, D., J. Brunner, A. Baker and G. Schatz. A 42 K outer-membrane protein is a component of the yeast mitochondrial import site. *Nature* 341: 205–209, 1989.
185. Vickers, L.P. Trimeric enzymes in carbamoylphosphate metabolism. *Trends Biochem. Sci.* 6: xi-xii, 1981.
186. Von Heijne, G. Transcending the impenetrable: how proteins come to terms with membranes. *Biochim. Biophys. Acta* 947: 307–333, 1988.
187. Von Heijne, G., J. Steppuhn and R.G. Herrmann. Domain structure of mitochondrial and chloroplast targeting peptides. *Eur. J. Biochem.* 180: 535–545, 1989.
188. Vorhaben, J.E., D.D. Smith and J.W. Campbell. Characterization of glutamine synthetase from avian liver mitochondria. *Int. J. Biochem.* 14: 747–756, 1982.
189. Vorhaben, J.E., L. Wong and J.W. Campbell. Assay for glutamine synthetase activity. *Biochem. J.* 135: 893–896, 1973.
190. Waley, S.G. Some aspects of the evolution of metabolic pathways. *Comp. Biochem. Physiol.* 30: 1–11, 1969.
191. Walsh, P.J., E. Danulat and T.P. Mommsen. Variation urea excretion in the Gulf toadfish, *Opsanus beta*. *Mar. Biol.*, 106: 323–328, 1990.
192. Watford, M., Y. Hod, Y. Chiao, M.F. Utter and R.W. Hanson. The unique role of the kidney in gluconeogenesis in the chicken. The significance of a cytosolic form of phosphoenolpyruvate carboxykinase. *J. Biol. Chem.* 256: 10023–10027, 1981.
193. Watts, D.C. and R.L. Watts. Carbamoyl phosphate synthetase in the Elasmobranchii: osmoregulatory function and evolutionary implications. *Comp. Biochem. Physiol.* 17: 785–798, 1966.
194. Webb, J.T. and G.W. Brown, Jr. Some properties and occurrence of glutamine synthetase in fish. *Comp. Biochem. Physiol.* 54B: 171–175, 1976.
195. Webb, J.T. and G.W. Brown, Jr. Glutamine synthetase: assimilatory role in liver as related to urea retention in marine Chondrichthyes. *Science* 208: 293–295, 1980.
196. Weiss, R.L. and R.H. Davis. Intracellular localization of enzymes of arginine metabolism in *Neurospora*. *J. Biol. Chem.* 248: 5403–5408, 1973.
197. Wilson, R.P. Nitrogen metabolism in channel catfish *Ictalurus punctatus* – II. Evidence for an apparent incomplete ornithine-urea cycle. *Comp. Biochem. Physiol.* 46B: 625–634, 1973.
198. Wilson, R.P. and P.L. Fowlkes. Activity of glutamine synthetase in channel catfish tissues determined by an improved tissue assay method. *Comp. Biochem. Physiol.* 54B: 365–368, 1976.
199. Wilson, R.P. and W.E. Poe. Nitrogen metabolism in channel catfish *Ictalurus punctatus* – III. Relative pool sizes of free amino acids and related compounds in various tissues of the catfish. *Comp. Biochem. Physiol.* 48B: 545–556, 1974.
200. Wong, T.M. and D.K.O. Chan. Physiological adjustments to dilution of the external medium in the lip-shark *Hemiscyllium plagiosum* (Bennett). Branchial, renal, and rectal gland function. *J. Exp. Zool.* 200: 85–96, 1977.
201. Wood, C.M., S.F. Perry, P.A. Wright, H.L. Bergman and D.J. Randall. Ammonia and urea dynamics in the Lake Magedi tilapia, a ureotelic teleost fish adapted to an extremely alkaline environment. *Resp. Physiol.* 77: 1–20, 1989.
202. Wood, J.D. Nitrogen excretion in some marine teleosts. *Canad. J. Biochem. Physiol.* 36: 1237–1242, 1958.
203. Woolfolk, C.A., B. Shapiro and E.R. Stadtman. Regulation of glutamine synthetase – I. Purification and properties of glutamine synthetase from *Escherichia coli*. *Arch. Biochem. Biophys.* 116: 177–192, 1966.
204. Wu, C. Glutamine synthetase – I. A comparative study of its distribution in animals and its inhibition by DL-allo-d-hydroxyllysine. *Comp. Biochem. Physiol.* 8: 335–351, 1963.
205. Wu, C. Glutamine synthetase. II. The intracellular localization in rat liver. *Biochim. Biophys. Acta* 77: 482–493, 1963.
206. Xiong, X. and P.M. Anderson. Purification and properties of ornithine carbamoyl transferase from liver of *Squalus acanthias*. *Arch. Biochem. Biophys.* 270: 198–207, 1989.
207. Zhaung, Z. and R. McCauly. Ubiquitin is involved in the *in vitro* insertion of monoamine oxidase B into mitochondrial outer membranes. *J. Biol. Chem.* 264: 14594–14596, 1989.

CHAPTER 4

Frontiers in the study of the biochemistry and molecular biology of vision and luminescence in fishes

MARGARET MCFALL-NGAI AND WESLEY TOLLER

Department of Biological Sciences, University of Southern California, Los Angeles, CA 90089–0371, U.S.A.

I. Introduction
 1. The aquatic light environment
 2. Adaptations of visual and luminescence systems as responses to the aquatic light field
II. Visual systems of fishes
 1. The ecology of fish vision
 2. The properties of the visual pigments
  2.1. The chromophore
  2.2. The opsin
 3. Horizons in the study of fish vision
III. Bioluminescence in fishes
 1. Bacterial bioluminescence in fishes
  1.1. Occurrence
  1.2. Culturable bacterial symbioses
  1.3. Uncultured bacterial symbioses
  1.4. The molecular biology, biochemistry and physiology of fish/bacterial light organ symbioses
 2. Non-bacterial bioluminescence in fishes
 3. Tissues supporting photogenic cells
IV. Conclusions
V. References

I. Introduction

This chapter highlights those portions of the fields of fish vision and fish bioluminescence that are at the horizons of our understanding of the biochemical and molecular bases for these phenomena. As such, the treatment is not intended to be an exhaustive review of the vast literature on the visual and luminescence systems of fishes, subjects that have been reviewed recently by a number of authors[49,60,123]. Instead we will focus specifically on the rapid advances in biochemistry and molecular biology that are making new techniques more readily accessible, and are permitting a more detailed view of how both vision and luminescence are mediated at the level of the gene. It is at this level that we are only now beginning to explore many areas of fish biology.

Because the visual system is shared by all vertebrate classes, while bioluminescence is restricted to fishes[60], it is not surprising that the frontiers for molecular research of vision in fishes are more clearly defined than are those of bioluminescence. For example, recent benchmark studies in the molecular biology of the rhodopsins[3,57,112,114,167], found in organisms as diverse as bacteria and terrestrial vertebrates, have provided great insight into how this class of molecules functions in sensory transduction. These data and experimental approaches can be, and are now being, directly applied to studies of how fishes are adapted to their visual world. Conversely, an understanding of the basic mechanisms by which rhodopsins work promises to be greatly enhanced by the study of the diversity of fish visual systems. The time scale over which adaptive selection has been operating on their evolution[17], and the vast array of light fields in which they occur[65], present unique opportunities for comparative studies using fishes. Such studies will be invaluable to an understanding of how evolutionary selection pressure acts on the rhodopsin molecule. Further, the well-studied subject of the visual physiology of fishes[1,123] affords a considerable volume of background information about how visual systems of fishes are adapted to their various aquatic habitats. The direction of future efforts is clear and exciting, and promises to be fruitful; the questions are defined, and answers await.

In contrast, directions in the study of the molecular biology and biochemistry of fish bioluminescence are not so clearly discernable, primarily due to the following three circumstances. Firstly, the subject has been less intensely studied, probably because luminescence does not occur in other vertebrate classes, but in addition, many bioluminescent fishes are less easily accessible and more difficult to maintain under laboratory conditions. Secondly, rather than the single evolutionary origin of the visual system, luminescence capability has arisen independently many times within the fishes[60]. Thirdly, some families of fishes are bioluminescent through symbiotic associations with luminous bacteria, while others directly control the substrates of luminescence, which are derived either from their gene products or from the gene products of other luminous animals[61]. Thus, while we have a broad understanding of the phylogenetic occurrence of luminescence in fishes, we have only begun to uncover the biochemical and molecular details characterizing the diverse array of luminescence systems that have evolved. These facts notwithstanding, the study of certain fish bioluminescence systems have seen remarkable advances in recent years, and their successes present intriguing and useful signposts for the future.

By way of introduction to research frontiers of fish vision and bioluminescence, a brief survey of the light fields in the oceans will be presented, to familiarize the reader with the suite of selection pressures imposed on these systems by the nature of the fishes' environments. Within this context, we will explore why it is useful to consider fish vision and bioluminescence, two light-related systems, simultaneously.

1. The aquatic light environment

The interaction of light with water, and with materials suspended and dissolved in water, creates a variety of photic environments in aquatic habitats markedly richer than that encountered on land[93]. This difference in the number of distinct light regimes has resulted both in a wider array of adaptations of the visual systems in aquatic vertebrates[1,123] and in the widespread occurrence of luminescence[60,61]. To understand the mechanisms by which selection pressure on vision and luminescence is directed by the quality of the photic environment, it is useful to consider the types of light fields present, and their biotic and abiotic determinants, comparing them with the more familiar light fields encountered on land.

The terrestrial biosphere occurs as a thin, heterogeneous veneer over the surface of the land. Within the relatively short distances this coverage represents, light is little affected by the medium itself (the air), nor is it affected by suspended biotic or abiotic particles for the protracted periods of time that would be necessary for these factors to influence evolutionary selection[93]. The day–night cycle and the seasons provide the dominant variables in the quality of environmental light, i.e., its intensity and spectrum[96].

While variations in the quality of the aquatic photic environment also result from diurnal and seasonal changes, significant compounding influences include: (1) the depth of the aquatic biosphere[13]; (2) the water itself[65,70,159]; and, (3) biotic and abiotic materials suspended in the water[65]. The ocean is several miles deep in places, creating a vast three-dimensional field that has no true analogue in the terrestrial environment. Most fishes spend the majority of their lives in the water column, often miles from any substrate. As light travels through these great distances, the water and materials in the water alter the light field. The variety and complexity of the photic environments of the ocean has led to detailed classifications of coastal and oceanic waters based on the quality and quantity of materials in the water column[65]. Broadly speaking, and of relevance to selection on visual and luminescence systems, the ocean can be divided into the open ocean and nearshore environments. In clear oceanic waters, medium effects predominate. The water itself strongly absorbs light, resulting in its logarithmic decay with depth, and the short-wavelength and long-wavelength components of downwelling sunlight are more strongly absorbed, resulting in the predominance of blue wavelengths below 50 m. These processes yield a relatively predictable light field that vertically divides the open ocean into the euphotic (= epipelagic), dysphotic (= mesopelagic) and aphotic zones (= bathypelagic)[85]. The euphotic zone, defined as that portion of the water column where enough light is present for photosynthesis, extends to about 200 m in the clearest ocean waters. The dysphotic zone begins at the lower limit of the euphotic zone, or at the photosynthetic compensation depth, where blue light is still present but at intensities insufficient for photosynthesis to balance respiration; it extends from approximately 200 to 1000 m. At depths greater than 1000 m, all solar light has been absorbed and, thus, the environment is aphotic. With the abyssal plains of the ocean averaging approximately 4000 m, the majority of the biosphere occurs in regions without sunlight.

The light quality of nearshore habitats is further affected by biotic (e.g., plankton blooms[4,69,70]) and abiotic (e.g., runoff from the land, sediment disturbances) influences[65,93]. These factors affect both the attenuation of the light as a function of depth, as well as the wavelengths that are preferentially absorbed. Such influences are often variable in time and space and are largely dependent on the types and concentration of the organic and inorganic load in these waters. Generally, coastal water is more highly attenuated and shifted toward greener wavelengths[65]. Also significant is the scatter of light by particulate material in the water. Such scattering is responsible for the production of 'veiling' light, or bright diffused light of the sunlit areas of the water column[81]. This phenomenon also occurs in terrestrial environments when sunlight is scattered by the fog or mist[91]. The importance of veiling light to the overall pattern of the light field is, of course, directly related to the concentration of particulates in the water.

While several other features of the light field have influenced the evolution of visual and luminescence systems of fishes (e.g., the polarization of light, or changes in angular distribution of downwelling light with depth[29]), intensity, wavelength and scatter are those features that have been most strongly implicated in the evolution of these two systems and their associated biochemistries[2,49,93].

2. Adaptations of visual and luminescence systems as responses to the aquatic light field

Although selection on a large number of fish visual systems, such as those characteristic of diurnal reef fishes[1,123], is not influenced by luminescence, a significant subset of fishes lives in environments where luminescence is either a prevalent component of the light field, as in the dysphotic zone, or the only available light, as in the aphotic zone. Considerable evidence exists for the coevolution of these systems in such habitats, and a number of authors have addressed this subject (e.g., refs. 2, 14, 84, 122).

Probably the most striking evidence for coevolution of vision and bioluminescence is their structural similarity at the anatomical level (Fig. 1; refs. 10, 76). The source of the light emission is endogenous in luminescence and exogenous in the visual system, but the supporting anatomy for the processing of the photons is analogous, and under some circumstances may even be homologous. In the vertebrate eye (for a review, see ref. 75), a photon from the environment passes first through the cornea and lens, to be bent and focussed on the small area of photoreceptor cells of the retina. A choroid behind the retina serves to capture any photons not absorbed by the visual cell outer segments. Further, in some animals a reflector, or tapetum, is present behind the retina to permit another pass of the photon over the retina, increasing efficiency of photon capture. The luminescence systems of most fishes are, in general, similar to the eye in construction. Photons produced by a localized photogenic area are directed out through tissues that act, depending on the system, as lenses or corneas, bending or diffusing the point source light. Further, a reflector layer insures efficient light emission and a pigmented layer absorbs stray photons[61].

The structural convergence of these two organ types is often accomplished by the recruitment of similar biochemical raw materials. The reflectors are composed of imbricated purine platelets[25,92,121,158] and the pigmented layers contain high concentrations of melanin[61,75]. At a higher level of organization, in some instances,

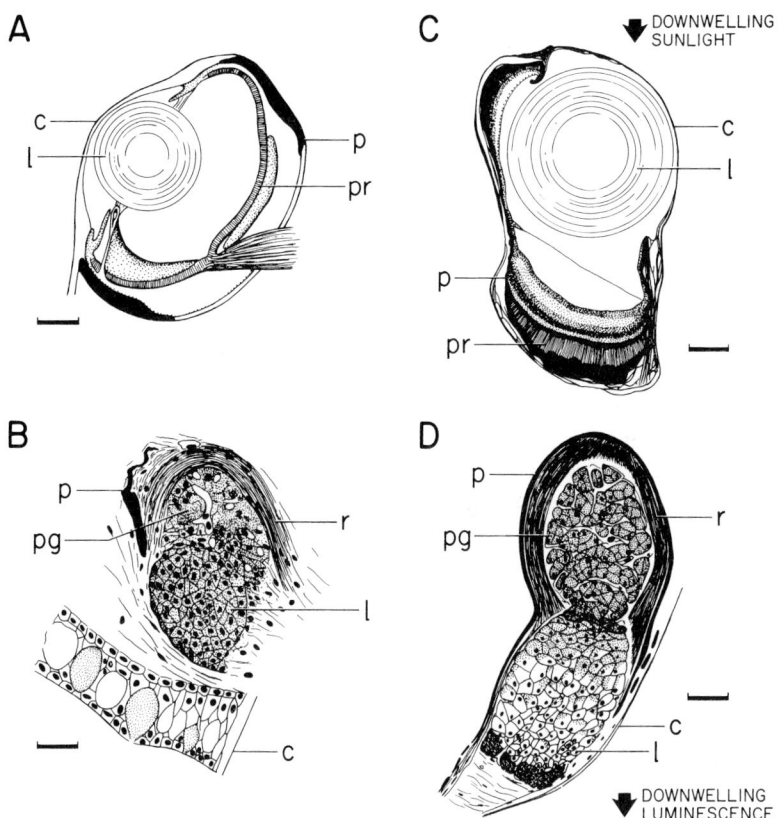

Fig. 1. Diagrammatic vertical sections of ocular and photogenic structures from shallow-water and mesopelagic fishes. A: a typical eye of a teleost such as *Porichthys notatus*, an inhabitant of shallow, nearshore waters of the Eastern Pacific (bar = 1 mm; modified from ref. 123). B: a ventral photophore from *P. notatus* (bar = 100 μm; modified from ref. 144). C: the upturned, 'tubular' eye of *Argyropelecus affinis*, an inhabitant of the mesopelagic zone (bar = 1 mm; modified from ref. 10). D: downturned, ventral photophore of *A. affinis* (bar = 0.5 mm, modified from ref. 10). c = cornea, l = lens, p = pigmented layer, pg = photogenic tissue, pr = photoreceptor layer, r = reflective layer.

the tissues that function as lenses and corneas may have similar properties to accomplish the same function. For example, the lenses and corneas of many of the bioluminescent fishes are formed from tissues that are embryologically similar to the eye lens and cornea[61]. It remains to be determined whether this homology extends to the biochemicals that are expressed within these tissues.

Physiological controls are also imposed on these systems to change the intensity and/or the spectrum of the light that either enters the organ, in the case of the eye[75], or leaves the organ, in the case of a light-emitting structure[61]. In the fish eye, photomechanical change in the retina partially controls the level of environmental light that is integrated by higher neuronal processes[1,123]. Similarly, bioluminescence intensity is controlled by the action of chromatophores in the tissues through which light passes. In addition, luminescent organs often have shutters, which can completely occlude light emission into the environment, that are analogous to the eye lid of terrestrial vertebrates[95,99]. The spectral quality of light transmitted can also be controlled physiologically. In both systems, pigments occur in intermediary tissues[56,162], such as the lens. The pigments can be either static within these tissues, or their concentration in relation to the light path can be subject to dynamic change under varying conditions.

The above-mentioned features of visual and bioluminescence systems indicate that because these systems interact with light they are physiologically and anatomically similar. However, the fine tuning of the convergence of these two systems has most certainly been accomplished through selection on the behavior and ecology of these fishes using bioluminescence. A large body of literature exists that supports the idea that the coevolution of luminescence and vision is central to the evolution of predator–prey interactions and intraspecific communication of animals in the dysphotic and aphotic zones[2,20,84,166]. For example, in the mesopelagic zone, the majority of the fishes use ventrally directed bioluminescent light to camouflage themselves from predators located below. The systems that have been studied in detail have revealed that the luminescence matches very precisely the intensity, spectrum and angular distribution of downwelling light, even to the extent that filters over the bioluminescent organs of some fishes narrow the spectrum of the bioluminescence to match that of ambient light[19,29,30]. Further, a number of examples have been reported of animals with specially coevolved visual systems that function to vitiate this camouflage[103,142,143]. In the aphotic zone, luminescence is the only light available for intra- and interspecific communication and, thus, must be tightly coupled with the evolution of the eyes of those fishes.

II. Visual systems of fishes

1. The ecology of fish vision

Research over the past twenty-five years has provided much information about visual transduction in fishes[1,123]. Particularly numerous are studies that have correlated the light quality of the fishes' environment with anatomy, physiology, and biochemistry of their visual systems[24,78,81,82,102]. Probably the most conspicuous adaptations are those associated with the differences in light intensity presented by the various habitats of the marine world. Generally, the eyes of fishes in the euphotic zones of the open ocean and in nearshore habitats have duplex retinas with both rods and cones, the two basic types of visual cells. Rods are associated with scotopic (dim light) vision, while cones play a greater role in photopic (bright light) vision. Various physiological and anatomical adaptations of the rods and cones optimize sensitivity and resolution, respectively, under the conditions presented over the day–night cycle. In contrast, in the low light regimes of the dysphotic zone, and in the aphotic zone where only bioluminescent light is available, the visual systems are adapted for maximum photon capture. Fishes in these habitats usually have pure rod retinas, and often improve sensitivity through

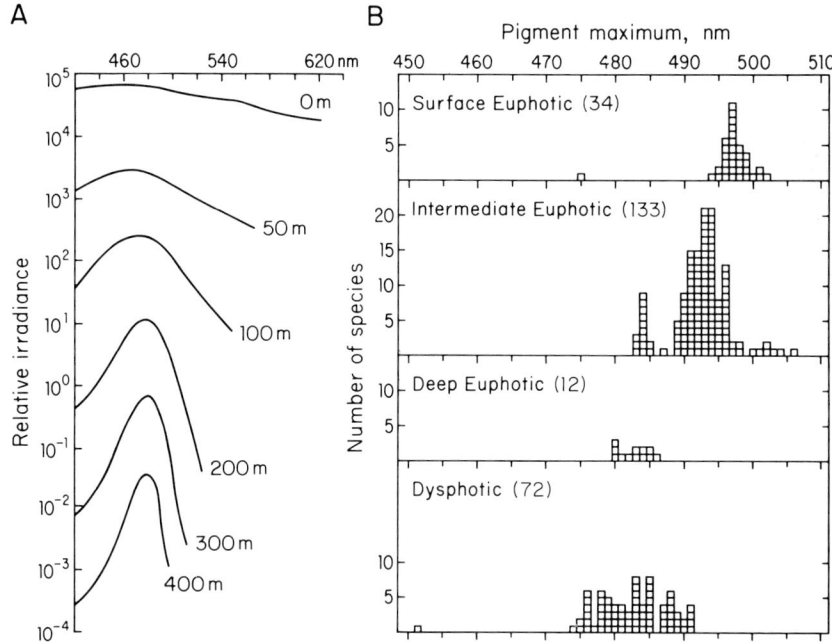

Fig. 2. A: the progressive attentuation and selective absorption of light with increasing depth for clear oceanic waters (from ref. 65). B: the trend towards blue-shifted rhodopsin absorption maxima with increasing depth of habitat[105,128,129].

elaborations of the shape and size of the eye, and peculiar arrangements of the ocular tissues, such as the lens[78,84,101].

Just as is the case with adaptations to different intensities, there is a large body of data suggesting that fishes are adapted to the spectral quality of their environment[24,82]. Visual pigments, the molecules responsible for transduction of photons to chemical signals, absorb light within defined regions of the spectrum. Several different cone pigments have been identified, including some that are UV sensitive[9,48]; similarly, the rod pigments vary in the maximum wavelength at which they absorb[27,104,156]. More data exist on rod pigments because a technique for the extraction and spectrophotometric analysis of rod pigments has been available for many years. This technique was not suitable for cone pigments; however, with the recent development of microspectrophotometric (MSP) techniques[160], which permit the analysis of visual pigments in the intact cell, measurement of the pigment absorption properties of cones has been made possible, providing valuable data on cone pigments. These techniques have also been applied to rods, and these data confirm rod pigment absorption characteristics that had been obtained by pigment extraction techniques.

The large body of literature on variation of rod pigments has permitted the correlation of absorbance maxima of the pigment of a given fish species with the properties of the light field of its natural environment (Fig. 2). Rod pigments most often absorb around 500 nm but, generally, deep-sea fishes have more blue-shifted pigments, while nearshore and freshwater fishes have rod pigments that absorb maximally at greener wavelengths. Early work indicated that in some cases the fish's visual pigment very closely matches the spectrum of environmental light[27,156], while in other species, the visual pigment absorption maximum is several nanometers away from the peak wavelength of the light in the habitat[24]. These findings have resulted in the formulation of two hypotheses to explain the relationship between the visual pigment and environmental light: the sensitivity hypothesis and the contrast hypothesis. The sensitivity hypothesis holds that, to maximize sensitivity, the fish's visual pigment absorbance will peak at the dominant wavelength of environmental light at the time of day or depth most important to the fish's use of vision in its behavior[106]. For example, the blue shift in the pigments of fishes in the dysphotic zone supports the sensitivity hypothesis. Further, Munz and McFarland[105] concluded that the rhodopsins of reef fishes match the spectrum of twilight underwater irradiance that is characteristic of the reef habitat at the time when visual sensitivity is most important to survival. The contrast hypothesis holds that, in environments where scatter is significant (i.e., where there is a great deal of veiling light), the peak wavelength of the visual pigment will be shifted away from the peak of ambient light to increase the contrast of objects against background light, thereby increasing resolution[79,80]. Because the shallow water environments present such a complex light field, the visual system of a fish in these habitats will certainly be optimized for sensitivity and contrast under different conditions, and it will be difficult, if not impossible, to discern which of the two is more important in driving selection on the visual pigment molecule.

While the precise relationship of the properties of fish visual pigments and fish ecology remains disputed, it is nonetheless clear that natural selection has acted upon the rod pigments of fishes to produce an extraordinary diversity of absorption maxima. The biochemical and molecular mechanisms underlying such variations are not yet understood. However, recent analyses and experimental manipulations of other rhodopsins, particularly those of bacteria[107,108,161], insects (*Drosophila*[36,130,167]) and mammals[111,112,113,114], provide exciting clues as to how spectral maxima are biochemically determined. These studies have been primarily concerned with the function of these rhodopsin molecules at the most basic biochemical level and not with their evolution. However, the resultant sequencing data have provided a molecular clock for the rhodopsins, and structural data on the molecule that suggest those areas most likely involved in the determination of spectral absorbance. An understanding of what these studies tell us of the basic function of rod pigments promises to permit the design of experiments to approach how selection pressure has acted on fish pigments.

2. The properties of the visual pigments

Visual pigments are integral membrane proteins composed of a chromophoric group covalently bound, via a protonated Schiff base, to a lysine residue of the protein opsin[155]. The disposition of the protein within the cell membrane positions the water-insoluble chromophore near the center of the lipid bilayer[148]. Absorp-

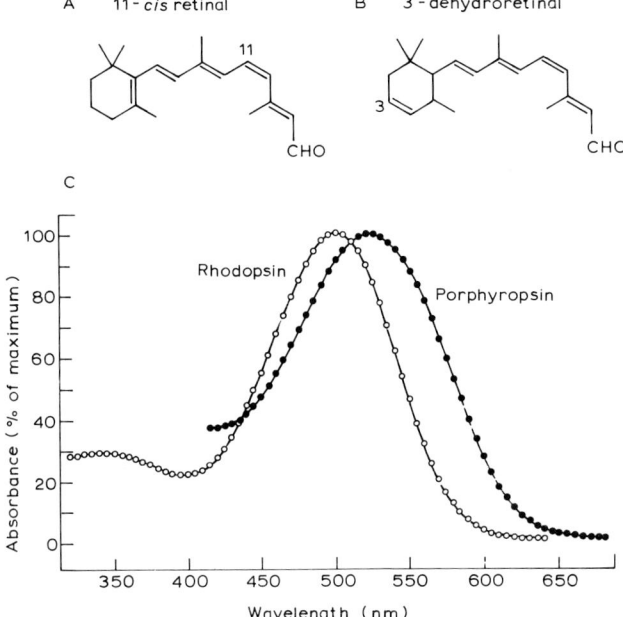

Fig. 3. A: the chemical structure of 11-*cis*-retinal. B: the chemical structure of 11-*cis*-3-dehydroretinal. C: typical absorption spectra for a rhodopsin (λ_{max} = 500 nm) and a porphyropsin (λ_{max} = 523 nm) (modified from ref. 12).

tion of a photon of light causes isomerization of the chromophore. The isomerized chromophore is expelled from the protein, presumably revealing an enzymatic surface. This enzymatic surface binds G-protein (guanosine triphosphate binding protein), initiating an enzymatic cascade that eventually leads to a neural response via cGMP[145]. Light-dependent phosphorylation of the opsin C-terminal residues appears to be instrumental in shutting-off the cascade triggered by an activated rhodopsin molecule[77]. The chromophore is enzymatically converted back to the initial isomer, which reassociates with the opsin to regenerate rhodopsin. An understanding of how the chromophore and opsin act in concert to bring about absorption of certain wavelengths has been achieved by extensive experimental studies of the chromophore and the opsin molecule.

2.1. The chromophore

Vertebrates have two chromophore types that associate with the opsin protein, both of which are aldehyde derivatives of vitamin A (Fig. 3A and B). Most commonly, the chromophore is 11-*cis*-retinal, which when bound to the opsin forms the molecule called rhodopsin. Certain fishes and a few other vertebrates utilize 11-*cis*-3-dehydroretinal, which binds with opsin to form porphyropsin. Visual pigments have an absorption spectrum much like a bell-shaped curve, the peak being the wavelength of maximum absorption, λ_{max}. Rhodopsin and porphyropsin differ in where they absorb in the visible spectrum and in the broadness of their absorption curves (Fig. 3C). Vertebrate rhodopsins typically have a λ_{max} of 500 nm. Porphyropsins absorb more effectively at slightly longer wavelengths, with a typical λ_{max} of around 540 nm. The rhodopsin absorption curve has a narrower half-band width than that of porphyropsin.

Spectral location of a visual pigment λ_{max} results from interaction between opsin and chromophore, as well as of the chromophore's intrinsic light absorption properties. The chromophore 11-*cis*-retinal dissolved in ethanol absorbs light maximally at 379 nm[72]. Schiff base protonation results in a shift of the chromophore λ_{max} to 440 nm[110]. However, when attached to the opsin protein by a Schiff base linkage, the spectral absorption of retinal is further shifted toward longer wavelengths (e.g., bovine rhodopsin has a λ_{max} of 498 nm). The cause of this phenomenon, known as the opsin-shift[108], has been attributed to the structure of the retinal binding pocket (see below) created by the opsin protein. The mechanism by which chromophoric absorption is modified by the protein is not yet understood; however, several hypotheses have been proposed[7,62,64]. Some current studies have employed analogues of 11-*cis*-retinal as artificial chromophores to probe the ligand–opsin interaction (e.g., ref. 161).

2.2. The opsin

Rhodopsins belong to a large family of membrane receptor proteins that includes adrenergic and muscarinic receptor proteins[164]. Members of this family are all characterized by having seven transmembrane segments (Fig. 4A) and by coupling with G-proteins. All rhodopsins, including those of bacteria, interact with the

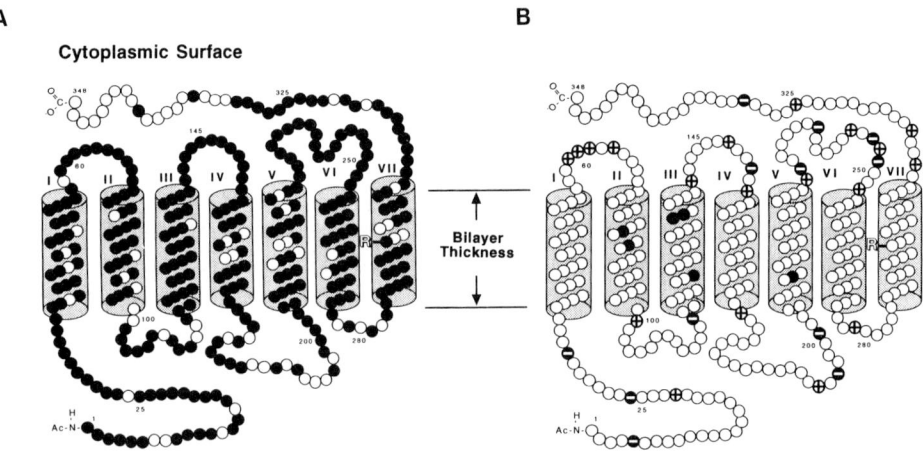

Fig. 4. A: amino acid sequence identities (black circles) among four vertebrate rhodopsins: human, bovine, ovine, and chicken. Non-conserved amino acids are indicated by open circles. Shaded cylinders represent α-helical regions that traverse the lipid bilayer; R = retinal attachment site (refs. 38, 112, 113, 146; illustration modified from ref. 3). B: mutagenesis studies of the binding pocket of rhodopsin[111]. Black circles indicate charges amino acids that were replaced by site-directed mutagenesis; open circles indicate uncharged amino acids. Also illustrated are positively charged (basic) residues (⊕), and negatively charged (acidic) residues (●) located outside the pocket that were not replaced.

ligand retinal[107], although there is no discernable sequence similarity between eukaryotic and prokaryotic rhodopsins[38]. However, a recent study of the three-dimensional structure of the bacteriorhodopsins has shown that the 21 amino acids that form the retinal binding pocket are highly conserved, and distributed among the seven transmembrane segments[57]. This indicates that the functioning of the molecule demands a certain environment surrounding the ligand, and that few substitutions within the retinal binding pocket are tolerated. Given the highly conserved nature of the vertebrate rhodopsin molecule (see below), it seems likely that similar constraints are operating.

Vertebrate rhodopsins have an apoprotein that is 348 amino acids in length. Hydropathy plots and patterns of proteolytic degradation[38] both indicate that the opsin consists of the characteristic seven trans-membrane, α-helical segments connected by six loops (see Fig. 4A). Three cytosolic loops, and the carboxy terminus, are enzymatically active. Retinal is bound to lysine 296 on the seventh transmembrane segment. The six remaining α-helices presumably surround the chromophore in a three dimensional array that forms a binding pocket[110]. As in bacterial rhodopsins, the binding pocket has received considerable study because the amino acids located here are in intimate contact with the chromophore and, therefore, have the potential to influence chromophore spectral absorption.

Primary structures of several visual pigments have been determined. A complete amino acid sequence was first accomplished for bovine rhodopsin[47,127], and subsequently for ovine rhodopsin[11]. Amino acid sequences have been deduced from gene sequences of human[113], bovine[112], chicken[146], fruit fly[167], and octopus[126] rhodopsins. The gene structure and amino acid sequence of human color visual pigments[114] and a pineal color pigment from the blind cavefish *Astyanax fasciatus*[163] have been determined as well.

Although human color visual pigments and invertebrate rhodopsins share only moderate sequence similarity with vertebrate rhodopsins, the vertebrate rhodopsins are themselves highly conserved (Fig. 4A). Over 93% of the residues are identical among the mammalian rhodopsins sequenced, and chicken rhodopsin approaches 90% amino acid sequence identity with the mammals. These determinations have permitted estimations of the rate of evolution of the molecule. For example, comparison of the amino acid sequences of human and bovine rhodopsins reveals 94.6% identity[113]. Primates diverged from other mammals 60–80 million years ago[120] during which time the rhodopsin sequences have diverged by approximately 7%. If substitutions have occurred at a roughly constant rate, then there has been a 1% change in the molecule every 10 million years. Comparisons between and among the vertebrates sequences support this estimate as a general value for neutral evolutionary change in vertebrate rhodopsin[114]. Extrapolating from these values to fishes, with much of neoteleostean radiation occurring within the last 60 millions years[17], a 6% difference in rhodopsin amino acid sequences would be predicted for more distantly related groups. Thus, it can be predicted that recently diverged genera will have very few neutral substitutions in the rhodopsin molecule.

Sequence determinations, in addition to providing a molecular clock for rhodopsin, have permitted the design of experiments to explore how the molecule

TABLE 1

Locations of substitutions from site-directed mutagenesis of the bovine rhodopsin molecule [a]

Site number	Original amino acid	Mutant amino acid	New λ_{max}
83	D	G	501
83	D	N	492
86	M	E	498
122	E	Q	498
122	E	I	497
134	E	L	479
134	E	Q	497
135	R	L	492
211	H	C	495
211	H	F	496
122, 83	E, D	I, G	498
122, 86	E, M	I, E	496
134, 135	E, R	L, L	498
134, 135	E, R	R, E	493

[a] Original λ_{max} = 498 nm; bovine rhodopsin numbering system[111]. Amino acid coding system: C = cysteine, D = aspartic acid, E = glutamic acid, F = phenylalanine, G = glycine, H = histidine, I = isoleucine, L = leucine, M = methionine, Q = glutamine, R = arginine.

functions. Nathans[111], through site-directed mutagenesis of bovine rhodopsin, has produced substitutions in six sites (Fig. 4B) within the membrane-spanning regions of bovine rhodopsin, a molecule with a native λ_{max} of 498 nm. To test the point charge model[62], which predicts that wavelength maxima are set by the interaction of charged residues with the chromophore, Nathans chose to manipulate charged residues located within the cell membrane (Fig. 4B). A total of 14 different mutations in 6 amino acid positions yielded rhodopsins with absorption maxima ranging from 479 to 501 nm (Table 1). While the data failed to support the point charge model for wavelength regulation, they did demonstrate the manipulative capabilities of molecular cloning techniques, and the ability of single amino acid substitutions in the opsin to alter rhodopsin spectral absorption properties.

3. Horizons in the study of fish vision

The studies by Nathans and coworkers[111] using site-directed mutagenesis have provided great insight into the role of the opsin protein in spectral absorption modulation. A potential limitation of this technique, however, is that it does not provide information on changes that have resulted from natural selection. Selection pressure must act upon the entire molecule in such a way as to preserve all functions, such as the interaction with G-proteins, not just the function of light absorption. Mutations that produce changes in λ_{max} might not be permitted in nature because they would compromise other functions of the molecule. For this reason, experimental studies with site-directed mutagenesis must be complemented by studying the results of 'successful experiments' that nature has performed on the rhodopsin molecule.

The fishes offer an ideal source of models for such an approach because of their diversity, their evolutionary history and the tremendous body of data available on their visual pigment absorption. These data support the idea that the special qualities of the aquatic photic environment have been important in the selection of functionally meaningful differences in the peak wavelengths of absorbance. Molecular cloning techniques can be readily applied to deduce amino acid sequence from the nucleotide sequence of the gene. Such methods can be used to compare the primary structure of the rhodopsin of fish species that exhibit different rhodopsin absorbance properties. More specifically, by comparing closely related species that live in different photic environments, the influence of neutral substitutions should be reduced in number, with functional implications for those amino acid substitutions that do occur. Alternatively, comparison of more distantly related species that live under similar light regimes, but who have presumably converged upon an advantageous λ_{max}, might give insight into alternative mechanisms by which pigment spectral modulation results.

Existing data on light absorption by marine fish rhodopsins suggest several phylogenetic groups in which the question of pigment spectral adaptation might be fruitfully addressed. Particularly interesting are mesopelagic fishes (Table 2). As inhabitants of the dysphotic zone, these phyletically diverse fishes have rhodopsins significantly shifted toward the shorter wavelengths, with absorptions ranging from 451–490 nm, and averaging about 475 nm, the peak wavelength of downwelling light (see Fig. 2A, and Table 2). The absorption of these pigments was so different from previously studied rhodopsins that they were thought to be another type of visual pigment entirely, and were called 'chrysopsins'[26]. Because many of these fishes are distantly related, a blue-shifted rhodopsin probably arose independently in several lineages (Table 2).

Divergent evolution of rhodopsins might also be studied in the deep-sea fishes. For example, the deep-sea smelts (Bathylagidae) exhibit a variety of rhodopsin absorption maxima within a single genus. *Bathylagus stilbius* has a single rhodopsin that absorbs at 491 nm[37]. However, other members of this family have within their retinas two rhodopsins, with disparate light absorbing properties (Table 3). *Bathylagus longirostris*, for example, has one rhodopsin absorbing at 474 nm and one at 502 nm[129]. The phenomenon of dual rhodopsins also occurs in *Bathylagus wesethi*[104] and *B. bercoides*[128]. The occurrence of dual rhodopsins is not restricted to bathylagids (Table 3). And, in no case where such dual pigments occur is the ecological significance of this capability known.

Divergence of visual pigment absorption (or spectral modulation) is perhaps best studied for closely related species that inhabit drastically different photic environments. For example, members of the family Scorpaenidae occur within a variety of photic environments. The lionfish, *Pterois radiata*, is largely restricted to shallow coral reefs and has a rhodopsin λ_{max} of 494 nm[105]. In contrast, the longspine thornyhead, *Sebastolobus altivelis*, is a deep-sea form found in the temperate eastern Pacific Ocean and has a λ_{max} of 484 nm[23]. In the latter case, the

TABLE 2

Rhodopsin absorbance maxima of mesopelagic fishes [a]

	Range (nm)	Number of species
Chondrichthyes		
Rhinochimaeridae (ratfishes)	481	(1)
Anguilliformes		
Notacanthidae (spiny eels)	481	(1)
Derichthyidae (longneck eels)	489	(1)
Synaphobranchidae (cutthroat eels)	476–478	(2)
Eurypharyngidae (gulpers)	474	(1)
Salmoniformes		
Alepocephalidae (slickheads)	476–480	(5)
Platytroctidae (Searsiidae)	475–480	(4)
Bathylagidae (deap-sea smelts) *	479–491	(5)
Opisthoproctidae (spookfishes)	477	(1)
Stomiiformes		
Gonostomatidae (bristlemouths)	481–484	(3)
Sternoptychidae (hatchetfishes)	475–485	(8)
Photichthyidae	478	(1)
Chauliodontidae (viperfishes)	484–488	(3)
Stomiidae (scaly dragonfishes)	489	(1)
Astronesthidae (snaggletooths)	480–483	(3)
Melanostomiidae (scaleless dragonfishes) *	483–485	(4)
Idiacanthidae (black dragonfishes)	483	(1)
Aulopiformes		
Scopelarchidae (pearleyes)	451	(1)
Notosudidae (waryfishes)	487	(1)
Myctophiformes		
Myctophidae (lanternfishes)	483–491	(10)
Gadiformes		
Moridae (morid cods)	485	(1)
Ophidiiformes		
Ophidiidae (cusk eels)	479	(1)
Beryciformes		
Melamphaeidae (bigscales)	482–488	(7)
Anoplogasteridae (fangtooths)	482	(1)
Scorpaeniformes		
Scorpaenidae (scorpionfishes)	484	(2)
Perciformes		
Zoarcidae (eelpouts)	485	(1)
Chiasmodontidae (swallowers)	483	(1)

[a] Data from ref. 129, and references cited therein.
* Members of this family with dual pigment not included here (see Table 3).

10 nm blue-shift of λ_{max} from 494 to 484 nm correlates with depth of habitat, and probably represents adaptation to more blue-shifted environmental light. Because the two fishes are closely related, however, the protein sequences are likely to be very similar.

TABLE 3

Mesopelagic fishes with two types of rod pigments

	λ_{max}	Pigment [a]
Salmoniformes		
Bathylagidae (deep-sea smelts)		
Bathylagus wesethi	478, 501	n.d.
B. bercoides	467, 497	n.d.
B. longirostris	474, 502	R, R
Stomiiformes		
Malacosteidae (loosejaws)		
Malacosteus niger	521, 538	R, P
Aristomstomias scintillans	526, 551	n.d.
Melanostomiidae		
Pachystomias microdon	513, 539	R, P
Gadiformes		
Macrouridae (rattails)		
Malacocephalus laevis	478, 485	n.d.
Perciformes		
Apogonidae (cardinalfishes)		
Howella sherborni	463, 492	R, R

[a] Type of visual pigment where data are available. R = rhodopsin (vitamin A_1) pigment, P = porphyropsin (Vitamin A_2), n.d. = not determined.

III. Bioluminescence in fishes

While luminescence is rare in terrestrial and freshwater environments, it is a commonly occurring feature of marine animals[60]. All bioluminescent fishes described thus far are marine[61]. Even in habitats (e.g., Lake Baikal in the Soviet Union) where freshwater fishes show remarkable adaptational convergences with marine fishes[73], there is no strong evidence that such convergence extends to the development of bioluminescence capability.

Although the rarity of bioluminescence in freshwater habitats is not well understood, the restricted occurrence of bioluminescence in marine fishes suggests that this capability would only be effective under certain light regimes. The amount of bioluminescent light that can be emitted by a fish is restricted by the maximum quantum yield of the luminescent reaction itself and by the number of reactive molecules that can be packed into a given area[89]. The brightest luminescence does not approach the intensity of the diurnal light field of the euphotic zones[166]. Thus, the utility of luminescence is restricted to the dysphotic and aphotic zones, or to the nocturnal realm. In those areas where luminescence does commonly occur, the light produced is most often in the blue wavelengths[59,122], indicating selection on molecular species that transmit well through the aquatic medium or mimic well the color of the ambient light.

The bioluminescence of the light organs of fishes is of one of two origins: either symbiotic luminous bacteria, or the fishes own tissue biochemistry[61]. The anatomical characteristics of both types of light organ systems, and the few observations of luminescence in fishes in nature, support the view that light production is used in both intra- and interspecific communication[98,166].

1. Bacterial bioluminescence in fishes

1.1. Occurrence

While most luminous fishes have evolved mechanisms of non-bacterial luminescence, more is known about the biochemistry and molecular biology of bacterially produced symbiotic luminescence[31,49,52,116,119]. Bacterial light organs occur in hundreds of species, representing 20 families of fishes, 9 of which belong to the suborder of deep-sea anglerfishes, the Ceratioidea[60,94]. Most of these fishes are either coastal or benthic, with the exception of the bacterially luminous spookfishes (Opisthoproctidae) and the deep-sea anglerfishes (Ceratioidea), which are pelagic species. These numerous examples of light-organ symbioses can be further divided into those fishes having bacterial symbionts that have been grown in laboratory culture, and those for whom the symbionts are as yet uncultured.

1.2. Culturable bacterial symbioses

Three species of culturable marine luminous bacteria have been identified from the light organs of fishes: *Photobacterium leiognathi*, *P. phosphoreum* and *Vibrio fischeri*[53,94]. As in the light organs of marine squids[8,42] the symbioses exhibit unilateral species specificity between hosts and bacteria; i.e., a given fish species will always harbor a given luminous bacterial species, but a given symbiotic luminous bacterial species may occur in more than one fish species or family. Three biological characteristics of the light organs and their bacteria have been invoked to explain the species-specific nature of the occurrence of these bacteria with different species of host fishes: (a) the relationship between the geographic range of the host and the different growth temperature optima of the bacterial species, (b) the anatomical location of the light organ, and (c) the phylogenetic position of the host (Fig 5; Table 4)[53,94].

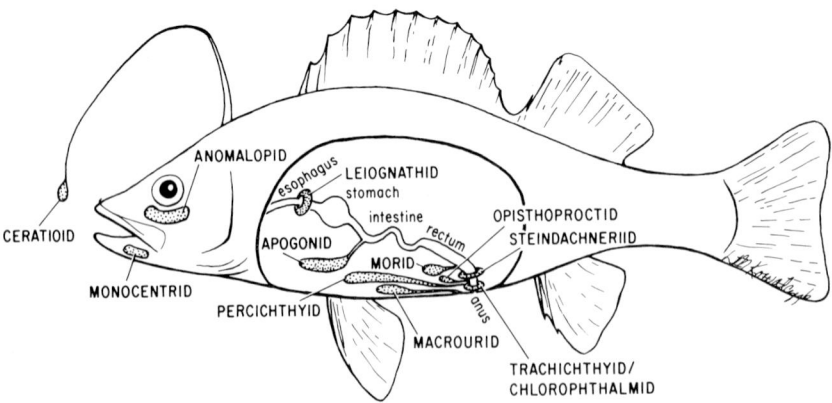

Fig. 5. Anatomical locations of bacterial light organs in fishes.

TABLE 4

Characteristics of fish/bacterial light organ associations

Host fish	Habitat	Light Organ Position	Bacterial Species
Aulopiformes			
Chlorophthalmidae	S-Bp	hindgut	n.d.
Salmoniformes			
Opisthoproctidae	M/B	hindgut	Pp
Gadiformes			
Steindachneriidae	S-Bp	hindgut	Pp
Moridae	S-Bp	hindgut	Pp
Macrouridae	S-Bp	hindgut	Pp
Lophiiformes			
Ceratioidei	B	barbel	n.d.
Beryciformes			
Trachichthyidae	S-Bp	hindgut	Pp
Anomalopidae	SH-Trop	under the eye	n.d.
Monocentridae	SH-Trop/Temp	lower jaw	Vf
Perciformes			
Leiognathidae	SH-Trop/Temp	esophageal	Pl
Apogonidae	SH-Trop	pyloric caecae	Pl
Percichthyidae	S-Bp	hindgut	Pl

S-Bp = shelf-benthopelagic, M = mesopelagic, B = bathypelagic, S-Bp = shelf-benthopelagic, SH = shallow, Trop = tropical, Temp = temperate, Pp = *Photobacterium phosphoreum*, Pl = *Photobacterium leiognathi*, Vf = *Vibrio fischeri*, n.d. = not determined.

The three species of culturable bacteria in fish light organs also occur as free-living components of the bacterioplankton, and their distribution and abundance in water masses of different temperatures often correlate with the range of different growth optima of the bacterial species[52,119,135,138]. In addition, the species of symbiotically luminous fishes[50] in these water masses also reflect those different growth optima. For example, *Photobacterium leiognathi* is characterized by a high growth temperature optimum[132,134], and is not only a common component of tropical waters, but also is the symbiont of leiognathids and apogonid luminous fishes. Similarly, *Photobacterium phosphoreum* is the species of symbiotic bacteria that is best adapted to growth under conditions characteristic of cold, deeper ocean waters[119,134], and most of the deep-sea fishes with bacterial light organs harbor this species. Finally, *Vibrio fischeri* is the luminous bacterial species that is most often found in the temperate zones, and occurs in monocentrids, or pinecone fishes, whose distribution includes the temperate zones[39,125,133].

Because current evidence suggests that the light organs of symbiotic fishes are inoculated with luminous bacteria during the larval stage[61,119], a theory has arisen that fish light organs evolved to utilize the predominant luminous bacterial species present in their ambient habitat[53,134,136]. The following four points suggest that this theory is too simplistic. Firstly, although one species of luminous bacteria may be most abundant in a given area, often several species of symbiotic luminous bacteria can be detected[52,136]. Secondly, bacteriologists have reasoned that selection should occur for a light organ symbiosis in which the bacteria grow slowly but luminesce brightly[53,117]. Thus, one could argue that the utilization of a bacterial species that is viable but not at its optimum growth temperature would be more suited for the efficient functioning of the association. Thirdly, the ontogeny and evolutionary history of the fishes must be considered. Many fishes, particularly deep-sea fishes, have larvae that develop in surface water masses that are at a considerably warmer temperature than the waters where the adults occur[100]. Thus, at the time of the larval fishes' inoculation, it would likely be exposed to a bacterial population dominated by a luminous bacterial species different from that present in the adults' environment. In addition, there is some evidence to indicate that the deep sea was considerably warmer at the time when most of these fishes evolved than it is today[137]. Fourthly, a closer examination of the geographical occurrence of certain host species does not support a correlation between the symbiotic bacterial species and the luminous bacteria present in the ambient water. For example, the percichthyid fishes occur in deep waters of tropics and temperate regions[86,141] yet have light organs containing *P. leiognathi*[43]. Similarly, while most of the *P. leiognathi*-containing leiognathid fish species occur in the tropical Indo-West Pacific, some are found in the colder waters of northern Japan[45]; conversely, while some monocentrids, which have the temperate-associated luminous bacterium *V. fischeri*[39,133], occur in these colder waters of Japan, they are spread throughout the Philippines and Indonesia in distinctly tropical seas[58]. Thus, the overriding imperative for fish/bacterial light organ associations appears to be that phylogenetically related fishes will obtain and maintain the same luminous bacterial species regardless of the relative abundance of the bacteria in the environment, or their relative growth temperature optima.

The anatomical location of the light organ also exhibits some patterns that are generally correlated with the species of luminous bacteria present (Fig. 5, Table 4; ref. 53). In fish families with *P. phosphoreum* as symbionts, the light organs are always associated with the hindgut[134]. In two of the three families with foregut associations (i.e., the leiognathids and apogonids), the symbiont is *P. leiognathi*[50,165]. However, the percichthyids once again are the exception. These fishes have a hindgut association with *P. leiognathi*[43]. So that although all teleost/*P. phosphoreum* light organs thus far studied are hindgut associations, not all hindgut light organs have this bacterial species[53], and *P. leiognathi* associations are quite variable with respect to which area of the gut is involved. Once again, the phylogeny of the host provides a better predictor of the bacterial symbiont's species than anatomy alone.

There also is some support for the idea that the type of bacterial symbiont correlates with the phylogenetic position and evolutionary history of the host fish[94]. All bacterially luminous perciform fishes have light organs harboring *P. leiognathi*, while the more primitive fishes have either *P. phosphoreum* or *V. fischeri*. All of the fishes within a certain family, regardless of their depth distribution or geographical occurrence, harbor one particular species of symbiont. This fact argues that rather than there being a more random, adventitious incorporation of the most prevalent luminous bacterial species, both the evolutionary history and ontogeny of these associations is characterized by specific signalling between partners.

While some of these biogeographic, anatomical and phylogenetic correlations may fit, and are of heuristic value in our study of these symbioses, we have insufficient data about the evolutionary history of these associations to make any strong conclusions. Further, the number of bacterially bioluminescent fishes is insufficient to confirm the presence of solid trends. How and why certain bacterial species associate with particular fish families is likely to remain an unresolved question until the mechanisms of initiation and development of the symbioses are made clear.

1.3. Uncultured bacterial symbioses

The light organ bacteria of the flashlight fishes (Anomalopidae[68]) and the deep-sea anglerfishes (Ceratioidea[124]) are as yet uncultured. However, biochemical and molecular techniques have been applied to the isolated bacteroids of both these groups indicating which species of culturable luminous bacterium is most similar to the species contained within these organs. Kinetics of the luminous reaction[74] and a sequence of one of the genes of the luciferase molecule (*lux*A, see below; M. Haygood, personal communication) indicate that the symbionts of two species of anomalopids (*Kryptophanaron alfredi* and *Anomalops katoptron*) are most closely related to *Vibrio harveyi*, a species of marine luminous bacterium that is common in the water column, but has not been shown to be a light organ symbiont in fishes[52]. While anomalopids, as nocturnal fishes occurring in the shallow waters of the tropics[66,99], are accessible for such studies, the ceratioids occur primarily in the bathypelagic zone[6], and we know somewhat less about the symbionts of these fishes. However, a comparison of luciferase enzyme biochemistry of the symbionts of four species of these fishes has suggested that they are more similar to the three culturable symbiotic species (*Photobacterium leiognathi*, *P. phosphoreum*, or *V. fischeri*) than to *V. harveyi*[74].

1.4. The molecular biology, biochemistry and physiology of fish / bacterial light organ symbioses

Knowledge of how light emission is produced and controlled in luminous bacteria is essential to an understanding of how fish light organs are adapted to promote luminescence of their symbionts. Bacteriologists have made great strides in understanding the molecular basis for luminescence in light-producing bacterial species. Basically, the heterodimeric enzyme luciferase is a mixed function oxidase that uses molecular oxygen to simultaneously oxidize the two reduced substrates $FMNH_2$ to FMN, and the aldehyde tetradecanal to tetradecanoic acid. This reaction normally competes with the bacterial respiratory pathway for these reduced substrates[83], and luciferase may act as the major terminal oxidase when the cytochrome system is compromised, such as in the absence of sufficient iron[54] or oxygen[67,118], conditions that may be manipulated in symbiotic associations within the light organ[53,117].

The genes responsible for luminescence have been well characterized in several species of luminous bacteria[5,97], and are generally organized into a single major operon (the *lux* operon). The two genes coding for the heterodimeric luciferase molecule, *lux* A and *lux* B, are flanked by the structural genes for several enzymes responsible for tetradecanal synthesis. In addition, in at least two species[34,140], there is a region of regulatory gene(s) that is influenced by the intracellular concentration of a molecule called autoinducer, identified as homoserine lactone in *V. fischeri*[35] or a related molecule in *V. harveyi*[16]. Because it is continuously excreted, when bacterial cells are in a confined environment, such as within the light organ, autoinducer builds up, leading to a dramatic induction of the synthesis of the proteins of the luciferase operon[97,115].

Because it has not been possible to experimentally generate aposymbiotic (= without symbionts) fishes, attempts to model these symbioses have been forced to rely upon an extrapolation from bacterial behavior under axenic culture conditions in laboratory media. Proceeding from the assumption that the fish would gain maximum benefit by a symbiotic relationship in which bacterial growth is slowed and luminescence enhanced[53], physiological conditions have been manipulated in culture to produce these dual effects. Four culture conditions have been found to result in decreased growth and enhanced luminescence in some isolates from light organ symbioses: oxygen[118] and iron limitation[54], nutritional source[41,133], and variations in medium osmolarity[32]. Because none of these conditions has been actually measured for any intact light organ, any test of their existence remains to be performed. In addition to the absence of direct evidence, several additional caveats should be considered when reviewing models of light organ physiology derived from bacterial culture studies. Firstly, the luminous bacteria are physiologically plastic. They are facultative anaerobes, each species occurring in a wide variety of niches — as free-living, saprophytic, parasitic, enteric, as well as light organ symbionts[52]. Thus, extrapolations of behavior in culture to that in symbiosis should be made cautiously, especially for the many studies that have been performed with bacteria maintained in culture for years after their isolation from the symbiosis. Because none of these bacteria can be tested to see if they still retain the ability to infect and survive as a symbiont, it is possible that they have lost important symbiosis-specific characteristics as has been shown for symbiotic *Rhizobium* spp.[40]. Secondly, under no culture conditions have researchers been able to duplicate the distinctive morphology of the bacteria as they exist in the light organs of fishes[53]. Thus, because in the light organ environment these bacteria are probably in some differentiated state induced by the symbiosis (as has been reported for other examples of microbial symbionts[109]), studies of bacteria grown in the laboratory cannot be expected to successfully mimic specific symbiotic physiological activities.

Although limitations such as these characterize current models of light organ physiology and biochemistry, there is compelling evidence that one of the four physiological factors studied in bacterial culture, the ambient concentration of oxygen, plays an important role in the intact association. *V. fischeri*, the luminous bacterial symbiont in monocentrid fishes, luminesces most brightly when grown under growth-limiting concentrations of oxygen[118]. Estimates of the rate of bacterial release from the light organ indicate that bacterial growth rate is between 1 and 10% that under optimal aerobic growth conditions[55]. Laboratory studies of strains of *V. fischeri* have shown that they excrete pyruvate during incomplete oxidation of carbohydrate substrates[133], and electron microscopic studies of the monocentrid light organ have demonstrated unusually mitochondria-rich cells in the animal tissue surrounding the bacteria[147]. These three observations have been used to suggest a model that describes the slow-growth and high luminescence within the monocentrid light organ. In this model, the bacterially produced pyruvate feeds the aerobic respiration of the adnate host cells, which in turn limits the availability of oxygen to the bacteria, thus resulting in their low growth rate and induced level of luminescence[117,133].

Oxygen availability has also been implicated as a controlling factor that can modulate luminescence levels in leiognathid light organs. The circumesophageal light organ of these fishes is poorly vascularized[92] and, unlike *V. fischeri*, induction of *P. leiognathi* luminescence is depressed by low ambient oxygen concentrations[118]. Evidence exists that the gas bladder, which is separated from the bacteria-containing tissue by a thin, gas-permeable tissue layer, serves as the primary source of oxygen for the bacterial culture[33,92,94]. In an anesthetized fish, in which the window remains open, raising or lowering the percentage of oxygen in the bladder lumen promptly leads to a corresponding increase or decrease in bacterial luminescence levels, and thus indicates that gas bladder gases are available, and important to, the symbionts' activity[94]. Because the fish can rapidly separate the bacterial culture from this window by covering it with a tissue layer which curtails gas diffusion[92], a natural mechanism exists for controlling the availability of oxygen to the luminescing cells.

2. Non-bacterial bioluminescence in fishes

Luminescence by fishes that do not harbor luminous bacterial symbionts has traditionally been referred to as 'autogenic', because the animal typically makes both the enzymes and the substrates of the bioluminescent reaction. However, recent evidence has shown that substrates may also be obtained through the animal's diet (see below), rendering the term 'autogenic' problematic; hence we will instead use the word 'non-bacterial' to refer to the luminescence systems of animals that do not involve the mixed-function oxidase system of bacteria.

The majority of the fishes in the dysphotic zone, both by species and by biomass, exhibit non-bacterial luminescence[61]. In addition, there are seven known families of coastal and benthic fishes that have non-bacterial photophores. While some of these luminescence systems have been described[21,139], the vast majority of them remains unstudied at the biochemical or molecular levels. However, the data that are available suggest some interesting trends in the occurrence of these systems.

Biochemical research, particularly in the 1970's, elucidated details of the luminescent reactions of certain marine animals (for reviews, see refs. 21, 49). Unlike the monophyletic origin of the clearly homologous family of bacterial luciferases,

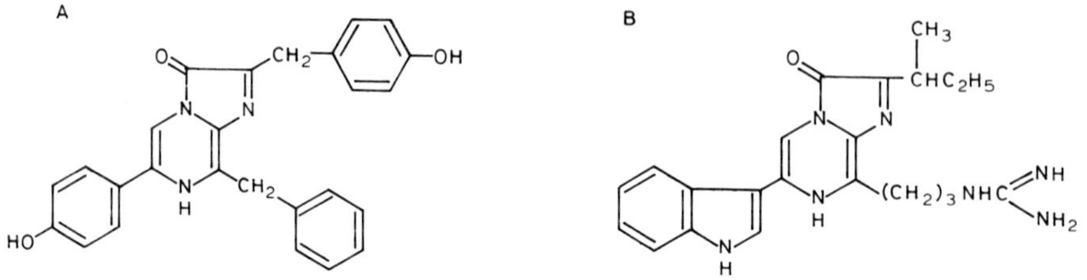

Fig. 6. The chemical structure of coelenterate-type (A) and *Vargula*-type (B) luciferins.

the luciferases of different taxonomic groups of animals, while analogous in function, are structurally and catalytically distinct, and are of a multiphyletic origin[15,49,90]. In addition, unlike bacterial luciferases, they do not oxidize either $FMNH_2$ or tetradecanal, but instead use one of a variety of different substrates, grouped generically as 'luciferins', for the luminescence reaction. One striking observation was that the luciferin molecules of many phylogenetically distant animals were one of two chemical types[90], referred to as coelenterate-type luciferin, and *Vargula* (previously *Cypridina*)-type luciferin, so named because they were originally described from a jellyfish and an ostracod, respectively. Both of these classes of luciferins are elaborations of a fused imidazole-pyrazine ring (Fig. 6; refs. 63, 71). Coelenterate-type luciferin occurs in a variety of coelenterates, a squid, several crustacean species and in two species of fishes (a neoscopelid and a myctophid)[21,139]; *Vargula*-type luciferin occurs in other species of crustacea and is

TABLE 5

Taxonomic occurrence of Coelenterate-type and *Vargula*-type luciferins

Coelenterate-type luciferin	*Vargula*-type luciferin
1. Teleosts	
F—Neoscopelidae	F—Batrachoididae
Neoscopelus microchir	*Porichthyes* (14 ssp.) [b]
F—Sternoptychidae [a]	F—Pempheridae
Argyropelecus hemigymnus	*Parapriacanthus ransonnetti*
F—Myctophidae	F—Apogonidae
Diaphus coeruleus	*Apogon ellioti*
Diaphus elucens	
F—Photichthyidae [a]	
Yarella illustris	
2. Other taxa	
P—Cnidaria	
P—Ctenophora	P—Arthropoda
P—Mollusca	
P—Arthropoda	

P = phylum; F = family; modified from refs. 21, 139 [a] and 157 [b].

used by some of the luminous species from three fish families; the Apogonidae, the Pempheridae and the Batrachoididae[21,46,90].

Because luminescence is believed to be a derived character in most of the above-mentioned species, the presence of a similar or identical substrate for the luminescent reaction has resulted in speculation about the biochemical origin of these luciferin molecules[49,90]. Two theories exist: (1) convergent evolution resulting in identical luciferin molecules; and, (2) evolution of a strategy to acquire a suitable luciferin through the ingestion of a bioluminescent coelenterate or *Vargula*. Evidence in support of convergence of the luciferins has derived from descriptions of the luminescent reaction and possible biosynthetic pathways for luciferin[49,88]. Not only are the luciferins often identical among these groups, but the luciferase is always an oxygenase, two features suggesting that a limited number of mechanisms are available for the generation of the electronically excited states that characterize luminescent reactions[49]. Further, McCapra[87] showed that these luciferins can be chemically synthesized from a tripeptide intermediate, supporting the idea that a common basis exists for luciferin biosynthesis. These features taken together provide some indirect evidence that the luciferins may have independently evolved several times in distantly related taxa.

The best evidence that bioluminescence can be supported by luciferin that has been obtained through ingestion of luminescent prey comes from studies of animals containing *Vargula*-type luciferin. Direct evidence for a utilization of luciferin obtained in the diet has been confined to studies of naturally occurring populations of the midshipman, *Porichthys notatus*[149]. The fishes of this genus have several hundred, superficial ventral photophores containing *Vargula*-type luciferin[22,154], and populations of the fish inhabit both areas where *Vargula* is present and areas where this ostracod is absent[149,150]. Midshipmen that co-occur with *Vargula* are luminescent, while fish that do not encounter *Vargula* do not luminesce. *P. notatus* has specific mechanisms for the binding and transport of the easily oxizided luciferin from the gut, where it is absorbed, to the superficial photophores[151]. Experimental manipulations of this system have demonstrated that a single dietary exposure to *Vargula* is sufficient to confer luminescence capability to an individual fish throughout its life history[153]. It has recently been shown that this is accomplished not by the triggering of luciferin biosynthesis in the fish, but by an efficient utilization and recycling of the ostracod-synthesized luciferin[153].

While these elegant studies of the luminescence system of *Porichthys notatus* provide strong evidence for acquisition of bioluminescent substrates through the diet, the presence of such a mechanism in other bioluminescent animals awaits further evidence. It is possible that examples of both independent convergent evolution and diet-derived sources of luciferin can be found among the animal species with either coelenterate or *Vargula*-type luciferin systems. Of use in these studies will be continued DNA and protein sequence data for phylogenetic analyses[152], and research on the mechanisms of control of luciferase gene expression, two areas of molecular biology that should be pursued to complement the well-developed field of animal luciferase biochemistry[21,49].

3. Tissues supporting photogenic cells

Accessory structures that serve to modify biologically produced light, occur in a variety of forms, reflecting the location (superficial or internal) and type (non-bacterial or bacterial) of the light organ system[61]. Superficial light organs are most similar in structure to the eyes of fishes, for all components of the luminescence system are localized. In contrast, internal light organs must recruit tissues that are often some distance from the photogenic tissue to propagate light into the environment. For example, light produced by the circumesophageal light organ of leiognathids is reflected by the silvery gas bladder and transmitted through translucent musculature of the entire ventrum of the fish[33,92,95]. Further, chromatophores throughout the body serve to modulate light intensity or preclude light emission in certain directions. Thus, rather than a discrete, localized set of tissues, much of the leiognathid's body is used in the light organ system. The involvement of such tissues in leiognathids, and in other fishes with internal light organs, has resulted in selection on the anatomy and biochemistry of the tissues to serve this auxiliary function. For example, to serve as an efficient reflector, the gas bladders of leiognathids have extremely high levels of purines — those biochemicals laid down in imbricated layers in fish tissue to form reflective layers[92,94].

Fishes with non-bacterial light organs, whether superficial or internal, have direct control over the intensity and color produced by the photogenic cells[19,28,29,30]. The nervous or endocrine systems modulate intensity, and selection pressure can act directly on the genome of fishes to alter the color of light emission resulting from the light-producing reaction[18]. In contrast, fishes with bacterial light organs must deal with the properties inherent to the luminescence of the bacterial symbiont[53]. Studies of bacterial light production indicate that, once induced, the bacteria are constantly luminous, unless the availability of a substrate, such as oxygen, is limiting[49,53]. In superficial bacterial organs, such as those in the flashlight fishes and the monocentrids, bacterial light emission is thought to be continuous[51,52], and fishes have evolved melanin rich shutters with which to indirectly control the level of light emission by occlusion of the organ[44,66,125]. The pattern of luminescent display in these fishes is controlled in a simple on-off fashion by the action of these shutters. Maximum efficiency of photon flux into the environment is controlled by reflective tissue with purine platelets backing the light organ[44,158]. The color of the luminescence is initially defined by the excited states of the bacterial luciferin–luciferase system, but the spectrum of light that leaves the animal can be modified by filters over the light organ tissue, such as the red filter covering the light organ of monocentrids[44].

The control of light emission in fishes with internal bacterial light organs is not as well understood. In such fishes, in addition to the use of other portions of the body to affect a variety of behaviors, bacterial luminescence levels may also be controlled, as suggested in leiognathids, where the availability of gas bladder oxygen is thought to be crucial[92]. Whether the bacterial light produced in the internal light organs of other fishes is in any way controlled by the fish host remains to be investigated.

IV. Conclusions

In this synthesis we have attempted to outline those areas of fish vision and bioluminescence that are at the forefront of research efforts in these fields of study. While there is good evidence that selection pressure on these two facilities is strongly linked under a variety of circumstances, the historical approach to these subjects and the current directions are somewhat different. For these reasons, our focus and perspective have not been the same for each of these areas. However, both fields are clearly at that crossroads where applications of current techniques in molecular biology and biochemistry will permit great strides in our understanding of how vision and luminescence integrate with the behavior and ecology of fishes. Explorations in these areas should also provide an understanding of how animals in general adapt to aquatic photic environments. Finally, as fishes offer rare opportunities as model systems for the study of basic biological phenomena[131], research efforts toward understanding such adaptations in fishes should prove invaluable, particularly in the case of the visual system.

V. References

1. Ali, M.A. *Vision in Fishes: New Approaches in Research*. New York: Plenum Press, 1974, 863 pp.
2. Anctil, M. Prospects in the study of interrelationships between vision and bioluminescence. In: *Vision in Fishes: New Approaches in Research*, edited by M.A. Ali. New York: Plenum Press, 1974, pp. 657–671.
3. Applebury, M.L. and P.A. Hargrave. Molecular biology of the visual pigments. *Vision Res.* 26: 1881–1895, 1986.
4. Atlas, D. and T.T. Bannister. Dependence of mean and spectral extinction coefficient of phytoplankton on depth, water color, and species. *Limnol. Oceanogr.* 25: 157–159, 1980.
5. Baldwin, T.O., J.H. Devine, R.C. Heckel, J.W. Lin and G.S. Shadel. The complete nucleotide sequence of the *lux* regulon of *Vibrio fischeri* and the *luxABN* region of *Photobacterium leiognathi* and the mechanism of control of bacterial bioluminescence. *J. Biolumin. Chemilumin.* 4: 326–341, 1989.
6. Bertelsen, E. Ceratioidei: development and relationships. In: *Ontogeny and Sytematics of Fishes*, edited by H.G. Moser. Spec. Publ. No. 1, Am. Soc. Ich. Herp., 1984, pp. 325–334.
7. Blatz, P.E., J.H. Mohler and H.V. Navangul. Anion-induced wavelength regulation of absorption maxima of Schiff bases of retinal. *Biochemistry* 11: 848–855, 1972.
8. Boettcher, K.J. and E.G. Ruby. Depressed light emission by symbiotic *Vibrio fischeri* of the sepiolid squid *Euprymna scolopes*. *J. Bacteriol.* 172: 3701–3706, 1990.
9. Bowmaker, J.K. and Y.W. Kunz. Ultraviolet receptors, tetrachromatic colour vision and retinal mosaics in the brown trout (*Salmo trutta*): age-dependent changes. *Vision Res.* 27: 2101–2108, 1987.
10. Brauer, A. 'Die Tiefsee-Fische' II Anatomische Teil. *Wiss. Ergebn. der Deutschen Tiefsee-Expedition, auf dem Dampfer 'Valdivia.'* G. Fisher, Jena, F.R.G., 15: 1–266, 1908.
11. Brett, M. and J.B.C. Findlay. Isolation and characterization of the CNBr peptides from the proteolytically derived N-terminal fragment of ovine opsin. *Biochem. J.* 211: 661–670, 1983.
12. Bridges, C.D.B. Spectroscopic properties of porphyropsins. *Vision Res.* 7: 349–369, 1967.
13. Briggs, J.C. *Marine Biogeography*. McGraw-Hill, New York, 1974.
14. Buck, J. Speculations on the interrelations and evolution of photic organs. In: *Light and Life*, edited by W.D. McElroy and B. Glass. Baltimore, MD: Johns Hopkins Press, 1961, pp. 754–770.
15. Buck, J.B. Functions and evolutions of bioluminescence. In: *Bioluminescence in Action*, edited by P.J. Herring. New York: Academic Press, 1978, pp. 419–460.
16. Cao, J.-G. and E.A. Meighen. Purification and structural identification of an autoinducer for the luminescence system of *Vibrio harveyi*. *J. Biol. Chem.* 264: 21670–21676, 1989.

17. Carrol, R. *Vertebrate Paleontology and Evolution*. New York: W.H. Freeman, 1988.
18. Case, J.F. and L.G. Strause. Neurally controlled luminescent systems. In: *Bioluminescence in Action*, edited by P.J. Herring. New York: Academic Press, 1978, pp. 331–366.
19. Case, J.F., J. Warner, A.T. Barnes and M. Lowenstein. Bioluminescence of lantern fish (Myctophidae) in response to changes in light intensity. *Nature* 265: 179–181, 1977.
20. Clarke, W.D. Function of bioluminescence in mesopelagic organisms. *Nature* 198: 1244–1246, 1963.
21. Cormier, M.J. Comparative biochemistry of animal systems. In: *Bioluminescence in Action*, edited by P.J. Herring, New York: Academic Press, 1978, pp. 75–108.
22. Cormier, M.J., J.M. Crane and Y. Nakano. Evidence for the identity of the luminescent systems of *Porichthys porosissimus* (fish) and *Cypridina hilgendorfii* (crustacean). *Biochem. Biophys. Res. Commun.* 29: 747–752, 1967.
23. CrescitelliI, F., M.J. McFall-Ngai and J. Horwitz. The visual pigment sensitivity hypothesis: further evidence from fishes of varying habitats. *J. Comp. Physiol.* A 157: 323–333, 1985.
24. Dartnall, H.J.A. Assessing the fitness of visual pigments for their photic environments. In: *Vision in Fishes: New Approaches in Research*, edited by M.A. Ali. New York: Plenum Press, 1974, pp. 543–563.
25. Denton, E.J. Reflectors in fishes. *Sci. Am.* 224: 65–72, 1971.
26. Denton, E.J. and F.J. Warren. Visual pigments of deep-sea fish. *Nature* 178: 1059, 1956.
27. Denton, E.J. and F.J. Warren. The photosensitive pigments in the retinae of deep-sea fish. *J. Mar. Biol. Ass. U.K.* 36: 651, 1957
28. Denton, E.J., J.P. Gilpin-Brown and P.G. Wright. On the 'filters' in the photophores of mesopelagic fish and on a fish emitting red light and especially sensitive to red light. *J. Physiol. (Lond.)*, 20: 872–873, 1970.
29. Denton, E.J., J.P. Gilpin-Brown and P.G. Wright. The angular distribution of the light produced by some mesopelagic fish in relation to their camouflage. *Proc. Roy. Soc. Lond.* B 182: 145–158, 1972.
30. Denton, E.J., P.J. Herring, E.A. Widder, M.A. Latz and J.F. Case. The roles of filters in the photophores of oceanic animals and their relation to vision in the oceanic environment. *Proc. Roy. Soc. Lond.* B 225: 63–97, 1985.
31. Devine, J.H., C. Countryman and T.O. Baldwin. Nucleotide sequence of the luxR and luxI genes and the structure of the primary regulatory region of the lux regulon of *Vibrio fischeri* ATCC 7744. *Biochemistry* 27: 837–842, 1988.
32. Dunlap, P.V. Osmotic control of luminescence and growth in *Photobacterium leiognathi* from ponyfish light organs. *Arch. Microbiol.* 141: 44–50, 1985.
33. Dunlap, P.V. and M.J. McFall-Ngai. Initiation and control of the bioluminescent symbiosis between *Photobacterium leiognathi* and leiognathid fish. In: *Endocytobiology III*. Ann. N.Y. Acad. Sci. 1987, pp. 260–283.
34. Dunlap, P.V. and E.P. Greenberg. Control of *Vibrio fischeri lux* gene transcription by a cyclic AMP receptor protein-*lux*R protein regulatory circuit. *J. Bacteriol.* 170: 4040–4046, 1988.
35. Eberhard, A., A.L. Burlingame, C. Eberhard, G.L. Kenyon, K.H. Nealson and N.J. Oppenheimer. Structural identification of autoinducer of *Photobacterium fischeri* luciferase. *Biochemistry* 20: 2444–2449, 1981.
36. Feiler, R., W.A. Harris, K. Kirschfeld, C. Wehrhan and C.S. Zucker. Targeted misexpression of a *Drosophila* opsin gene leads to altered visual function. *Nature*, 333: 737–741, 1988.
37. Fernandez, H.R. Visual pigments of bioluminescent and nonbioluminescent deep-sea fishes. *Vision Res.* 19: 589–592, 1978.
38. Findlay, J.B. and D.J. Pappin. The opsin family of proteins. *Biochem. J.* 238: 625–642, 1986.
39. Fitzgerald, J.M. Classification of luminous bacteria from the light organ of the Australian pinecone fish, *Cleidopus gloriamaris*. *Arch. Microbiol.* 112: 153–156, 1977.
40. Flores, M., V. Gonzalez, M.A. Pardo, A. Leija, E. Martinez, D. Romero, D. Pinero, G. Davila and R. Palacios. Genetic instability in *Rhizobium phaseoli*. *J. Bacteriol.* 170: 1191–1196, 1988.
41. Friedrich, W.F. and E.P. Greenberg. Glucose repression of luminescence and luciferase in *Vibrio fischeri*. *Arch. Microbiol.* 134: 87–91, 1983.
42. Fukasawa, S. and P.V. Dunlap. Identification of luminous bacteria from the light organ of the squid, *Doryteuthis kensaki*. *Agric. Biol. Chem.* 50: 1645–1646, 1986.
43. Fukasawa, S., T. Suda and S. Kubota. Identification of luminous bacteria isolated from the light organ of the fish, *Acropoma japonicum*. *Agric. Biol. Chem.* 52: 285–286, 1988.
44. Haneda, Y. On a luminous organ of the Australian pinecone fish, *Cleidopus gloria-maris* De Vis. In: *Bioluminescence in Progress*, edited by F.H. Johnson and Y. Haneda. Princetown, NJ: Pinceton University press, 1966, pp. 547–555.

45. Haneda, Y. and F.I. Tsuji. The luminescent system of pony fishes. *J. Morphol.* 150: 539–552, 1976.
46. Haneda, Y., F.H. Johnson and O. Shimomura. The origin of luciferin in the luminous ducts of *Parapriacanthus ransonneti, Pempheris klunzingeri*, and *Apogon ellioti*. In: *Bioluminescence in Progress*, edited by F.H. Johnson and Y. Haneda. Princeton, NJ: Princeton University Press, 1966, pp. 533–545.
47. Hargrave, P.A., D.H. McDowell, D.R. Curtis, J.K. Wang, E. Jusczak, S.L. Fong, J.K. Mohanna Rao and P. Argos. The structure of bovine rhodopsin. *Biophys. Struct. Mech.* 9: 235–244, 1983.
48. Harosi, F.I. and Y. Hashimoto. Ultraviolet visual pigments in a vertebrate: a tetrachromatic cone system in the dace. *Science* 222: 1021–1023, 1983.
49. Hastings, J.W. Biological diversity, chemical mechanisms, and the evolutionary origins of bioluminescent systems. *J. Mol. Evol.* 19: 309–321, 1983.
50. Hastings, J.W. and G. Mitchell. Endosymbiotic bioluminescent bacteria from the light organ of pony fish. *Biol. Bull.* 141: 261–268, 1971.
51. Hastings, J.W. and K.H. Nealson. Endosymbiotic luminous bacteria occurring in luminous organs of higher animals. In: *Endocytobiology*, edited by W. Schwemmler and H.E.A. Schenk. Berlin: Walter de Gruyter, 1980, pp. 467–471.
52. Hastings, J.W. and K.H. Nealson. The symbiotic luminous bacteria. In: *The Prokaryotes*, edited by M.P. Starr, H. Stolp, H.G. Truper, A. Balows and H.G. Schlegel. Berlin: Springer-Verlag, 1981, pp. 1332–1345.
53. Hastings, J.W., J. Makemson and P.V. Dunlap. How are growth and luminescence regulated independently in light organ symbionts? *Symbiosis* 4: 3–24, 1987.
54. Haygood, M.G. and K.H. Nealson. Mechanisms of iron regulation of luminescence in *Vibrio fischeri*. *J. Bacteriol.* 162: 209–216, 1985.
55. Haygood, M.G., B.M. Tebo and K.H. Nealson. Luminous bacteria of a monocentrid fish (*Monocentris japonicus*) and two anomalopid fishes (*Photoblepharon palpebratus* and *Kryptophanaron alfredi*): population sizes and growth within the light organs, and rates of release into the seawater. *Mar. Biol.* 78: 249–254, 1984.
56. Heinermann, P.H. Yellow intraocular filters in fishes. *Exp. Biol.* 43: 127–147.
57. Henderson, R., J.M. Baldwin and T.A. Ceska. Model for the structure of bacteriorhodopsin based on high-resolution electron and cryo-microscopy. *J. Mol. Biol.* 213: 899–929, 1990.
58. Herre, A.W. Check list of Phillipine fishes. *Res. Rep. 20, Fish and Wildlife Serv., U.S. Dept. Int., U.S. Gov. Printing Office*, 1953.
59. Herring, P.J. The spectral characteristics of luminous organisms. *Proc. Roy. Soc. Lond. B* 220: 183–217, 1983.
60. Herring, P.J. Systematic distribution of bioluminescence in living organisms. *J. Biolumin. Chemilumin.* 1: 147–163, 1987.
61. Herring, P.J. and J.G. Morin. Bioluminescence in fishes. In: *Bioluminescence in Action*, edited by P.J. Herring. New York: Academic Press, 1978, pp. 273–329.
62. Honig, B., U. Dinur, K. Nakanishi, V. Balough-Nair, M.A. Gawinowicz, M. Arnaboldi and M.G. Motto. An external point charge model for wavelength regulation in visual pigments. *J. Am. Chem. Soc.* 101: 7084–7086, 1979.
63. Hori, K., H. Charbonneau, R.C. Hart and M.J. Cormier. Structure of native *Renilla reniformis* luciferin. *Proc. Natl. Acad. Sci. U.S.A.* 74: 4285–4287, 1977.
64. Irving, C.S., G.W. Byers and P.A. Leermakers. Spectroscopic model for the visual pigments: influence of microenvironmental polarizability. *Biochemistry* 9: 858–864, 1970.
65. Jerlov, N.G. *Marine Optics*. New York: Elsevier, 1976, 231 pp.
66. Johnson, G.D. and R.H. Rosenblatt. Mechanisms of light organ occlusion in flashlight fishes, family Anomalopidae (Teleostei: Beryciformes), and the evolution of the group. *Zool. J. Linn. Soc.* 94: 65–96, 1988.
67. Karl, D.M. and K.H. Nealson. Regulation of cellular metabolism during synthesis and expression of the luminous system in *Beneckea* and *Photobacterium*. *J. Gen. Microbiol.* 117: 357–368, 1980.
68. Kessel, M. The ultrastructure of the relationship between the luminous organ of the teleost fish *Photoblepharon palpebratus* and its symbiotic bacteria. *Cytobiology*. 15: 145–158, 1977.
69. Kiefer, D.A. and R.W. Austin. The effect of varying phytoplankton concentration on submarine light transmission in the Gulf of California. *Limnol. Oceanogr.* 19: 55–64, 1974.
70. Kirk, J.T.O. *Light and Photosynthesis in Aquatic Ecosystems*. Cambridge, U.K.: Cambridge University Press, 1983, 401 pp.
71. Kishi, Y, T. Goto, Y. Hirata, O. Shimomura and F.H. Johnson. *Cypridina* bioluminescence I. Structure of *Cypridina* luciferin. *Tet. Lett.* 3427–3436, 1966.
72. Koutalos, Y., T.G. Ebrey, M. Tsuda, K. Odashima, T. Lien, M.H. Park, N. Shimizu, F. Derguini,

K. Nakanishi, H.R. Gilson and B. Honig. Regeneration of bovine and octopus opsins *in situ* with natural and artificial retinals. *Biochemistry* 28: 2732–2739, 1989.
73. Kozhov, M. In: *Lake Baikal and its Life*, edited by W. Junk, The Hague, Netherlands: Martinus Nijhof, 1963, 344 pp.
74. Leisman, G., D. Cohen and K.H. Nealson. Bacterial origin of luminescence in marine animals. *Science* 208: 1271–1273.
75. Levine, J.S. The vertebrate eye. In: *Functional Vertebrate Morphology*, edited by M. Hildebrande, D.M. Bramble, K.F. Liem and D.B. Wake. Harvard University Press, 1985, pp. 317–337.
76. Leydig, F. *Die Augenähnlichen Organe der Fische*. Bonn: E. Strauss, 1881.
77. Liebman, P.A., K.R. Parker and E.A. Dratz. The molecular mechanism of visual excitation and its relation to the structure and composition of the rod outer segment. *Annu. Rev. Physiol.* 49: 765–791, 1987.
78. Locket, N.A. Some problems of deep-sea fish eyes. In: *Vision in Fishes: New Approaches in Research*, edited by M.A. Ali. New York: Plenum Press, 1974, pp. 645–655.
79. Lythgoe, J.N. Visual pigments and visual range underwater. *Vision Res.* 8: 997–1012, 1968.
80. Lythgoe, J.N. Vision in fishes: ecological adaptations. In: *Environmental Physiology of Fishes*, edited M.A. Ali. New York: Plenum Press, 1979, pp. 431–445.
81. Lythgoe, J.N. *The Ecology of Vision*. Oxford: Clarendon Press, 1980, 244 pp.
82. Lythgoe, J.N. Visual pigments and environmental light. *Vision Res.* 24: 1539–1550, 1984.
83. Makemson, J.C. Luciferase-dependent oxygen consumption by bioluminescent vibrios. *J. Bacteriol.* 165: 461–466, 1986.
84. Marshall, N.B. *Explorations in the Lives of Fishes*. Cambridge, MA: Harvard University Press, 1971, 204 pp.
85. Marshall, N.B. *Deep-Sea Biology: Developments and Perspectives*. New York: Garland STPM Press, 1980, 566 pp.
86. Matsubara, K. Revision of the japanese serranid fish, referable to the genus *Acropoma*. *Mem. Coll. Agric. Kyoto Univ.* 66: 21–29, 1953.
87. McCapra, F. The chemistry of bioluminescence. *Endeavor* 32: 139–145, 1973.
88. McCapra, F. Chemical mechanisms in bioluminescence. *Acc. Chem. Res.* 9: 201–209, 1976.
89. McCapra, F. The chemistry of bioluminescence. In: *Bioluminescence in Action*, edited P.J. Herring. New York: Academic Press, 1978, pp. 49–73.
90. McCapra, F. and R. Hart. The origins of marine bioluminescence. *Nature* 286: 660–661, 1980.
91. McCartney, E.J. *Optics of the Atmosphere*. New York: John Wiley, 1976.
92. McFall-Ngai, M.J. Adaptations for reflection of bioluminescent light in the gas bladder of *Leiognathus equulus* (Perciformes: Leiognathidae). *J. Exp. Zool.* 227: 23–33, 1983.
93. McFall-Ngai, M.J. Crypsis in the pelagic environment. *Am. Zool.* 30: 175–188, 1990.
94. McFall-Ngai, M.J. Luminescent bacterial symbioses and adaptive radiation in fishes. In: *Symbiosis as a Source of Innovation, Speciation, and Morphogenesis*, edited L. Margulis and R. Fester. Cambridge, MA: MIT Press, 1991.
95. McFall-Ngai, M.J. and P.V. Dunlap. Three new modes of luminescence in the leiognathid fish *Gazza minuta*: discrete projected luminescence, ventral body flash and buccal luminescence. *Mar. Biol.* 73: 227–237, 1983.
96. McFarland, W.N. Light in the sea – Correlations with behaviors of fishes and invertebrates. *Am. Zool.* 26: 389–401, 1986.
97. Meighen, E.A. Enzymes and genes from *lux* operons of bioluminescent bacteria. *Annu. Rev. Microbiol.* 42: 151–176, 1988.
98. Morin, J.G. Coastal bioluminescence: patterns and functions. *Bull. Mar. Sci.* 33: 787–817, 1983.
99. Morin, J.G., A. Harrington, K. Nealson, N. Krieger, T.O. Baldwin and J.W. Hastings. Light for all reasons: versatility in the behavioral repertoire of the flashlight fish. *Science* 190: 74–76, 1975.
100. Moser, H.G. (ed.). *Ontogeny and Systematics of Fishes*. Spec. Publ. No. 1, Am. Soc. Ich. Herp., 1984, 760 pp.
101. Munk, O. Ocular anatomy of some deep-sea teleosts. *Dana Rep.* 70: 1–62, 1966.
102. Muntz, W.R.A. Visual pigments and the environment. In: *Vision in Fishes: New Approaches in Research*, edited by M.A. Ali. New York: Plenum Press, 1974, pp. 565–577.
103. Muntz, W.R.A. On yellow lenses in mesopelagic animals. *J. Mar. Biol. Assoc. U.K.* 56: 963–976, 1976.
104. Munz, F.W. Photosensitive pigments from the retinae of certain deep-sea fishes. *J. Physiol. (Lond.)* 140: 220–235, 1958.
105. Munz, F.W. and W.N. McFarland. The significance of spectral position in the rhodopsins of tropical marine fishes. *Vision Res.* 13: 1829–1874, 1973.

106. Munz, F.W. and W.N. McFarland. Evolutionary adaptations of fishes to the photic environment. In: *Handbook of Sensory Physiology, Vol. VII 5.*, edited by F. Crescitelli. Berlin: Springer-Verlag, 1977, pp. 193–274.
107. Nakanishi, K. Why 11-*cis* retinal? *Am. Zool.* in press.
108. Nakanishi, K., M. Balogh-Nair, M. Aniboldi, K. Tsujimoto, and B. Honig. An external point-charge model for bacterio-rhodopsin to account for its purple color. *J. Am. Chem. Soc.* 102: 7945–7947, 1980.
109. Nardon, P., V. Gianinazzi-Pearson, A.M. Grenier, L. Margulis and D.C. Smith. *Endocytobiology IV*. Paris: INRA Service des Publications, 1990.
110. Nathans, J. Molecular biology of visual pigments. *Annu. Rev. Neurosci.* 10: 163–194, 1987.
111. Nathans, J. Determinants of visual pigment absorbance: Role of charged amino acids in putative transmembrane segments. *Biochemistry* 29: 937–942, 1990.
112. Nathans, J. and D. Hogness. Isolation, sequence analysis and intron-exon arrangement of the gene encoding bovine rhodopsin. *Cell* 34: 807–814, 1983.
113. Nathans, J. and D. Hogness. Isolation and nucleotide sequence of the gene encoding human rhodopsin. *Proc. Natl. Acad. Sci. U.S.A.* 81: 4851–4855, 1984.
114. Nathans, J., D. Thomas and D.S. Hogness. Molecular genetics of human color vision: the genes encoding blue, green, and red pigments. *Science* 232: 193–202, 1986.
115. Nealson, K.H. Autoinduction of bacterial luciferase: occurrence, mechanism and significance. *Arch. Microbiol.* 112: 73–79, 1977.
116. Nealson, K.H. Isolation, identification and manipulation of luminous bacteria. *Methods Enzymol.* 57: 153–166, 1978.
117. Nealson, K.H. Alternative strategies of symbiosis of marine luminous fishes harboring light-emitting bacteria. *Trends Biochem. Sci.* 4: 105–110, 1979.
118. Nealson, K.H. and J.W. Hastings. Low oxygen is optimal for luciferase synthesis in some bacteria. *Arch. Microbiol.* 112: 9–16, 1977.
119. Nealson, K.H. and J.W. Hastings. Bacterial bioluminescence: its control and ecological significance. *Microbiol. Rev.* 43: 469–518, 1979.
120. Nei, M. *Molecular Evolutionary Genetics*. New York: Columbia University Press, 1987.
121. Nicol, J.A.C. Guanine strata argentea of fish eyes. *Can. J. Zool.* 58: 488–491, 1980.
122. Nicol, J.A.C. Bioluminescence and vision. In: *Bioluminescence in Action*, edited by P.J. Herring. New York: Academic Press, 1978, pp. 367–398.
123. Nicol, J.A.C. *The Eyes of Fishes*. Oxford: Clarendon Press, 1989, 308 pp.
124. O'Day, W.T. Bacterial bioluminescence in the deep-sea anglerfish *Oneirodes acanthias* (Gilbert 1915). *Contr. Sci.* 255: 1–12, 1974.
125. Okada, Y.K. On the photogenic organ of the knightfish (*Monocentrus japonicus* (Houttuyn)). *Biol. Bull. Mar. Biol. Lab. Woods Hole* 50: 365–373, 1926.
126. Ovchinnikov, Y.A., N.S. Abdulaev, A.S. Zolotarev, I.D. Artamonov, I.A. Bespalov, A.E. Dergachev and M. Tsuda. Octopus rhodopsin: amino acid sequence deduced from cDNA. *FEBS Lett.* 232: 69–72, 1988.
127. Ovchinnikov, Y.A., N.G. Abdulaev, M.Y. Feigina, I.D. Artamonov, A.S. Bogachuk, A.S. Zolotarev, E.R. Eganyan and P.V. Kostetskii. Visual rhodopsin III: complete amino acid sequence and topography in the membrane. *Bioorg. Khim.* 9: 1331–1340, 1983.
128. Partridge, J.C., S.N. Archer and J.N. Lythgoe. Visual pigments in the individual rods of deep-sea fishes. *J. Comp. Physiol. A* 162: 543–550, 1988.
129. Partridge, J.C., J. Shand, S.N. Archer, J.N. Lythgoe and W. Van Groningen-Luyben. Interspecific variation in the visual pigments of deep-sea fishes. *J. Comp. Physiol. A* 164: 513–529, 1989.
130. Pollock, J.A. and S. Benzer. Transcript localization of four opsin genes in the three visual organs of *Drosophila*: RH2 is ocellus specific. *Nature* 333: 779–782, 1988.
131. Powers, D.A. Fish as model systems. *Science* 246: 352–358.
132. Reichelt, J.L. and P. Baumann. Taxonomy of the marine luminous bacteria. *Arch. Mikrobiol.* 94: 283–330, 1973.
133. Ruby, E.G. and K.H. Nealson. Symbiotic associations of *Photobacterium fischeri* with the marine luminous *Monocentris japonica*: a model of symbiosis based on bacterial studies. *Biol. Bull.* 151: 574–586, 1976.
134. Ruby, E.G. and J.G. Morin. specificity of symbiosis between deep-sea fishes and psychrotrophic luminous bacteria. *Deep-Sea Res.* 25: 161–167, 1978.
135. Ruby, E.G. and K.H. Nealson. Seasonal changes in the species composition of luminous bacteria in nearshore seawater. *Limnol. Oceanogr.* 23: 530–533, 1978.

136. Ruby, E.G., E.P. Greenberg and J.W. Hastings. Planktonic marine luminous bacteria: species distribution in the water column. *Appl. Environ. Microbiol.* 39: 302–306, 1980.
137. Shackleton, N.J. and J.P. Kennett. Paleotemperature history of the Cenozoic and the initiation of Antarctic glaciation-oxygen and isotope analyses in DSDP sites 277, 279, and 281. *Initial Reports of the Deep-Sea Drilling Project*, Washington, DC: U.S.G.P.O., V. 29, 1975, pp. 743–755.
138. Shilo, M. and T. Yetinson. Physiological characteristics underlying the distribution patterns of luminous bacteria in the Mediterranean Sea and the Gulf of Elat. *Appl. Environ. Microbiol.* 38: 577–584, 1979.
139. Shimomura, O., S. Inoue, F.H. Johnson and Y. Haneda. Widespread occurrence of coelenterazine in marine bioluminescence. *Comp. Biochem. Physiol. B* 65: 435–437, 1980.
140. Showalter, R.E., M.O. Martin and M.R. Silverman. Cloning and nucleotide sequence of *lux*R, a regulatory gene controlling bioluminescence in *Vibrio harveyi*. *J. Bacteriol.* 172: 2946–2954, 1990.
141. Smith, J.L.B. *The Sea Fishes of South Africa*. Cape Town, South Africa: Cape and Transvaal Printers, 1965.
142. Somiya, H. 'Yellow lens' eyes and luminous organs of *Echiostoma barbatum* (Stomiatoidei: Melanostomiatidae). *Jpn. J. Ichthyol.* 25: 269–272, 1979.
143. Somiya, H. 'Yellow lens' eyes of a stomiatoid deep-sea fish, *Malacosteus niger*. *Proc. Roy. Soc. Lond. B* 215: 481–489, 1982.
144. Strum, J. Fine structure of the dermal luminescent organs in the fish *Porichthys notatus*. *Anat. Rec.* 164: 433–461, 1969.
145. Stryer, L. Cyclic GMP cascade of vision. *Annu. Rev. Neurosci.* 9: 87–119, 1986.
146. Takao, M., A. Yasui and F. Tokunaga. Isolation and sequence determination of the chicken rhodopsin gene. *Vision Res.* 28: 471–480, 1988.
147. Tebo, B.M., D.S. Linthicum and K.H. Nealson. Luminous bacteria and light-emitting fish: ultrastructure of the symbiosis. *Biosystems* 11: 269–280, 1979.
148. Thomas, D.D. and L. Stryer. The transverse location of the retinal chromophore of rhodopsin in the outer segment disc membranes. *J. Mol. Biol.* 154: 145–157, 1982.
149. Thompson, E.M. and F.I. Tsuji. Two populations of the marine fish *Porichthys notatus*, one lacking luciferin essential for bioluminescence. *Mar. Biol.* 102: 161–165, 1989.
150. Thompson, E.M., B.G. Nafpaktitus and F.I. Tsuji. Latitudinal trends in size-dependence of bioluminescence in the midshipman fish *Porichthys notatus*. *Mar. Biol.* 98: 7–13, 1988.
151. Thompson, E.M., B.G. Nafpaktitus and F.I. Tsuji. Dietary uptake and blood transport of *Vargula* (crustacean) luciferin in the bioluminescent fish, *Porichthys notatus*. *Comp. Biochem. Physiol. A* 89: 203–209, 1988.
152. Thompson, E.M., S. Nagata and F.I. Tsuji. Cloning and expression of cDNA for the luciferase from the marine ostracod *Vargula hilgendorfii*. *Proc. Natl. Acad. Sci. U.S.A.* 86: 6567–6571, 1989.
153. Thompson, E.M., T. Yoshiaki, B.G. Nafpaktitus, T. Goto, and F.I. Tsuji. Induction of bioluminescence capability in the marine fish, *Porichthys notatus*, by *Vargula* (crustacean) [C14]luciferin and unlabelled analogues. *J. Exp. Biol.* 137: 39–51, 1988.
154. Tsuji, F.I., Y. Haneda, R.V. Lynch III and N. Sugiyama. Luminescence cross-reactions of *Porichthys* luciferin and theories on the origin of luciferin in some shallow-water fishes. *Comp. Biochem. Physiol. A* 40: 163–179, 1971.
155. Wald, G.W. The molecular basis of visual excitation. *Nature* 219: 800–807, 1968.
156. Wald, G.W., P.K. Brown and P.S. Brown. Visual pigments and depths of habitat of marine fishes. *Nature* 180: 969–971, 1957.
157. Walker, H.J. and R.H. Rosenblatt. Pacific toadfishes of the genus *Porichthys* (Batrachoididae) with descriptions of three new species. *Copeia* 1988: 887–904, 1988.
158. Watson, M., E.L. Thurston and J.A.C. Nicol. Reflectors in the light organs of *Anomalops* (Anomalopidae: Teleostei). *Proc. Roy. Soc. Lond. B*, 202: 339–351, 1978.
159. Wheeler, W.N. and M. Neushul. The aquatic environmnent. In: *Physiological Plant Ecology I*, edited by O.L. Lange, P.S. Nobel, C.B. Osmond and H.S. Ziegler. Berlin Springer-Verlag, 1981.
160. Widder, E.A., P. Hiller-Adams and J.F. Case. A multichannel microspectrophotometer for visual pigment investigation. *Vision Res.* 27: 1047–1055, 1987.
161. Yan, B., T. Takahashi, D.A. McCain, V. Jayathirtha Rao, K. Nakanishi and J.L. Spudich. Effects of modifications of the retinal beta-ionone ring on archaebacterial sensory rhodopsin I. *Biophys. J.* 57: 477–483, 1990.
162. Yasaki, Y. On the nature of the luminescence of the knight fish, *Monocentrus japonicus* (Houttuyn). *J. Exp. Biol.* 50: 495–505, 1928.
163. Yokoyama R. and S. Yokoyama. Isolation, DNA sequence, and evolution of a color visual pigment gene of the blind cave fish *Astyanax fasciatus*. *Vision Res.* 30: 807–816, 1990.

164. Yokoyama, S., K.E. Isenberg and A.F. Wright. Adaptive evolution of G-protein coupled receptor genes. *Mol. Biol. Evol.* 6: 342–353, 1989.
165. Yoshiba, S. and Y. Haneda. Bacteriological study on the symbiotic luminous bacteria cultivated from the luminous organ of the apogonid fish *Siphamia versicolor* and the Australian pinecone fish *Cleidopus gloria-maris. Sci. Rep. Yokosuka Cy. Mus.* 13: 82–84, 1967.
166. Young, R.E. Oceanic bioluminescence: an overview of general functions. *Bull. Mar. Sci.* 33: 829–845, 1983.
167. Zucker, C.S., A.F. Cowman and G.M. Rubin. Isolation and structure of rhodopsin gene from *D. melanogaster. Cell* 40: 851–859, 1985.

CHAPTER 5

Function and evolution of fish hormonal pheromones

NORM STACEY * AND PETER SORENSEN **

* Zoology Department, University of Alberta, Edmonton, Alta., Canada T6G 2E1, and ** Department of Fisheries and Wildlife, University of Minnesota, St. Paul, MN 55108, U.S.A.

I.   Introduction
II. Evidence for hormonal pheromones in fish
     1. Gobiidae
     2. Cottidae
     3. Clariidae
     4. Cyprinidae
        4.1. Zebrafish
        4.2. Goldfish
III. Functions of hormonal pheromones in fish
IV. Are these compounds true pheromones?
V.   Evolution of hormonal pheromones in fish: the origins of a vertebrate communication system
VI. Future directions
VII. References

I. Introduction

Fish have well developed chemical senses[20], a predictable evolutionary consequence of life in a medium where chemical cues abound and visual cues are often limited. It perhaps is not surprising, then, that a great variety of fish are keenly sensitive to the odor of conspecifics which they employ for functions as diverse as individual, kin, sex, and species recognition, and the promotion of reproductive synchrony[1,8,30,48,63,64]. Although these systems are in general poorly understood, exciting discoveries over the past decade suggest that fish commonly use hormones and their metabolites as chemical cues to synchronize reproduction[48]. As with other odors affecting conspecifics, these released hormonal compounds have been loosely termed *pheromones*, and *communicative* functions attributed to them. However, it is important to appreciate that these are not straightforward issues, and that the chemical nature and function of pheromones is a poorly understood and controversial subject in vertebrate biology. The purpose of this chapter is to consider what the identified *hormonal pheromones* of fish really are. After briefly reviewing what is known about these compounds and their functions in a few model species, we discuss questions of function (particularly communicative function), whether hormonal pheromones should be considered true pheromones, and how they might have evolved.

II. Evidence for hormonal pheromones in fish

Although a relationship between endocrine activity and sex pheromone function in fish was first established by Tavolga's[68,69] classical studies of pheromones in a goby (*Bathygobius soporator*), studies of crab sex pheromones[23] led to the first clear indication that aquatic organisms might commonly use released hormones as pheromones. In many marine decapod Crustaceans, the odor of premolt females stimulates males to search for them and to carry them until moulting, when copulation is possible[16]. Suspecting a close link between endocrine and pheromone function, Kittredge et al.[23] discovered that the decapod steroidal moulting hormone, crustecdysone, elicited behavioral responses at concentrations as low as 10^{-13} M, and suggested that this steroid was likely to be a mating pheromone in many crustaceans. Extrapolating further, Kittredge and Takahashi[22] speculated that hormones are likely to function as pheromones in many species of aquatic animals because they represent pre-existing chemical signals produced and released at biologically meaningful times, whose recognition might simply require a mutation by which hormonal receptors are expressed on chemosensory membranes. This preadaptation hypothesis appeared to answer the evolutionary dilemma of how a sex pheromone system might simultaneously evolve the abilities to synthesize and detect a specialized chemical cue. Ironically, although the findings of Kittredge et al.[23] have been harshly criticized by others studying Crustacean pheromones[18], they appear to have served as a strong theoretical stimulus for fish pheromone research[8,12,53]. Presently, there is good evidence for hormonal sex pheromones in four teleost families (Gobiidae, Cottidae, Clariidae, Cyprinidae) (Table 1); these examples account for virtually all identified vertebrate sex pheromones[33,36]. Here, we review only the stronger evidence and suggest readers also consult more complete reviews[48,55,63].

1. Gobiidae

The gobies, global in distribution and comprising perhaps 10% of extant teleost species[34], are excellent subjects for sex pheromone research. The common gobiid reproductive pattern, in which territorial males defend a nest-site which ovulated females must locate, presents a favorable situation for the evolution of a male attractive pheromone. Studies of the black goby (*Gobius jozo*) bear out this prediction and indicate that the male pheromone is a steroid hormone metabolite[6-8].

As in many Gobiids[34], the testes of *G. jozo* are unusual in that they contain a mesorchial gland rich in interstitial Leydig cells. Based on *in vitro* studies showing that the major steroid product of the mesorchial gland is etiocholanolone glucuronide (EG; a 5β-reduced androgen conjugated to glucuronic acid), Colombo et al.[6,7] tested the behavioral actions of this compound and found that ovulated females were attracted to artificial nests associated with EG, whereas non-ovulated females were not.

Colombo et al.[8] noted that this EG system bears striking similarities to the reduced androgen pheromone of pigs[33]; both are specialized hormonal metabolites produced by specialized tissues, both compounds have no known endocrine function, and both are specially suited to their medium of dispersal, EG being water soluble and the pig pheromone being volatile. Colombo et al.[8] cite evidence that male urine elicits behavioral responses in females, but have not determined whether EG is present in urine, or whether the male has control of pheromone release (control of the signal). Finally, Colombo et al.[8] assert that EG likely acts as a pheromone in several fish including the goldfish (*Carassius auratus*) and the guppy (*Poecilia reticulata*). However, we have been unable to measure any behavioral, endocrine, or chemosensory response to EG in goldfish[51,62].

2. Cottidae

Reproductive behavior of many Cottids is comparable to that of Gobiids in that territorial males defend a nest-site to which females are attracted for spawning. Results from a combination of techniques (chromatographic characterization, behavioral bioassay, electrophysiological recording) have led Dmitrieva and co-workers (refs. 10, 11, and unpublished results) to propose that sexual interactions of the yellowfin Baikal sculpin (*Cottocomephorus grewingki*) are synchronized by the release of three distinct male sex pheromones, two of which are steroidal. The first of these is a non-steroidal attractant, suggested to be a polyenic alcohol. The second pheromone, tentatively identified as 11β-hydroxytestosterone, induces ovulation of attracted females whereas the third, consisting of δ_4-3-keto-C19 steroids such as testosterone and androstenedione (hereafter referred to as 'androgens' for simplicity) trigger spawning behavior in ovulated females which are at the nest-site. Unlike the situation with *G. jozo*, the sex pheromones of male *C. grewingki* have been isolated from urine.

3. Clariidae

The African catfish (*Clarias gariepinus*), a potentially valuable aquaculture species in the Middle East and Africa, has been the subject of an elegant series of pheromone studies[29,40,41,42,43,76]. *Clarias* spawn in the summer, evidently in response to factors associated with flooding; males compete aggressively for a female, and the resulting pair then swims into submerged vegetation where the eggs are scattered[29]. The evolution of chemical cues synchronizing sexual activities presumably has been favored by the fact that spawning often occurs in turbid water and at night.

Although there is good evidence that female *Clarias* release a sex pheromone[41], research has focussed on a male pheromone produced by the seminal vesicle, an accessory gland of the reproductive tract known to synthesize and store a variety of free and conjugated steroids[46]. Using a Y-maze in which female *Clarias* choose between two odor sources, Resink et al.[40] demonstrated that ovulated females prefer male water to female water, whereas non-ovulated and anosmic females do

TABLE 1

Identified hormonal pheromones in teleosts

Species	Donor sex	Proposed pheromone	Response	Evidence [1] Res.[2]	Syn.[3]	Rel.[4]	Olf.[5]	References
Black goby (*Gobius jozo*)	M	etiocholanolone glucuronide	female is attracted and stimulated to oviposit	+	+	–	–	7, 8
Baikal sculpin (*Cottocomephorus grewingki*)	M	11β-hydroxytestosterone	induces ovulation	+	+	U	–	PC
	M	Δ_4-3-keto-C19 steroids (e.g. testosterone)	induces female behavior	+	+	U	–	10, 11
African catfish (*Clarias gariepinus*)	M	5β-pregnane-3α, 17α-diol-20-one-3α-glucuronide	attracts ovulated female	+	+	–	+	40–44
Zebrafish (*Brachydanio rerio*)	F	testosterone and estradiol glucuronides	attracts male and stimulates male courtship	+	+	–	–	29, 71
	M	steroid glucuronides	induces ovulation	+	+	–	–	29, 72, 73
Goldfish (*Carassius auratus*)	F	17α,20β-dihydroxy-4-pregnen-3-one (17,20β-P)	increases milt volume and induces ovulation	+	+	+	+	15, 48, 51, 54 56, 62, 66
	F	prostaglandin F2α metabolites	stimulates male courtship	+	+	+	+	48, 49, 52, 53 54, 59
	F/M	Δ_4-3-keto-C19 steroids (e.g. androstenedione)	antagonizes effects of 17,20β-P	+	+	–	+	67, and unpublished results

[1] Evidence relating to the isolated pheromone;
[2] Response demonstrated by exposure to pure pheromone;
[3] Synthesis of proposed pheromone demonstrated;
[4] Release of proposed pheromone demonstrated;
[5] Olfactory responses to proposed pheromone demonstrated.
+ = published evidence;
– = no evidence available;
U = in urine;
PC = personal communication.

not. Further choice tests, using the odor of seminal vesicle fluid[41] or the odor of males in which either the testes or the seminal vesicle had been removed[40], clearly indicated that the seminal vesicle is the source of the male pheromone.

Both behavioral and electrophysiological studies clearly demonstrate that steroid glucuronides are active components of the male *Clarias* attractant pheromone synthesized by the seminal vesicle. Ovulated *Clarias* are attracted to a fraction of seminal vesicle fluid containing steroid glucuronides, but not to fractions containing free steroids or proteins, and β-glucuronidase treatment of the glucuronide fraction destroys its activity[41]. These behavioral findings are consistent with the results of electro-olfactogram (EOG) recording[43], a technique that measures voltage gradients, believed to reflect multi-unit generator potentials, from the surface of the olfactory epithelium[37]. As measured by EOG, the olfactory potency of the odor of intact males is greater than that of males from which the seminal vesicle has been removed, and less than that of males which have been castrated, an operation which causes seminal vesicle hypertrophy[43] (Fig. 1). Seminal vesicle fluid also is a potent olfactory stimulant and, although all seminal vesicle fractions induce olfactory responses (Fig. 1), the activity of the steroid glucuronide fraction is significantly reduced by β-glucuronidase treatment[43].

In vitro studies[41,46] have shown that the seminal vesicle can synthesize 17 free steroids and 8 glucuronated steroids. A mixture of 7 of these glucuronides attracted ovulated *Clarias* in choice tests[41]. When the olfactory potency of these glucuronides was tested by EOG, two were found to have the low detection thresholds expected of a potential pheromone: 5β-pregnane-3α,17α-diol-20-one-3α-glucuronide (10^{-11} M); 5β-androstane-3α,11β-diol-17-one-3α-glucuronide (10^{-9} M)[43] (Fig. 1). Although these studies clearly indicate that male African catfish produce specialized steroid glucuronides in the seminal vesicle, and that some of these glucuronides are potent olfactory stimulants, the specificity of the neurological and behavioral responses to these compounds have yet to be ascertained. Similarly, it has not been determined whether these steroids are synthesized in significant quantities *in vivo*, and whether and how they might be released.

4. Cyprinidae

Hormonal sex pheromones have been identified in two cyprinids, the zebrafish (*Brachydanio rerio*) and the goldfish. Although both species employ the egg-scattering reproductive behaviors common in Cyprinids, they appear to use very different sex pheromones.

4.1. Zebrafish

Female zebrafish exhibit a 4- to 5-day ovulatory cycle when held in mixed-sex groups, and there is evidence that both ovulation and attraction of males to ovulated females are induced by steroidal pheromones[29]. Females fail to ovulate when isolated from males, and resume ovulation when exposed to male holding water, unless they are made anosmic[72]. Testis homogenate and a testicular fraction containing steroid glucuronides also induce ovulation in isolated females, whereas

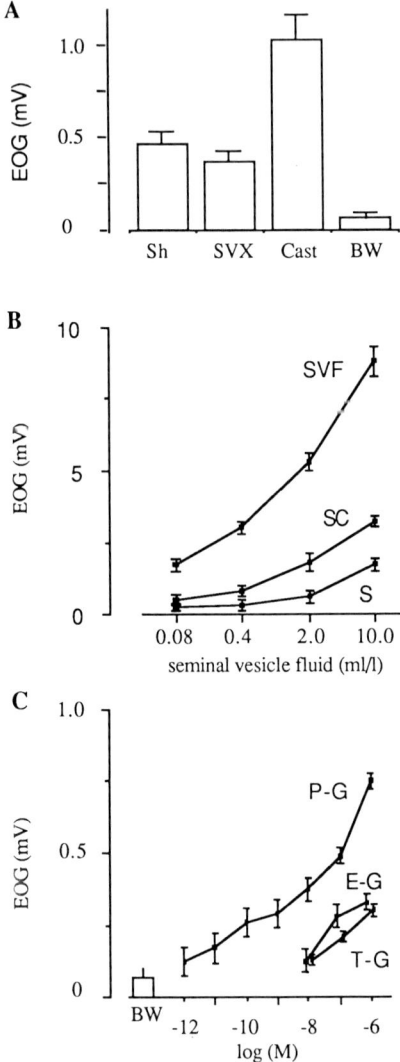

Fig. 1. Electro-olfactogram (EOG) recordings from the olfactory epithelium of the African catfish (*Clarias gariepinus*) demonstrating sensitivities to steroid glucuronide synthesized by the *in vitro* seminal vesicle (redrawn from Resink et al.[43]). A: EOG responses of female *Clarias* to blank water (BW) and to water which held a sham- operated (Sh) male, a male from which the seminal vesicle has been removed (SVX), or a castrated male (Cast). Castration causes seminal vesicle hypertrophy. B: EOG dose–responses of female *Clarias* to seminal vesicle fluid (SVF) and to the steroid conjugate (SC) and free steroid (S) fractions of SVF. C: EOG dose–responses to 5β-pregnane-3α,17α-diol-20-one-3α-glucuronide (P-G), the most potent steroidal olfactory stimulant isolated from the *Clarias* SV, and to less potent SVF glucuronides (E-G; etiocholanolone-glucuronide: T-G; testosterone glucuronide). E-G and T-G have been proposed as steroidal pheromones in the goby, *Gobius jozo*[8], and the zebrafish, *Brachydanio rerio*[71], respectively.

testicular fractions that have been treated with β-glucuronidase, or that contain free steroids, are ineffective[72]. Unfortunately, the active testicular component of the pheromone has not been identified. *In vitro* studies have identified a number of testicular steroid glucuronides, one of which (5α-androstane-3α,17β-diol-glucuronide) also has been identified in male holding water[72]; however, attempts to induce ovulation with synthetic glucuronides have so far been unsuccessful [73].

Using attraction to an odor source as a bioassay, Van Den Hurk and Lambert[71] provide strong evidence that steroid glucuronides also are components of a female zebrafish pheromone. Males are attracted to an aqueous ovarian extract from mid-cycle females, but exhibit stronger responses to ovarian extracts from ovulated fish; neither females nor anosmic males are attracted[71]. Testing a number of fractions prepared from the aqueous extract demonstrated that only the fraction containing steroid glucuronides was attractive[71]. Finally, a mixture of estradiol and testosterone glucuronides, which were ineffective alone, attracted males, suggesting that these identified ovarian compounds[29] likely are components of the female pheromone. However, it is not yet known whether these glucuronides are released, what specific behavioral function they might perform, or if they are detected by the olfactory system.

4.2. Goldfish

At present, more is known about sex pheromone function in goldfish than any other teleost fish[48,55]. However, it is clear that our understanding of this species is far from complete.

Goldfish reproduction is typical of many temperate cyprinids, gonadal growth taking place over winter and spawning coinciding with spring warming. Ovulation, which can occur several times each spawning season, is triggered by a periovulatory gonadotropin (GtH) surge of approximately 15 h duration which commences in the afternoon and rapidly stimulates ovarian synthesis of 17α,20β-dihydroxy-4-pregnen-3-one (17,20β-P), a progesterone derivative which induces oocyte final maturation (i.e., the resumption and completion of meiosis)[19,66]. Ovulation (rupture of the follicles and release of the mature oocytes to the ovarian lumen) occurs early the following dawn. The presence of ovulated eggs in the reproductive tract then stimulates synthesis and release of prostaglandin $F_{2\alpha}$ ($PGF_{2\alpha}$), which enters the circulation and acts within the brain to trigger female spawning behavior[19,58,59]. Groups of males pursue the ovulated female(s) for the hour or so required for the female to release her ovulated eggs, during which time the males compete vigorously for position beside the female each time she enters aquatic vegetation to oviposit[58]. In addition to performing critical hormonal functions in the female, 17,20β-P and prostaglandins (PGs) are released to the water where they function as potent chemical cues eliciting dramatic changes in the reproductive physiology and behavior of males[48,50].

Our discovery of hormonal pheromones in the goldfish was the serendipitous result of an experiment to determine whether 17,20β-P, which also is produced by the teleost testis[17,70], stimulates milt (sperm and seminal fluid) production. Although 17,20β-P injection increased milt volume the following day, similar re-

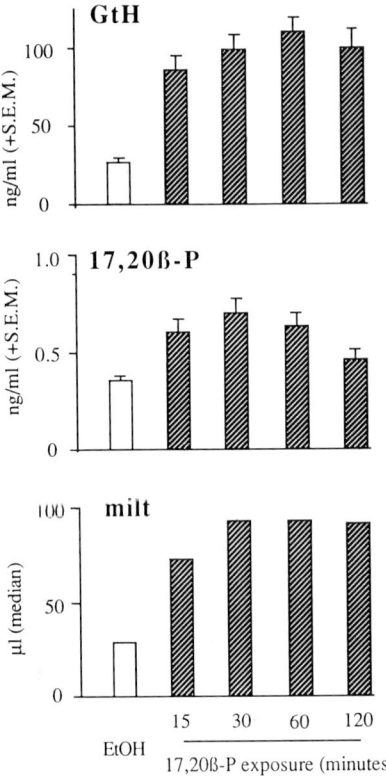

Fig. 2. Effect of water-borne 17α,20β-dihydroxy-4-pregnen-3-one (17,20β-P) on blood gonadotropin (GtH), blood 17,20β-P, and milt volume in goldfish. Blood samples were taken following 15 to 120 min exposure to 17,20β-P (5×10^{-10} M) or ethanol vehicle (EtOH); males were then moved to clean aquaria and stripped of milt 8 h later. Redrawn from Dulka et al.[15].

sponses were observed in uninjected males held in the same aquaria. Suspecting that the milt responses of uninjected males were due to 17,20β-P released by injected males, we added 17,20β-P directly to tank water and observed increased milt at the lowest dose tested (10^{-10} M)[62]. The response is specific to 17,20β-P and related progestogens — a variety of free and glucuronated estrogens and androgens are completely ineffective — and is blocked by sectioning the olfactory tracts[62]. The milt response to water-borne 17,20β-P is mediated by a rapid release of GtH from the pituitary[15,54] (Fig. 2); recently hypophysectomized males do not respond, whereas intact males exhibit increased blood GtH levels within 15 min of exposure.

The finding that water-borne 17,20β-P stimulates milt volume by increasing blood GtH[15] suggested that this steroid might be the female ovulatory pheromone shown by Kobayashi et al.[24,25] to increase GtH levels in male goldfish held with periovulatory females. To test this possibility, males were placed either in contact with a mature female, or separated from her by an opaque barrier which permitted water flow[66]. Aquatic vegetation was then added to promote ovulation. At 3 h intervals throughout the periovulatory period, groups of fish were removed for blood sampling, measurement of milt volume, and collection of female holding water. The periovulatory profile of blood 17,20β-P in ovulating females was remarkably similar to the profile of 17,20β-P released to the water (Fig. 3), indicating rapid clearance from the blood to the water. The GtH levels and milt volumes of males exposed to the odor of ovulating females increased relative to those of males with non-ovulatory females, and were equivalent to those of males in direct contact with ovulatory females, supporting the possibility that water-borne 17,20β-P has a pheromonal function (Fig. 3). Although a comprehensive biochemical analysis of 17,20β-P-like compounds released by goldfish has not been conducted, we know from RIA of water samples that ovulatory females release approximately as much glucuronated 17,20β-P (17,20β-P-G) (Fig. 3) and 17α-hydroxyprogesterone (17-P)[66,75] as free 17,20β-P. However, these compounds are much less potent than 17,20β-P in stimulating either EOG or endocrine responses (refs. 51, 56, and unpublished results).

At ovulation, when 17,20β-P release is rapidly decreasing, female goldfish start releasing another hormonal pheromone which triggers male behavior. Partridge et al.[37], the first to describe this postovulatory pheromone, provided evidence that it is associated with ovarian fluid and detected by olfaction. Several lines of evidence support the possibility that this pheromone is comprised of metabolites of $PGF_{2\alpha}$. Males show equal sexual interest in ovulated females and in females in which sexual activity has been induced by $PGF_{2\alpha}$ injection[57]. This lack of discrimination appears attributable to chemical cues because anosmic males pursue neither ovulated nor $PGF_{2\alpha}$-injected females[61], and because males exhibit a similar set of behavioral responses when exposed to the odors of ovulated and $PGF_{2\alpha}$-injected females in an open-field maze[49]. Finally, immunoreactive PGF is released both by $PGF_{2\alpha}$-injected females and by recently ovulated females, provided they have ovulated eggs in the ovarian lumen[48,53]. However, because an amount of $PGF_{2\alpha}$ — sufficient to release a stimulatory odor when injected into females — failed to affect male behavior when added directly to aquarium water[49], it appears likely that injected $PGF_{2\alpha}$ is metabolized prior to release. This possibility is supported by EOG and behavioral studies; 15-keto-prostaglandin $F_{2\alpha}$ (15-K-$PGF_{2\alpha}$), a mammalian metabolite of $PGF_{2\alpha}$ (PG metabolism has not been studied in fish), is a more potent olfactory stimulant (10^{-12} M threshold) than $PGF_{2\alpha}$ (10^{-10} M threshold)[53], and induces male sexual behaviors at lower concentrations than $PGF_{2\alpha}$ when added directly to aquarium water (cf. Fig. 4; ref. 54).

Whereas unmodified hormonal 17,20β-P appears to function as a pheromone on release, $PGF_{2\alpha}$, which stimulates female spawning behavior more effectively than 15-K-$PGF_{2\alpha}$[52], is almost certainly modified for pheromonal function. Preliminary studies indicate that female goldfish injected with radiolabelled $PGF_{2\alpha}$ release at least three labelled olfactory stimulants; unfortunately, none of these appears to be $PGF_{2\alpha}$ or 15-K-$PGF_{2\alpha}$ (P.W. Sorensen and F.W. Goetz, unpublished results).

Although it now is clear that water-borne metabolites of $PGF_{2\alpha}$ exert a rapid

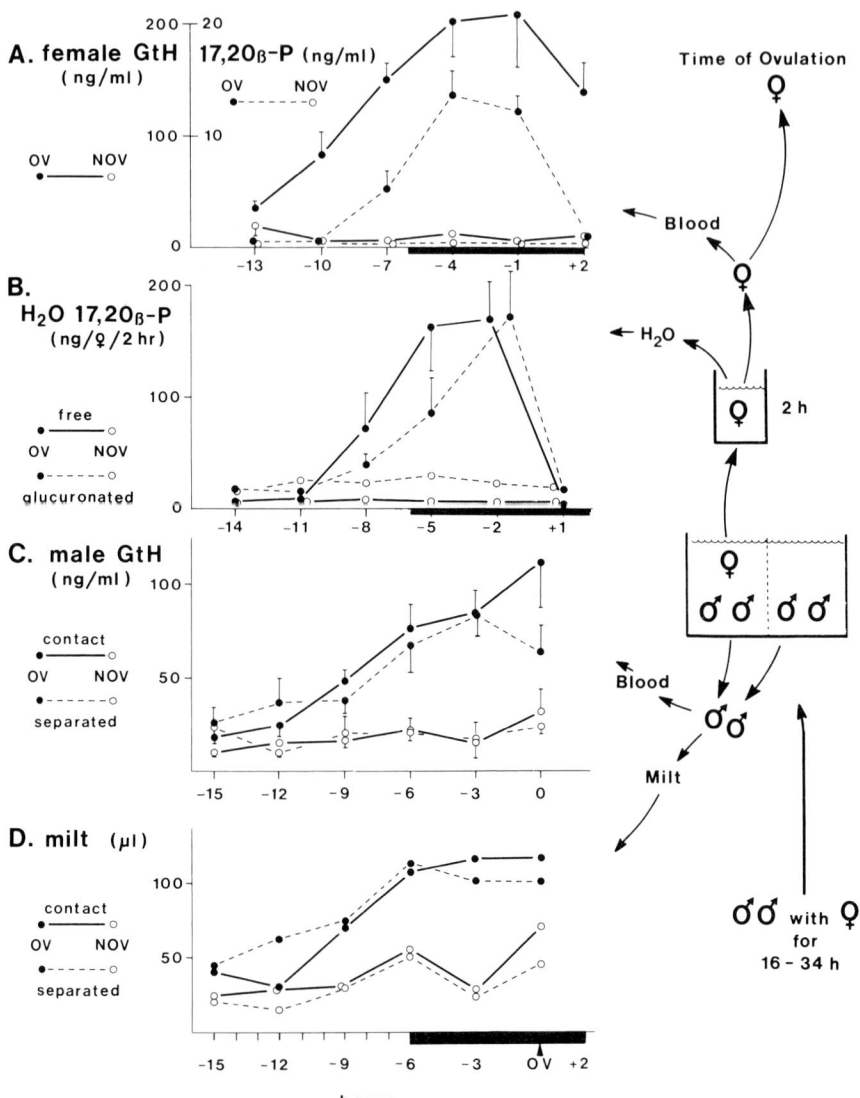

Fig. 3. Correlation between periovulatory release of free and glucuronated 17α,20β-dihydroxy-4-pregnen-3-one (17,20β-P) by female goldfish and increase of blood gonadotropin (GtH) and milt volume in males. Male parameters were not influenced by whether the males were in contact with the female (contact) or separated from the female by an opaque barrier allowing water flow (separated). Redrawn from Stacey et al.[66].

effect on male sexual behaviors, it is not clear what role this pheromone has in eliciting the rapid GtH and milt increases which occur during sexual interactions[28]. Water-borne 17,20β-P is thought to increase GtH simply by triggering a neuroendocrine reflex, and therefore can increase GtH levels[54] and milt volume[62] in

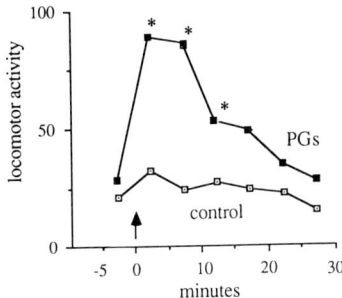

Fig. 4. Locomotor activity (median number of aquarium areas entered) of groups of five male goldfish prior to and during continuous infusion arrow of either ethanol vehicle (control; 16 groups) or a mixture (PGs; 16 groups) of 3×10^{-8} M prostaglandin $F_{2\alpha}$ ($PGF_{2\alpha}$) and 3×10^{-8} M 15-keto-prostaglandin $F_{2\alpha}$ (15-K-$PGF_{2\alpha}$). In addition to increased locomotory activity, prostaglandin-exposed groups exhibited cloacal nudging and reduced feeding, behaviors typical of males interacting with ovulated females[54]. *Significantly different from control (Friedman's test and Tukey's method for pairwise comparisons).

isolated males. In contrast, water-borne PGs appear to trigger these endocrine and gonadal responses indirectly, by stimulating social interactions with conspecifics[54]. Furthermore, whereas several hours are required for water-borne 17,20β-P to increase milt volume by stimulating the pituitary-gonadal axis[15], sexual interaction can increase milt volume in less than 1 h[54] and without evident GtH increase[28]. These rapid milt responses to sexual interaction may be mediated by contractions of testis and sperm duct muscles initiated by input from the preoptic area[14].

Although our initial studies examined the pheromonal effects of 17,20β-P independently from those of PG pheromones, more recent work has shown that these two odors can have synergistic actions, as might be predicted from their sequential release during the periovulatory period. Males which have been exposed to 17,20β-P, and then stripped of milt the following morning, are able to replace a larger milt volume during sexual interaction than males which have not been exposed to 17,20β-P[65]. Similarly, males exposed simultaneously to 17,20β-P and $PGF_{2\alpha}$-injected females have higher blood GtH levels than do males exposed to only one of these stimuli[54]. Finally, Defraipont and Sorensen[9] have recently shown that during group spawning, males recently exposed to 17,20β-P are more successful in competing for access to females than are males which have not been exposed; whether this effect is mediated by the endocrine responses induced by 17,20β-P is not known. Similarly, whether pheromones might affect sperm quality as well as quantity has yet to be determined.

EOG studies have contributed substantially to our present understanding of hormonal pheromone function in goldfish by demonstrating the sensitivity and specificity of the olfactory system. In the case of PG pheromones, EOG studies[53] have shown that F_2-series prostaglandins are more stimulating than others (e.g., E and F_1 series), and that 15-K-$PGF_{2\alpha}$ is the most potent known $PGF_{2\alpha}$ metabolite. EOG studies[51,56] of pheromonal 17,20β-P demonstrate extreme olfactory specificity for this steroid and illustrate the close correlation between olfactory potency

A. Cross-adapt to $^{-7}$ M 15-K-PGF$_{2o}$

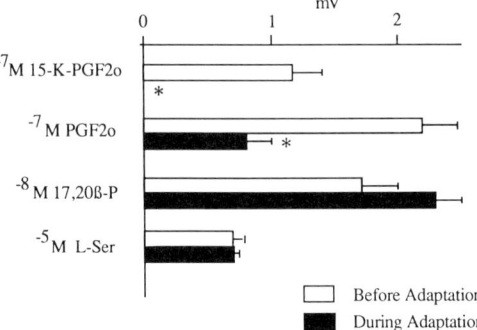

B. Cross-adapt to $^{-8}$ M 17,20ß-P

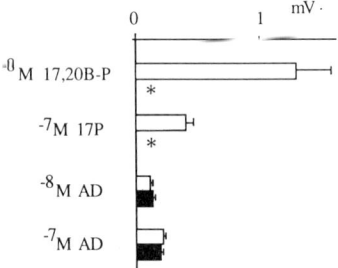

Fig. 5. Cross-adaptation experiments indicating separate olfactory receptor mechanisms for prostaglandins and 17α,20β-dihydroxy-4-pregnen-3-one (17,20β-P). A: EOG responses (in mV) prior to and during adaptation to 10^{-7} M 15-keto-prostaglandin F$_{2α}$ (15-K-PGF$_{2α}$). Results indicate that prostaglandin F$_{2α}$ (PGF$_{2α}$) interacts with the 15-K-PGF$_{2α}$ receptor mechanism, but that 17α,20β-dihydroxy-4-pregnen-3-one (17,20β-P) and L-serine (L-ser) do not. B: EOG responses prior to and during adaptation to 10^{-8} M 17,20β-P. Results indicate that 17α-hydroxyprogesterone (17P) interacts with the 17,20β-P receptor but that androstenedione (AD) does not. Redrawn from Sorensen et al.[53,56].

of 17,20β-P-like steroids and their effectiveness in stimulating endocrine response. EOG cross-adaptation studies, in which the sensory epithelium is first adapted to one odor and then exposed to a second odor, have provided preliminary information about the number of olfactory receptor mechanisms mediating pheromonal responses. For example, PGF$_{2α}$ and 15-K-PGF$_{2α}$ appear to preferentially activate separate receptor mechanisms, which in turn are distinct from those mediating amino acid or 17,20β-P responses[53,56] (Fig. 5A). In vitro studies indicate that 17,20β-P[45] and L-amino acids[39] act via high affinity receptors associated with olfactory epithelium membrane (Fig. 6); the intriguing possibility that gonadal 17,20β-P receptors[31] have been externalized to the olfactory epithelium has not been explored. Finally, EOG studies intended to demonstrate the specificity of the goldfish olfactory system to 17,20β-P have surprisingly shown that androgens such as androstenedione stimulate the goldfish olfactory system (10^{-10} M threshold) through receptor mechanisms distinct from those which detect 17,20β-P (ref. 56,

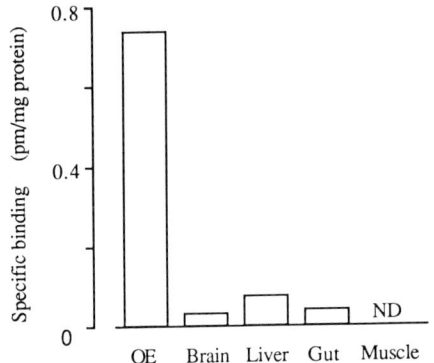

Fig. 6. Specific binding of ^{3}H-17α,20β-dihydroxy-4-pregnen-3-one to membranes isolated from goldfish olfactory epithelium (OE), brain, liver, intestine, and skeletal muscle after a 60 min incubation at 4°C. ND; not detectable. Redrawn from Rosenblum et al.[45].

and unpublished results) (Fig. 5B). These unexpected responses to androgens likely would not have been uncovered had endocrine bioassay been the only technique employed.

The most straightforward interpretation of the effects of released female 17,20β-P and PGs on the reproductive physiology and behavior of males is that they enhance male reproductive success. For instance, it seems reasonable that males exposed to 17,20β-P have a competitive advantage at the time of spawning because of their increased sperm stores, while males detecting prostaglandin metabolites are more readily able to locate the ovulated female in a spawning group (Fig. 7). It is worth noting that this simple model has not been directly tested, and that there are reasons to believe that it may omit important functions of other hormonal pheromones in goldfish.

For example, there is preliminary evidence that females respond to 17,20β-P and that males release it. Exposure to water-borne 17,20β-P increases the incidence of spontaneous ovulation[50], suggesting that release of 17,20β-P by ovulatory females might trigger ovulation in other females. Such a primer pheromone function, which may be the explanation for anecdotal and experimental reports[27] of synchronous ovulation in goldfish, might confer advantages on both spawners and their spawn through 'predator swamping', a reproductive strategy whereby highly synchronized reproduction reduces the proportion of a population that can be ingested by predators. That water-borne 17,20β-P affects both sexes perhaps is not surprising, considering that male and female olfactory systems exhibit similar sensitivities to this steroid[51], whereas $PGF_{2\alpha}$, which is only known to elicit responses in males, is a more potent olfactory stimulant in males than in females (unpublished results). Despite equivalent male and female olfactory responses to 17,20β-P, their GtH responses appear to be quite different; 17,20β-P exposure consistently induces GtH increase even in males beginning testicular recrudescence, whereas females show no GtH increase unless capable of ovulation (unpublished results).

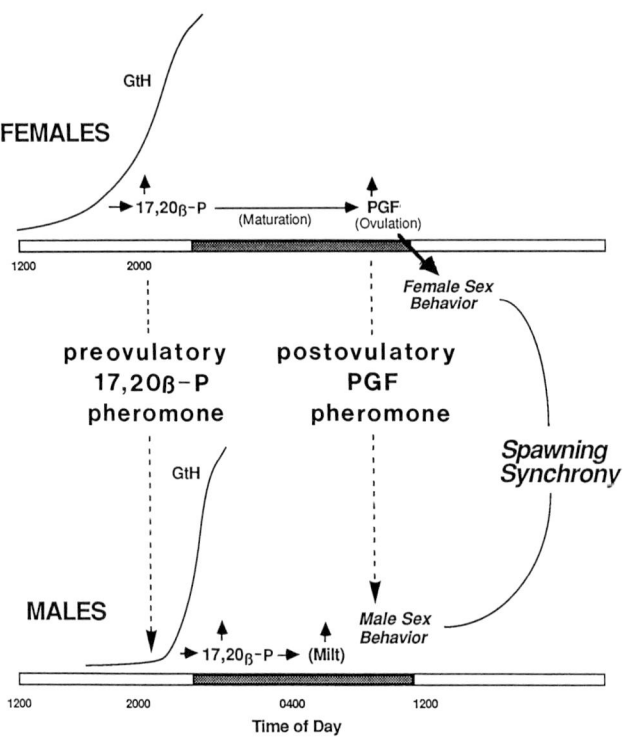

Fig. 7. The dual hormonal pheromone system of the goldfish. The female preovulatory gonadotropin (GtH) surge stimulates synthesis of ovarian 17α,20β-dihydroxy-4-pregnen-3-one (17,20β-P) which in turn induces oocyte final maturation. 17,20β-P is released throughout the preovulatory period and triggers a surge of male GtH which, by stimulating testicular 17,20β-P synthesis, increases milt volume in time for spawning. At ovulation, increased prostaglandin $F_{2\alpha}$ (PGF) synthesis stimulates female spawning behavior; PGF is then metabolized and released to the water where it stimulates male sexual behavior. Redrawn from Sorensen et al.[53].

A pheromone function of 17,20β-P released by ovulatory females and triggerring reproductive responses in both sexes would be analogous to that seen in Pacific herring (*Clupea harengus pallasi*), in which a pheromone released in milt rapidly triggers both male and female spawning behaviors[60]. However, unlike the situation in herring, where only one sex produces the bisexual releaser pheromone, there is reason to believe that both female and male goldfish can serve as sources of pheromonal 17,20β-P. Although 17,20β-P release by males has yet to be demonstrated, it is clear that males rapidly increase their blood 17,20β-P levels if exposed to ovulatory females[24] or to 17,20β-P[15] (Figs. 2 and 7). Provided that 17,20β-P is released at high rates by males only following interaction with ovulatory females, such 17,20β-P should carry the same information (imminent spawning opportunity) and thus warrant the same GtH responses as 17,20β-P released by females. As yet, however, the only evidence that males might respond to 17,20β-P released by other males is our recent finding[67] that milt production is increased when males are exposed to a male injected with human chorionic gonadotropin, a treatment known to increase 17,20β-P levels in male blood[26].

A second omission from the model presented in Fig. 7 is the potential pheromonal role of water-borne androgens. Having found that androgens (especially androstenedione, AD) are potent olfactory stimulants (10^{-10} M threshold)[56] which do not stimulate milt production[62], we examined the possibility that they might instead perform some inhibitory function. In clear support of this proposal, overnight milt volume increases to threshold 17,20β-P concentrations (10^{-11} to 10^{-10} M) were blocked in fish which were simultaneously exposed to 17,20β-P and a 100-fold excess of either AD or testosterone (T) (unpublished results). The anomalous overnight milt volume increase observed when males are isolated[67] also can be reduced by water-borne AD (unpublished results). Although it is assumed that androgens inhibit milt production by reducing blood GtH levels, this remains to be shown. There presently is no information as to which sex releases pheromonal androgens, what specific androgen compounds are released, how males benefit by reducing milt production, and whether the female's ovulatory response to water-borne 17,20β-P[50] also might be inhibited by androgens. Nonetheless, we feel the most parsimonious interpretation of the observed androgen effects is that both sexes normally produce a dual-component pheromone consisting of stimulatory progestins and inhibitory androgens, and that this pheromone stimulates GtH increase only when above a threshold progestin:androgen ratio. We believe such a pheromone is necessary because low levels of 17,20β-P are measurable in the blood of both non-ovulatory females[66] and non-stimulated males[15]; non-ovulatory females are known to release 17,20β-P[66] and males are presumed to do so. Androgens could mask this basal 17,20β-P release, thus restricting GtH increases to interactions with conspecifics releasing high 17,20β-P levels.

III. Functions of hormonal pheromones in fish

Demonstrating that an individual's odor elicits a reproductive *response* in conspecifics is quite distinct from demonstrating what reproductive *function* that response might serve. For example, although it is clear that male goldfish increase releasable sperm stores when exposed to 17,20β-P, there is no evidence that this response increases either the male's reproductive success, or the reproductive success of the female releasing the 17,20β-P. This is not a trivial point, for we cannot claim to understand the function of a sex pheromone until we understand how it affects the reproduction of sender and receiver. No study has experimentally addressed this problem in fish, function being deduced from observed behavioral or endocrine response. Nonetheless, the compelling evidence for the existence of fish hormonal pheromones suggests a range of functional types. At one extreme, both sender and receiver may benefit (bilateral benefit); at the other, benefit appears restricted to the receiver (unilateral benefit).

The case of the steroidal pheromone, etiocholanolone glucuronide (EG), released by the male black goby (*G. jozo*) provides the clearest indication of bilateral reproductive benefits. Presumably, this pheromone benefits females by providing a distinctive odor with which to locate the male's cryptic nest, and benefits males by attracting receptive females. Moreover, the fact that the male has evolved a specialized pheromone-producing tissue, the testicular mesorchial gland[8], is strong evidence for signal specialization, an important criterion of true communication[30]. Whether the male employs additional means to increase the efficiency of his chemical signalling (controlled pheromone release, associated behaviors) is not known. The finding that ovulated females are attracted to male urine (obtained from the bladder after ligating the urogenital papilla), raises the possibility that EG is stored for pulsatile release[8], although there unfortunately is no evidence that EG is present in urine of *G. jozo*. However, Dmitrieva and Ostrovmov [10] find that urine of male *C. grewingki* contains the androgenic pheromone eliciting female spawning behavior, and that the capacity of the male's urinary bladder increases approximately five fold during the spawning season, possibly a reflection of its function in pheromone storage.

In zebrafish, goldfish, and *Clarias* it seems clear that hormonal releaser pheromones benefit the receivers by enabling them to locate and/or identify sexually active conspecifics. These pheromones also should benefit the senders by indicating their reproductive status to potential spawning partners. However, if the release of steroids or PGs normally produced for endogenous hormonal function is sufficient to maximize such benefits to the sender, there may be no mechanism whereby the receiver's response can exert selective pressure for signal specialization. In this respect, it is interesting that neither zebrafish nor goldfish are known to possess structural specializations for pheromone production or transmission such as those seen in *G. jozo* and *C. grewingki*. In *Clarias*, pheromone production by the seminal vesicle[40] could be considered indicative of a signalling specialization, but a more straightforward interpretation would be that any pheromonal effects of released seminal vesicle products are incidental to their primary functions of regulating sperm storage and fertility[74]. Unfortunately, the entire question of signal specialization as it relates to pheromone function in fish is poorly understood, although it seems likely that whatever specializations have evolved will be less overt than the elaborate pheromone distribution systems of terrestrial species simply because soluble products are easily dispersed in water. It is possible that zebrafish, goldfish, and *Clarias* do possess specializations for signalling, but that these are subtle metabolic changes (e.g., an increase in hormone production over that required for endogenous hormonal function, or a modified enzymatic pathway for hormone metabolism) rather than structural modifications.

The goldfish pheromone, 17,20β-P, is similar to the steroidal pheromones of the catfish and zebrafish, and the PG pheromone of goldfish, in that it shows no evidence of signal specialization, but is different in that it is difficult to understand how the male's endocrine and gonadal responses could directly benefit the individual female. Our laboratory studies showing that spawning involves a number of males, which can rapidly increase their milt volume during spawning, indicate that female fertility is unlikely to be limited by availability of sperm. Furthermore, even if sperm were limiting, the facts that most 17,20β-P is released in darkness hours before ovulation and spawning, that goldfish do not exhibit mate fidelity, and that females may ovulate in synchrony, all would tend to reduce the likelihood that a female will benefit through the increased fertility of those males she stimulates prior to ovulation. However, the question of female benefit through 17,20β-P release can only be properly addressed through studies of feral goldfish spawning under natural conditions.

Despite the fact that virtually nothing is known concerning the precise functions of fish hormonal pheromones, functional diversity is indicated by the nature of the pheromones (modified or unmodified hormone), presence or absence of signalling specializations, and likelihood of bilateral reproductive benefits. Although these differences might suggest that there is little or no relationship among the hormonal pheromones of fish, we believe that the differences illustrate distinct stages in an evolutionary sequence common to all hormonal pheromones.

IV. Are these compounds true pheromones?

The term pheromone was coined by Karlson and Lüscher[21] who derived it from the Greek terms *pherin* (to transfer) and *hormon* (to excite) and proposed that it replace the older term ectohormone introduced by Bethe in 1932[3]. Karlsson and Lüscher defined pheromones as 'substances that are excreted to the outside by an individual and received by a second individual of the same species in which they release a specific reaction for example a definite behavior or developmental process.' Wilson and Bossert[78] later categorized pheromones either as primers, whose principal effects are physiological, or as releasers, whose principal effects are behavioral. Although the pheromone concept has won widespread popular acceptance, it encompasses important issues which have drawn considerable criticism and stimulated attempts to develop new terminologies[2,13]. One issue of particular relevance is Karlson and Lüscher's[21] statement that 'the substance is...secreted...outside the body [to] serve communication between individuals.' Unfortunately, Karlson and Lüscher do not define what they mean by communication, an omission which has created confusion because of recent controversy about the definition of this term[5].

Reviewing chemical communication in fish, Liley[30] adopts Wilson's[77] definition of communication: '...an action on the part of another organism or cell that alters the probability pattern of behavior of another organism (or cell) in an adaptive fashion.' Liley then deduces that the term adaptive implies that the 'signal, or the response, or both, have been genetically programmed to some extent by natural selection.' Taking this a step further, Liley argues that this implies that 'the signaller benefits by influencing the behavior of other individuals' (a point also made by Burghardt[5] in his definition), and 'that performance of an appropriate response benefits the receiver.' In other words, Liley states: 'the key feature that distinguishes a communication signal is that there is some degree of evolutionary development or specialization of the stimulus (ritualization) making it a more effective signal.' The problem of applying rigorous definitions of pheromones and chemical communication to fish is summarized by Liley[30]: 'even if one individual emits a chemical that influences the behavior of a conspecific..should the chemical be considered a pheromone — a... component in the species' communication system — or...a metabolic product...which elicits responses in conspecifics in much the same way that abiotic...stimuli may elicit specific adaptive responses?' Liley's question strikes at the heart of the matter when one questions whether the hormonally derived cues in fish can be categorized as communicative chemicals or 'true pheromones'. At the present time, with the possible exception of *Gobius jozo*, we know too little about pheromone synthesis and release mechanisms to state unequivocally whether hormonal pheromones are specialized byproducts of specialized tissues (or biochemical pathways) which have evolved because they benefit the donor, or merely unspecialized metabolites which benefit only the receiver. In spite of these ambiguities about communication function, we believe that fish hormonal pheromones should be considered true pheromones because at the very least they appear to represent necessary steps in the evolution of true communicative signals. The following section addresses this hypothesis.

V. Evolution of hormonal pheromones in fish: the origins of a vertebrate communication system

Because the hormonal functions of the identified hormonal sex pheromones of fish (androgens, estrogens, progestins, prostaglandins) are highly conserved not just in fish, but throughout vertebrates, it seems reasonable to assume that these functions predate the evolution of the more specialized pheromonal functions. If so, what are the evolutionary selective pressures through which critical reproductive responses of some fish have come to be regulated by released hormones and hormone metabolites of conspecifics?

Early theoretical arguments as to why aquatic organisms either do use[22] or should use[12] hormones and hormone metabolites as pheromones have focussed on physiological preadaptation (see II. Evidence for hormonal pheromones in fish). Basically, a potential signal (released hormone) which already is in place could readily become a functional signal for a conspecific with a mutation which externalizes its existing hormone receptors on chemosensory membranes. Interestingly, both hormonal and olfactory receptors for $17,20\beta$-P in goldfish appear to be membrane bound, although whether these receptors are biochemically related is not known[45,48].

The preadaptation hypothesis appears to clarify the conceptual problem of the communication function of hormonal pheromones. For example, the hypothesis implies that a communicative pheromone system (in which both sender and receiver benefit) need not arise *de novo*, but rather can be derived from a system which unilaterally benefits a receiver which responds adaptively to unspecialized hormonal products from a sender. The transition from a system of unilateral benefit (*spying*) to one of bilateral benefit involving signal specialization (*communication*) presumably is determined by whether the receiver's response suffi-

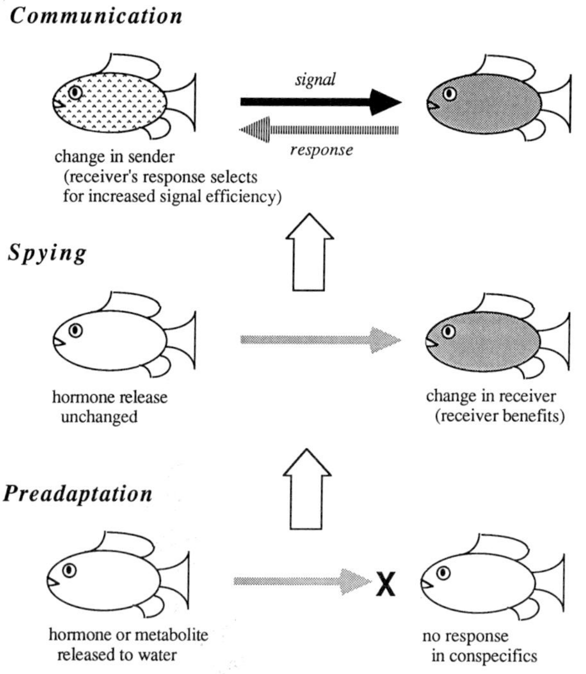

Communication

Spying

Preadaptation

Fig. 8. Proposed stages in the evolution of a communication function by hormonal pheromones.

ciently increases the reproductive success of the sender (Fig. 8). From this perspective, quibbling about whether specific hormonal pheromones should be considered to have communicative functions, and thus represent true pheromones, seems misguided, since they meet other criteria informally required of pheromones (relative species specificity, innate recognition, and clear and adaptive functions). Rather, the biologically relevant issue about these chemical cues is the nature of the evolutionary processes which have led to their present functions, and whether or not they could eventually come to assume the communicative functions expected of pheromones. We believe hormonal pheromones not only represent true pheromones but they may in fact constitute the clearest examples of pheromonal function among the vertebrates. Hormonal pheromones of fish, which appear to exemplify a range of conditions between spying and communication, should provide excellent opportunities to understand how chemical communication can evolve.

If the sender's reproductive success is unaffected by the receiver's adaptive response, there is no selective pressure for signal specialization and the interaction will remain one of intraspecific spying; as discussed above, the endocrine response of male goldfish to 17,20β-P released by females[66] may be of this nature. But even in some situations where the sender does benefit from the receiver's response, signalling may fail to evolve. In the case of the prostaglandin releaser pheromone of goldfish, for example, intense selective pressure on males to identify sexually active females evidently has led to a situation where female reproductive success is now dependent on the ability to release prostaglandins – anosmic males do not spawn. Despite this benefit to females, evolution of female signalling would be expected only if increasing prostaglandin production over that required for female behavior could increase fertility by attracting additional males. Given the competitive nature of male courtship, however, this seems unlikely. Thus, while spying can create selective pressure favoring the evolution of communication, a pheromone system need not progress beyond the spying stage.

In other cases where the sender's reproductive success is increased because of pheromone release, there is a possibility that communication can evolve. The male pheromone of *G. jozo*, for example, presumably originated with a mutation enabling females to detect and approach EG fortuitously released by territorial males, and led to a signalling or communicative function through sexual selection by females based on differential male production of EG. In general, territorial, nest guarding species (in which one sex can control reproductive success and fecundity is limited) would be expected to evolve relatively sophisticated bilateral communication systems operating in mate selection. Released hormones and hormone metabolites may readily be incorporated into these communication systems because they can convey information about reproductive status of the sender. Indeed, even in non-territorial species such as goldfish, females may have the opportunity to identify reproductively active (recently spawning) males on the basis of altered levels of released steroids. In this situation, however, females may avoid spawning with such males if, as in the lemon tetra, *Hyphessobrycon pulchipinnus*[35], their fertility is reduced by recent spawning activity.

The concept that fish hormonal pheromones originate with a spying stage which may then progress to a communication stage not only offers a unifying theoretical framework for comparing this process among diverse species but also focusses attention on the potential evolutionary problems inherent in the use of hormones as pheromones. One potential problem is that having reached the spying stage, progression to the signalling stage might be blocked if increasing the strength of the signal (i.e., increasing hormone production) adversely affects the sender's physiology. Two solutions might be envisaged. First, high plasma hormone levels could be prevented by rapid metabolism or release. The extremely rapid clearance of plasma 17,20β-P by goldfish may be an example. Second, adverse effects would be less likely if the pheromone were a hormonally inactive metabolite whose synthesis could be separated from the sender's endocrine system, as has been suggested for the mesorchial gland of *G. jozo*[8]. Indeed, selection for a separation of hormonal and pheromonal function in the sender might account for the fact that all but one of the known hormonal pheromones of fish (17,20Bβ-P in goldfish) are hormone metabolites; on the other hand, the apparent prevalence of metabolite pheromones might simply reflect the fact that hormones are commonly metabolized prior to release.

A second potential problem with the use of hormones as pheromones is that the structural diversity of those hormones proposed to function as fish pheromones (steroids, prostaglandins) is too limited to always allow species-specificity. This problem would be reduced, of course, if the pheromones were odor mixtures, either exclusively hormonal[43,71], or containing non-hormonal components[4]. Both the PG and 17,20β-P pheromone systems of goldfish appear to be comprised of more than one component, each of which is recognized by separate receptor mechanisms[53,56]. Another factor favoring the evolution of multi-component chemical signals is that they should facilitate orientation to an odor source. For example, many fish may have difficulty locating the source of a single-component pheromone because olfactory receptors saturate within several log units of threshold concentration (e.g., 17,20β-P and PG pheromones in goldfish); by using a multi-component pheromone containing components with a variety of dose-response relationships, fish could extend the range of pheromone concentrations over which they could detect the gradients required for orientation. However, even with a single-component hormonal pheromone, there are several reasons for believing that the potential problem of lack of species specificity should not prevent widespread evolution of these pheromonal systems.

First, lack of species specificity is a problem only if an individual enters the 'active space' (region of detection) of a heterospecific pheromone. Unfortunately, the active space of any fish pheromone has yet to be determined. However, in goldfish, the only fish species for which both the olfactory detection threshold[56] and pheromone release rate[66] have been measured, it appears that males must be within several body-lengths of an ovulatory female in order to detect 17,20β-P[55]. This surprisingly small active space, which likely is an overestimate because it does not consider the potential inhibitory effects of water-borne androgens[67], contradicts the prevailing intuition that aquatic pheromones have large active spaces and long transmission times, and raises the theoretical possibility of information being encoded through pulsatile release in urine.

A second factor which should reduce the potential problems of using hormonal pheromones with limited chemical diversity is that even minor variations in hormone or metabolite structure may be distinguishable by fish olfactory systems, which appear to be extraordinarily specific. The 17,20β-P system in goldfish is the best understood. Of nearly two dozen compounds closely related to 17,20β-P (Fig. 9), only the metabolite 17α,20β,21-trihydroxy-4-pregnen-3-one (17,20β,21-P) elicited appreciable EOG responses[56]. Electrical recording from the olfactory tracts, *in vitro* binding studies, and endocrine bioassays confirm this olfactory specificity (evidently the most extreme example demonstrated in a vertebrate) and suggest that a single population of olfactory receptors is responsible. If olfactory systems of other fishes are equally specific, effective species specificity could be achieved with only small changes in hormone metabolic pathways.

Although there is strong behavioral evidence for species-specific reproductive pheromones in fish (e.g., belontiids[32]), there unfortunately is no information about the chemical identities of these pheromones. Evidence for species specific pheromones also comes from EOG studies of *Clarias* and goldfish; the steroids most stimulatory for the goldfish olfactory system[51,56] are not those most stimulatory for the African catfish[43]. These EOG studies also show that the steroid glucuronides proposed as pheromones in *G. jozo*[8] and zebrafish[71] are poor

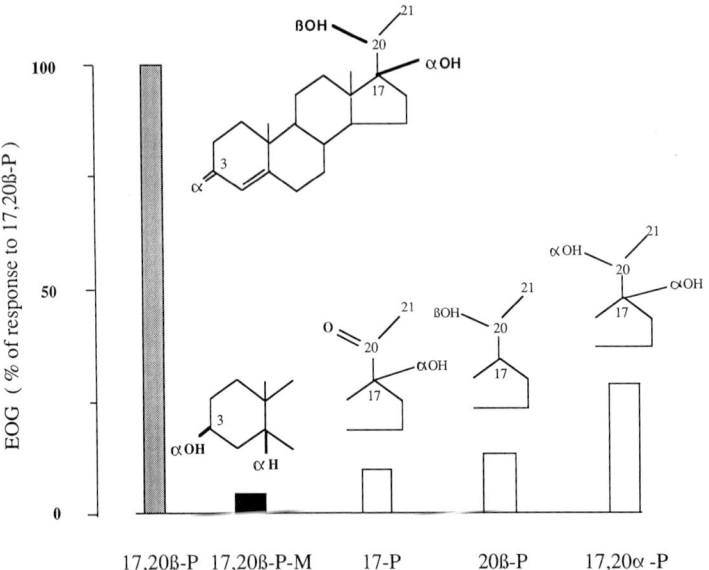

Fig. 9. Olfactory specificity of goldfish to 17α,20β-dihydroxy-4-pregnen-3-one (17,20β-P) as measured by electro-olfactogram (EOG) recording. Responses to 17,20β-P and related C21 steroids, all tested at 10^{-8} M. 17,20β-P-M; 5α-pregnan-3α,17α,20β-triol: 17-P; 17α-hydroxyprogesterone: 20β-P; 20β-hydroxyprogesterone: 17,20α-P; 17α,20β-dihydroxy-4-pregnen-3-one. Redrawn from Sorensen et al.[56].

olfactory stimulants in African catfish[43] (Fig. 1) and are not detected by the goldfish olfactory system[51]. Hopefully, future studies will examine the question of species specific hormonal pheromones by using a wider variety of species, and by focussing on related, reproductively sympatric species in which the opportunities for heterospecific interactions would be greatest.

Species specificity might originate via two mechanisms. In the simplest case, species specific pheromones could evolve passively as a result of chance differences in hormone metabolism. For example, prior to evolving hormonal pheromones, females of two species might have metabolized the maturational steroid, 17,20β-P, in different ways and therefore released different metabolites. Later, if it became advantageous for males to anticipate ovulation, natural selection would have favored those males capable of detecting and responding to the major 17,20β-P metabolite of conspecific females. Species specific pheromones would have resulted, but selection would have operated independently within each species. Such a scenario probably accounts for the different steroidal sensitivities of goldfish and African catfish olfactory systems[43,56].

A more complex and interesting scenario is anticipated in cases where species using the same hormonal pheromone run the risk of responding inappropriately to heterospecific odors. Such a situation could arise if females of two allopatric species had released the same progestin metabolite, and males of each species had evolved a pheromonal response to this common odor. If these species then experienced reproductive sympatry, reduced reproductive performance resulting from heterospecific responses would create selective pressure for species specificity, although the precise nature of these selective forces would depend on whether the benefits of pheromonal responses are unilateral or bilateral. In the case of unilateral benefit (spying), selection might favor male genotypes in each species which respond selectively to a secondary progestin metabolite (or metabolites) released only by conspecific females; thus species-specific pheromones would result even though female odors remain unchanged. In cases where bilateral benefits have led to signalling by females (i.e., 'true' communication), selection might favor both female genotypes which release species specific secondary metabolites and male genotypes capable of responding to them.

The evolution of a species specific pheromone will be determined not only by the nature of the benefits risked through responding to heterospecific pheromones, but also by the costs of ensuring that responses to conspecifics occur and that responses to heterospecifics do not. Although difficult to estimate, these costs are related to reproductive strategy and physiology and expressed as reduced reproductive potential due to wasted gametes, missed spawning opportunities, or increased risk of predation. In the above case of two sympatric species using pheromonal 17,20β-P, for example, responses to heterospecifics might be avoided by selection for sensitivity to a species specific secondary metabolite; but if that secondary metabolite were released in much lower quantities than 17,20β-P (and therefore were more difficult to detect), the benefit of specificity might well be outweighed by the cost of reduced fertility. In general, the cost of responding to heterospecifc pheromones should be insignificant if such mistakes occur only infrequently. Thus, it is interesting to speculate that non-specific hormonal pheromones may be one factor responsible for the widespread occurrence of natural hybridization between related fish species[47]. We presently are exploring this possibility by examining EOG responses to steroids and prostaglandins in a variety of cyprinids and catostomids known to hybridize under natural conditions.

VI. Future directions

Although the study of fish hormonal pheromones is only beginning, the tremendous diversity of fish and their reproductive strategies suggest that further research will continue to produce exciting findings of significance to a variety of scientific disciplines. Our present knowledge of fish hormonal pheromones emphasizes how little is understood about their function and evolution. In not one species, for example, is it known to what extent hormone release has been specialized for pheromone function, or how responses to hormonal pheromones influence reproductive success. Similarly, the important question of species specific hormones has yet to be examined in a group of closely related species. Answers to these questions not only should have practical application in fish culture and fisheries management, but also will provide key insights into the biology of fish and how they have evolved.

VII. References

1. Bardach, J.E. and J.H. Todd. Chemical communication in fish. In: *Communication by Chemical Signals*, edited by J.W. Johnson, Jr., D.G. Moulton and A. Turk. New York: Appelton-Century-Crofts, 1970, pp. 205–240.
2. Beauchamp, G.K., R.L. Doty, D.G. Moulton and R.A. Mugford. The pheromone concept in chemical communication: a critique. In: *Mammalian Olfaction, Reproductive Processes and Behavior*, edited by R.L. Doty. New York: Academic Press, 1976. pp. 143–160.
3. Bethe, A. Vernachlässigte Hormone. *Naturwiss.* 11:177–181, 1932.
4. Bryant, B.P. and J. Atema. Diet manipulation affects social behavior of catfish: the importance of food odor. *J. Chem. Ecol.* 13: 1645–1662, 1987.
5. Burghardt, G.M. Defining "Communication". In: *Advances in Chemoreception*, edited by D.G. Moulton and A. Turk. New York: Appleton Century Crofts, 1970, pp. 5–18.
6. Colombo, L., P.C. Belvedere and A. Marconato. Biochemical and functional aspects of gonadal biosynthesis of steroid hormones in teleost fish. *Proc. Indian Natl. Acad. Sci. B* 45: 226–234, 1979.
7. Colombo, L., A. Marconato, P.C. Belvedere and C. Frisco. Endocrinology of teleost reproduction: a testicular steroid pheromone in the black goby, *Gobius jozo* L. *Boll. Zool.* 47: 355–364, 1980.
8. Colombo, L., P.C. Belvedere, A. Marconato and F. Bentivegna. Pheromones in teleost fish. In: *Proceedings of the Second International Symposium on the Reproductive Physiology of Fish*, edited by C.J.J. Richter and H.J.Th. Goos. The Netherlands: Pudoc, 1982, pp. 84–94.
9. Defraipont, M. and P.W. Sorensen. Effects of the steroidal priming pheromone, $17\alpha,20\beta$-dihydroxy-4-pregnen-3-one, on behavioral spawning success and sperm quality and quantity in male goldfish. *Anim. Behav.* submitted.
10. Dmitrieva, T.M. and V.A. Ostroumov. The role of chemical communication in the organization of spawning behavior of the yellowfin Baikal sculpin (*Cottocomephorus grewingki*). *Biol. Nauki* 10: 38–42, 1986.
11. Dmitrieva, T.M., P.L. Katsel, R.B. Valeyev, V.A. Ostroumov and U.P. Kozlov. The isolation of the sexual pheromone of the male yellowfin Baikal sculpin (*Cottocomephorus grewingki*). *Biol. Nauki* 6: 39–44, 1988.
12. Doving, K.B. Evolutionary trends in olfaction. In: *The Structure-Activity Relationships in Chemoreception*, edited by G. Benz. London: IRL Press, 1976, pp. 149–159.
13. Drickhammer, L.C. Pheromones: behavioral and biochemical aspects. In: *Advances in Comparative and Environmental Physiology, Vol. III*, edited by J. Balthazart. New York: Springer-Verlag, 1989, pp. 270–356.
14. Dulka, J.G. and L.S. Demski. Sperm duct contractions mediate centrally evoked sperm release in goldfish. *J. Exp. Zool.* 237: 271–279, 1986.
15. Dulka, J.G., N.E. Stacey, P.W. Sorensen and G.J. Van Der Kraak. A sex steroid pheromone synchronizes male-female spawning readiness in the goldfish. *Nature* 325: 251–253, 1987.
16. Dunham, P.J. Sex pheromones in Crustacea. *Biol. Rev.* 53: 555–583, 1978.
17. Fostier, A., F. Le Gac and M. Loir. Steroids in male reproduction. In: *Proceedings of the Third International Symposium on the Reproductive Physiology of Fish*, edited by D.R. Idler, L.W. Crim and J.M. Walsh. St. John's, Canada: Memorial University Press, 1987, pp. 239–245.
18. Gleeson, R.A., M.A. Adams and A.B. Smith. Characterization of a sex pheromone in the blue crab, *Callinectes sapidus*: Crustecdysone studies. *J. Chem. Ecol.* 10: 913–921, 1984.
19. Goetz, F.W. Hormonal control of oocyte final maturation and ovulation in fishes. In: *Fish Physiology, Vol. IXB*, edited by W.S. Hoar, D.J. Randall and E.M. Donaldson. New York: Academic Press, 1983, pp. 117–170.
20. Hara, T.J. Role of olfaction in fish behavior. In: *Behavior of Teleost Fishes*, edited by T.J. Pitcher. London: Croom Helm, 1986, pp. 152–176.
21. Karlson, P. and M. Lüscher. "Pheromones": a new term for a class of biologically active substances. *Nature* 183: 55–56, 1959.
22. Kittredge, J.S. and F.T. Takahashi. The evolution of sex pheromone communication in the Arthropoda. *J. Theor. Biol.* 35: 467–471, 1972.
23. Kittredge, J.S., M. Terry and F.T. Takahashi. Sex pheromone activity of the moulting hormone, crustecdysone, on male crabs (*Pachygrapsus crassipes, Cancer antennarius,* and *C. anthonyi*). *Fish.Bull.* 69: 337–343, 1971.
24. Kobayashi, M., K. Aida and I. Hanyu. Gonadotropin surge during spawning in male goldfish. *Gen. Comp. Endocrinol.* 62: 70–79, 1986.

25. Kobayashi, M., K. Aida and I. Hanyu. Pheromone from ovulatory female goldfish induces gonadotropin surge in males. *Gen. Comp. Endocrinol.* 62: 70–79, 1986.
26. Kobayashi, M., K. Aida and I. Hanyu. Effects of hCG on milt amount and plasma levels of steroid hormones in male goldfish. *Bull. Jap. Soc. Sci. Fish.* 52: 755, 1986.
27. Kobayashi, M., K. Aida and I. Hanyu. Hormone changes during the ovulatory cycle in goldfish. *Gen. Comp. Endocrinol.* 69: 301–307, 1988.
28. Kyle, A.L., N.E. Stacey, R.E. Peter and R. Billard. Elevations in gonadotropin concentrations and milt volumes as a result of spawning in the goldfish. *Gen. Comp. Endocrinol.* 57: 10–22, 1985.
29. Lambert, J.G.D., R. Van Den Hurk, W.G.E.J. Schoonen, J.W. Resink and P.G.W.J. Van Oordt. Gonadal steroidogenesis and the possible role of steroid glucuronides as sex pheromones in two species of teleosts. *Fish Physiol. Biochem.* 2: 101–107, 1986.
30. Liley, N.R. Chemical communication in fish. *Can. J. Fish. Aquatic Sci.* 39: 22–35, 1982.
31. Maneckjee, A., M. Weisbart and D.R. Idler. The presence of $17\alpha,20\beta$-dihydroxy-4-pregnen-3-one receptor activity in the ovary of the brook trout, *Salvelinus fontinalis*, during terminal stages of oocyte maturation. *Fish. Physiol. Biochem.* 6: 19–38, 1989.
32. McKinnon, J.S. and N.R. Liley. Assymetric species specificity in response to female pheromone by males of two species of *Trichogaster* (Pisces: Belontidae). *Can. J. Zool.* 65: 1129–1134, 1987.
33. Melrose, D.R., H.C.B. Reed and R.L.S. Patterson. Androgen steroids associated with boar odour as an aid to detection of oestrous in pig artificial insemination. *Br. Vet. J.* 127:497–502, 1971.
34. Miller, P.J. The tokology of Gobiid fishes. In: *Fish Reproduction: Strategies and Tactics,* edited by G.W. Potts and R.J. Wootton. London: Academic Press, 1984, pp. 119–153.
35. Nakatsuru, K. and D.L. Kramer. Is sperm cheap? Limited male fertility and female choice in the lemon tetra (Pisces, Characidae). *Science* 216: 753–754, 1982.
36. Novotny, M., B. Jemiolo, S. Harvey, D. Weisler and A. Marchlewska-Koj. Adrenal mediated androgenous metabolites inhibit puberty in mice. *Science* 231: 722–725, 1986.
37. Ottoson, D. The electro-olfactogram. In: *Handbook of Sensory Physiology*, edited by L.M. Biedler. New York: Springer-Verlag, 1971, pp. 95–131.
38. Partridge, B.L., N.R. Liley and N.E. Stacey. The role of pheromones in the sexual behavior of the goldfish. *Anim. Behav.* 24: 291–299, 1976.
39. Rehnberg, B.G. and C.B. Schreck. The olfactory L-serine receptor in coho salmon: biochemical specificity and behavioral response. *J. Comp. Physiol. A* 159: 61–67, 1986.
40. Resink, J.W., R. Van Den Hurk, R.F.O. Groeninx Van Zoelen and E.A. Huisman. The seminal vesicle as source of sex attracting substances in the African catfish, *Clarias gariepinus. Aquaculture* 63: 115–127, 1987.
41. Resink, J.W., W.G.E.J. Schoonen, P.C.H. Albers, D.M. File, C.D. Notenboom, R. Van Den Hurk and P.G.W.J. Van Oordt. The chemical nature of sex attracting pheromones from the seminal vesicle of the African catfish, *Clarias gariepinus. Aquaculture* 83: 137–151, 1989.
42. Resink, J.W., T.W.M. Van Den Berg, R. Van Den Hurk, E.A. Huisman and P.G.W.J. Van Oordt. Induction of gonadotropin release and ovulation by pheromones in the African catfish, *Clarias gariepinus. Aquaculture* 83: 167–177, 1989.
43. Resink, J.W., P.K. Voorthuis, R. Van Den Hurk, R.C. Peters and P.G.W.J. Van Oordt. Steroid glucuronides of the seminal vesicle as olfactory stimuli in African catfish, *Clarias gariepinus. Aquaculture* 83: 153–166, 1989.
44. Resink, J.W., P.K. Voorthuis, R. Van Den Hurk, R.C. Peters and P.G.W.J. Van Oordt. Pheromone detection and olfactory pathways in the brain of the African catfish, *Clarias gariepinus. Cell Tissue Res.* 256: 337–345, 1989.
45. Rosenblum, P.M., P.W. Sorensen, N.E. Stacey and R.E. Peter. Binding of the steroidal pheromone $17\alpha,20\beta$-dihydroxy-4-pregnen-3-one to goldfish, *Carassius auratus,* olfactory epithelium membrane preparations. *Chem. Senses* 16: 143–154, 1991.
46. Schoonen, W.G.E.J., J.G.D. Lambert and P.G.W.J. Van Oordt. Quantitative analysis of steroids and steroid glucuronides in the seminal vesicle fluid of feral spawning and feral and cultivated non-spawning African catfish, *Clarias gariepinus. Gen. Comp. Endocrinol.* 70: 91–100, 1988.
47. Schwartz, F.J. World literature to fish hybrids with an analysis by family, species, and hybrid. *Suppl. 1, NOAA Tech. Rept.* NMFS-SSRF-750, 507 pp., 1981.
48. Sorensen, P.W. Hormones, pheromones and chemoreception. In: *Fish Chemoreception*, edited by T.J. Hara. London: Croom Helm, in press.
49. Sorensen, P.W., N.E. Stacey and P. Naidu. Release of spawning pheromone(s) by naturally-ovulated and prostaglandin-injected, non-ovulated female goldfish. In: *Chemical Signals in Vertebrates, Vol. IV,*, edited by D. Duvall, D. Muller-Schwarze, and R.M. Silverstein. New York: Plenum Press, 1986, pp. 149–154.

50. Sorensen, P.W. and N.E. Stacey. 17α,20β-dihydroxy-4-pregnen-3-one functions as a bisexual priming pheromone in goldfish. *Am. Zool.* 27: abstr. 412, 1987.
51. Sorensen, P.W., T.J. Hara and N.E. Stacey. Extreme olfactory sensitivity of mature and gonadally-regressed goldfish to a potent steroidal pheromone, 17α,20β-dihydroxy-4-pregnen-3-one. *J. Comp. Physiol. A* 160: 305–313, 1987.
52. Sorensen, P.W., K.J. Chamberlain, N.E. Stacey and J.G. Dulka. Differing roles of prostaglandin $F_{2\alpha}$ and its metabolites in goldfish reproductive behavior. In: *Proceedings of the Third International Symposium on the Reproductive Physiology of Fish*, edited by D.R. Idler, L.W. Crim and J.M. Walsh. St. John's, Canada: Memorial University Press, 1987, p. 164.
53. Sorensen, P.W., T.J. Hara, N.E. Stacey and F.W. Goetz. F prostaglandins function as potent olfactory stimulants that comprise the postovulatory female sex pheromone in goldfish. *Biol. Reprod.* 39: 1039–1050, 1988.
54. Sorensen, P.W., N.E. Stacey and K.J. Chamberlain. Differing behavioral and endocrinological effects of two female sex pheromones on male goldfish. *Horm. Behav.* 23: 317–332, 1989
55. Sorensen, P.W. and N.E. Stacey. Identified hormonal pheromones in the goldfish: the basis for a model of sex pheromone function in teleost fish. In: *Chemical Signals in Vertebrates, Vol. V.*, edited by D. MacDonald, D. Muller-Schwarze and S.E. Natynczuk. Oxford: Oxford University Press, 1990, pp. 302–311.
56. Sorensen, P.W., T.J. Hara, N.E. Stacey and J.G. Dulka. Extreme olfactory specificity of male goldfish to the preovulatory steroidal pheromone 17α,20β-dihydroxy-4-pregnen-3-one. *J. Comp. Physiol. A* 166: 373–383, 1990.
57. Stacey, N.E. Hormonal regulation of female reproductive behavior in fish. *Am. Zool.* 21: 305–316, 1981.
58. Stacey, N.E. Roles of hormones and pheromones in fish reproductive behavior. In: *Psychobiology of Reproduction: an Evolutionary Perspective*, edited by D. Crews. Englewood Cliffs, NJ: Prentice-Hall, 1987, pp. 28–69.
59. Stacey, N.E. and F.W. Goetz. Role of prostaglandins in fish reproduction. *Can. J. Fish. Aquatic Sci.* 39: 92–98, 1982.
60. Stacey, N.E. and A.S. Hourston. Spawning and feeding behavior of captive Pacific herring, *Clupea harengus pallasi. Can. J. Fish. Aquatic Sci.* 39: 489–498, 1982.
61. Stacey, N.E. and A.L. Kyle. Effects of olfactory tract lesions on sexual and feeding behavior in the goldfish. *Physiol. Behav.* 30: 621–628, 1983.
62. Stacey, N.E. and P.W. Sorensen. 17α,20β-dihydroxy-4-pregnen-3-one: a steroidal primer pheromone which increases milt volume in the goldfish, *Carassius auratus. Can. J. Zool.* 64: 2412–2417, 1986.
63. Stacey, N.E., A.L. Kyle and N.R. Liley. Fish reproductive pheromones. In: *Chemical Signals in Vertebrates, Vol. IV,* edited by D. Duvall, D. Muller-Schwarze and R.M. Silverstein. New York: Plenum Press, 1986, pp. 117–133.
64. Stacey, N.E., P.W. Sorensen, J.G. Dulka, G.J. Van Der Kraak and T.J. Hara. Teleost sex pheromones: recent studies on identity and function. In: *Proceedings of the the Third International Symposium on the Reproductive Physiology of Fish*, edited by D.R. Idler, L.W. Crim and J.M. Walsh. St. John's, Canada: Memorial University Press, 1987, pp. 150–154.
65. Stacey, N.E., K.J. Chamberlain, P.W. Sorensen and J.G. Dulka. Milt volume increase in goldfish: interaction of pheromonal and behavioral stimuli. In: *Proceedings of the Third International Symposium on the Reproductive Physiology of Fish*, edited by D.R. Idler, L.W. Crim and J.M. Walsh. St. John's, Canada: Memorial University Press, 1987, p. 165.
66. Stacey, N.E., P.W. Sorensen, J.G. Dulka and G.J. Van Der Kraak. Direct evidence that 17α,20β-dihydroxy-4-pregnen-3-one functions as a goldfish primer pheromone: preovulatory release is closely associated with male endocrine responses. *Gen. Comp. Endocrinol.* 75: 62–70, 1989.
67. Stacey, N.E., P.W. Sorensen, J.G. Dulka, J.R. Cardwell and A.S. Irvine. Fish sex pheromones: current status and potential applications. *Bull. Inst. Zool.*, Academia Sinica. in press.
68. Tavolga, W.N. The effects of gonadectomy and hypophysectomy on the pre-spawning behavior in males of the gobiid fish, *Bathygobius soporator. Physiol. Zool.* 28: 218–233, 1955.
69. Tavolga, W.N. Visual, chemical and sound stimuli as cues in the sex discriminatory behavior of the gobiid fish, *Bathygobius soporator. Zoologica* 41: 49–65, 1956.
70. Ueda, H., A. Kambegawa and Y. Nagahama. Involvement of gonadotropin and steroid hormones in spermiation in the amago salmon, *Oncorhynchus rhodurus*, and goldfish, *Carassius auratus. Gen. Comp. Endocrinol.* 59: 24–30, 1985.
71. Van Den Hurk, R. and J.G.D. Lambert. Ovarian steroid glucuronides function as sex pheromones for male zebrafish, *Brachydanio rerio. Can. J. Zool.* 61: 2381–2387, 1983.
72. Van Den Hurk, R., W.G.E.J. Schoonen, G.A. Van Zoelen and J.G.D. Lambert. The biosynthesis of

steroid glucuronides in the testis of the zebrafish, *Brachydanio rerio*, and their pheromonal function as ovulation inducers. *Gen. Comp. Endocrinol.* 68: 179–188, 1987.
73. Van Den Hurk, R., G.A. Van Zoelen, W.G.E.J. Schoonen, J.W. Resink, J.G.D. Lambert and P.G.W.J. Van Oordt. Do testicular steroidglucuronides of zebrafish, *Brachydanio rerio*, evoke ovulation in female conspecifics? *Gen. Comp. Endocrinol.* 66: 19, 1987.
74. Van Den Hurk, R., J.W. Resink and J. Peute. The seminal vesicle of the African catfish, *Clarias gariepinus*. A histological, histochemical, enzyme-histochemical, ultrastructural and physiological study. *Cell Tissue Res.* 247: 573–582, 1987.
75. Van Der Kraak, G.J., P.W. Sorensen, N.E. Stacey and J.G. Dulka. Periovulatory goldfish release three potential pheromones: $17\alpha,20\beta$-dihydroxyprogesterone, $17\alpha,20\beta$-dihydroxyprogesterone glucuronide, and 17α-hydroxyprogesterone. *Gen. Comp. Endocrinol.* 73: 452–457, 1989.
76. Van Oordt, P.G.W.J. and H.J.Th. Goos. The African catfish, *Clarias gariepinus*: a model for the study of reproductive endocrinology in teleosts. *Aquaculture* 63: 15–26, 1987.
77. Wilson, E.O. *Sociobiology: The New Synthesis*. Cambridge, MA: Harvard University Press, 1975, 697 p.
78. Wilson, E.O. and W.H. Bossert. Chemical communication among animals. *Rec. Prog. Horm. Res.* 19: 673–716, 1975.

CHAPTER 6

Urea synthesis in fishes: evolutionary and biochemical perspectives

THOMAS P. MOMMSEN * AND PATRICK J. WALSH **

*Department of Biochemistry and Microbiology, University of Victoria, Victoria, B.C., Canada V8W 3P6, and **Rosenstiel School of Marine and Atmospheric Science, University of Miami, Miami, FL 33149, U.S.A.*

I. Introduction
II. The ornithine–urea cycle
III. Urea in fishes
IV. Evolutionary perspectives
V. Intracellular and intercellular transport phenomena
   1. Urea, arginine, ornithine and citrulline
   2. Glutamine and Ammonia
   3. Alanine
   4. Aspartate
   5. Carbon dioxide
VI. Detoxification
VII. Zonation
VIII. Regulation
   1. Hormonal control
   2. Metabolic control
   3. Feedback inhibition?
Acknowledgements
IX. References

I. Introduction

Metabolic turnover of nitrogenous substances, such as amino acids, proteins, nucleotides and related compounds, leads to the necessity for animals to rid the body of the resulting end products. Since the immediate end product is usually ammonia, which is highly toxic to all vertebrates, mechanisms are invoked to convert ammonia to innocuous nitrogenous compounds such as urea or uric acid. The ability to detoxify ammonia is especially important under situations of limited water availability, and thus it is usually considered a prerequisite to terrestrial vertebrate life. However, while the direct excretion of ammonia is straightforward for animals living in an aqueous environment and involves little metabolic expenditure (28,90,113), the synthesis of urea or uric acid necessitates considerable metabolic cost. The synthesis of urea by the ornithine–urea cycle, for instance, requires the input of at least 2 ATPs for each ammonia nitrogen assimilated. The relative advantages or disadvantages of one product over the other two have been documented numerous times[45], but the ultimate choice of end product or end

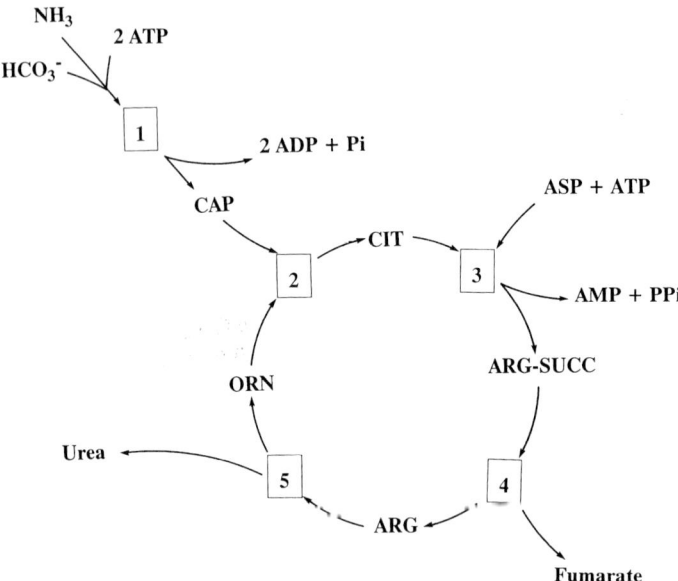

Fig. 1. Urea synthetic pathway in mammalian liver with ammonia as substrate. CAP = carbamoylphosphate; CIT = citrulline; ASP = aspartate; ARG-SUCC = argininosuccinate; ARG = arginine. Enzymes: (1) carbamoylphosphate synthetase I (ammonia and N-acetylglutamate-dependent); (2) ornithine carbamoyl transferase; (3) argininosuccinate synthetase; (4) argininosuccinate lyase; (5) arginase. For subcellular compartmentation of enzymes, cf. Fig. 4.

products is somewhat dependent on different parameters, including diet, specific metabolic situations, water availability, stage in the lifecycle or genetic capacity.

II. The ornithine-urea cycle

In mammals, the liver is the main site for urea synthesis with extra-hepatic tissues generally serving as the principal sources of nitrogenous compounds for hepatic urea synthesis. Both alanine and glutamine formed in extrahepatic locations are considered the two most important intertissue vehicles for [nitrogen] transport, assuming pivotal roles in the alanine-glucose cycle and the intertissue glutamine cycle, respectively. Urea synthesis in the mammalian liver proceeds by the so-called ornithine–urea cycle, first deduced in part by Krebs and Henseleit[55], involving the following five enzymes (Fig. 1): (1) carbamoylphosphate synthetase I (CPS); (2) ornithine carbamoyl transferase; (3) argininosuccinate synthetase; (4) argininosuccinate lyase; and (5) arginase.

Since the two entry points for nitrogen into the cycle, i.e. CPS I and argininosuccinate synthetase, utilize ammonia and aspartate, respectively, a number of additional reactions are required for the cycle to exploit alanine or glutamine as nitrogen sources. Transamination of alanine into aspartate, for example, requires the concurrent availability of alanine and aspartate aminotransferases, while the production of ammonia from alanine for the synthesis of carbamoyl phosphate necessitates the presence of alanine aminotransferase and glutamate dehydrogenase. Glutaminase, a mitochondrial enzyme in hepatic tissue, is a prerequisite for the use of glutamine as ammonia donor. Other ancillary reactions may play critical roles in the regulation of the cycle. These include carbonic anhydrase, which may be essential to furnish bicarbonate from carbon dioxide for CPS and N-acetylglutamate synthase, the enzyme responsible for synthesis of N-acetylglutamate, an essential allosteric modulator for CPS I catalysis. At times, ornithine aminotransferase may be instrumental in controlling the availability of ornithine[61] for the first step of the cycle *per se*.

The ornithine–urea cycle of the mammalian liver is a compartmentalized pathway: CPS I and ornithine carbamoyl transferase are localized in the mitochondrial matrix, while the synthesis and degradation of argininosuccinate, and the ensuing enzymatic hydrolysis of arginine are catalyzed by cytosolic enzymes. However, the last three enzymes are not randomly distributed over the cytosolic compartment. Instead, they appear to be tightly associated with the outer mitochondrial membrane in such a fashion that neither argininosuccinate nor arginine exchange freely with their cytosolic pools not actively involved in urea synthesis[16].

Carbamoyl phosphate can also be synthesized by a glutamine-dependent CPS II. However, although related to CPS I[25], CPS II is not dependent on N-acetyl-L-glutamate and is exclusively localized in the cytosol where it forms part of a multifunctional protein responsible for the first three steps in the synthesis of pyrimidines.

III. Urea in fishes

With few notable exceptions, freshwater fish and marine teleosts excrete predominantly ammonia/ammonium *via* the gills by diffusion of NH_3 or NH_4^+ or exchange of ammonium ions for Na^+ (ref. 28). Although gill tissue exerts an extremely high metabolic rate accounting for almost 10% of the entire oxygen demand of teleosts for osmoregulatory purposes[68,87], the overall metabolic expenditures for the release of ammonia and ammonium ions appear to be minimal[30]. In addition to ammonia and a few minor nitrogenous compounds such as creatinine, uric acid or allantoic acid, these fish void a small, albeit highly variable, amount of urea, usually ranging around 20% of total nitrogen[9]. Since, at least initially, these fish were also thought to be devoid of a functional ornithine–urea cycle and thus *de novo* synthesis of urea was ruled out, other sources for excretory urea were proposed. Biochemical pathways leading to a potential supply of urea for excretion are the degradation of purines *via* uric acid (Fig. 2) and the hydrolysis of arginine. Both enzyme systems required for these processes are active in teleost livers[21,22,44,100,101,111], with key enzymes of purine degradation localized in peroxisomes[33,44]. The wide-spread occurrence of arginase points towards dietary arginine as a prime source for excretory urea observed in fish species[37]. However, arginase appears to scale somewhat with the degree of air-breathing in teleosts, and, indeed, highest levels

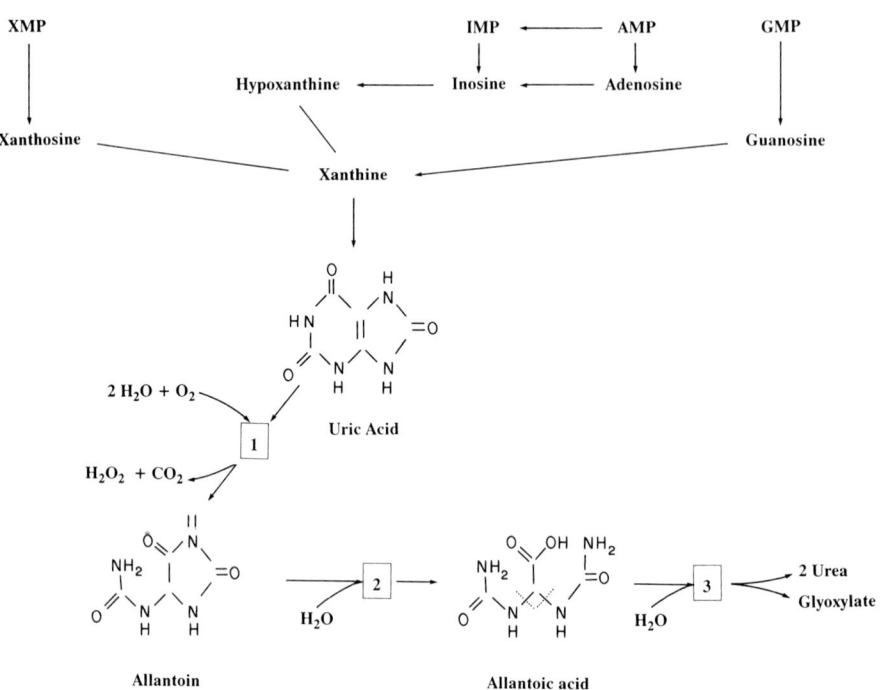

Fig. 2. Urea production through degradation of purine nucleotides in vertebrate hepatic tissue. Key enzymes: (1) urate oxidase; (2) allantoinase; (3) allantoicase.

of arginase are found in species which also contain the full complement of urea cycle enzymes[97,100]. It is a general observation that the teleostean gill is highly permeable to urea[60] and, as a consequence, endogenously generated urea or experimentally administered urea are quickly lost across the gills into the surrounding water.

Although it seemed from previous evidence on the absence of one or more ornithine–urea cycle enzymes in fish liver that the appropriate genes were deleted[8], other data supported the notion that the genes were present. Nevertheless, the degree of expression of ornithine–urea cycle enzyme genes and enzyme concentrations maintained were minute[23,46]. In agreement with the latter interpretation are data from radiotracer studies on the rainbow trout (*Oncorhynchus mykiss*, formerly *Salmo gairdneri*) and the guppy (*Poecilia reticulata*). These two teleostean species incorporate labelled bicarbonate into urea[17,24], but actual rates are about two orders of magnitude below those observed for elasmobranchs[74], and thus almost three orders of magnitude below mammalian rates[51]. Further, rainbow trout failed to significantly bolster their excretion of urea in response to increasing environmental concentrations of ammonia[83]. The recognition that arginine constitutes an essential amino acid for teleostean fishes[107] is a further indirect confirmation that the endogenous rate of arginine synthesis from ornithine via the urea cycle is clearly insufficient to supply the animals' needs.

For a number of years, the description of high urea cycle enzyme activity in the liver of the teleost oyster toadfish[92] had remained a deviation from the accepted picture outlined above. Recently, an extensive reassessment of urea cycle activity in teleostean fishes has made it abundantly clear that the toadfish is not a captivating oddity, but an example that *de novo* synthesis of urea by the ornithine–urea cycle may indeed occur in teleostean fishes at substantial rates. In the course of the last few years, Read's original observation on the toadfish (*Opsanus tau*[92])) has been confirmed[75] and expanded to a congener (*O. beta*[75,114]), a tilapia (*Oreochromis alcalicus grahami*) living in an extremely alkaline environment[89] and several air-breathing teleosts from the Indian subcontinent, namely *Heteropneustes fossilis*[96], *Clarias batrachus*, *Anabas testudineus* and *Amphipnous cuchia*[97]. Enzyme titers in all of these species are considerably higher than those for wholly aquatic teleosts living near neutral pHs[2,8,42,89,118]. While these enzyme data supply evidence for the *potential* for urea synthesis, additional corroborating documentation for hepatic urea synthesis in teleosts comes from the measurements of generally high proportions and rates of urea excretion by these species. In fact, when exposed to experimentally elevated concentrations of ammonia, the Gulf toadfish (*O. beta*) and an air-breathing catfish (*H. fossilis*) significantly increase the proportion of urea in their nitrogenous excretory products[95,112]. The extreme case appears to be the alkaline lake tilapia which excretes urea exclusively under the specific alkaline conditions of its environment (pH > 9.5) which preclude the excretion of ammonia. However, to date the only *direct* evidence for hepatic urea synthesis in teleosts has been supplied by studies on isolated hepatocytes of the two toadfish species[75,114]. When supplied with appropriate sources of nitrogen and carbon, such as aspartate, glutamine, ammonia or alanine and bicarbonate, isolated liver cells of *O. tau* and *O. beta* synthesize urea and release it into the surrounding medium at higher rates than hepatocytes obtained from a ureogenic marine skate (*Raja erinacea*)[74]. Rates for the warm-water Gulf toadfish measured at 24°C were noticed to fluctuate considerably with the circannual cycle or the subpopulation sampled[75,112,114], and can reach up to one third of the rate routinely found in rat hepatocytes[51].

For a long time it has been known that an entirely different relationship between ammonia and urea exists in marine elasmobranchii, such as sharks, rays and skates, the holocephalans (chimaeras) and the coelacanth (*Latimeria*). These fishes, whose *milieu intérieur* is isotonic or slightly hypertonic to seawater, are distinguished by accumulating high levels of urea (up to 400 mM) in their blood and tissues. This urea, which serves as the animals' most abundant non-ionic osmolyte, is largely derived from endogenous sources through *de novo* synthesis by the ornithine–urea cycle in liver. On the one hand, elasmobranchs and related fishes — for obvious reasons no data are as yet available for *Latimeria* — employ ammonia and not urea as their primary vehicle for nitrogen transport and excretion[56,86]. While the primary data on this relationship are of a rather indirect nature, insignificant levels of ornithine–urea cycle enzymes together with the abandonment of urea as an osmolyte in freshwater elasmobranchs[36,38] supports this notion. Thus these fishes expend little energy to void nitrogen. Instead, they undergo considerable metabolic expense to synthesize urea at the hepatic level for the purpose of osmoregulation by somewhat indirect means. Osmoregulatory emphasis in teleostean fishes is put on the gill, with its substantial endogenous metabolic rate, which is the main site for direct osmoregulation. On the other hand, all marine elasmobranch fishes mentioned above, as well as the coelacanth, display substantial activities of the entire set of ornithine–urea cycle enzymes[11,13,38–40,46,91]. Actual rates of urea synthesis, however, are relatively modest when contrasted with rates achieved by mammals[38,73]. It appears that the hepatic capacity to synthesize urea in marine elasmobranchs and by association the coelacanth, must merely match routine losses in body urea. Such losses are incurred through passive leakage of urea at gill, and body surface, or through active and passive processes in kidney[38,39]. Other routes for the potential loss of body urea are its destruction by endogenous microbial processes in tissues or in the gastrointestinal tract[52]. The permeability for urea in the elasmobranch gill is extremely low, compared with the 'urea-leaky' gill tissue of teleostean fishes.

Although the urea synthetic capability of elasmobranch fishes was documented a long time ago[102], it is only comparatively recently that the biochemical characteristics of their hepatic urea cycle were scrutinized. While ornithine carbamoylphosphate transferase, argininosuccinate synthetase, and argininosuccinate lyase show little, if any, difference to their counterparts in mammalian liver, important difference are apparent for CPS and arginase. Shark, skate and ratfish CPSs all use glutamine and not ammonia as the nitrogen donor. However, just as mammalian CPS I, the so-called CPS III of sharks is an enzyme of the mitochondrial matrix and dependent on the presence of N-acetyl-L-glutamate as a modulator[2]. Thus, the enzyme is clearly distinct from CPS II, the glutamine-dependent enzyme of the orotate pathway located in the cytosol of all vertebrate animals. Arginase, the enzyme actually liberating urea, is localized in the mitochondrial compartment in shark, skate, and chimaera, while the enzyme is cytosolic in mammalian liver.

With regard to urea synthesis, a special situation is evident for the lungfishes. While the inveterately aquatic lungfish *Neoceratodus forsteri* and the non-estivating *Lepidosiren paradoxa* contain only minute activities of ornithine–urea cycle enzymes in their livers[34,41], their estivating African equivalents *Protopterus aethiopicus* and *P. annectens* display considerable hepatic activities of all ornithine–urea cycle enzymes[48,75].

IV. Evolutionary perspectives

Interestingly, the pathway for urea synthesis in teleostean fishes appears to be identical to the elasmobranch situation and thus quite different from the pathway used in mammals. Just as in the elasmobranch fishes, teleost CPS is characterized by a preference for glutamine as nitrogen donor over ammonia and dependence on N-acetyl-L-glutamate and thus has been classified as a CPS III. Ironically, among the vertebrates, CPS III was first described for *Micropterus salmoides*, an ammoniotelic teleost[1], before discovery of its widespread occurrence in elasmobranchs[2]

TABLE 1

Characteristics of arginase and mitochondrial carbamoylphosphate synthetase (CPS) and distribution of urea synthetic activity in liver of 'fishes' and tetrapods

	Species analyzed	CPS	Arginase	Urea synthesis
Hagfish	(*Eptatretus stouti*)		Mitochondrial	No
Lampreys				No
Shark	(*Squalus acanthias*)	III	Mitochondrial	Yes
Skate	(*Raja erinacea*)	III	Mitochondrial	Yes
Chimaera	(*Hydrolagus colliei*)	III	Mitochondrial	Yes
Bichir	(*Polypterus sp.*)	III	Mitochondrial	No
Sturgeon	(*Scaphirhynchus platorhynchus*)			No
Paddlefish	(*Polyodon spathula*)			No
Gar	(*Lepisosteus platostomus*)			No
Bowfin	(*Amia calva*)	III	Mitochondrial	No
Rainbow Trout	(*Oncorhynchus mykiss*)	III	Mitochondrial	No
Carp	(*Cyprinus carpio*)	III	Mitochondrial	No
Bass	(*Micropterus salmoides*)	III		No
Midshipman	(*Porichthys notatus*)	III		No
Gulf Toadfish	(*Opsanus beta*)	III	Mitochondrial	Yes
Oyster Toadfish	(*Opsanus tau*)	III	Mitochondrial	Yes
Coelacanth	(*Latimeria chalumnae*)	III	Mitochondrial	Yes
Lungfish	(*Protopterus aethiopicus*)	I	Cytosolic	Yes
Lungfish	(*Protopterus annectens*)	I	Cytosolic	Yes
Tetrapods		I	Cytosolic	Yes

CPS III = carbamoylphosphate synthetase III (glutamine-dependent); CPS I = carbamoylphosphate synthetase I (ammonia-dependent). For individual references, see ref. 75. Livers of additional species of teleostean fishes show urea synthetic activity, but information about specific characteristics of CPS or arginase is not available (see text).

and ureogenic teleosts[75,97]. Glutamine-dependent synthesis of carbamoyl phosphate in elasmobranch liver had been described earlier[116], but dependence on N-acetyl-L-glutamate had not been analyzed and thus the crucial distinction between CPS II and CPS III could not be made. The other important difference between teleosts and mammals is the subcellular localization of arginase. In all teleosts analyzed, just as in the elasmobranchs, arginase clearly behaves like an enzyme associated with the mitochondrial matrix[75]. In contrast, the enzyme of mammalian liver is localized in the cytosolic compartment, albeit associated with the outer mitochondrial membrane due to its functional role of channelling ornithine[16].

When tabulating the specific properties of CPS and arginase in a plethora of fishes spanning a wide area of fish systematics, an interesting pattern was recognized (Table 1). This pattern leads to a number of important conclusions, not only about the evolution of urea synthesis in the vertebrate line *per se* but also about possible systematic relationships among different groups of fishes. Clearly, as Table 1 suggests, the synthesis of urea appears to by a monophyletic trait within the vertebrates. According to the data compiled in Table 1, only two key changes have to be postulated in the course of the evolution of urea synthesis in the vertebrates. As the table implies, these changes were probably not temporally

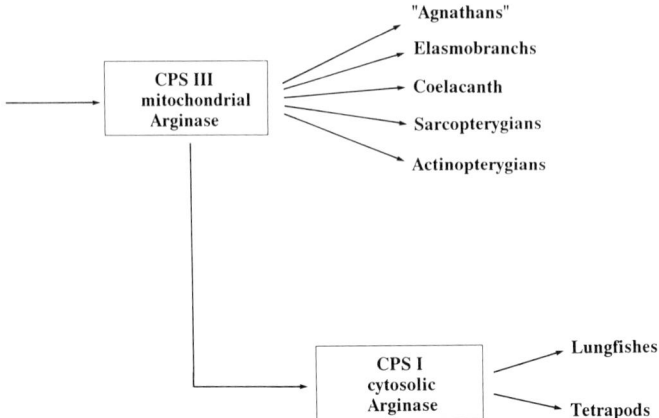

Fig. 3. Putative evolutionary trends in hepatic carbamoylphosphate synthetase and arginase in vertebrates. CPS III = carbamoylphosphate synthetase III (glutamine-dependent); CPS I = carbamoylphosphate synthetase I (ammonia-dependent).

independent of each other and occurred before the evolution of the extant lungfishes (Fig. 3). One of these changes is the replacement of mitochondrial CPS III with ammonia-dependent CPS I, which biochemically represents only a minor switch. First, these two enzymes are immunologically very closely related[25]. Second, a single point mutation in the glutamine-binding region of CPS III, leading to the replacement of a cysteine with a serine, may be at the root of the enzyme losing its glutamine-binding capability and acquiring a decreased Michaelis-Menten constant for ammonia (CPS I)[81,82,93,94]. A more detailed description of the putative evolution and relationships between all CPSs is presented in another chapter[9]. The other biochemical alteration indispensable to the conclusion from Table 1 is the replacement of a mitochondrial arginase in liver with a cytosolic enzyme. Again, the molecular events required are minor, merely involving the loss of the leader sequence responsible for the post-synthesis transport of this protein into the mitochondria[9].

A key role in this analysis is played by two groups of fishes whose relative systematic positions has been under debate for a long time, namely the coelacanth, the only living representative of the crossopterygians, and the lungfishes. As mentioned above, both groups are potentially ureogenic and contain all hepatic enzymes required for *de novo* synthesis of urea. But this is where the biochemical similarities end. The coelacanth, which utilizes urea as an osmolyte, is not biochemically distinct from the other fishes covered in this chapter, since it, too, possesses CPS III and an arginase which behaves like glutamate dehydrogenase or citrate synthase (Table 1), two commonly used marker enzymes for the mitochondrial matrix. The lungfishes, on the other hand, are characterized by CPS I and a cytosolic arginase. Both of these traits are familiar from the mammalian-type ornithine–urea cycle. Therefore, when only these two biochemical traits are employed, it appears that the lungfishes lead the way toward the evolution of the other vertebrate groups (Fig. 3). A more recent analysis of the relatedness of mitochondrial DNA in the coelacanth, lungfish and tetrapods[66] unequivocally supports our conclusions drawn from the analysis of urea cycle characteristics. While our analysis decidedly rules out the positioning of the coelacanth in such a pivotal role, it does not supply any detailed information about the exact relationship of the coelacanth within the fishes, and other parameters have to be considered in its classification[31,66,80].

With the present state of knowledge about the urea synthetic machinery in vertebrate fishes, a rather cogent picture is emerging. First, all fishes appear to contain the genes for all ornithine–urea cycle enzymes, but the degree of expression varies, probably depending largely on the physiological circumstance or stage during the individual life cycle[24]. Molecular biology techniques will furnish information of whether the genes are merely silenced or actually lost in some species of fishes. Existing data on enzyme presence, but lack of physiological role, in a number of teleosts suggest that the fish genome still contains the entire set of genes for a functional ornithine–urea cycle but the reason why these genes were not lost in the course of the evolution of at least some of our aquatic friends are not self-evident. Both CPS I and CPS III, for instance, are relatively inefficient enzymes with specific activities of about 3.5 μmol of carbamoyl phosphate produced per min (at 37°C) per mg of highly purified CPS I[59] and one tenth of this value or less for elasmobranch CPS III (at 26°C)[3] and teleost CPS III[12], respectively. Hepatic pyruvate kinase, for comparison, catalyzes specific turnover rates of some 520 μmol of product per min per mg of enzyme protein, while the real champions among the enzymes, such as catalase or carbonic anhydrase, are more than 100 times faster than pyruvate kinase and may only be limited by the rate by which substrate is available to them. Therefore, to achieve the high rates of urea synthesis realized in mammalian liver, CPS I may account for some 20% of the entire mitochondrial protein[18] at an enzyme concentration approaching 1.5 mM![59].

It is possible that in spite of the high metabolic cost of *de novo* urea formation, the complete set of urea cycle genes has been conserved in teleosts, because, as some data on juvenile teleosts suggest[24], urea synthesis may play an important, as yet unappreciated, role during the ontogenetic development of teleosts. There is some indication from viviparous teleosts that the ovarian fluid may contain appreciable amounts of urea[26] which appears to be contributed by the embryo. Although it seems most likely that high turnover of nucleotides by the developing embryo, and thus a catabolic process (cf. Fig. 2), is the source of urea found in the ovary, comparative data on the developing, ureogenic dogfish pup suggest a route relevant to the present discussion. Since the pup is being bathed in high concentrations of ammonia (up to 22 mM) in the uterine fluid, contributed by the maternal system[53], ammonia detoxification appears to be the prime objective of the pup urea cycle. Indeed, urea accounts for 98% of the [nitrogen] excreted/lost by the pup in the early stages of development, decreasing slightly to around 93% in the later stages[54]. In either case, production of urea is more predominant in the pup than in the adult dogfish where urea accounts for less than 80% of total excretion of nitrogenous compounds[86]. Although the ovarian fluid in viviparous teleosts contains much less ammonia than in the dogfish[26], extrapolation from the elasmobranch system to teleostean models with higher urea cycle activity during embryogenesis will supply ready reasons for the persistence of urea cycle enzymes in teleosts. It may also explain the occurrence of urea cycle activity in the liver of *Potamotrygon sp.*[38], a freshwater elasmobranch which does not employ urea as an osmolyte even under conditions of increased environmental salinity[36].

Further, since all fishes, with the exception of the lungfishes, display identical biochemical characteristics for ornithine–urea cycle enzymes and compartmentation, the urea cycle is undoubtedly a monophyletic trait among the vertebrates. Thus the physiological divergence of the product into osmolyte, vehicle for nitrogen recycling as in ruminants and hibernating mammals, or excretory vehicle, is secondary to given biochemical constraints; many selective factors may have summed to strongly favour retention of these genes.

The difference in the above two factors of the urea synthetic pathway between lungfish and higher vertebrates on one side and 'fishes' on the other has a number of important biochemical repercussions, ones that can be exploited by comparative biochemists and physiologists. Parameters affected are, among others, transport phenomena, especially for mitochondria, detoxification mechanisms, metabolic zonation, and the role of glutamine and general nitrogen transport. In an attempt to simplify, when we mention 'fishes', in the following we refer to all fishes possessing CPS III and mitochondrial arginase (cf. Fig. 3). Similarly, when we mention mammalian-type urea cycle, we mean all mammals and other ureogenic animals, including lungfish, amphibians and some reptiles, comprising hepatic CPS I and a cytosolic arginase (cf. Fig. 3).

V. Intracellular and intercellular transport phenomena

1. Urea, arginine, ornithine and citrulline

Since hepatic arginase is a cytosolic enzyme in mammals, urea in these animals is liberated in the cytosol. In fishes, in contrast, urea is generated within the mitochondrial compartment and subsequently has to cross into the cytosol before transport/diffusion into the bloodstream (Fig. 4). While no research has been done on the intracellular transport phenomena for urea in teleosts, it is evident that the mitochondrial membrane of elasmobranchs does not pose a significant diffusion barrier to urea. Urea added to a suspension of isolated mitochondria equilibrates instantaneously across the membrane (ref. 78, and J. Ballantyne and C. Moyes, unpublished results).

Since ornithine, too, is produced from arginine in the cytosol in mammals, it has to cross into the mitochondrion before it can be utilized by ornithine carbamoyl transferase located in the mitochondrial matrix. In turn, citrulline exits from the mitochondrion. Experimental evidence supports the notion that an ornithine/citrulline antiport is responsible for the transport of these two compounds in the mammals and that cytosolic arginase is somehow spatially associated

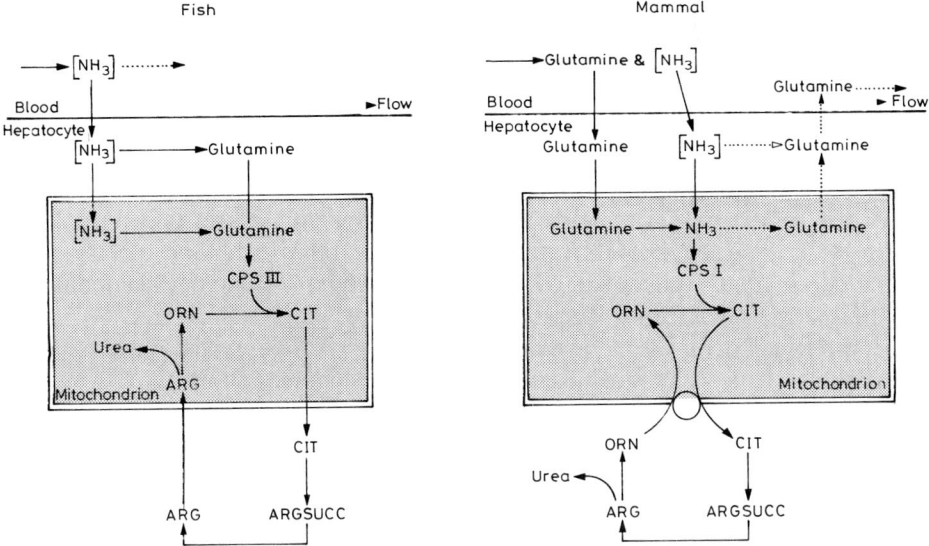

Fig. 4. Subcellular compartmentation of urea synthesis in fish and mammalian livers, integrated with the preferred plasma vehicle for 'nitrogen'. CPS III = carbamoylphosphate synthetase III (glutamine-dependent); CPS I = carbamoylphosphate synthetase I (ammonia-dependent). Other abbreviations as in Fig. 1.

with ornithine entry since ornithine appears to be channelled directly from arginase to ornithine carbamoyl transferase. Conversely, in the piscine situation, ornithine is generated within the mitochondrion, and thus a specific ornithine transporter is not required[11], while channelling could still be realized through spatial association of the two enzymes within the mitochondrial matrix. Such channelling would not only directly supply ornithine carbamoyl transferase with its substrate but also allow higher concentrations to be reached in the microenvironment around the active site than would be possible if the enzyme and its substrate were suspended freely in the bulk water — should it exist — inside of the mitochondrion.

It is clear that ornithine does enter the mitochondrion of elasmobranch fishes, but due to the nature of the existing data, no conclusions about velocity or energy-dependence of the uptake process can be drawn. While isolated dogfish (*Squalus acanthias*) mitochondria may utilize exogenously supplied ornithine for citrulline synthesis[5], experiments on isolated skate (*Raja erinacea*) mitochondria revealed no oxidative capacity for ornithine[78]. It is well possible that ornithine oxidation is hampered by the lack of ornithine aminotransferase, or an inadequate level of this enzyme in skate mitochondria. Since in our own experiments on isolated skate hepatocytes ornithine was an essential requirement for urea synthesis (T. Mommsen and T. Moon, unpublished results), it is clear that ornithine readily crosses through the hepatocyte membrane and into the mitochondrion. The fact that the addition of ornithine is a *sine qua non* for urea synthesis further implies that intramitochondrial ornithine concentrations are low, possibly due to losses incurred during hepatocyte isolation. Because of the cyclical nature of the urea synthetic process it is conceivable that only minute amounts of ornithine have to cross into the mitochondrion, merely sufficient to 'spark' urea synthesis. The latter consideration, however, applies only to hepatocyte suspensions and not to isolated mitochondria. The apparent inability of isolated elasmobranch mitochondria to metabolize ornithine through oxidation renders them a simple, as yet unexploited, model system to study the uptake of this amino acid into mitochondria. In rat liver mitochondria, for comparison, the rapid oxidation of ornithine makes it necessary to study uptake of ornithine under non-physiological conditions which curtail the secondary utilization of ornithine.

Keeping in mind that isolated skate mitochondria fail to oxidize ornithine, in spite of the fact that they can take it up from the surrounding medium, it is not surprising that the mitochondria will not oxidize exogenously administered arginine. Although skate mitochondria contain large levels of arginase[11,13], these results confirm the contention that no obvious metabolic route is operating for intramitochondrial oxidation of ornithine. At any rate, with the intramitochondrial arrangement of arginase, CPS III and ornithine carbamoyltransferase, a possible depletion of hepatic ornithine into polyamine synthesis through ornithine decarboxylase, an enzyme usually located in the cytosol[19], is minimized, even if channelling between these three urea cycle enzymes should not exist in the fishes. From this it appears that two mechanisms available to rat liver mitochondria to regulate the rate of urea synthesis, i.e. the regulation of ornithine levels through ornithine decarboxylase or ornithine transaminase, do not apply to the piscine liver. Unfortunately, the other important regulatory site for urea synthesis identified for the rat liver mitochondria, i.e. control of N-acetyl-L-glutamate levels by N-acetyl-L-glutamate synthase[14,79,88], remains to be analyzed for the fishes (see below).

In ureogenic fishes, arginine must cross into the mitochondrion to undergo hydrolysis by mitochondrial arginase, but to date no direct analysis for arginine uptake is available. Further, exogenous application of arginine has yet to be employed as an experimental route in studying citrulline synthesis in isolated elasmobranch or toadfish mitochondria.

2. Glutamine and ammonia

In contrast to ornithine or arginine, isolated skate mitochondria will actively oxidize glutamine[78]. This observation implies two different facts. First, glutamine is taken up by the mitochondria — the mechanisms is not known — and second, it is converted to glutamate. Once inside of the mitochondrion, glutamine can be used by CPS III for the synthesis of carbamoyl phosphate and citrulline[5], or it can be deaminated by a mitochondrial glutaminase. Since the isolated skate mitochondria were not supplied with exogenous bicarbonate for carbamoyl phosphate synthesis nor with carbamoyl phosphate itself for citrulline synthesis, it seems likely that skate mitochondria do indeed contain glutaminase. It is this enzyme, then, which was generating intramitochondrial glutamate as an excellent oxidative substrate to be used by glutamate dehydrogenase for production of a Krebs cycle intermediate.

The above experiments further indicate that skate liver glutamate dehydrogenase may function in the oxidative direction and not assume a role in reducing the concentration of intramitochondrial ammonia.

Also, since teleost liver glutaminase is localized in the mitochondrial matrix (ref. 10 and T. Mommsen and P. Walsh, unpublished results), as it is in mammals, the possibility of futile cycling of glutamine between glutaminase and glutamine synthetase may exist. This is especially relevant in species, such as the oyster toadfish, which contain most or all of their glutamine synthetase in the same compartment as glutaminase. Again, possible channelling of glutamine from its site of production to CPS III may eliminate this possibility, and thus also prevent the potential competition between glutaminase and CPS III. It seems likely that glutaminase is spatially separated from glutamine synthetase and CPS III and deals with glutamine from other sources or serves as an anaplerotic reaction generating intramitochondrial glutamate and 2-oxoglutarate.

The catalytic properties of glutaminase in fish liver may give additional insight into the evolution of CPS III *versus* I. In the mammalian liver, these two enzymes are fine-tuned not only through channelling of ammonia from glutaminase to CPS I[63], but also through *N*-acetylglutamate. This modulator which is synthesized from acetyl-coenzyme A and glutamate, is a strong activator of both enzymes[65]. In fishes, in contrast, CPS III and glutaminase are potential competitors for glutamine and therefore reciprocal, and not parallel, control mechanisms can be expected.

The overall importance of glutamine as intertissue vehicle or extrahepatic substrate in fishes is difficult to assess from the selected topics covered in piscine literature. For mitochondria isolated from the red muscle of *Salvelinus namaycush* or *Amia calva*, glutamine serves as the best oxidative amino acid substrate *in vitro*[14a], an observation that is supported by corresponding levels of mitochondrial, phosphate-dependent glutaminase is this tissue[14a]. Glutamine-derived carbon appears to be funnelled into the Krebs cycle via glutamate dehydrogenase and complete oxidation of glutamine carbon in red muscle may also involve mitochondrial malic enzyme. In addition, *in situ* synthesis of glutamine may be relevant, especially in the bowfin (*A. calva*) red muscle, which contains at least as much glutamine synthetase as the liver[14a] of this ammoniotelic species[23]. In *Salvelinus* red muscle synthesis of glutamine, in contrast, appears insignificant, but obviously, muscular synthesis of glutamine, carefully weighed against its oxidation, should be considered for ureogenic species.

At this point, the actual contribution of ammonia *versus* urea to nitrogen excretion in the elasmobranch fishes is not clear. Although it is always assumed that these vertebrates are not only ureogenic but also ureotelic, some experimental and theoretical evidence does not support the notion of ureotelism. Of course, this may be more of a semantical problem than a biochemical distinction. However, due to rather limited velocity and virtual independence from experimental manipulations it appears that actual rates of urea synthesis are primarily fine-tuned to compensate for passive losses of urea through the gills or *via* the urine. Therefore, this rate of leakage/synthesis will merely set a baseline for urea excretion, and not, in fact, determine the bulk of nitrogen excretion. The fact that elasmobranch urea synthesis is generally rather slow and not subject to rapid changes suggests that most nitrogen excretion, which is particularly relevant after a meal, is *via* ammonia and not the metabolically costly urea. Further, it is likely that a major portion of excretory urea observed following a meal, is derived from other sources, such as argininolysis or uricolysis. After all, arginine is one of the most abundant amino acids in fish protein. It would be interesting to study relative rates of urea and ammonia excretion after a meal in marine elasmobranchs with different dietary preferences.

Additional evidence from teleosts points toward the fact that ammonia rather than an amino acid is the preferential vehicle for nitrogen transport. In most ammoniotelic fishes, capacity to generate ammonia in gill tissue – the main site for ammonia/ammonium exchange into the surrounding water – is minimal, and it appears instead that intrahepatically generated ammonia accounts for most of the ammonia lost[29,68]. It is not clear, how much of a contribution ammonia derived from white muscle AMP during strenuous exercise[73] may make to periodical nitrogen excretion by fish.

Even the ureogenic, albeit considerably ammoniotelic, dogfish fails to employ glutamine as a prevalent vehicle for nitrogen and seems to favour ammonia instead and possibly alanine during periods of starvation[56]. However, the bulk of nitrogen is transported as ammonia. As a consequence, the importance of glutamine as a substrate, transport vehicle or metabolic regulator for the fishes must not be overstated. It is intriguing, however, that fishes, as a rule, use a glutamine-dependent CPS while at the same time transporting nitrogen in the form of ammonia. Consequently, a hepatic glutamine synthetase is required, not primarily to detoxify ammonia, but rather to convert ammonia into a form which is more readily acceptable to CPS III. Elasmobranch glutamine synthetase, with its extremely high affinity for ammonia, is thus ideally suited to control the synthesis of the animal's most important osmolyte. In the mammal, the situation is completely reversed: here, ammonia is the exclusive substrate for CPS I, but glutamine serves as the intertissue transport vehicle. A mitochondrial glutaminase must be in place to deliver, in fact channel, ammonia to CPS I[63]. In view of this somewhat counterintuitive arrangement, it would be interesting to analyze the intertissue nitrogen transport in ureotelic teleosts under extreme physiological conditions, such as in the alkaline lake tilapia or in air-breathing fish exposed to high environmental ammonia or with restricted access to water.

Taking the entire route of hepatic ammonia into urea in fish liver into consideration, the following stoichiometry is obtained (Eqn. 1):

$$2 \text{ NH}_4^+ + \text{HCO}_3^- + 4 \text{ ATP} \rightarrow \text{urea} + 3 \text{ ADP} + \text{AMP} + 3 \text{ P}_i + \text{PP}_i \qquad (1)$$

Hereby, the three ATP-utilizing reactions are glutamine synthetase, CPS III (2 ATP) and argininosuccinate synthetase. Since the production of AMP by the latter comprises the hydrolysis of two high energy phosphates, a total of 5 ATP are hydrolyzed to ADP for each molecule of urea synthesized, corresponding to 2.5

ATP used for each nitrogen assimilated. While at first inspection urea synthesis from ammonia in the mammalian liver may appear energetically less costly at 4 ATP/urea[105] this view ignores two crucial facts. Once the mandatory delivery of [nitrogen] to CPS I via glutaminase *and* the extrahepatic synthesis of glutamine through glutamine synthetase have been included, identical stoichiometries for urea synthesis in fishes and mammals are realized. At any rate, the above stoichiometry excludes the energy required to transport aspartate into the mitochondrion. In case the malate/aspartate shuttle is utilized, energy demand may amount to about one third of an ATP for each aspartate delivered. Further, the shuttling of citrulline/ornithine in mammals and that of citrulline/arginine in fishes, respectively (cf. Fig. 4), may have some admittedly minor bearing on the overall stoichiometry of urea synthesis in the vertebrates.

3. Alanine

Since alanine is one of the most abundant amino acids in fish tissues and blood and is a highly favoured amino acid substrate for gluconeogenesis, it is this particular amino acid that is most likely to be supply nitrogen essential for urea production. In the case of the oyster toadfish, which contains only cytosolic glutamine synthetase, alanine-nitrogen has a long itinerary before it can serve as a substrate for mitochondrial CPS III. Alanine (or aspartate or glutamate) nitrogen must enter the mitochondrion and undergo the glutamate dehydrogenase reaction before exiting into the cytosol for glutamine synthesis. By the time alanine-derived nitrogen has arrived at the active site of CPS III, it has crossed the mitochondrial membrane three times.

Considering this somewhat convoluted route, the question suggests itself, whether CPS III may not employ ammonia and thus that the detoxification function of glutamine synthetase and the ornithine–urea cycle are separate rather than obligatorily linked metabolic processes. In this context it should be reiterated that CPS III can utilize both glutamine and ammonia as substrates, although maximum velocities with saturating concentrations of glutamine are 4 to 5-fold higher than with ammonia. In this respect, isolated toadfish hepatocytes behave similar to purified fish CPS III (3,5,12). Indeed, when isolated hepatocytes from *O. beta* or *O. tau* were supplied with ammonia instead of glutamine, they were able to synthesize urea albeit at drastically reduced rates (Table 2). Interestingly, when Gulf toadfish liver cells are exposed simultaneously to alanine and glutamate, rates of urea synthesis (in the presence of aspartate) are severely depressed compared with hepatocytes which had been supplied with glutamine as substrate. Since the concurrent presence of methioninesulfoximine, an inhibitor of glutamine synthetase, decreases the rate of urea synthesis by isolated gulf toadfish hepatocytes even further (Table 2), it can be concluded, that, at least in this species, the mitochondrial glutamine synthetase is not bypassed and glutamine, indeed, serves as the *in vivo* substrate for carbamoyl phosphate synthesis.

On the one hand, alanine is instrumental in supplying carbon backbones for *de novo* synthesis of glucose and, in fact, it is likely to be the most important

TABLE 2

Urea synthesis in isolated toadfish (*Opsanus beta*) hepatocytes and its dependence on different nitrogenous substrates

Glutamine	100
Glutamine plus methionine sulfoximine	104.5 ± 5.9 (9)
Ammonia	19.6 ± 2.9 (6)
Ammonia plus glutamate	31.1 ± 4.2 (6)
Ammonia plus glutamate plus methionine sulfoximine	20.8 ± 2.7 (6)
Alanine	9.4 ± 2.2 (3)
Alanine plus glutamate	11.1 ± 2.4 (3)
Alanine plus glutamate plus methionine sulfoximine	5.6 ± 1.0 (3)
Serine plus glutamate	11.3 ± 2.8 (3)

Rates of urea synthesis are expressed in % of control in the presence of glutamine ± SEM, with the number of independent observations in parentheses. From ref. 75 and T. Mommsen and P. Walsh, unpublished results. Actual rates of urea synthesis were 17.26 ± 3.16 μmol of urea synthesized per gram of packed hepatocytes in one hour ($n = 10$). Isolated Gulf toadfish hepatocytes were incubated with the listed substrates, as well as with aspartate and ornithine and ^{14}C labelled bicarbonate, for 90 min. Carbon dioxide was collected after enzymatic hydrolysis of urea. Concentrations were 5 mM for all substrates, except for methioninesulfoximine, an inhibitor of glutamine synthetase, which was present at 10 mM. For procedural and other experimental details, see ref. 75.

gluconeogenic substrate *in vivo*. On the other hand, the existence of a glucose–alanine cycle, which tightly links extrahepatic alanine synthesis with muscular activity is doubtful for piscine systems for two reasons. First, rates of hepatic gluconeogenesis in fishes are generally insufficient to make the same important contribution to glucose turnover as in mammals. Second, although to date little attention has been devoted to amino acid *versus* ammonia efflux from exercising fish muscle, it appears that ammonia and not alanine is the predominant product of muscular activity in fish muscle[67,73,119].

Alanine production by extrahepatic tissues, especially the white muscle, is important, however, during periods of extended starvation in fishes. Under these specific conditions, which lead to extensive muscular proteolysis[32,56,72], alanine appears to be the main amino acid exported from the muscles and available to the liver. As a consequence, alanine is among the best gluconeogenic substrates for the fish liver, generally approaching rates achieved with lactate and supplying further evidence that liver is the main site of ammonia production in resting fish.

4. Aspartate

Since the enzymatic reaction feeding aspartate into the urea cycle, i.e., argininosuccinate synthetase, is cytosolic in both fishes and mammals, no class-specific differences have to be invoked. However, some recent experiments with respect to the regulation of urea synthesis in the toadfish indicate that aspartate delivery may be an important site for control. Interestingly, both aspartate aminotransferase and malic enzyme were among the few enzymes increased following long-term exposure of toadfish to glucagon or the synthetic glucocorticoid dexamethasone[70]. Since at the same time, the activity of glucose 6-phosphate dehydrogenase was markedly curtailed by the hormonal treatment, it seems reasonable to postulate that in this teleost, malic enzyme functions anaplerotically to increase the pool of 4-carbon intermediates, especially aspartate. Thus the enzyme does not assume its normal role of contributing reducing power for fatty acid synthesis.

5. Carbon dioxide

In both mammals and the Gulf toadfish, carbonic anhydrase hydrates metabolically generated CO_2, mainly from the Krebs cycle, to yield HCO_3^- for urea synthesis[27,114]. Although it appears likely that the intramitochondrial isozyme is the most important in this regard, the role of cytosolic carbonic anhydrase is not clear. Its degree of association with the outer mitochondrial membrane could control rates of CO_2 diffusion from the mitochondrion, and hence the availability for ureogenesis. Association of intramitochondrial carbonic anhydrase with CPS III or Krebs cycle enzymes could further enhance bicarbonate delivery (cf. ref. 113).

VI. Detoxification

Since unionized ammonia exerts numerous toxic effects on tissues, rapid and efficient removal of this species is of paramount importance to fish survival. When fish are exposed experimentally to increased concentrations of NH_3, the tissue undergoing the largest changes appears to be the brain. The most pronounced change relevant to the present discussion is a drastic increase in the concentration of glutamine at the expense of the glutamate pool of the tissue[6,57], due to the presence of substantial amounts of glutamine synthetase[117]. Minor increases in glutamine levels are observed for liver and intestine[6,85]. In line with generally low levels of this enzyme in the teleost liver[117], only a small (less than 20%) increase in hepatic glutamine is noted, mirrored by a corresponding decrease in glutamate[6]. Most teleosts mistreated this way alter their intermediary metabolism, but are not able to detoxify ammonia through the ornithine urea cycle and thus, urea excretion experiences only minor readjustments[83]. Although exposure to ammonia in ureogenic teleosts, such as air-breathing fishes or the marine toadfish, increases the rate of urea excretion[95,112] and may even enlarge titers of hepatic urea cycle enzymes[98], concurrent biochemical events in cerebral tissues of these fishes have been ignored. Judging by equally high levels of glutamine synthetase activities in all teleost brains[117], we can assume that the primary line of defense in all teleost brain is the synthesis of glutamine, but secondary events are likely quite different in solely ammoniotelic and ureogenic species. First, it seems that the ureogenic species are able to withstand much higher concentrations of ambient unionized ammonia than exclusively ammoniotelic species such as the rainbow trout. To rainbow trout, a concentration of 36 μmolar NH_3 is lethal[6], while Gulf toadfish can be chronically exposed to 20 μmolar NH_3 without any apparent ill effects[112].

To date two champions of ammonia tolerance have left their mark in the comparative literature. One is an air-breathing catfish (*H. fossilis*) from India which survives in water supplemented with 75 mM NH_4Cl (98); assuming that ambient pH was around neutrality, these fish can withstand 240 μmolar NH_3. In the other, the alkaline lake tilapia (*O. alcalicus grahami*), exposure to experimentally augmented ammonia levels was associated with increased urea output; otherwise the fish were unfazed by 500 μmolar concentrations of unionized ammonia (at pH 10)[120]!

These observations may lead to interesting hypotheses concerning the relative roles of ammonia and glutamine in ammoniotelic fishes on one side and ureotelic species on the other. If we accept the contention that ammonia and alanine are the most important transport vehicles for [nitrogen] in fishes[32,56,72], exogenous glutamine, although abundant in teleost plasma[85,115], does not seem readily available for hepatic urea synthesis. Therefore the fate of glutamine synthesized in brain of ammonia-exposed teleosts and potential sources of ammonia for intrahepatic glutamine synthesis in ureogenic species become important considerations. While it is possible that glutamine synthesis provides the brain with a momentary ammonia sink to be emptied when a temporary ammonia insult has subsided, catalyzed by brain glutaminase[110], glutamine is likely to be released from the brain under conditions of chronic ammonia exposure. Possible target tissue for circulating glutamine are the liver, kidney and gill of ammoniotelic teleosts by virtue of their ability to rapidly deaminate glutamine[10,104,109] and their high content in glutaminase[110,115]. Unfortunately, comparable data are not available for ureogenic teleosts, but the apparent absence of a glutamine shuttle in ureogenic elasmobranchs[56] also suggests intrahepatic sources of ammonia and that *in situ* glutamine synthesis delivers [nitrogen] to the urea cycle.

VII. Zonation

In the mammalian liver, clear heterogeneity exists among parenchymal cells with regard to their ability to synthesize urea. Cells located in the well-oxygenated periportal zone are characterized by higher activities of anabolic pathways, including urea synthesis, while cells of the perivenous area tend to be more catabolic in scope[50]. In fact, the intensity of staining for CPS I decreases from the portal to the venule side of the liver and staining is entirely absent from the cell layers surrounding the hepatic venules. Thus clear-cut metabolic zonation is observed. Glutamine synthetase, which in mammals assumes the function of curtailing spillover of hepatic ammonia into the venous circulation (cf. Fig. 4), is exclusively localized in the two or three rows of parenchymal cells surrounding the hepatic venule[35] and spatially separated from cells with urea cycle activity. While enzymatic heterogeneity appears to occur in fish liver[43,69,76,84], possibly rendering parenchymal heterogeneity a basic trait in vertebrate evolution, it does not apply to urea synthesis in the toadfish liver[76]. In addition, glutamine synthetase appears evenly dissipated among different subpopulations of toadfish parenchymal cells which are easily separated by gradient centrifugation. Together with the presence of glutaminase activity within the mitochondrial compartment of the toadfish, these data also imply that neither glutaminase nor glutamine synthetase can function vectorially as they do in the mammalian liver (cf. Fig. 4). Homogeneous distribution of glutamine synthetase in the toadfish is in line with the functional role of this enzyme in this ureogenic teleost[76], i.e. as feeder enzyme to hepatic CPS III (cf. Fig. 4, Table 3). In this context, it would be interesting to analyze parenchymal heterogeneity and hepatic zonation in African lungfishes, especially in regard to urea synthesis and the relative spatial distribution of glutamine synthetase. From the initial data on the biochemical set-up it seems reasonable to expect mammalian-type separation of glutamine synthetase and urea cycle. Taking this line of thought a step further, the African lungfish could be an ideal experimental system to access trends in the evolution of transport mechanisms for 'nitrogen' and the intriguing shift in the specific role of glutamine.

VIII. Regulation

1. Hormonal control

While we prefer not to delve into the ongoing discussion on the importance of pH regulation and bicarbonate removal *versus* ammonia detoxification to urea synthesis (cf. refs. 7, 15, 64, 112), in mammals a number of different mechanisms are actuated to alter urea cycle activity. One of the best examples is probably the rise in urea synthetic activity following the exposure of hepatic tissue to increased concentrations of glucagon. In the rat liver, for instance, glucagon directly affects the activity of CPS I by altering the concentration of its essential allosteric effector N-acetyl-L-glutamate[58,103] and the availability of intramitochondrial ATP through increased oxidative throughput. In addition, glucagon's effects persist in subsequently isolated mitochondria which show increased rates of citrulline production[88,108]. In the short term, the hormone also triggers an activation of the mitochondrial glutaminase, possibly through glucagon-dependent changes in enzyme interaction with the mitochondrial membrane[62]. In the longer term, glucagon is known to augment the titer of all five urea cycle enzymes through enzyme induction[47,58]. While most of the effects of glucagon are correlated with increases in cAMP, an alternative route of increasing urea synthesis in the rat liver is through α-adrenergic mechanisms, as invoked through exposure to epinephrine, vasopressin or angiotensin[20]. Under closer scrutiny, it becomes apparent that none of these common mechanisms regulating urea cycle activity in the mammalian liver can apply to the piscine situation. A few of them can be dismissed on grounds of the biochemical set-up alone, while others can be eliminated by experimental evidence. For an overview, the specific characteristics of hepatic urea synthesis in fishes *versus* mammals are presented in Table 3. Hepatic glutaminase, for instance, occupies an entirely different position in the fish liver and thus is quite dissociated from the urea cycle *per se*, and its pivotal role is assumed by glutamine

TABLE 3

Characteristics of urea synthesis in vertebrate liver

Fish liver	Mammalian liver
1. Glutamine synthetase feeder enzyme for CPS III	1. Glutamine synthetase – scavenger enzyme preventing hepatic NH_3 efflux
2. Carbamoyl phosphate synthetase III mitochondrial matrix glutamine dependent N-acetylglutamate essential cofactor	2. Carbamoyl phosphate synthetase I mitochondrial matrix ammonia dependent N-acetylglutamate essential cofactor
3. Mitochondrial arginase	3. Cytosolic arginase
4. Urea produced in mitochondrion	4. Urea produced in cytosol
5. Arginine enters mitochondrion	5. Ornithine enters mitochondrion via ornithine-citrulline antiport
6. Citrulline exits	6. Citrulline exits
7. Glutamine synthesized intrahepatically	7. Glutamine synthesis extrahepatically
8. Glutaminase not required for urea synthesis	8. Glutaminase essential for urea synthesis
9. Nature of CPS III and glutamine synthetase regulation unknown	9. Reciprocal control of CPS I and glutamine synthetase

synthetase. Further, other transporters are involved in shuttling intermediates between the cytosol and the mitochondrial matrix of fish liver and in most cases their specific regulation is not known. Alterations in the availability of intramitochondrial ATP which are subjugated by the mammals as a means to regulate CPS I activity, do not seem to be exploited by the fishes either, since none of the common hormones increasing oxidative performance of mitochondria (glucagon, epinephrine, vasoactive peptides), exert this effect in the fish liver[49,77]. When isolated toadfish (*O. beta*) hepatocytes were exposed to physiological or pharmacological concentrations of numerous hormones, including glucagon, glucagon-like peptide, epinephrine, and vasoactive peptides, no acute effects on the rate of urea synthesis were noted[71]. Since generally α-adrenergic phenomena are lacking in fish hepatic tissue[49,77], the failure of epinephrine to acutely influence toadfish urea synthesis is not surprising. In further agreement with the above conclusions, longer term exposure of toadfish to glucagon or the glucocorticoid analogue dexamethasone – treatments resulting in multifold changes in mammalian urea cycle activity – failed to alter urea synthesis *in vitro* (with exogenous glutamine added to the incubates) or the activities of two urea cycle enzymes assayed. Hormonal treatment, however, significantly increased the activities of three enzymes ancillary to the cycle: glutamine synthetase, malic enzyme and aspartate aminotransferase[70].

2. Metabolic control

While these results confirm the short-term unresponsiveness of the toadfish urea cycle *per se* to hormonal regulation, they point towards two new concepts in the regulation of the urea cycle: the pivotal role of glutamine synthetase and the availability of aspartate. The existing evidence for the central regulatory role of glutamine synthetase in the toadfish liver, which are likely to apply to most, if not

TABLE 4

The pivotal role of hepatic glutamine synthetase in piscine urea synthesis

1. Ammonia – main transport form of [nitrogen]
2. Low K_m of glutamine synthetase for ammonia
3. Glutamine synthetase likely saturated with glutamate
4. Glutamine likely *in vivo* substrate for CPS III
5. High K_m of CPS III for ammonia
6. Non-induction of toadfish urea cycle enzymes by numerous hormones effective in mammalian liver
7. Inducibility of hepatic glutamine synthetase by dexamethasone
8. Lack of zonation for glutamine synthetase in toadfish hepatocytes
9. Lack of zonation for urea synthetic activity in toadfish hepatocytes

all teleosts, is provided in Table 4. The only remaining factor of the above list compiled for mammals which may be relevant for regulating urea synthesis in fish liver is therefore the synthesis and degradation of *N*-acetylglutamate and its transport out of the mitochondrion. Unfortunately, to date the knowledge about *N*-acetylglutamate in fishes is restricted to the observation that it is essential for CPS III catalysis, while actual occurrence of the compound has yet to be reported. Similarly, occurrence, distribution or kinetics of the corresponding synthase (cf. Eqn. 2) await analysis.

$$\text{Acetyl-coenzyme A} + \text{glutamate} \rightarrow N\text{-acetylglutamate} + \text{coenzyme A-SH} \quad (2)$$

From a comparative point of view, two specific properties of *N*-acetylglutamate synthase in the mammalian liver, make this a rather intriguing enzyme. First, *N*-acetylglutamate synthase is localized in the mitochondrial matrix and thus potentially a competitor for intramitochondrial glutamate. Second, the enzyme is strongly activated by micromolar concentrations of arginine[99]. A similar setting in fish liver would lead to potential removal of this important activator of CPS III by mitochondrial arginase, a counterintuitive situation which, again, suggests spatial separation of *N*-acetylglutamate synthase, CPS III and arginase as well as channelling of arginine from its cytosolic production site directly to arginase. A different pool of intramitochondrial arginine has to be postulated should fish *N*-acetylglutamate synthase be subject to regulation by this amino acid.

3. Feedback inhibition?

Although, on the surface, all fishes may possess similar CPS and arginases, minor differences in regulatory mechanisms are already emerging. In the elasmobranch dogfish (*S. acanthias*), for instance, the rate of citrulline synthesis by isolated mitochondria is inhibited by the presence of urea[4] as is the reaction of CPS III *in vitro*[3]. In contrast, the hepatic CPS III of smallmouth bass reveals less powerful inhibition by urea[12], which is in agreement with results on toadfish (*O. beta*) hepatocytes where small, albeit supraphysiological, concentrations of urea failed to alter the rate of urea synthesis *in vitro*[75].

Our scheme of urea synthesis touches on a couple of additional concepts in evolution of metabolic pathways and their regulation. The two congeneric toadfish are an interesting case in point. In one, the Gulf toadfish, glutamine synthetase is localized in the mitochondrion, while in the other, the enzyme is clearly cytosolic[75]. Thus, in the Gulf toadfish, the enzyme is localized in the 'right' compartment to serve as a feeder enzyme for mitochondrial CPS III, and consequently contiguous channelling of substrate is a distinct possibility, while across-membrane channelling of glutamine has to be postulated for the congener. If we assume that in most teleost fishes ammonia is the primary transport vehicle for [nitrogen] in plasma or blood — a view that is supported by limited experimental evidence detailed above — hepatic and mitochondrial mechanisms must have evolved for the efficient uptake of ammonia/ammonium. In ureogenic species, hepatic glutamine synthetase may be the essential ammonia trap in this scheme. In case alanine serves as the prevailing intertissue vehicle for nitrogen shuttling in fishes — a concept sustained by some experiments but at odds with others (cf. ref. 106) — different avenues must be in place to funnel [nitrogen] *via* glutamine into the ornithine–urea cycle or *via* glutamate dehydrogenase into ammonia for excretion. In a third — at this point merely theoretical — scenario, glutamine functions as an important intertissue vehicle and delivers [nitrogen] promptly to hepatic CPS III in ureogenic species or to mitochondrial glutaminase and glutamate dehydrogenase in the liver for liberation of ammonia destined for excretion by the gills.

Acknowledgements. We thank Drs. B. Korsgaard, T. Moon and E. Plisetskaya for stimulating discussions and Dr. D. Siebert for helpful comments on one of our previous publications. Research in our laboratories is supported by grants from NSERC (Canada) and NSF (U.S.A.).

IX. References

1. Anderson, P.M. A glutamine- and N-acetyl-L-glutamate-dependent carbamyl phosphate synthetase activity in the teleost *Micropterus salmoides*. *Comp. Biochem. Physiol.* 54B: 261–263, 1976.
2. Anderson, P.M. Glutamine- and N-acetylglutamate-dependent carbamoyl phosphate synthetase in elasmobranchs. *Science* 208: 291–293, 1980.
3. Anderson, P.M. Purification and properties of the glutamine- and N-acetyl-L-glutamate-dependent carbamoyl phosphate synthetase from liver of *Squalus acanthias*. *J. Biol. Chem.* 256: 12228–12238, 1981.
4. Anderson, P.M. Effects of urea, trimethylamine oxide, and osmolarity on respiration and citrulline synthesis by isolated hepatic mitochondria from *Squalus acanthias*. *Comp. Biochem. Physiol.* 85B: 783–788, 1986.
5. Anderson, P.M. and C.A. Casey. Glutamine-dependent synthesis of citrulline by isolated hepatic mitochondria from *Squalus acanthias*. *J. Biol. Chem.* 259: 456–462, 1984.
6. Arillo, A., C. Margiocco, F. Medlodia, P. Mensi and G. Schenone. Ammonia toxicity mechanism in fish: studies on rainbow trout (*Salmo gairdneri* Rich.). *Ecotoxicol. Environ. Safety* 5: 316–328, 1981.
7. Bean, E.S. and D.E. Atkinson. Regulation of the rate of urea synthesis in liver by extracellular pH. A major factor in pH homeostasis in mammals. *J. Biol. Chem.* 259: 1552–1559, 1984.

8. Brown, G.W. Jr. and P.P. Cohen. Comparative biochemistry of urea synthesis. 3. Activities of urea cycle enzymes in various higher and lower vertebrates. *Biochem. J.* 75: 82–91, 1960.
9. Campbell, J.W. and P.M. Anderson. Evolution of mitochondrial enzyme systems in fish: the mitochondrial synthesis of glutamine and citrulline. In: *Biochemistry and Molecular Biology of Fishes*, edited by P.W. Hochachka and T.P. Mommsen. Amsterdam, New York: Elsevier, 1991, pp. 43–75.
10. Campbell, J.W., P.L Aster and J.E. Vorhaben. Mitochondrial ammoniagenesis in liver of the channel catfish *Ictalurus punctatus*. *Am. J. Physiol.* 244: R709–R717, 1983.
11. Casey, C.A. and P.M. Anderson. Subcellular location of glutamine synthetase and urea cycle enzymes in liver of spiny dogfish (*Squalus acanthias*). *J. Biol. Chem.* 257: 8449–8453, 1982.
12. Casey, C.A. and P.M. Anderson. Glutamine- and N-acetyl-L-glutamate-dependent carbamoyl phosphate synthetase from *Micropterus salmoides*. *J. Biol. Chem.* 258: 8723–8732, 1983.
13. Casey, C.A. and P.M. Anderson. Submitochondrial localization or arginase and other enzymes associated with urea synthesis and nitrogen metabolism, in liver of *Squalus acanthias*. *Comp. Biochem. Physiol.* 82B: 307–315, 1985.
14. Cathelineau, L., D. Rabier, F. Petit and P. Kamoun. Physiological and hormonal variations of acetylglutamate and citrullinogenesis in rat liver mitochondria. *Enzyme* 26: 245–253, 1981.
14. (a) Chamberlin, M.E., H.C. Glemet and J.S. Ballantyne. Glutamine metabolism in a holostean fish (*Amia calva*) and a teleost (*Salvelinus namaycush*). *Am. J. Physiol.* 260: R159–R166, 1991.
15. Cheema-Dhadli, S., R.L. Jungas and M.L. Halperin. Regulation of urea synthesis by acid-base balance in vivo: role of NH_3 concentration. *Am. J. Physiol.* 252: F221–F225, 1987.
16. Cheung, C.-W., N.S. Cohen and L. Raijman. Channeling of urea cycle intermediates *in situ* in permeabilized hepatocytes. *J. Biol. Chem.* 264: 4038–4044, 1989.
17. Chiu, Y.N., R.E. Austic and G.L. Rumsey. Urea cycle activity and arginine formation in rainbow trout (*Salmo gairdneri*). *J. Nutr.* 116: 1640–1650, 1986.
18. Clarke, S. A major polypeptide component of rat liver mitochondria: carbamyl phosphate synthetase. *J. Biol. Chem.* 251: 950–961, 1976.
19. Corti, A., P.L. Tartoni, S. Astancolle, P. Davalli and R. Viviani. Liver ornithine decarboxylase activity in the European sea bass (*Dicentrarchus labrax* L.): effects of protein composition of the diet, environmental conditions and age. *Comp. Biochem. Physiol.* 89B: 137–141, 1988.
20. Corvera, S. and J.A. Garcia-Sainz. Vasopressin and angiotensin II stimulate ureogenesis through increased mitochondrial citrulline production. *Life Sci.* 31: 2493–2498, 1982.
21. Cvancara, V.A. Studies on tissue arginase and ureogenesis in freshwater teleosts. *Comp. Biochem. Physiol.* 30: 489–496, 1969.
22. Cvancara, V.A. Comparative study of liver uricase activity in freshwater teleosts. *Comp. Biochem. Physiol.* 28: 725–732, 1969.
23. Cvancara, V.A. Liver carbamyl phosphate synthetase in the primitive freshwater bony fishes (Chondrostei, Holostei). *Comp. Biochem. Physiol.* 49B: 785–787, 1974.
24. Dépêche, J., R. Gilles, S. Daufresne and H. Chapello. Urea content and urea production via the ornithine–urea cycle pathway during development of two teleost fishes. *Comp. Biochem. Physiol.* 63A: 51–56, 1979.
25. Devaney, M.A. and S.G. Powers-Lee. Immunological cross-reactivity between carbamyl phosphate synthetases I, II, and III. *J. Biol. Chem.* 259: 703–706, 1984.
26. de Vlaming, V., D. Baltz, S. Anderson, R. Fitzgerald, G. Delahunty and M. Barkley. Aspects of embryo nutrition and excretion among viviparous embiotocid teleosts: potential endocrine involvements. *Comp. Biochem. Physiol.* 76A: 189–198, 1983.
27. Dodgson, S.J., R.E. Forster and B.T. Storey. The role of carbonic anhydrase in hepatocyte metabolism. *Ann. N.Y. Acad. Sci.* 429: 516–524, 1984.
28. Evans, D.H. and J.N. Cameron. Gill ammonia transport. *J. Exp. Zool.* 239: 17–23, 1986.
29. Evans, D.H. and K.J. More. Modes of ammonia transport across the gill epithelium of the dogfish pup (*Squalus acanthias*). *J. Exp. Biol.* 138: 375–397, 1988.
30. Evans, D.H., K.J. More and S.L. Robbins. Modes of ammonia transport across the gill epithelium of the marine teleost fish *Opsanus beta*. *J. Exp. Biol.* 144: 339–356, 1989.
31. Forey, P.L. Golden jubilee for the coelacanth *Latimeria chalumnae*. *Nature* 336: 727–732, 1988.
32. French, C.J., P.W. Hochachka and T.P. Mommsen. Metabolic organization of liver during spawning migration of sockeye salmon. *Am. J. Physiol.* 245: R827–R830, 1983.
33. Fujiwara, S., S. Hayashi, T. Noguchi, N. Hanada and T. Takehara. Subcellular distribution of hepatic allantoinase varies among fishes. *Comp. Biochem. Physiol.* 93B: 213–215, 1989.
34. Funkhouser, D., L. Goldstein and R.P. Forster. Urea biosynthesis in the South American lungfish, *Lepidosiren paradoxa*: relation to its ecology. *Comp. Biochem. Physiol.* 41A: 439–443, 1972.

35. Gaasbeek Janzen, J.W., W.H. Lamers, A.F.M. Moorman, A. De Graf, J.A. Los and R. Charles. Immunohistochemical localization of carbamoyl-phosphate synthetase (ammonia) in adult rat liver. *J. Histochem. Cytochem.* 32: 557–564, 1984.
36. Gerst, J.W. and T.B. Thorson. Effects of saline acclimation on plasma electrolytes, urea excretion, and hepatic urea biosynthesis in a freshwater stingray, *Potamotrygon sp.* Garman, 1877. *Comp. Biochem. Physiol.* 56A: 87–93, 1977.
37. Goldstein, L. and R.P. Forster. The role of uricolysis in the production of urea by fishes and other aquatic vertebrates. *Comp. Biochem. Physiol.* 14: 567–576, 1965.
38. Goldstein, L. and R.P. Forster. Urea biosynthesis and excretion in freshwater and marine elasmobranchs. *Comp. Biochem. Physiol.* 39B: 415–421, 1971.
39. Goldstein, L. and R.P. Forster. Osmoregulation and urea metabolism in the little skate *Raja erinacea.* *Am. J. Physiol.* 220: 742–746, 1971.
40. Goldstein, L., S. Harley-DeWitt and R.P. Forster. Activities of ornithine–urea cycle enzymes and of trimethylamine oxidase in the coelacanth, *Latimeria chalumnae.* *Comp. Biochem. Physiol.* 44B: 357–362, 1973.
41. Goldstein, L., P.A. Janssens and R.P. Forster. Lungfish *Neoceratodus forsteri*: activities of ornithine–urea cycle and enzymes. *Science* 157: 316–317, 1967.
42. Gregory, R.B. Synthesis and total excretion of waste nitrogen by fish of the Periophthalmus (mudskipper) and Scartelaos families. *Comp. Biochem. Physiol.* 57A: 33–36, 1977.
43. Hampton, J.A., R.C. Lantz, P.J. Goldblatt, D.J. Lauren and D.E. Hinton. Functional units in rainbow trout (*Salmo gairdneri*, Richardson) liver: II. The biliary system. *Anat Rec.* 221: 619–634, 1988.
44. Hayashi, S., S. Fujiwara and T. Noguchi. Degradation of uric acid in fish liver peroxisomes. *J. Biol. Chem.* 264: 3211–3215, 1989.
45. Hoar, W.S. *General and Comparative Physiology.* Englewood Cliffs, NJ: Prentice-Hall, 1983, edn. 3, pp. 1–851.
46. Huggins, A.K., G. Skutch and E. Baldwin. Ornithine–urea cycle enzymes in teleostean fish. *Comp. Biochem. Physiol.* 28: 587–602, 1969.
47. Husson, A., C. Buquet and R. Vaillant. Induction of the five urea-cycle enzymes by glucagon in cultured foetal rat hepatocytes. *Differentiation* 35: 212–218, 1987.
48. Janssens, P.A. and P.P. Cohen. Ornithine–urea cycle enzymes in the African lungfish *Protopterus aethiopicus.* *Science* 152: 358–359, 1966.
49. Janssens, P.A. and P. Lowrey. Hormonal regulation of hepatic glycogenolysis in the carp, *Cyprinus carpio.* *Am. J. Physiol.* 252: R653–R660, 1987.
50. Jungermann, K. and N. Katz. Functional specialization of different hepatocyte populations. *Physiol. Rev.* 69: 708–764, 1989.
51. Kashiwagura, T., M. Erecinska and D.F. Wilson. pH Dependence of hormonal regulation of gluconeogenesis and urea synthesis from glutamine in suspensions of hepatocytes. *J. Biol. Chem.* 260: 407–414, 1985.
52. Knight, I.T., D.J. Grimes and R.R. Colwell. Bacterial hydrolysis of urea in tissues of carcharinid sharks. *Can. J. Fish Aquat. Sci.* 45: 357–360, 1988.
53. Kormanik, G. Time course of the establishment of uterine seawater conditions in late-term pregnant spiny dogfish (*Squalus acanthias*). *J. Exp. Biol.* 137: 443–456, 1988.
54. Kormanik, G. Nitrogen budget in developing embryos of the spiny dogfish *Squalus acanthias.* *J. Exp. Biol.* 144: 583–587, 1989.
55. Krebs, H.A. and K. Henseleit. Untersuchungen über die Harnstoffsynthese im Tierkörper. *Hoppe-Seyler's Z. physiol. Chem.* 210: 33–66, 1932.
56. Leech, A.R., L. Goldstein, C.J. Cha and J.M. Goldstein. Alanine biosynthesis during starvation in skeletal muscle of the spiny dogfish, *Squalus acanthias.* *J. Exp. Zool.* 207: 73–80, 1979.
57. Levi, G., G. Morisi, A. Coletti and R. Catanzaro. Free amino acids in fish brain: normal levels and changes upon exposure to high ammonia concentrations in vivo, and upon incubation of brain slices. *Comp. Biochem. Physiol.* 49A: 623–636, 1974.
58. Lin, R.C., P.J. Snodgrass and D. Rabier. Induction of urea cycle enzymes by glucagon and dexamethasone in monolayer cultures of adult rat hepatocytes. *J. Biol. Chem.* 257: 5061–5067, 1982.
59. Lusty, C.J. Carbamoylphosphate synthetase I of rat-liver mitochondria. *Eur. J. Biochem.* 85: 373–383, 1978.
60. Masoni, A. and P. Payan. Urea, inulin and para-amino-hippuric acid (PAH) excretion by the gills of the eel, *Anguilla anguilla* L. *Comp. Biochem. Physiol.* 47A: 1241–1244, 1974.
61. McGivan, J.D., N.M. Bradford and A.D. Beavis. Factors influencing the activity of ornithine aminotransferase in isolated rat liver mitochondria. *Biochem. J.* 162: 147–156, 1977.

62. McGivan, J.D., M. Vadher, J. Lacey and N.M. Bradford. Rat liver glutaminase. *Eur. J. Biochem.* 148: 323–327, 1985.
63. Meijer, A.J. Channeling of ammonia from glutaminase to carbamoyl-phosphate synthetase in liver mitochondria. *FEBS Lett.* 191: 249–251, 1985.
64. Meijer, A.J., W.H. Lamers and R.A.F.M. Chamuleau. Nitrogen metabolism and ornithine cycle function. *Physiol. Rev.* 70: 701–748, 1990.
65. Meijer, A.J. and A.J. Verhoeven. Regulation of hepatic glutamine metabolism. *Biochem. Soc. Trans.* 14: 1001–1004, 1986.
66. Meyer, A. and A.C. Wilson. Origin of tetrapods inferred from their mitochondrial DNA affiliation to lungfish. *J. Mol. Evol.* 31: 359–364, 1990.
67. Milligan, C.L. and C.M. Wood. Muscle and liver intracellular acid-base and metabolite status after strenuous activity in the inactive benthic starry flounder *Platichthys stellatus*. *Physiol. Zool.* 60: 54–68, 1987.
68. Mommsen, T.P. Metabolism of the fish gill. In: *Fish Physiology*, edited by W.S. Hoar and D.J. Randall. New York: Academic Press, 1984, p. 203–238.
69. Mommsen, T.P., E. Danulat, M.E. Gavioli, G.D. Foster and T.W. Moon. Separation of enzymatically distinct populations of trout hepatocytes. *Can. J. Zool.* 69: 420–426, 1991.
70. Mommsen, T.P., E. Danulat and P.J. Walsh. Effects of glucagon and dexamethasone on metabolism in the ureogenic teleost *Opsanus beta*. *Gen. Comp. Endocrinol.* in press.
71. Mommsen, T.P., E. Danulat and P.J. Walsh. Hormonal regulation of metabolism in hepatocytes of the ureogenic teleost *Opsanus beta*. *Fish Physiol. Biochem.* in press.
72. Mommsen, T.P., C.J. French and P.W. Hochachka. Sites and pattern of protein and amino acid utilization during the spawning migration of salmon. *Can. J. Zool.* 58: 1785–1799, 1980.
73. Mommsen, T.P. and P.W. Hochachka. The purine nucleotide cycle as two temporally separated metabolic units: a study on trout muscle. *Metabolism* 37: 552–556, 1988.
74. Mommsen, T.P. and T.W. Moon. The metabolic potential of hepatocytes and kidney tissue in the little skate, *Raja erinacea*. *J. Exp. Zool.* 244: 1–8, 1987.
75. Mommsen, T.P. and P.J. Walsh. Evolution of urea synthesis in vertebrates: the piscine connection. *Science* 243: 72–75, 1989.
76. Mommsen, T.P. and P.J. Walsh. Metabolic and enzymatic heterogeneity in liver of the ureogenic teleost *Opsanus beta*. *J. Exp. Biol.* 156: 407–418, 1991.
77. Moon, T.W. and T.P. Mommsen. Vasoactive peptides and phenylephrine actions in isolated teleost hepatocytes. *Am. J. Physiol.* 259: E644–E649, 1990.
78. Moyes, C.D., T.W. Moon and J.S. Ballantyne. Oxidation of amino acids, Krebs cycle intermediates, fatty acids and ketone bodies by *Raja erinacea* liver mitochondria. *J. Exp. Zool.* 237: 119–128, 1986.
79. Natale, P.J. and G.C. Tremblay. Studies on the availability of intramitochondrial carbamoylphosphate for utilization in extramitochondrial reactions in rat liver. *Arch. Biochem. Biophys.* 162: 357–368, 1974.
80. Nelson, J.S. The next 25 years: vertebrate systematics. *Can. J. Zool.* 65: 779–785, 1987.
81. Nyunoya, H., K.E. Broglie and C.J. Lusty. The gene coding for carbamyl-phosphate synthetase I was formed by fusion of an ancestral glutaminase gene and a synthetase gene. *Proc. Acad. Sci. U.S.A.* 82: 2244–2246, 1985.
82. Nyunoya, H., K.E. Broglie, E.E. Widgren and C.J. Lusty. Characterization and derivation of the gene coding for mitochondrial carbamyl phosphate synthetase I of rat. *J. Biol. Chem.* 260: 9346–9356, 1985.
83. Olson, K.R. and P.O. Fromm. Excretion of urea by two teleosts exposed to different concentrations of ambient ammonia. *Comp. Biochem. Physiol.* 40A: 999–1007, 1971.
84. Ottolenghi, C., D. Ricci, M.E. Gavioli, A.C. Puviani, E. Fabbri, A. Capuzzo, L. Brighenti and E.M. Plisetskaya. Separation of catfish hepatocytes by digitonin infusion: some metabolic patterns and hormonal responsiveness. *Can. J. Zool.* 69: 427–435, 1991.
85. Pequin, L. and A. Serfaty. La régulation hépatique et intestinale de l'ammoniémie chez la Carpe. *Arch. Sci. Physiol.* 22: 449–459, 1968.
86. Perlman, D.F. and L. Goldstein. Nitrogen Metabolism. In: *Physiology of Elasmobranch Fishes*, edited by T.J. Shuttleworth. Berlin: Springer-Verlag, 1988, pp. 253–275.
87. Perry, S.F. and P.J. Walsh. Metabolism of isolated fish gill cells: contribution of epithelial chloride cells. *J. Exp. Biol.* 144: 507–520, 1989.
88. Rabier, D., P. Briand, F. Petit, P. Parvy and P. Kamoun. Acute effects of glucagon on citrulline biosynthesis. *Biochem. J.* 206: 627–631, 1982.
89. Randall, D.J., C.M. Wood, S.F. Perry, H. Bergman, G.M.O. Maloiy, T.P. Mommsen and P.A.

Wright. Urea excretion as a stategy for survival in a fish living in a very alkaline environment. *Nature* 337: 165–166, 1989.
90. Randall, D.J. and P.A. Wright. Ammonia distribution and excretion in fish. *Fish Physiol. Biochem.* 31987.
91. Read, L.J. Enzymes of the ornithine–urea cycle in the chimaera *Hydrolagus colliei*. *Nature* 215: 1412–1413, 1967.
92. Read, L.J. The presence of high ornithine–urea cycle enzyme activity in the teleost *Opsanus tau*. *Comp. Biochem. Physiol.* 39B: 409–413, 1971.
93. Rubino, S.D., H. Nyunoya and C.J. Lusty. Catalytic domains of carbamyl phosphate synthetase. *J. Biol. Chem.* 261: 11320–11327, 1986.
94. Rubino, S.D., H. Nyunoya and C.J. Lusty. In vivo synthesis of carbamyl phosphate from NH3 by the large subunit of Escherichia coli carbamyl phosphate synthetase. *J. Biol. Chem.* 262: 4382–4380, 1987.
95. Saha, N. and B.K. Ratha. Effect of ammonia stress on ureogenesis in a freshwater air-breathing teleost, *Heteropneustes fossilis*. In: *Contemporary Themes in Biochemistry. Proceedings of the Fourth Federation of Acian and Oceanian Biochemists Congress*, edited by O.L. Lon, M.C.-M. Chung, P.L.H. Hwang, S.-F. Leong, K.H. Loke, P. Thiyagarajah and P.T.-H. Wong. Cambridge: Cambridge University Press, 1987, pp. 342–343.
96. Saha, N. and B.K. Ratha. Active ureogenesis in a freshwater air-breathing teleost, *Heteropneustes fossilis*. *J. Exp. Zool.* 241: 137–141, 1987.
97. Saha, N. and B.K. Ratha. Comparative study of ureogenesis in freshwater, air-breathing teleosts. *J. Exp. Zool.* 252. 1–8, 1989.
98. Saha, N. and B.K. Ratha. Alterations in excretion pattern of ammonia and urea in a freshwater air-breathing teleost, *Heteropneustes fossilis* (Bloch) during hyper-ammonia stress. *Ind. J. Exp. Biol.* 28: 597–599, 1990.
99. Shigesada, K. and M. Tatibana. N-acetylglutamate synthetase from rat-liver mitochondria. Patial purification and catalytic properties. *Eur. J. Biochem.* 84: 285–291, 1978.
100. Singh, R.A. and S.N. Singh. Liver arginase in air-breathing and non-air-breathing freshwater teleost fish. *Biochem. Syst. Ecol.* 14: 239–241, 1986.
101. Singh, R.A. and S.N. Singh. Tissue distribution, effect of starvation and seasonal variation of arginase in the freshwater teleost, *Clarias batrachus* (L.). *Biochem. Arch.* 4: 329–334, 1988.
102. Smith, H.W. The retention and physiological role of urea in the Elasmobranchii. *Biol. Rev. Camb. Phil. Soc.* 11: 49–82, 1936.
103. Staddon, J.M., N.M. Bradford and J.D. McGivan. Effects of glucagon *in vivo* on the N-acetylglutamate, glutamate and glutamine contents of rat liver. *Biochem. J.* 217: 855–857, 1984.
104. Stieber, S.F. and V.A. Cvancara. Tissue deamination of L-amino acids in the teleost *Stizostedion vitreum* (Mitchill). *Comp. Biochem. Physiol.* 56B: 285–287, 1977.
105. Stryer, L. *Biochemistry*. New York: W.H.Freeman and Company, 1988, edn. 3, pp. 1–1089.
106. Suarez, R.K. and T.P. Mommsen. Gluconeogenesis in teleost fishes. *Can. J. Zool.* 65: 1869–1882, 1987.
107. Tacon, A.G.J. and C.B. Cowey. Nutrition. In: *Fish Energetics*, edited by P. Tytler and P. Calow. Baltimore, MD: Johns Hopkins University Press, 1985, pp. 155–183.
108. Titheradge, M.A. and R.C. Haynes. The hormonal stimulation of ureogenesis in isolated hepatocytes through increases in mitochondrial ATP production. *Arch. Biochem. Biophys.* 201: 44–55, 1980.
109. Van Waarde, A. and F. Kesbeke. Nitrogen metabolism in goldfish, *Carassius auratus* L. Influence of added substrates and enzyme inhibitors on ammonia production by isolated hepatocytes. *Comp. Biochem. Physiol.* 70B: 499–507, 1981.
110. Van Waarde, A. and F. Kesbeke. Nitrogen metabolism in goldfish, *Carassius auratus* L. Activities of amidases and amide synthetases in golfdish tissues. *Comp. Biochem. Physiol.* 71B: 593–603, 1981.
111. Vellas, F. and Y. Creac'h. Importance of purine metabolism in the biosynthesis of urea in the common carp (*Cyprinus carpio* L.). *Arch. Sci. Physiol.* 26: 207–218, 1972.
112. Walsh, P.J., E. Danulat and T.P. Mommsen. Variation in urea excretion in the Gulf toadfish, *Opsanus beta*. *Mar. Biol.* 106: 323–328, 1990.
113. Walsh, P.J. and R.P. Henry. Carbon dioxide and ammonia metabolism and exchange. In: *Biochemistry and Molecular Biology of Fishes*, edited by P.W. Hochachka and T.P. Mommsen. Amsterdam, New York: Elsevier, 1991 pp. 181–207.
114. Walsh, P.J., J.J. Parent and R.P. Henry. Carbonic anhydrase supplies bicarbonate for urea synthesis in toadfish (*Opsanus beta*) hepatocytes. *Physiol. Zool.* 62: 1257–1272, 1989.

115. Walton, M.J. and C.B. Cowey. Aspects of ammoniogenesis in rainbow trout, *Salmo gairdneri*. *Comp. Biochem. Physiol.* 57B: 143–149, 1977.
116. Watts, D.C. and R.L. Watts. Carbamoyl phosphate synthetase in the Elasmobranchii: osmoregulatory function and evolutionary implications. *Comp. Biochem. Physiol.* 17: 785–798, 1966.
117. Webb, J.T. and G.W. Brown. Glutamine synthetase: assimilatory role in liver as related to urea retention in marine chondrichtyes. *Science* 208: 293–295, 1980.
118. Wilson, R.P. Nitrogen metabolism in channel catfish *Ictalurus punctatus* – II. Evidence for an apparent incomplete ornithine–urea cycle. *Comp. Biochem. Physiol.* 46B: 625–634, 1973.
119. Wood, C.M. Acid-base and ionic exchanges at gills and kidney after exhaustive exercise in the rainbow trout. *J. Exp. Biol.* 136: 461–481, 1988.
120. Wood, C.M., S.F. Perry, P.A., Wright, H.L. Bergman and D.J. Randall. Ammonia and urea dynamics in the Lake Magadi tilapia, a ureotelic teleost fish adapted to an extremely alkaline environment. *Resp. Physiol.* 77: 1–20 (1989).

CHAPTER 7

The interface of animal and aqueous environment: strategies and constraints on the maintenance of solute balance

STEPHEN H. WRIGHT

Department of Physiology, University of Arizona, Tucson, AZ 85724, U.S.A.

I. Introduction
II. Categories of transport
IV. Transepithelial solute exchange: transport in series
V. Fish in sea water — mechanisms for active secretion of Na and Cl
   1. Basolateral Na,K,2Cl-cotransport
   2. Apical Cl^- channel
   3. Paracellular Na^+ flux
   4. Basolateral Na,K-ATPase
VI. Fish in fresh water — mechanisms for active absorption of Na and Cl
   1. Apical Na/H exchange
   2. Apical Cl/HCO_3 exchange
   3. Apical electrogenic H-pump and a parallel Na-channel
   4. Basolateral Cl^- efflux
VII. Conclusion
    Acknowlegdements
VIII. References

I. Introduction

Claude Bernard's hallmark of living systems was the maintenance of a *'milieu intérieur'* that is strikingly different from the surrounding environment. For an aquatic animal, this poses some unique problems. Fish are immersed in the same solvent (i.e., water), that is the basis for the solutions found in the internal compartments (cells, intercellular spaces, and blood spaces), comprising the several, separate milieus of the animal. Consequently, the aqueous environment represents an infinite sink that constantly threatens to pull the animal from its routine, steady-state condition, toward an equilibrium condition (i.e., equal solute concentrations inside and out) that represents death. For a marine teleost, whose body fluids are dilute compared to its aqueous habitat, the threat is the accumulation of salt, primarily Na^+ and Cl^- (and a loss of water). The freshwater fish is faced with the opposite problem, a constant loss of NaCl (and gain of water). The maintenance of steady-state, homeostatic conditions within the internal compartments of fish (and these are remarkably similar for marine and fresh-water fish) is the result of the constant expenditure of energy to either lose or gain salts. The interface between animal and the aqueous environment is the front where these exchanges must take place, and they occur under the same set of energetic constraints, whether the issue is excreting salts from a marine teleost or absorbing salts from dilute pond water into a freshwater fish.

To meet each of these challenges, animals make use of a limited, but marvelously adaptable, set of cellular transport processes. In this chapter I will introduce the armamentarium of transport processes available for the exchange of solutes between fish and environment, and current models that have been proposed to account for the maintenance of salt balance in marine and freshwater fish, focusing on exchanges across branchial epithelia (The regulation of salt balance in fish has received considerable attention and has been reviewed elsewhere[26,33].) I will then discuss the thermodynamic constraints of each model. It is worth noting at the outset that adequate information does not exist to provide a rigorous, quantitative test of any of these models. Indeed, it is not my intent to 'prove' or 'disprove' the adequacy of the models under examination, but rather to provide a focus for the kinds of information that will prove useful in the future when (hopefully) appropriate data do exist. In addition, I hope this discussion will prove useful as an overview of the physical factors that influence the strategies used by fish to deal with some of the problems imposed by immersion in their aquatic habitat.

II. Categories of transport

There are a limited number of means by which solutes move across biological membranes (Fig. 1). The simplest is dictated by the lipid-solubility/diffusion characteristics of the solute (Fig. 1B). For inorganic ions, passive diffusion across the lipid bilayer of cell membranes is very slow, owing to the poor solubility of these electrolytes in lipid[24]. The presence of ion-selective channels in cell membranes (Fig. 1C) can increase the permeability of a membrane to ions by many orders of magnitude, thereby permitting a rapid flux of electrolyte, although a net flux is dependent upon the presence of a favorable electrochemical gradient.

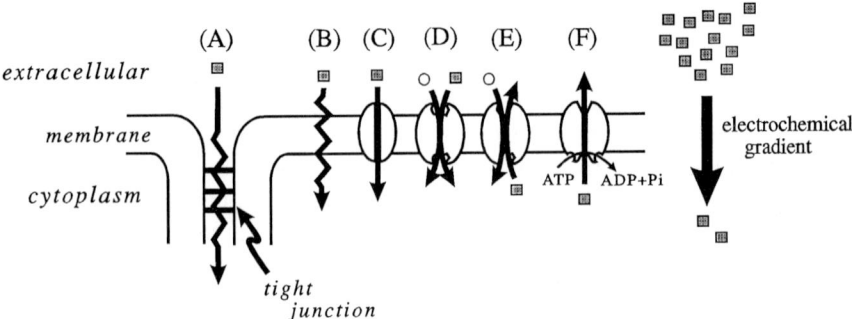

Fig. 1. Schematic representation of modes of solute transport across a cell plasma membrane (B–F), or across an epithelial tight junction (A). Arrows associated with these processes indicate the direction of net solute flux. The electrochemical gradient of solute represented by the square symbols is directed from outside to inside of the cell. Refer to the text for a discussion of the individual processes.

The maintenance of salt balance in fish requires that the animal be capable of moving solute against electrochemical gradients; active transport of Na^+ and Cl^- must form the basis for strategies of salt balance. If the energy source involves a direct coupling of a metabolic energy substrate (i.e., ATP) to the transport event, the process is referred to as primary active transport (Fig. 1F). Primary active transporters are ubiquitous; every cell possess several distinct primary active transporters in the plasma and intracellular membranes. Nevertheless, there are relatively few specific examples of primary active transport. Indeed, a conservative list would include only five unambiguous examples[30]: (1) Na,K-ATPase of the plasma membrane; (2) H-ATPase of selected plasma membranes; (3) the H-ATPase of the lysosomal membrane; (4) Ca-ATPase of the plasma membrane; and (5) Ca-ATPase of the endoplasmic reticulum membrane. There is increasingly compelling evidence for the presence of anion-dependent ATPases in some systems (including the branchial epithelia of some fish[6]), although, these data are controversial. Although limited in number, these processes represent the link between cell metabolism and all forms of active, transmembranous solute flux. It will become evident that primary active transport is the basis for every model of salt balance in fish.

The transmembranous ion gradients established by primary active transporters serve as the immediate source of energy to drive secondary active transport. The flux of the *activator* ion down its electrochemical gradient (the gradient established by primary active transport) provides the energy to move the *object* solute against its electrochemical gradient. There is an amazingly diverse set of such processes (literally dozens have been characterized in different cell types[30]) and it has become clear that secondary active transport is the routine strategy employed by animal cells to move both organic and inorganic solutes against electrochemical gradients.

There are two classes of secondary active transport, distinguished by the relative directions of the 'coupled fluxes' of the activator and object solutes. In a cotransport (symport) process, the activator and object solutes move in the same direction (Fig. 1D). In a countertransport (antiport or exchange) process, the activator and object solutes move in opposite directions (Fig. 1E). It should be emphasized that, like all chemical reactions, secondary active transport is a reversible process; appropriate changes in the relative magnitude of the gradient for an object solute can result in driving the uphill flux of the activator substrate. Indeed, many secondary active transport processes routinely change direction as external forces manipulate the gradients of the cotransported substrates. Thus the designation of activator *vs.* object solutes is an artificial one, but typically represents the relative roles of transported substrates under routine, physiological conditions.

III. Thermodynamic constraints on transport processes

The extent to which a transport process can move substrate against an electrochemical gradient, and the criteria which dictate the direction of an active transport process, become evident upon examination of the thermodynamic relationships that underlie transmembranous transport processes (see ref. 1 for a detailed discussion of these relationships). Consider a solute, A, that moves by means of one of the passive transport processes depicted in Fig. 1A–C. At equilibrium (i.e., when net flux of A by the passive pathways is zero), the distribution of A across the membrane must satisfy the following limiting relationship:

$$\frac{[A]_i}{[A]_o} \leq \exp\frac{-\psi Z_A F}{RT} \quad \text{(Eqn. 1)}$$

in which the subscripts i and o refer to the chemical concentrations (activities) of solute A, ψ is the electrical potential difference, in volts, across the membrane (outside considered to be 0 V); Z_A is the valence of solute A; and F/RT has its usual value (39.6 V^{-1} at 20°C). Inspection of Eqn. 1 reveals that it is simply a rearrangement of the familiar Nernst equation from which it is possible to calculate the size of opposing chemical and electrical gradients for a solute distributed across a membrane at equilibrium. For instance, the unidirectional flux of a monovalent cation arising from a 10-fold chemical gradient is exactly balanced (i.e., net flux = 0) by a backflux driven by an electrical force of -58 mV (at 20°C).

In primary active transport, the equilibrium distribution of a solute across the membrane can be shifted from the simple relationship represented in Eqn. 1 by adding the free energy of hydrolysis of ATP to the overall reaction. For a hypothetical primary active process in which ATP is hydrolyzed to support the transport of solute A from outside to inside of a cell, the limiting equilibrium distribution ratio is:

$$\frac{[A]_i}{[A]_o} \leq \exp\left(\frac{-\psi Z_A F}{RT} - \frac{\Delta G_{ATP}}{n_A RT}\right) \quad \text{(Eqn. 2)}$$

where ΔG_{ATP} is the free energy of hydrolysis of ATP (approximately -65 kJ · mol^{-1} under prevailing intracellular conditions); and n_A is the number of molecules of A transported per molecule of ATP hydrolyzed.

A portion of the free energy of hydrolysis of ATP is stored in the solute gradient generated by a primary active transporter. By tapping this store of potential energy, secondary active transport can drive the uphill transport of a second (and possibly third) solute. The following relationship describes the limiting distribution ratio for a solute, A, whose flux is coupled to the symport of solute B, and/or to the antiport of solute C:

$$\frac{[A]_i}{[A]_o} \leq \left(\frac{[B]_o}{[B]_i}\right)^{n_B/n_A} \left(\frac{[C]_i}{[C]_o}\right)^{n_C/n_A} \exp\left(\frac{-\psi F}{RT}\left(Z_A + \frac{n_B}{n_A}Z_B - \frac{n_C}{n_A}Z_C\right)\right)$$

$$\text{(Eqn. 3)}$$

where the subscripted values of n represent the numbers of molecules co- or countertransported during each cycle of the transport process.

Ion exchanges between fish and the aqueous medium occur at the branchial epithelial interfaces alluded to earlier. Therefore, before we consider the relationships presented in Eqns. 1, 2 and 3 as a means to test the energetic feasibility of models proposed to account for active ion transport in fish, it is appropriate to consider the unique structural aspects of epithelia that permit net, uphill fluxes between animal and environment.

IV. Transepithelial solute exchange: transport in series

Epithelial cell layers comprise the physiological interface between an animal and the aqueous environment. It is the nature of epithelia to separate adjacent compartments (e.g., the blood and extracellular spaces of an animal *vs.* the surrounding, aqueous environment). But rather than serve merely as a barrier that prevents mixing of the contents of these different compartments, epithelia also play an active role in defining and maintaining the chemical differences that are the basis of Bernards' concept of the *milieu intérieur*. This capability arises as a result of the unique, polarized structure of the epithelial cell that has a plasma membrane face exposed to one compartment, and a second, biochemically and physiologically distinct plasma membrane domain exposed to the opposing compartment ((3); consider the models presented in Figs. 2 and 3). Adjacent cells are connected to one another by a continuous band of junctional material called a tight junction (*zonula occludens*), which is localized at the apical face of the epithelium and forms a barrier between cells thereby limiting free diffusion between the opposing compartments *via* a paracellular pathway. With respect to the branchial epithelium, the apical pole is that aspect of the cell exposed to the environment. The basolateral pole is exposed to the extracellular/blood space of the animal. The branchial epithelium of fish contains several different cell types and these have been described in detail elsewhere (e.g., ref. 25). Current evidence

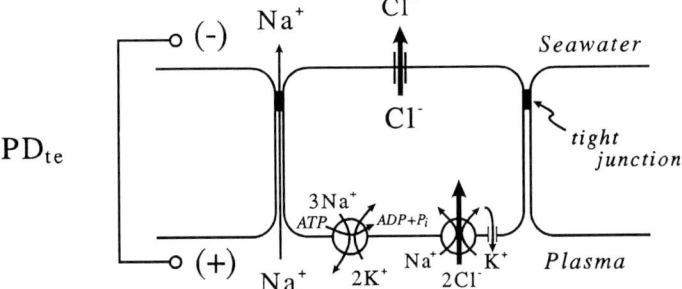

Fig. 2. Schematic representation of a current model for active, transepithelial secretion of Na^+ and Cl^- by marine fish. Refer to the text for a discussion of the individual processes involved.

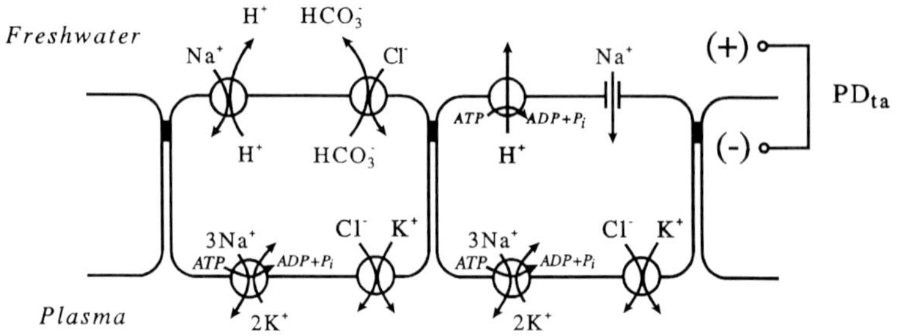

Fig. 3. Schematic representation of current models for active, transepithelial absorption of Na^+ and Cl^- by freshwater fish. Refer to the text for a discussion of the individual processes involved.

strongly supports the contention that a subpopulation of cells found in the gill and opercular epithelium, termed chloride cells, is the site of active salt extrusion from fish in seawater[13,14]. The site(s) of active salt accumulation in freshwater fish is not as clear[25].

V. Fish in sea water – mechanisms for active secretion of Na and Cl

Key's initial studies[22] of salt secretion by marine fish implicated Cl^- as the ion actively secreted by branchial epithelia. This suggestion was later contested (reviewed in ref. 33), but more recent data[13,14] has clearly shown that 'chloride cells' in branchial epithelia of saltwater adapted fish are the site for an active extrusion of Cl^-. Fig. 2 presents a working model of the most widely accepted model for active secretion of Na^+ and Cl^- from sea water-adapted fish. There are four basic elements to the active extrusion of Cl^- in this model. (1) Cl^- enters the cell across the basolateral membrane by a cotransporter that couples the flux of Na, K and 2Cl ions. This process is inhibitable by the renal diuretics, furosemide and bumetinide and has characteristics similar to those noted in other salt-transporting epithelia, including medullary thick limbs of Henle's loop in the mammalian kidney[15], and rectal salt glands of elasmobranch fish[17]. This process serves to drive Cl^- into the cytoplasm of the cell to a level well above the predicted equilibrium distribution for this ion, and thus produces an outwardly directed electrochemical gradient for Cl^-. (2) This outwardly directed gradient supports the passive, downhill exit of Cl^- from the cell, which is limited to a *trans*-apical flux by means of a Cl-specific channel. (3) The resulting net transepithelial Cl^- flux is the basis of a transepithelial electrical potential difference (PD_{te}) in which the seawater aspect of the epithelium is electrically negative compared to the blood space compartment. The PD_{te} supports a passive paracellular flux of the positively charged Na; i.e., Na^+ moves from a relatively low concentration in the blood to a high concentration in sea water drawn by the electrical force generated by the Cl^- diffusion potential. (4) The overall process is driven by the primary active Na,K-ATPase which is located in the basolateral membrane. Continued activity of this process maintains the inwardly directed Na^+ gradient which drives Cl^- uptake

TABLE 1

Concentration of major inorganic monovalent ions in seawater and the plasma and branchial cytoplasm of saltwater fish

Ion	Concentration (mM)	Reference
$[Na^+]_{plasma}$	170	27
$[Na^+]_{cell}$	37	16
$[Na^+]_{seawater}$	450	
$[K^+]_{plasma}$	3	32
$[K^+]_{cell}$	110	17
$[Cl^-]_{plasma}$	150	27
$[Cl^-]_{seawater}$	550	

into the cells; and the electrical profile of the cell layer, which helps support the paracellular flux of Na^+.

With respect to our discussion of the energetic feasibility of this model of salt secretion, let us consider each of these individual processes in turn.

1. Basolateral Na,K,2Cl-cotransport

This process has the task of raising intracellular Cl^- to a chemical concentration in excess of chloride's equilibrium distribution. Eqn. 3 can be rearranged to provide the appropriate limiting activity ratio for Cl^-:

$$\frac{[Cl]_{cell}}{[Cl]_{plasma}} \leq \left(\frac{[Na]_{plasma}}{[Na]_{cell}}\right)^{1/2} \left(\frac{[K]_{plasma}}{[K]_{cell}}\right)^{1/2} \qquad (Eqn. 4)$$

The coupled flux of two monovalent cations with two anions causes the electrical term to drop out this relationship, leaving an equilibrium distribution of solutes that is simply the product of opposing chemical gradients. Obviously, the key elements in determining the maximum intracellular concentration of Cl^- is knowledge of intracellular activities of Na^+ and K^+; these are not known for any branchial epithelial cell. Nevertheless, it is instructive to insert 'generic' values for these parameters, taken from studies of other salt-secreting epithelia (Table 1), thereby permitting the estimate of a maximal concentration of $[Cl^-]_{cell}$; the values in Table 1 suggest a maximal value for intracellular Cl^- of 53 mM, which agrees well with measured values for Cl^- distribution driven by the Na,K,2Cl-cotransporter in other salt secreting epithelia[17].

Recognizing the risk of constructing a teleological argument, it is intriguing to consider 'why' a Na,K,2Cl-cotransporter appears to be the mechanism of choice for accumulation of Cl^- in salt-secreting epithelia, in contrast to a NaCl cotransporter (which has been found in some absorptive tissues[28]). Both processes are capable of using the Na-electrochemical gradient to generate an outwardly directed electrochemical gradient for Cl^-. However, by coupling the Cl^- flux to an obligate uphill flux of K^+, the Na,K,2Cl-cotransporter 'splits' the energy in the Na gradient thereby limiting the maximum intracellular Cl^- concentration to approximately 50 mM. In contrast, an NaCl-cotransporter would be in equilibrium with an intracellular Cl^- concentration of almost 700 mM. This degree of transport potential would have to be tightly regulated to avoid compromising the routine maintenance of normal cell volume and the activity of cytoplasmic enzymes[29]. In addition, the concomitant accumulation of K^+ with Cl^- provides an 'assist' to the Na,K-ATPase (see upcoming discussion) in terms of maintaining an outwardly directed K^+ gradient. The electrogenic efflux of K *via* a basolateral K-channel (see ref. 16; Fig. 2), helps maintain the inside-negative membrane potential that is required to drive Cl^- out of the cell through a channel in the apical membrane.

2. Apical chloride channel

The Cl^- channel permits the exchange of Cl^- across the apical membrane between the cell and seawater. For this to represent a net secretory flux from the cell, the distribution of Cl^- across the apical membrane must reflect a non-equilibrium condition. The limiting condition describing this relationship is determined from Eqn. 1:

$$\frac{[Cl]_{cell}}{[Cl]_{sw}} \leq \exp\frac{-\psi Z_{Cl} F}{RT} \qquad (\text{Eqn. 5})$$

At the present time there is only one set of measurements for the apical membrane potential in a saltwater adapted fish. Zadunaisky and colleagues[32] reported a value of 18 mV (cell negative) across the apical membrane of Cl^- cells in the short-circuited, opercular epithelium of saltwater adapted *Fundulus heteroclitus*. Using this value for apical PD, the minimum value of $[Cl^-]_{cell}$ that will satisfy the relationship presented in Eqn. 5 is 269 mM, well in excess of the maximum value of 53 mM for cell Cl^- predicted to occur *via* the cotransporter. Indeed, if we assume that cell Cl^- does not exceed an upper level of approximately 70 mM[17], then the apical potential difference must be on the order of 52 mV (inside negative) to drive this anion into a sea water chloride concentration almost eight times larger than that of the cell! Potentials of this size have been measured in salt secreting epithelia of other fish (shark rectal glands[17]).

3. Paracellular Na flux

Like the apical flux of Cl^-, the net secretory flux of Na^+ depends upon an adequate electrical driving force to move this compound from a relatively low concentration in the blood (170 mM) into the high Na^+ concentration of seawater (450 mM). In this case, the driving force comes from the transepithelial potential of the branchial cell layer. There have been many measurements of transepithelial PD in fish epithelia (see ref. 27), and for most marine teleosts the PD_{te} is approximately 25 mV, blood side positive (see Fig. 2). Given the steady-state transepithelial distribution of Na (450 mM out vs. 170 mM in), a 25 mV potential represents an equilibrium potential for transepithelial Na^+; i.e., the opposing chemical and electrical forces acting on Na^+ would result in no net transepithelial flux. However, Potts[27] has suggested that the concentration of Na^+ in the intercellular spaces of the branchial epithelium is likely to be higher than the value measured in plasma, owing to the localized activity of the basolateral Na,K-ATPase. This would provide the imbalance in the opposing chemical and electrical forces required to produce a net, passive secretory flux of Na^+ from the blood to seawater via the paracellular route.

It is worth noting that, despite my use of a generic PD_{te} value of +25 mV, the literature includes studies of fish with PD_{te} values that are near 0 or that are actually negative (see ref. 27). Clearly, such results are inconsistent with the model proposed in Fig. 2, and emphasize the need to consider the possibility of multiple strategies for the maintenance of ion balance in saltwater fish.

4. Basolateral Na,K-ATPase

Elegant studies have shown that the Na,K-ATPase is localized to the basolateral aspect of branchial epithelia and, furthermore, is enriched in the chloride cells[21]. The routine cycling of this process results in the extrusion of three Na^+ ions and the accumulation of two K^+ ions at the expense of one molecule of ATP. This, then, represents the component of the secretory cycle in which metabolic energy is transduced into a form used directly in the actual secretory events discussed above. As it is the Na electrochemical gradient that serves to drive Cl^- transport via the basolateral cotransporter, it is instructive to consider the thermodynamic limit on the size of this gradient that can be maintained by activity of the primary Na,K-pump. Use of the thermodynamic concepts presented in Eqns. 2 and 3 results in an equation that sets a lower limit on the activity of cytoplasmic Na^+ from which active, uphill transport can occur:

$$\frac{[Na]_{plasma}}{[Na]_{cell}} \leq \left(\frac{[K]_{plasma}}{[K]_{cell}}\right)^{n_K/n_{Na}} \exp\left(\frac{1/RT(\psi F - \Delta G_{ATP})}{n_{Na}}\right) \quad \text{(Eqn. 6)}$$

In solving for $[Na]_{cell}$, I have used the Table 1 values for $[Na]_{plasma}$, $[K]_{cell}$, and $[K]_{plasma}$. I have also used a value of -75 mV for the electrical potential difference across the basolateral membrane. This choice reflects (1) the average transepithelial PD of 25 mV (blood side negative); and (2) the 'predicted' value of 50 mV (cell negative) for the apical membrane potential that will support an active Cl^- secretory flux. Finally, a value of -65 kJ·mol^{-1} represents a reasonable estimate of ΔG_{ATP} under the typical conditions of low intracellular ADP and inorganic phosphate. The resulting estimate for minimum $[Na]_{cell}$ is 0.7 mM. This value is substantially lower than typical measured values for total cell Na in marine animals, and is lower than estimates of cell Na$^+$ activity derived from studies with Na-sensitive microelectrodes (e.g., ref. 16).

It is also instructive to consider the relationship between the energetic potential of ATP (i.e., -65 kJ·mol^{-1}) and the observed steady state conditions for transepithelial distribution of the actively transported ions, Na and Cl. As noted earlier, the Na$^+$ distribution between plasma and seawater is virtually at equilibrium, i.e., the balance of chemical and electrical forces represents a ΔG of approximately zero. Instead, it is the chloride distribution that represents the 'uphill' movement, and the free energy cost per mole of Cl$^-$ transported is given by the relationship:

$$G = F\psi_{te} + RT \ln \frac{[Cl]_{seawater}}{[Cl]_{plasma}} \qquad (Eqn.\ 7)$$

Thus, the steady-state distribution of Cl$^-$ between a marine teleost and seawater 'costs' only 5.6 kJ mol^{-1}.

It is evident from this analysis that there is adequate energy in the hydrolysis of ATP to support the solute gradients that exist between fish and seawater, and between epithelial cells and the blood. Furthermore, the basolateral Na,K,2Cl-cotransporter should be able to sustain a non-equilibrium distribution of Cl$^-$, required if a net, *trans*-apical Cl$^-$ flux is to occur. However, it is also clear that the *trans*-apical PD (PD$_{ta}$) of chloride cells must be larger than current measurements suggest, if an apical chloride channel is to be effective in serving as the avenue for efflux in Cl$^-$ secretion. Indeed, this is likely the case. The technical difficulties associated in measuring intracellular potentials in chloride cells using conventional microelectrodes are extreme; the cells are small and the complex network of basolateral infoldings make it difficult to secure unambiguous recordings. Furthermore, it is not known to what extent the membrane potential will change when exposed to the 'open-circuit' conditions of the *in situ* system, including the asymmetric distribution of ions represents by seawater *vs.* plasma. Thus, until such measurements are made, using conventional or new technologies (e.g., potential-sensitive fluorescent dyes), the model of salt secretion in branchial epithelia presented in Fig. 2 must be accepted with the *caveats* noted here.

VI. Fish in fresh water – mechanisms for active absorption of Na and Cl

The environmental interface of a freshwater fish is the site of a constant, outwardly directed set of chemical forces that, if left unchecked, would lead to a depletion of solutes to the dilute external solution *. Stemming from the early studies of Krogh (reviewed in ref. 23), the accumulation of NaCl by freshwater-

---

* Unlike 'seawater', the chemical composition of 'freshwater' is not well-defined. We can, however, use an operational definition of freshwater as a medium having a NaCl concentration of <1 mM, and, for the purposes of this discussion, will use a value for ambient Na$^+$ and Cl$^-$ of 0.5 mM.

adapted fish has been linked to the excretion of H^+ and HCO_3^-. The results of a large number of studies (reviewed in ref. 25) lead to a general model for NaCl absorption that involves the action of two antiporters, Na/H exchange and Cl/HCO_3^- exchange, working in parallel within the apical membrane of branchial cells (Fig. 3). The net activity of this model involves four steps: (1) an 'uphill' entry of Na^+ from the external solution across the apical membrane, driven by an outwardly directed gradient for H^+; (2) an uphill movement of Cl^- across the apical membrane, driven by an outwardly directed gradient for HCO_3^-; (3) the uphill exit of Na^+ across the basolateral membrane via the Na,K-ATPase; and (4) the exit of Cl^- across the basolateral membrane by an unknown mechanism. Unfortunately, assessment of the energetic feasibility of each of these steps is currently even more limited by a lack of information pertaining to the relevant forces than is the situation for saltwater fish. We can, however, make some use of data from studies with freshwater-adapted frogs, as the frog skin shares many of the same tasks and faces the same energetic constraints as fish branchial epithelia.

1. Apical Na / H exchange

The evidence supporting the presence of the mediated exchange of Na^+ for H^+ is based primarily upon the observed inhibitory effect of the diuretic, amiloride, on uptake of Na^+ in freshwater fish[10]. The process must be capable of drawing Na^+ from an external concentration (activity) of 0.5 mM (or perhaps lower) into a cellular concentration that probably exceeds 1 mM *. The limiting ratio is:

$$\frac{[Na]_{cell}}{[Na]_{fw}} \leq \frac{[H]_{cell}}{[H]_{fw}} \qquad \text{(Eqn. 8)}$$

The outwardly directed gradient for H^+ must be larger than 2:1 (i.e., at least 0.3 pH units) to sustain the minimal Na gradient one can calculate from Table 2. Cytoplasmic pH in branchial epithelia has not been measured directly, but, based upon measurements in other cell types, and given the necessity to maintain an appropriate environment for cytoplasmic metabolism[29], pH_{cell} is likely to be on the order of 7.2–7.6. The pH of freshwater can be extremely variable; in an incomplete review of the literature I found reports in which trout (*Salmo gairdneri*) were acclimated in water with pH that varied from 8.0[2] to 6.8[31]. Under the former conditions, a proton driving force would seem capable of supporting active accumulation of Na^+, whereas under the latter conditions, the energetic constraints of the process make it an unlikely candidate as the primary mechanism for sodium absorption. There has been speculation that Na^+ absorption could also be driven

---

* Intracellular Na concentration/activity is not known for any freshwater teleost epithelium. However studies of freshwater-adapted frog skin employing X-ray microanalysis suggest a cellular Na content of less than 5–10 mM[7]. Ion sensitive microelectrode studies suggest an intracellular activity as low as 1 mM in low Na-adapted (i.e., 10 μM Na) animals[20]. The use of 1 mM for intracellular Na^+ is a reasonably conservative lower limit.

TABLE 2

Concentration of major inorganic monovalent ions in freshwater and the plasma and branchial cytoplasm of freshwater fish

Ion	Concentration	Reference
$[Na^+]_{plasma}$	138 mM	4
$[Na^+]_{cell}$	1–5 mM	7, 20
$[Na^+]_{freshwater}$	0.5 mM	
$[K^+]_{plasma}$	6 mM	4
$[K^+]_{cell}$	–	
$[Cl^-]_{plasma}$	146 mM	4
$[Cl^-]_{cell}$	5–20 mM	20
$[Cl^-]_{seawater}$	0.5 mM	
$[H^+]_{plasma}$	25 nM (pH 7.6)	4
$[H^+]_{cell}$	50 nM (pH 7.3)	19
$[H^+]_{freshwater}$	10–169 nM (pH 8–6.8)	2, 31

by NH_4^+ as a replacement ion for H^+ (see ref. 6). However, the evidence linking NH_4^+ secretion with Na^+ absorption is based primarily upon whole-animal studies and alternative explanations for this linkage have been proposed[11]. At this time, therefore, the energetic requirements for Na/H exchange make it a feasible strategy for Na^+ accumulation in branchial epithelia, although the process would be extremely sensitive to environmental pH.

2. Apical chloride / bicarbonate exchange

This process must also drive its object solute (Cl^-) from a low external concentration into an intracellular concentration which, based upon studies with freshwater-adapted frog skin[20] may be on the order of 5 to 20 mM. Therefore, a mechanism for the active accumulation of Cl^- must be able to sustain a minimum of a 10-fold, outwardly directed gradient. As with the preceding process, there is little information concerning the cytoplasmic concentrations of the relevant ions. We can estimate cytoplasmic HCO_3^-, however, from the Henderson-Hasselbalch relationship for a CO_2–bicarbonate buffer system. Using an arterial PCO_2 of 3 mmHg (solubility coefficient of 0.056), a plasma pH (at 20°C) of 7.6, and a pK for carbonic acid of 6.1, we can estimate a plasma bicarbonate concentration of approximately 5 mM. Although cytoplasmic pH is probably lower than 7.6 due to metabolic production of CO_2, cytoplasmic bicarbonate is still likely to be on the order of 5 mM. Owing, at least in part, to the low salt concentration and P_{CO_2} of 'freshwater', the ambient bicarbonate concentration is probably < 1 mM; i.e., an outwardly directed HCO_3^- gradient of between 5 and 10:1 is possible. Following from the relationship shown in Eqn. 8, apical Cl/HCO_3^- exchange is feasible as a means for sustaining intracellular Cl^- concentrations, although, without a mechanism for elevating cytoplasmic HCO_3^-, it would seem to be at the edge of energetic credibility.

3. Apical electrogenic H-pump and a parallel Na-channel

Several years ago evidence began to accumulate suggesting that accumulation of Na^+ from dilute external solution across frog skin involved the parallel activity of an electrogenic, H-ATPase and an amiloride-inhibitable Na-channel, located in the apical membrane of the 'mitochondria-rich' cells[8,9,19]. These data offered an alternative mechanism for the observed link between Na^+ absorption and acid secretion in frog skin, a link similar to that noted for fish. Despite a current lack of direct evidence for the presence of such a collection of transporters in the branchial epithelia of freshwater fish, this model (depicted in the righthand cell of Fig. 3) is consistent with many of the observations made using intact fish and preparations of perfused fish heads. Therefore, it seems appropriate to offer it here as a possible strategy used by freshwater fish, as well as by freshwater amphibians, to absorb Na^+ (see ref. 2).

The energetic feasibility of the combined activity of these processes is readily apparent (and discussed with respect to frog skin in ref. 8). The stoichiometric characteristics of the postulated pump are unknown, but, based upon the activity of electrogenic H-pumps in other systems[5], it could involve the electrogenic transfer of two H^+ per ATP hydrolyzed, against a steady-state apical membrane potential of approximately 50 mV (inside negative)[19]. From Eqn. 2 it can be shown that a primary active pump with these postulated characteristics could generate an H^+ gradient of approximately 5 pH units!

The electrogenic activity of this pump results in a net extrusion of positive charge from the cell across the apical membrane, thereby hyperpolarizing that membrane. Indeed, inhibition of this process in frog skin depolarizes the PD_{ta} by approximately 20 mV (from 47 to 25 mV, inside negative). While this process is active, the inside negative PD can serve as an attractive force to draw external Na up a six-fold chemical gradient; e.g., from an external concentration of 0.5 mM into a cellular concentration of 3 mM (refer to Eqn. 1), well within reported levels for freshwater-adapted epithelial cells[19].

A significant 'byproduct' of an apical H-pump is the generation of cytoplasmic HCO_3^- (by means of mass action). It is attractive to speculate that this could be the means by which the branchial epithelium increases the outwardly directed gradient for bicarbonate and thus increases the driving force for accumulation of Cl^-.

4. Basolateral chloride efflux

Current models for branchial absorption of NaCl in freshwater fish do not include a specific mechanism for the efflux of Cl^- from cells to blood. Although the electrical gradient (an inside negative PD) will favor Cl^- efflux, we cannot say without knowing the PD_{bl} and $[Cl]_{cell}$ whether the combined electrochemical gradient will favor a net efflux of Cl^- from the cell. A favorable gradient would be required if Cl^- efflux were limited to a conductive pathway (e.g., a Cl^- channel). Another possibility is a cotransport of K^+ with Cl^- (Fig. 3); evidence for such a process has been found in several absorptive epithelia (e.g., refs. 12, 18). The K^+

gradient directed outwardly from the branchial cell probably exceeds 30:1 and so has a potential energy equivalent to a membrane potential of approximately 85 mV (refer to Eqn. 1). Thus, linking Cl^- flux to the K^+ gradient, rather than to the membrane potential, provides an energetically feasible means to actively secrete Cl^- from an epithelial cell. I should stress, however, that there is currently no evidence concerning the specific mode of Cl^- efflux from the branchial cells of freshwater fish.

VII. Conclusion

The 'strategies' used in the maintenance of normal solute balance in fish are quite simple: marine fish must secrete NaCl, whereas freshwater fish must accumulate salt from the environment. The 'tactics' whereby these required tasks are carried out are less clear. There have proven to be a limited number of general categories of transport mechanism by which solutes cross membranes, although there are an enormous number of variations on these few themes. Here I have considered thermodynamic constraints common to all transport phenomena and used these as a means to test the feasibility of some of the models currently proposed to account for transepithelial NaCl transport in fish gills and opercular epithelia. It is clear that primary active transport processes (e.g., the Na,K-ATPase and, possibly, an H-ATPase) have ample potential to develop and sustain ion gradients which can serve to drive active salt transport by means of an appropriately arrayed set of secondary active transporters. At issue, then, is whether the collection of transporters proposed to be involved in salt-secretion or absorption are associated with steady-state gradients of chemical and electrical force capable of sustaining observed solute gradients between fish and aqueous environment. In most of the cases examined, the answers have been rather equivocal; tests of current models are frustrated by a paucity of data for fish branchial epithelia on such crucial issues as cell ions activities and electrical membrane potentials, the forces which influence net transepithelial solute flux. Nevertheless, I hope this exercise has focused attention on the need to consider the value of such measurements in providing, ultimately, the data required for rigorous tests of salt secretion models. In addition, I hope this discussion will prove useful for those interested in considering in a quantitative fashion energetic aspects of membrane transport.

Acknowledgements. Preparation of this review was supported through NSF Award DCB88-19367.

VIII. References

1. Aronson, P.S. Identifying secondary active solute transport in epithelia. *Am. J. Physiol.* 240: F1–F11, 1981.
2. Avella, M. and M. Bornancin. A new analysis of ammonia and sodium transport through the gills of the freshwater rainbow trout (*Salmo gairdneri*). *J. Exp. Biol.* 142: 155–175, 1989.

3. Berridge, M.J. and Oschman, J.L. *Transporting Epithelia*, New York: Academic Press, 1972, pp. 1–91.
4. Bornancin, M., I. Isia and A. Masoni. A re-examination of the technique of isolated perfused trout head preparation. *Comp. Biochem. Physiol.* 81A: 35–41, 1985.
5. Carty, S.E., R.G. Johnson and A. Scarpa. H^+-translocating ATPase and other membrane enzymes involved in the accumulation and storage of biological amines in chromaffin granules. In: *The Enzymes of Biological Membranes. Vol. 3*, edited by A.N. Martonosi. New York: Plenum Press, 1985, p. 449–495.
6. de Renzis, G. and M. Bornancin. Ion transport and gill ATPases. In: *Fish Physiology. Vol. X, Part B: Ion and Water Transfer*, edited by W.S. Hoar and D.J. Randall. New York: Academic Press, 1984, pp. 65–104.
7. Dörge, A., R. Rick, R. Bauer, Ch. Roloff, E. Arnim and K. Thurau. An X-ray microanalysis of cellular electrolytes in frog skin epithelium: the site of action of amiloride and the pathway of Na. In: *Amiloride and Epithelial Sodium Transport*, edited by A.W. Cuthbert, G.M. Fanelli, Jr. and A. Scriabine. Baltimore: Urban & Schwarzenberg, 1979, pp. 101–112.
8. Ehrenfeld, J., F. Garcia-Romeu and B.J. Harvey. Electrogenic active proton pump in *Rana esculenta* skin and its role in sodium ion transport. *J. Physiol. (Lond.)* 359: 331–355, 1985.
9. Ehrenfeld, J., I. Lacoste and B.J. Harvey. The key role of the mitochondria-rich cell in Na^+ and H^+ transport across the frog skin epithelium. *Pflügers Arch.* 414: 59–67, 1989.
10. Evans, D.H. The roles of gill permeability and transport mechanisms in euryhalinity. In: *Fish Physiology. Vol X. Gills, Part B: Ion and Water Transfer*, edited by W.S. Hoar and D.J. Randall. New York: Academic Press, 1984, pp.239–283.
11. Evans, D.H. and J.N. Cameron. Gill ammonia transport. *J. Exp. Zool.* 239: 17–23, 1986.
12. Eveloff, J. and D.G. Warnock. K-Cl transport systems in rabbit renal basolateral membrane vesicles. *Am. J. Physiol.* 252: F883–F889, 1987.
13. Foskett, J.K. and T.E. Machen. Vibrating probe analysis of teleost opercular epithelium: correlation between active transport and leak pathways of individual chloride cells. *J. Membrane Biol.* 85: 25–35, 1985.
14. Foskett, J.K. and C. Scheffey. The chloride cell: definitive identification as the salt-secretory cell in teleosts. *Science* 215: 164–166, 1982.
15. Greger, R. Ion transport mechanisms in thick ascending limb of Henle's loop of mammalian nephron. *Physiol. Rev.* 65: 760–797, 1985.
16. Greger, R. and E. Schlatter. Mechanism of NaCl secretion in rectal gland tubules of spiny dogfish (*Squalus acanthias*) II. Effects of inhibitors. *Pflügers Arch.* 402: 364–375, 1984.
17. Greger, R. and E. Schlatter. Mechanism of NaCl secretion in the rectal gland of the spiny dogfish (*Squalus acanthias*) I. Experiments in isolated in vitro perfused rectal gland tubules. *Pflügers Arch.* 402: 63–75, 1984.
18. Guggino, W.B. Functional heterogeneity in the early distal tubule of the *Amphiuma* kidney: evidence for two modes of Cl^- and K^+ transport across the basolateral membrane. *Am. J. Physiol.* 250: F430-F440, 1986.
19. Harvey, B.J. and J. Ehrenfeld. Regulation of intracellular sodium and pH by the electrogenic H^+ pump in frog skin. *Pflügers Arch.* 406: 362–366, 1986.
20. Harvey, B.J. and R.P Kernan. Intracellular ionic activities in frog skin in relation to external sodium concentration and in the presence of amiloride and/or ouabain. *J. Physiol. (Lond.)* 349: 501–517, 1984.
21. Karnaky, K.J, Jr., L.B. Kinter, W.B. Kinter and C.E. Stirling. Teleost chloride cell. II. Autoradiographic localization of gill Na,K-ATPase in killifish *Fundulus heteroclitus* adapted to low and high saline environments. *J. Cell Biol.* 70: 157–177, 1976.
22. Keys, A.B. The heart-gill preparation of the eel and its perfusion for the study of a natural membrane *in situ*. *Z. Vergl. Physiol.* 15: 352–363, 1931.
23. Krogh, A. *Osmotic Regulation in Aquatic Animals*, Cambridge: Cambridge University Press, 1939.
24. Pappone, P.A. and M.D. Cahalan. Ion permeation in cell membranes. In: *Physiology of Membrane Disorders*, edited by T.E. Andreoli, J.F. Hoffman, D.D. Fanestil and S.G. Schultz. New York: Plenum Press, 1986, pp. 249–272.
25. Payan, P. and J.P Girard. Branchial ion movements in teleosts: the roles of respiratory and chloride cells. In: *Fish Physiology. Vol. X, Part B: Ion and Water Transfer*, edited by W.S. Hoar and D.J. Randall. New York: Academic Press, 1984, pp. 39–63.
26. Pequeux, P., R. Gilles and W.S. Marshall. NaCl transport in gills and related structures. In: *Advances in Comparative and Environmental Physiology. Vol. 1. NaCl Transport in Epithelia*, edited by R. Greger. New York: Springer-Verlag, 1988, pp. 2–73.

27. Potts, W.T.W. Transepithelial potentials in fish gills. In: *Fish Physiology. Vol X, Part B: Ion and Water Transfer*, edited by W.S. Hoar and D.J. Randall. New York: Academic Press, 1984, pp. 105–128.
28. Powell, D.W. Ion and water transport in the intestine. In: *Physiology of Membrane Disorders*, edited by T.E. Andreoli, J.F. Hoffman, D.D. Fanestil and S.G. Schultz. Plenum Press, New York: 1986, pp. 559–596.
29. Somero, G.N. Protons, osmolytes, and fitness of internal milieu for protein function. *Am. J. Physiol.* 251: R197–R213, 1986.
30. Stein, W.D. *Transport and Diffusion across Cell Membranes*, New York: Academic Press, 1986, pp. 1–685.
31. Williams, E.M. and F.B. Eddy. Chloride uptake in freshwater teleosts and its relationship to nitrite and toxicity. *J. Comp. Physiol. B* 156: 867–872, 1986.
32. Zadunaisky, J., S. Curci, T. Schettino and J.I. Scheide. Intracellular voltage recordings in the opercular epithelium of *Fundulus heteroclitus*. *J. Exp. Zool.* 247: 126–130, 1988.
33. Zadunaisky, J.A. The chloride cell: the active transport of chloride and the paracellular pathways. In: *Fish Physiology. Vol. X. Gills, Part B: Ion and Water Transfer*, edited by W.S. Hoar and D.J. Randall. New York: Academic Press, 1984, pp. 129–176.

CHAPTER 8

Carbon dioxide and ammonia metabolism and exchange

PATRICK J. WALSH * AND RAYMOND P. HENRY **

*Division of Marine Biology and Fisheries, Rosenstiel School of Marine and Atmospheric Science, University of Miami, Miami, FL 33149–1098, U.S.A., and **Department of Zoology and Wildlife Science, Auburn University, Auburn, AL 36849–4201, U.S.A.*

I. Introduction
II. Metabolism of carbon dioxide and ammonia
   1. Were carbon dioxide and ammonia always waste products?
   2. Reactions involving carbon dioxide and ammonia
      2.1. Features of carbon dioxide metabolism
      2.2. Features of ammonia metabolism
      2.3. Urea synthesis
III. Production and transport of carbon dioxide and ammonia
   1. Carbon dioxide
   2. Ammonia
IV. Epilogue
V. References

I. Introduction

Carbon dioxide and ammonia * are both gases that are unique in physiological and biochemical systems because of their high solubilities in water; the solubility, or capacitance of carbon dioxide in water is roughly 28 times that for O_2 and for ammonia it is 23,000-fold higher[9,10]. The total capacities of water for carbon dioxide and ammonia are even greater as both CO_2 and NH_3 undergo chemical reactions with water and thus exist as both gaseous and ionic species. Other gases are highly soluble in water (e.g., nitrogen) but are chemically inert and thus have less biochemical consequence. CO_2 in water forms an acid, and because the pK' for the carbon dioxide system is about 6.1 (see ref. 9, for the rationale of defining a pK' for carbon dioxide *system*), at many environmental pHs (e.g., seawater = 8.0), and at *resting* physiological pH in fish, over 95% of the total carbon dioxide is in the form of HCO_3^-. Conversely, NH_3 in water acts as a base, and since the pK for ammonia is high (pK = 9–10, ref. 10) at typical environmental and physiological pHs about 99% of ammonia is in the protonated form, NH_4^+.

    The aqueous solubility and pK of both carbon dioxide and ammonia are dependent on physical factors such as temperature, salinity, and pH. Furthermore,

---
* We refer to the total quantities of all chemical forms as 'carbon dioxide' and 'ammonia', but we will refer to specific chemical forms by their chemical formulae.

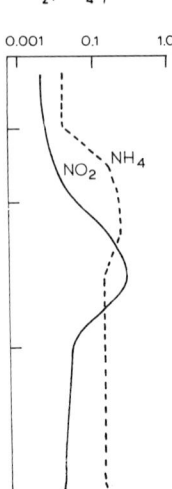

Fig. 1. Vertical profile of [ammonia] and [nitrite] (note logarithmic scale) *vs.* depth at a station in the Gulf of Maine. The peak of ammonia is attributed to a combination of physical and biological factors (from ref. 51).

the total amounts of both substances are influenced by biological activity such as photosynthesis and denitrification (e.g., Fig. 1, ref. 51). The end result is that, even in a large and ostensibly homogeneous environment such as the world ocean, ambient carbon dioxide and ammonia are highly variable both spatially and temporally. This variability is higher in estuarine waters, and it reaches extremes in environments such as tide pools. In such environments diurnal fluctuations in pH can be as great as 2 units (7.6 to 9.7), pCO_2 can vary by three orders of magnitude (10^{-3} to 2 Torr), and total carbon dioxide fluctuates between 0.8 and 2.1 mM[103]. Variability of the chemical species of carbon dioxide are also apparent in freshwater environments. In streams values for pH fall between 7.2 and 7.6, pCO_2 between 0.1 and 0.8 Torr, and total carbon dioxide between 0.2 and 0.4 mM; in rivers the corresponding ranges are 7.4 to 8.5, and 0.5 to 9 Torr with no available data on total carbon dioxide[36,59]. Typical lake pH values can range from 6 to 9.5 and total CO_2 up to 2 mM[125]. In standing water (e.g., stagnant ponds rich in vegetation), however, pCO_2 values as high as 60 Torr have been reported[45,106].

Ironically, the internal concentrations of carbon dioxide and ammonia in aquatic organisms (i.e., fish) have less to do with ambient carbon dioxide than they do with ambient O_2. Oxygen, by virtue of its much lower solubility in water, is present in relatively low concentrations. Consequently, for a fish to obtain sufficient amount of O_2 to sustain aerobic metabolism, it must ventilate large volumes of water over the gills. The end result is that carbon dioxide and ammonia, with their high water solubility (capacitance), are stripped from the blood in a highly efficient manner. Thus, in fully aquatic, resting fish, blood pCO_2 and pNH_3 are very low, 1–3 Torr[9] and ca. 40 μTorr[11], respectively, as are total carbon dioxide and total ammonia (3–10 mM,[78] and well below 1 mM,[85], respectively). Transport and excretion of carbon dioxide and ammonia by fish have been the subjects of a great deal of study at the whole animal and organ levels, and these topics are reviewed regularly (e.g., for carbon dioxide, see refs. 7, 76, 78, 83; for ammonia, see refs. 11, 32, 85, 86). In this chapter we will focus more on the molecular and biochemical interactions of carbon dioxide and ammonia. Although transport of these substances per se will not be our primary concern, there are some aspects of transport that impact on the utilization of these molecules in biochemical pathways, especially for carbon dioxide, that should be emphasized. The most abundant chemical species of carbon dioxide, and the one used as a substrate in most biochemical reactions, is HCO_3^-, but it is generally considered to be impermeant to biological membranes[29,40,77]. Molecular CO_2, the chemical species produced by most biochemical pathways, freely diffuses across membranes. Thus, the utilization of HCO_3^- in many ways depends on both the transport of CO_2 between subcellular compartments and its interconversion to HCO_3^-. This will be discussed in detail below, but first we shall consider carbon dioxide and ammonia as *bona fide* metabolic substrates and products.

II. Metabolism of carbon dioxide and ammonia

1. Were carbon dioxide and ammonia always waste products?

In modern day fishes (and aquatic species in general) it is apparent that carbon dioxide and ammonia are waste products of catabolism that undergo a net elimination from the animal. The metabolic reactions and pathways are, therefore, on net balance, producers of these molecular substances. It is possible, however, that the evolutionary ancestors of primitive fish inherited their metabolic organization from organisms that evolved in a completely different environment, one in which carbon dioxide and ammonia were more abundant. In such a case these molecules would have served as the chemical building blocks of amino acids, carboxylic acids, and other simple biological molecules, and the metabolic machinery of primitive organisms would have been organized for net utilization of these substances.

The pre-biotic earth atmosphere was a reducing one with high pressures of H_2. It is difficult to know the exact concentrations of CO_2 and NH_3, since their interconversion with CO and N_2, respectively, and the relative abundances of these four molecules depended largely on the partial pressure of H_2 at the time, an unknown quantity. However, it is clear that concentrations were substantially higher than today. Estimates place ammonia levels at 10 to 100 mM in the primordial seas, atmospheric CO_2 levels at 100–200 times higher than present, and oceanic pH levels at only slightly lower than present[50]. With ammonia (primarily NH_4^+) and carbon dioxide (mostly as HCO_3^-) in relative abundance, the biochemical reactions were poised in the direction of ammonia and carbon dioxide consumption through the utilization of external energy sources (UV light, electrical discharges, etc.) and primitive inorganic catalysts[61]. It is possible that primitive biological (organic) catalysts evolved under similar circumstances.

Early unicellular organisms (i.e., those present ⩾ 3.3 billion years ago (bya)) were anaerobic heterotrophs that utilized organic molecules from the primordial broth in simple pathways like glycolysis and the hexose monophosphate shunt. There is evidence of photosynthetic unicells as early as 3.2 bya, and their biochemical machinery was likely similar to modern-day purple photosynthetic bacteria. Carbon dioxide was assimilated using energy derived from light and reducing power from donors other than H_2O (i.e., H_2A). These organisms probably reduced environmental carbon dioxide and ammonia levels to some extent, but the appearance of photosynthetic unicells similar to modern-day cyanobacteria (earliest fossils appear at 2.3 bya) marked the beginning of very different environmental concentrations for ammonia and carbon dioxide. These organisms used water as their reducing agent and liberated molecular oxygen. Subsequent accumulation of oxygen changed the atmosphere from a reducing one and ensured that atmospheric ammonia levels would be lowered in favor of N_2 (further evidence of this shift lies in the ability of these organisms to assimilate nitrogen by N_2 fixation). They also probably enhanced trends towards reduction of environmental carbon dioxide levels by adding to the consumption of carbon dioxide in the dark reactions of photosynthesis.

Interestingly, contemporary photosynthetic organisms use molecular CO_2 in the biochemical reaction that fixes carbon, ribulose bisphosphate carboxylase. Although CO_2 is the species of carbon dioxide that is more easily diffusible through biological structures (i.e., membranes) aquatic photosynthetic organisms appear to have chosen to transport the form which is more abundant, namely HCO_3^-[87]. While the condition of the ancestral autotrophs is not known, extant members of the cyanobacteria possess the enzyme carbonic anhydrase (CA), the function of which has been shown to supply the photosynthetic pathway with molecular CO_2[57,93,104,133]. In primitive organisms, possessing metabolic pathways that are presumably most closely related to those that evolved in the primordial environment, CA supplies CO_2 for metabolism. It is therefore quite conceivable that this enzyme, which is often considered to have evolved in response to carbon dioxide excretion in animals, may have in fact initially evolved as a metabolic enzyme as part of a metabolic pathway of CO_2 utilization (synthesis).

This is an important concept because by the time multicellular organisms appeared (0.8 bya), it is likely that their environmental concentrations of carbon dioxide and ammonia were very similar to today. Ammonia concentrations were by and large low, with only local microenvironmental fluctuations. Atmospheric and dissolved carbon dioxide may have remained slightly higher than present (perhaps as high as 2-fold) until the appearance of vascular plants[50]. It is believed that present day global carbon dioxide levels were attained by the early-to-mid Paleozoic (0.6 to 0.4 bya) as vascular plants made their impact. Oxygen levels were probably still changing markedly at this time (cf., ref. 3), but it is reasonable to assume that fish evolved in an environment similar to that found presently with respect to carbon dioxide and ammonia. As such, carbon dioxide and ammonia were, on net balance, waste products to be excreted. It is important to glean from this possible scenario, however, that the fish metabolic systems which *produce* carbon dioxide and ammonia probably evolved from metabolic systems in which these molecules were *consumed* for synthetic purposes. Thus, these substances which were once abundant, can now be viewed as scarce precursors in fish metabolism.

No doubt that, over evolutionary time, there were considerable, and at times competing, selective pressures shaping the regulatory mechanisms that govern the use in synthesis and/or the accumulation and excretion of these substances. In this chapter we will discuss what we feel are several key features of carbon dioxide and ammonia metabolism that have responded to selective pressure resulting in the current roles of these molecules in the biochemical pathways of fish. These are: (1) bicarbonate is the preferred synthetic substrate, while CO_2 is the usual catabolic product; (2) CA is a necessary vital link in the rapid interconversion between these two metabolite pools (thus, CA is as much a metabolic pathway-level enzyme as it is an enzyme of CO_2 excretion, and it may have evolved primarily as the former); (3) NH_4^+ is freely liberated by few reactions, rather it is transferred whenever possible; and (4) NH_3 is produced by even fewer reactions (i.e., deaminations), but ones that may be quantitatively important to excretion and also contribute to intracellular pH (pH_i) stabilization under some physiological circumstances (e.g., exercising muscle).

2. Reactions involving carbon dioxide and ammonia

Carbon dioxide and ammonia participate in a variety of metabolic pathways. The major enzymes that consume or produce these compounds are listed in Tables 1 and 2, and the appropriate chemical species is highlighted. This information comes from a variety of sources on enzymes from other taxa (largely mammals) where the exact chemical species has been carefully documented (e.g., ref. 6) or deduced from information on the chemical/catalytic mechanism[113]. In some cases, simple but elegant direct approaches were used by investigators to determine if CO_2 or HCO_3^- is the reactive substrate (Fig. 2; ref. 19). While activities of all of the enzymes listed in Tables 1 and 2 have been reported in many fish species, detailed documentation of the exact chemical species consumed or generated by purified fish enzymes is sorely lacking (for review, see refs. 66–68, 101, 108, 109, 123). While it is likely that basic catalytic mechanisms are the same in fish and both more primitive and more advanced taxa, the exact chemical species should be established for selected fish enzymes; it is possible that water-breathers and air-breathers may show some fundamental differences in this regard due to the fact that air-breathers have higher carbon dioxide contents, and typically lower ammonia contents. As recently pointed out by Mommsen and Walsh[68] (see also Table 2), a marked change in carbamoylphosphate synthetase (CPS) nitrogen-donor specificities from glutamine to ammonia occurred between fish and higher vertebrates. It should be directly determined if other enzymes have kept their basic substrate/product preferences. For example, Suarez and Hochachka[100] purified pyruvate carboxylase (PC) from trout liver and examined the effects of $[HCO_3^-]$ on activity at constant pH. Indeed, typical hyperbolic saturation kinetics were ob-

TABLE 1

Enzymes that produce or consume carbon dioxide or bicarbonate

---

Ethanol fermentation
Pyruvate decarboxylase
pyruvate + H_2O ↔ acetaldehyde + CO_2

Tricarboxylic acid cycle
Pyruvate dehydrogenase
pyruvate + NAD^+ + CoA ↔ acetylCoA + NADH + H^+ + CO_2
Isocitrate dehydrogenase
isocitrate + $NAD(P)^+$ ↔ 2-oxoglutarate + $NAD(P)H$ + H^+ + CO_2
2-oxoglutarate dehydrogenase
2-oxoglutarate + NAD^+ + CoA ↔ succinylCoA + NADH + CO_2

Gluconeogenesis
Pyruvate carboxylase
HCO_3^- + pyruvate + ATP ↔ oxaloacetate + ADP + P_i
Phosphoenolpyruvate carboxy kinase
oxaloacetate + GTP ↔ phosphoenolpyruvate + GDP + CO_2
PropionylCoA carboxylase (also in fatty acid oxidation of odd #C)
HCO_3^- + propionylCoA + ATP ↔ D-methylmalonylCoA + AMP + PP_i + H^+

Pentose phosphate shunt and NADPH generation
6-Phosphogluconate dehydrogenase
6-phosphogluconate + $NADP^+$ ↔ D-ribulose-5-P + NADPH + H^+ + CO_2
Malic enzyme
malate + $NADP^+$ ↔ pyruvate + NADPH + H^+ + CO_2

Ketone body formation
Acetoacetate decarboxylase
acetoacetate + H^+ ↔ acetone + CO_2

Fatty acid synthesis
AcetylCoA carboxylase
HCO_3^- + acetylCoA + ATP ↔ malonylCoA + ADP + P_i
3-Ketoacyl-ACP synthase
acetyl-S-Cys-E-ACP-S-malonyl + H^+ ↔ HS-Cys-E-ACP-S-acetoacetyl + CO_2

Nitrogen metabolism
Uricase
uric acid + O_2 + $2H_2O$ ↔ allantoin + H_2O_2 + CO_2
Glycine synthase
glycine + FH_4 + NAD^+ ↔ N^5,N^{10}-methyleneFH_4 + NADH + NH_4^+ + CO_2
Carbamoylphosphate synthetase (CPS I)
HCO_3^- + 2ATP + NH_4^+ ↔ carbamoylphosphate + 2 ADP + P_i + H^+
Urease (microbial)
urea + $2H_2O$ ↔ $2NH_4^+$ + CO_3^{2-}

Hydration/dehydration
Carbonic anhydrase
CO_2 + H_2O ↔ H_2CO_3 ↔ H^+ + HCO^{3-}

---

served, and given available data for mammalian PC, Suarez and Hochachka could tentatively conclude that HCO_3^- is the true substrate in rainbow trout liver. However, in Suarez and Hochachka's measurements, pCO_2 must have also varied (at constant pH), so molecular CO_2 can not be completely ruled out as a substrate. Multifactorial experiments, in which kinetics of purified enzymes are examined

TABLE 2

Enzymes that produce or consume ammonia or ammonium

Glutamate dehydrogenase
glutamate + NAD(P)$^+$ + H$_2$O ↔ 2-oxoglutarate + NAD(P)H + H$^+$ + **NH$_4^+$**
Dehydratases
serine or threonine ↔ pyruvate or 2-oxobutyrate + **NH$_4^+$**
Glycine synthase
glycine + FH$_4$ + NAD$^+$ ↔ N^5, N^{10}-methyleneFH$_4$ + NADH + CO$_2$ + **NH$_4^+$**
L-Amino acid oxidases
L-amino acid + FAD + 1/2O$_2$ ↔ 2-oxoacid + FADH$_2$ + **NH$_4^+$**
Amide deaminases
asparagine + H$_2$O ↔ aspartate + **NH$_3$**
glutamine + H$_2$O ↔ glutamate + **NH$_3$**
Carbon-nitrogen lyases
histidine ↔ urocanate + **NH$_4^+$**
Glutamine synthetase
NH$_4^+$ + glutamate + ATP ↔ glutamine + ADP + P$_i$ + H$^+$
Carbamoylphosphate synthetase (CPS)
CPS I (lungfish and higher vertebrates)
NH$_3$ + 2ATP + HCO$_3^-$ ↔ carbamoylphosphate + 2 ADP + P$_i$ + H$^+$
CPS III (other fish)
glutamine + 2ATP + HCO$_3^-$ ↔ carbamoylphosphate + 2 ADP + P$_i$ + H$^+$ + glutamate
Cystathione-γ-lyase
cystathione + H$^+$ ↔ 2-oxobutyrate + cysteine + **NH$_4^+$**
Purine deaminases
adenosine(monophosphate) + H$_2$O ↔ Inosine(monophosphate) + **NH$_3$**
guanine + H$_2$O ↔ xanthine + **NH$_3$**
Urease (microbial)
urea + 2H$_2$O ↔ CO$_3^{2-}$ + 2**NH$_4^+$**

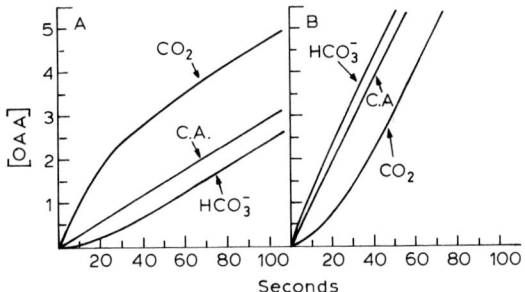

Fig. 2. Theoretical rates of product formation (OAA = oxaloacetic acid) from an enzyme if the active substrate species is CO$_2$ (A) or HCO$_3^-$ (B). The lines labelled CO$_2$ are when an acidic solution (rich in CO$_2$) is added to start the reaction, whereas the lines labelled HCO$_3^-$ are when a basic solution (rich in HCO$_3^-$) is added. CA (= carbonic anhydrase) represents a case when CA is added prior to the addition of either substrate. In the absence of CA, if CO$_2$ is the substrate (A), the reaction first proceeds at an initially high rate until the CO$_2$ and HCO$_3^-$ pools are equilibrated. Thereafter, the rate is similar to the rates for the CA curve or the later part of the HCO$_3^-$ curve. When bicarbonate is added, there is an initial lag in rate. Similar interpretations can be applied to B, where HCO$_3^-$ is the substrate. However, the addition of CA causes a larger difference when CO$_2$ is the added substrate because the equilibrium is ultimately poised towards HCO$_3^-$, leading to much larger changes in concentrations of both compounds when CO$_2$ is the starting material. (from ref. 19).

while each of the three appropriate variables (pH, pCO_2, [HCO_3^-] or pH, pNH_3, [NH_4^+]) are held constant while the other two vary (both in the presence or absence of CA in the case of carbon dioxide, Fig. 2) would simplify interpretations of enzyme function *in vivo*. These experiments would have the added benefit of being very useful in determining the mechanistic bases of observed independent cellular effects of these variables and potential regulatory roles of the enzymes involved (see below).

2.1. Features of carbon dioxide metabolism

Several features of enzyme reactions involving carbon dioxide (Table 1) emerge on closer examination. First, many of the biosynthetic pathways consume HCO_3^-, at least in the initial steps, by virtue of biotin-linked enzymes (e.g., PC, propionyl CoA carboxylase, acetyl CoA carboxylase, carbamoylphosphate synthetase). In some cases they may also be linked to steps later in the pathway that produce CO_2 (see below). Using the largest pool of inorganic carbon, HCO_3^-, for priming of synthetic reactions makes thermodynamic sense in that mass action effects are poised in favor of synthesis.

A second feature of enzyme reactions involving carbon dioxide is that most catabolic pathways evolve molecular CO_2. Since CO_2 is typically present in low concentrations, and biological membranes have high permeability constants for CO_2 (0.35 cm/s through a lipid bilayer[40,95]), generation of CO_2 by catabolic pathways ensures that catabolic pathway 'back-up' by mass action effects of accumulated CO_2 are minimized. Also, since CO_2 is highly diffusible through biological membranes, at least in unicellular organisms, its rapid and direct elimination was theoretically possible in a primordial sense. This may not be the case, however, in actuality. The transport of carbon dioxide may not necessarily become diffusion-limited only in complex, multicellular organisms. In a system in which CO_2 and HCO_3^- are in chemical equilibrium (e.g., the ocean), transport of CO_2 across a lipid membrane (e.g., a unicellular organism) is facilitated by the presence of CA[40]. The mechanism is straightforward: since CO_2 diffuses through the membrane and HCO_3^- does not, the maintenance of the CO_2 gradient ultimately depends upon the rate of dehydration of HCO_3^- (a relatively slow process). The presence of CA (particularly in the boundary layer at the surface of the membrane) results in the virtual instantaneous conversion of HCO_3^- to CO_2, preventing that step from becoming the limiting step in the overall process. Thus, because of permeability differences between HCO_3^- and CO_2, the rate-limiting step in carbon dioxide transport occurs at the site of the membrane. As a result, the selective pressure for facilitated CO_2 transport is great, even in unicellular organisms, and the presence of CA even in primitive cyanobacteria bears this out. It is not known if any organism, unicellular or multicellular, relies solely on the diffusion of CO_2 and the uncatalyzed hydration/dehydration reactions.

These two pools of carbon, anabolic and catabolic, could have remained relatively independent because uncatalyzed rates of interconversion are rather slow. The uncatalyzed hydration of CO_2 has a half-time of 14 s at 37°C[40], and the uncatalyzed rate of dehydration is even slower[28]; these values would be longer still at lower temperatures common in fish. This did not happen, and it is probably a direct result of two physiological and biochemical constraints. First, as discussed above, rapid interconversion of HCO_3^- to CO_2 is necessary for carbon dioxide transport even on the cellular level. The limitations of CO_2 transport and excretion would be even greater on the organ-system and whole organism levels and would require additional transport capacity (e.g., rapid interconversion in the red blood cell as catalyzed by erythrocytic CA). Second, there were likely circumstances in which bicarbonate became limiting to synthetic processes. Both constraints would be ameliorated by the presence and action of CA, and indeed the enzyme appears to be present wherever these processes take place. In mammals and various species of lower vertebrates CA has been shown to be necessary in providing HCO_3^- for the initial steps in the synthesis of glucose[25,26,48], glutamine[20,41,47], certain other amino acids[20], fatty acids[49] and urea[25,26,42,49].

It is quite plausible that the metabolic requirements for CA are similar, or even more pronounced, in fish. CA has already been shown to be important in urea synthesis (discussed below), and it could be equally necessary in gluconeogenesis (Fig. 3). During strenuous exercise, the pH_i of the muscle compartment in rainbow trout and other fish can be depressed to values as low as 6.5, and can remain at these low levels for at least 2 h (e.g., ref. 64). Although strenuous exercise increases blood pCO_2 levels, these are usually returned to near normal by 1 h post-exercise[129]. Furthermore, at least in mammals, tissues do not appear to have a high capacitance for carbon dioxide; increases in CO_2 production usually translate to increases in blood carbon dioxide[16]. Using pK' values for rainbow trout plasma (values for pK' in muscle tissue are not available) computed from algorithms in Boutilier et al.[5] for 15°C, and data from Milligan and Wood[64] on tissue pH_i and plasma pCO_2 (assuming equilibrium between plasma and tissue), calculated total tissue bicarbonate levels drop from approximately 2.0 mM at rest ($pH_i = 7.2$, $pK' = 6.135$, $pCO_2 = 3$ Torr) to as low as 0.6 mM at 2 h ($pH_i = 6.5$, $pK' = 6.2$, $pCO_2 = 4$ Torr). This decrease could impair biosynthetic ability. For example, recent evidence indicates that much of the post-exercise gluconeogenesis from lactate to replenish muscular glycogen must take place within the muscle mass of lower vertebrates[37,65,115], and it is clear at least in rainbow trout white muscle that glycogen replenishment is delayed for several hours post-exercise[64]. Initially it was speculated that glycogen replenishment must await correction of intramuscular pH_i[120]. In light of the discussion above, it is possible that HCO_3^- availability may also be a limiting factor for this pathway. However, one final caveat is important to note in this context: fish white muscle has low activities of PC and PEPCK[56], and it has been speculated that the initial steps of gluconeogenesis might proceed in this tissue by other means, i.e., actual reversal of the reaction catalyzed by PK (T.P. Mommsen, personal communication).

The above scenario for fish (and other aquatic species), at least relative to air-breathers, suggests a picture of potentially low HCO_3^- availability for synthesis due to the efficiency with which it is removed to the environment. Two important adaptations may ameliorate these potential problems. First, many initial steps in synthetic pathways are localized in the mitochondria (e.g., PC in gluconeogenesis,

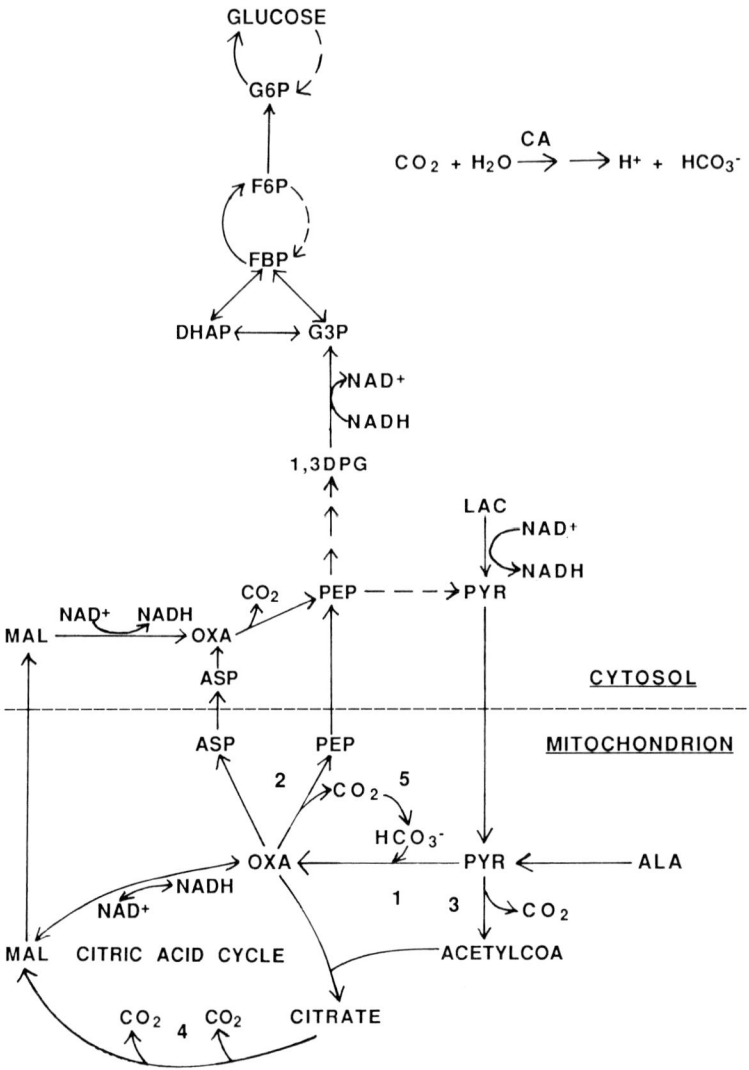

Fig. 3. Simplified schematic of pathways of gluconeogenesis from lactate and alanine shown in relation to opposing glycolytic (broken arrows) reactions and the Krebs cycle. Several key enzymes are noted by the following numbers: 1, pyruvate carboxylase; 2, phosphoenolpyruvate carboxykinase; 3, pyruvate dehydrogenase; 4, 2-oxoglutarate dehydrogenase and isocitrate dehydrogenase; 5, carbonic anhydrase (CA) (modified from ref. 101).

CPS in ureagenesis, etc.) and, at least *in vitro*, intramitochondrial pH remains considerably higher than external pH even at low absolute extramitochondrial pH values (for carp red muscle[72]). Second, mitochondria in at least one fish species (in toadfish liver) contain relatively high levels of CA[119] which should insure that any CO_2 generated by the tricarboxylic acid (TCA) cycle can be rapidly shunted to synthesis in times of low HCO_3^- availability. Gluconeogenesis can proceed in rainbow trout and toadfish hepatocytes even in the nominal absence of external HCO_3^- [115,118]. Also, the effects of the CA inhibitor acetazolamide on ureagenesis and gluconeogenesis were evaluated in toadfish hepatocytes *in vitro* and it was clear that CA is vital to high rates of biosynthesis, even under ample amounts of externally supplied HCO_3^- [119,120].

Obviously the initial step of a pathway like gluconeogenesis does not act in isolation, and as pointed out above (Table 1), a subsequent step catalyzed by PEPCK removes a molecule of CO_2 (Fig. 3). However the intracellular localization of PEPCK in fish liver (the organ of routine gluconeogenesis) is rather plastic, with some species exhibiting exclusive localization in the cytoplasm or mitochondria, some showing activity in both compartments, and some exhibiting modification of localization by dietary influences[101]. The plasticity of PEPCK localization can have important implications for carbon availability for synthesis, chemiosmotic synthesis of ATP and acid–base regulation. Intramitochondrial respiratory CO_2 will be converted to HCO_3^- by CA which will be consumed by the PC reaction. If the next step, via PEPCK, is mitochondrial, CO_2 is then regenerated in the mitochondria which CA should convert to HCO_3^-. This arrangement requires only a catalytic amount of CO_2 for the intramitochondrial steps of gluconeogenesis, but a disadvantage is that it does continue to generate H^+ intramitochondrially, a situation that is potentially at odds with mitochondrial ejection of protons in the chemiosmotic model. Alternatively, if PEPCK is cytoplasmic, oxaloacetate must move to the cytoplasm (via malate in the malate-aspartate shuttle), where CO_2 is released by PEPCK and converted by cytoplasmic CA to HCO_3^- and H^+. From the standpoint of proton ejection and energetic considerations, cytoplasmic PEPCK is more advantageous. However, since the inner mitochondrial membrane in mammals is impervious to HCO_3^- [30], HCO_3^- can only move back into the mitochondria for use in synthesis via counter-transporters or by cytosolic conversion to CO_2, diffusion, and reconversion to mitochondrial HCO_3^-. Therefore it would seem that from the standpoint of efficient incorporation of carbon, mitochondrial PEPCK localization is superior.

In addition to the potential (perhaps largely primordial) problems of synthetic substrate scarcity, the multiple metabolic forms of carbon dioxide imply that metabolic pathways can potentially be rather sensitive to more typical changes in the acid–base variables pCO_2 and $[HCO_3^-]$, even apart from the well-documented pH effects on enzyme activities. In different fish species (toadfish, rainbow trout) and experimental systems *in vitro* (hepatocytes and red blood cells) we have noted *independent* effects of pH, pCO_2, and $[HCO_3^-]$ on metabolic pathway activity[115,119,120,128]. However, mechanistic explanations for these effects remain largely speculative in the absence of firm data on forms of substrates/products for fish enzymes, effects of acid–base variables on transport and metabolism in isolated mitochondria, and identification of affected enzymes (by methods such as cross-over plots). The effects of these variables on rainbow trout hepatocyte metabolism[118] offer some interesting insights nonetheless. Increased pCO_2 markedly *inhibits* both oxidation of lactate and gluconeogenesis from lactate in these cells, independently from pH and $[HCO_3^-]$ effects. Based on the predominance of molecular CO_2 release as the product from TCA cycle enzymes (Table 1), simple mass-action effects could account for the depression of lactate oxidation. Although slowing of the TCA cycle by increased pCO_2 should lead to accumulation of pyruvate and priming of the PC reaction and gluconeogenesis (Fig. 3), it is probable that this is offset by inhibition of CO_2 release from PEPCK, and that this causes the observed slowdown of gluconeogenesis by high pCO_2. Such parsimonious explanations are not as apparent for independent stimulatory effects of bicarbonate on both pathways in these cells. Bicarbonate could potentially act in at least two ways: (1) since it is a counter-ion in mitochondrial transport, increased bicarbonate could stimulate release of mitochondrial products; or (2) it could directly stimulate as a substrate. One final intriguing means of potential regulation for both CO_2 and HCO_3^- is only hinted at in the literature. Bicarbonate appears to directly stimulate rat kidney mitochondria pyruvate dehydrogenase (PDH)[90], despite the fact that the enzyme produces CO_2, suggesting it can be an allosteric modifier. In the case of CO_2, given the likely covalently modification of fish hemoglobin to form carbamino compounds[88,124], perhaps metabolic enzymes are subject to the same types of covalent modifications leading to changes in their functional properties.

One final interesting body of information deserves comment before we leave carbon dioxide, i.e., the regulation of acid–base status during hypercapnia. Cameron and Iwama[13] recently reviewed this topic and noted that when aquatic organisms are subjected to elevated pCO_2, their blood pH initially decreases along the buffer line (i.e., by purely physicochemical means) and in subsequent hours blood pH approaches normal values by elevations of $[HCO_3^-]_e$. The adjustments are rarely complete, and an upper limit to the amount of bicarbonate is apparent (Fig. 4). In attempting to answer the question 'why can't (or don't) animals continue compensating until the original pH is restored?', Cameron and Iwama[13] suggested that (1) an upper limit to plasma $[HCO_3^-]$ is reached (see also ref. 43), or that (2) a new pH set-point is reached that represents a compromise between pH regulation and ion homeostasis. Few data are available for the intracellular compartment. Wood et al.[127] exposed skates to moderate environmental hypercapnia (7.5 Torr) and found that intracellular pH depressions were compensated by increased $[HCO_3^-]_i$ to values up to about 13 mM, but values were typically lower than arterial blood. Given the pervasive metabolic effects of hypercapnia, and the additional metabolic compromises that must also ensue, studies of the intracellular limits of bicarbonate accumulation are clearly in order.

Much research remains in the area of carbon dioxide metabolism. Several fruitful lines are apparent. (1) Although CA appears to be ubiquitous in fish tissues, its localization in the liver mitochondria of *Opsanus beta* could be related to a special function, urea synthesis. A wider systematic screening of fish tissue compartments for CA activity is called for. Furthermore, the regulatory properties of CA in fish tissues should be examined, notably its pH-dependency. (2) Enzymes that utilize and produce HCO_3^- and CO_2 should be purified and examined kinetically to determine especially (i) the precise species used/formed, (ii) their K_m–[substrate] relationships, and (iii) whether allosteric or covalent modifications

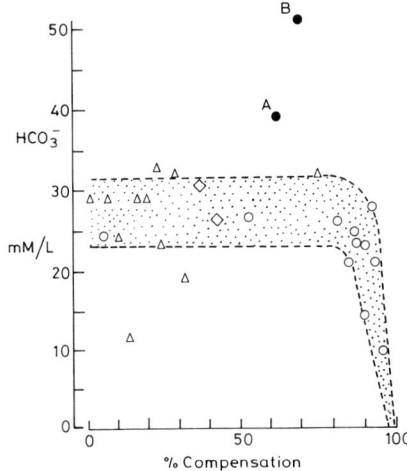

Fig. 4. A summary diagram of the responses of a variety of lower vertebrates to hypercapnia. The stippled area represents a postulated bicarbonate threshold. Circles (○) represent fish species, triangles (△) represent amphibians, and diamonds (◇) represent reptiles. Points A and B are for blue crabs and catfish, respectively, exposed to progressive hypercapnia[12] (diagram originally assembled by Heisler[43] and modified by Cameron and Iwama[13]).

of enzyme function can be brought about by HCO_3^- and CO_2. (3) Studies which determine the sites of regulation of pathway flux by CO_2 and HCO_3^- should be performed with particular attention to cross-over studies of [metabolite] following perturbations by changes in acid–base variables in isolated cells and mitochondria. Since only one variable of pCO_2, $[HCO_3^-]$, and pH can be held constant at a time, carefully controlled multi-factorial experimental designs will be necessary. (4) More attention should be paid to acid–base studies of intracellular compartments. Of course more caution in methodology is required (see refs. 43, 62,79–81, for discussions of methodological considerations), but further such carefully performed studies will be invaluable.

2.2. Features of ammonia metabolism

Several features of enzyme reactions involving ammonia (Table 2) are apparent on closer examination, and all appear to relate to its toxic effects. In contrast to carbon dioxide metabolism, where CO_2 is freely liberated and the total free pool is relatively large, the number of free ammonia molecules generated in the tissues is more severely limited by virtue of the heavy reliance on transaminases. A second feature is that nearly all ammonia-producing reactions yield NH_4^+. This is because the already protonated amine group of amino acids is in large measure the source of liberated NH_4^+. These reactions often do not consume a proton directly from solution but add a hydrogen from a reduced source (H_2O, NADH, etc.) to an already charged — NH_3^+ moiety. Since one aspect of ammonia toxicity relates to the strong tendency for NH_3 to combine with a proton (at physiological pH_i) to raise pH[14], metabolism appears to be already adapted to minimize these effects by releasing mostly NH_4^+.

Unionized ammonia (NH_3) does appear to be the product of a limited number of reactions, specifically when unprotonated — NH_2 groups (as in the amide of glutamine or the nitrogens of purine rings) are removed. Fish do not rely heavily on purine degradation to uric acid for excretion of nitrogen, so generation of large quantities of NH_3 by these means would not appear to be a problem in fish. However, muscle ammonia levels do increase markedly in fish following burst exercise[24,130] or anoxia[107,112]. Mommsen and Hochachka[67] have demonstrated that, at least during exercise in rainbow trout, generation of NH_4^+ occurs by NH_3 production from AMP deaminase and subsequent protonation. Their theoretical treatments suggest that NH_3 can remove a significant quantity of protons by subsequent formation of NH_4^+. This would clearly be advantageous vis á vis the large pH_i changes discussed earlier. The net increase in ammonium that they measured was 6.61 mmol/kg, and given a typical buffering capacity of -51.32 mmol/pH/kg for rainbow trout white muscle[64], one can calculate that the net decrease in pH_i would be 0.13 units greater if the metabolic base, NH_3 was not liberated. This value may not seem large, but it represents 20% of the observed pH_i depression (about 0.6 units). Also, as Heisler[44] points out, one potential limit to the duration of burst exercise is the extent of the pH_i depression as it relates to acid-inactivation of phosphofructokinase (PFK-1); any measure that delays this inactivation may have profound effects on survival.

Only one other major enzyme reaction appears to liberate NH_3, namely glutaminase. However, sufficient biochemical data are not available to conclusively state this. Many references give the product of glutaminase as either NH_3 or NH_4^+, depending upon pH (e.g., refs. 79, 109, 113). Standard uncatalyzed nucleophilic substitutions of OH for NH_2 in amides yield a $RCOO^-$ and a NH_4^+ under acidic conditions, or a $RCOO^-$ and a NH_3 under basic conditions[71], but the pH dependency of the species of ammonia produced does not depend on the pH and pK relationships of ammonia *per se*, but on the chemical species that initially attacks the amido carbon (H^+ *vs.* OH^-) and how the reaction mechanism proceeds (via a positively charged or negatively charged intermediate). Since glutaminase is a mitochondrial enzyme, usually working under basic conditions, it would appear that it has the potential to produce NH_3, perhaps under most circumstances.The species liberated has important implications for excretion and it should be carefully documented for fish glutaminases. Furthermore, although the number of enzymatic reactions that release NH_3 are few (i.e., the various deaminases), it is apparent that they must be quantitatively important in generating the species of ammonia destined for excretion, especially following acid–base disturbances. Data from studies which examine acid fluxes between fish and environment utilizing the 'δ-HCO_3^-' approach[43] and the 'titratable acidity' approach (e.g., ref. 63) validate the underlying assumption that metabolism produces largely NH_3.

2.3. Urea synthesis

Urea synthesis in fish is reviewed in detail elsewhere in this volume[69], but it deserves special mention in the present context because it utilizes both carbon dioxide and ammonia. It has recently been documented that rat liver CPS I directly uses NH_3^{18}. Since NH_3 is the toxic form of ammonia (see for example ref. 21), it makes sense from a regulatory standpoint that the pathway is sensitive to this particular molecular form. Urea synthesis in the majority of fish species takes place via CPS III which utilizes glutamine, synthesized from glutamate by glutamine synthetase (GS), as a nitrogen donor. Studies of purified mitochondrial GS from *Squalus acanthias* demonstrated that it has an extremely low K_m for ammonia (15 μM), but the species of ammonia utilized was not determined[92]. It is important to know the substrate preference of GS since it will determine whether the enzyme is saturated or not *in vivo*, and consequently how well it can respond to increased ammonia in a regulatory manner. If fish GS utilizes NH_4^+ as appears to be the case in mammals[113], it will likely be saturated most of the time, and regulatory responses would be limited to longer term changes in enzyme quantities. It already is apparent in the toadfish that urea synthesis responds to $[HCO_3^-]$ within the physiological range (pathway $K_m = 1.3$ mM) and that the pathway is sensitive to HCO_3^- supply by CA[119]. Since rates of NH_3–NH_4^+ interconversion are rapid (with a $t_{1/2}$ of < 50 ms[99]), and fish appear to tolerate higher total ammonia concentrations than mammals[85], it is tempting to speculate that urea synthesis in fish is controlled (at least in the short term) largely by the availability of HCO_3^-, rather than ammonia.

In recent years considerable controversy has surrounded the theory of Atkinson and co-workers that bicarbonate consumption by urea synthesis is an important pathway of acid–base balance in terrestrial vertebrates. Meijer et al.[60] have recently detailed a consensus viewpoint that recognizes a role for urea synthesis in stabilizing $[HCO_3^-]$, but one that also does not trivialize the contribution of other organs to acid–base balance. Most importantly, the view of Meijer *et al.* is consistent with much of the experimental findings for mammals. However, direct application of this framework to piscine systems in general is doubtful, with the possible exception of the Lake Magadi tilapia. In the toadfish, urea synthesis does not respond to changes in pH[119], a key requirement of the hypothesis of Atkinson and colleagues. Furthermore, unlike air breathers, aquatic fish can directly excrete HCO_3^- to their environment at the gill (see below). In species that do undergo periodic air-exposure and limitation to gill HCO_3^- excretion, it is probable that, within limits, HCO_3^- retention on top of the normally low levels found in fish does not pose a particular metabolic hardship (see for example Fig. 4). Viewed from this perspective, it is probable that ammonia would reach toxic levels (i.e., in the sub-millimolar range) long before bicarbonate posed a threat if air-exposed fish did not make urea. Additionally, air-exposure in fish is characterized by the rapid onset of severe respiratory acidosis (decreased pH, increased pCO_2), (e.g., ref. 52). If an air-exposed fish in acidosis were to manipulate pH upward by increasing bicarbonate content, it is possible to envision a scenario where bicarbonate consumption by urea synthesis would be counter-productive if viewed strictly from an acid–base perspective. The ureagenic toadfish (*Opsanus beta*) survives air-exposure readily[121], so it appears to be a prime system to evaluate the applicability of Atkinson's hypothesis to fish by simultaneously performing traditional acid–base analyses and measurements of urea synthesis. The ureagenic Lake Magadi tilapia is another such system. The high bicarbonate content of its environment probably subjects the fish to continual bicarbonate entrance, a substantial osmotic, ionic and acid–base threat, and bicarbonate consumption by urea synthesis might be an important adaptive trait of this pathway in this species[84].

Recent data obtained from permeabilized rat hepatocytes demonstrated that intermediates of urea synthesis are 'channelled', or transferred efficiently from enzyme to enzyme, with very low release of free substrates to the surrounding intracellular environment[17]. For this and other pathways (e.g., glycolysis) it is suggested that grouping of enzyme-enzyme complexes is an important feature in the organization of metabolism[17,94]. This concept states that metabolites do not move through a pathway by random diffusion through the aqueous cytoplasm to and from the active sites of enzymes; rather, metabolic intermediates are transferred directly from active site to active site of enzyme complexes. Given that enzyme active sites are, for the most part, hydrophobic environments, these studies have two potentially important implications. First, water may be a more limiting substrate in synthetic pathways (e.g., in the hydration of CO_2) than traditionally believed. If so, it is likely that intracellular CA would also be associated with these multi-enzyme complexes (e.g., with CPS or PC), in line with its role as a *bona fide* metabolic enzyme. Furthermore, if such multi-enzyme complexes can also transfer small molecules like carbon dioxide and ammonia, as opposed to obtaining them from the general intracellular pool, much of the consternation by us and others over the exact metabolic forms of these substrates and products may prove to be irrelevant!

III. Production and transport of carbon dioxide and ammonia

1. Carbon dioxide

Carbon dioxide and ammonia, as discussed above, are on net balance waste products for elimination. Where are they generated and how are they transported? Much of the CO_2 generation will take place in mitochondria via the TCA cycle. Carbon dioxide not utilized for synthesis (in tissues like liver) can readily leave mitochondria in the molecular form by diffusion. Is it possible for carbon dioxide to also leave mitochondria as HCO_3^-? Permeability of the inner mitochondrial membrane to bicarbonate is restricted to transport through a limited number of sites. For rat kidney mitochondria, Robinson et al.[90] have suggested that elevated *extramitochondrial* HCO_3^- can substitute for OH^- on the $H_2PO_4^{2-}/OH^-$ transporter. This reverses the normal inwardly directed flux of $H_2PO_4^{2-}$, and via a series of linked carriers, ultimately leads to the ejection of citrate and 2-oxoglutarate from the mitochondria. Their interpretation implies that if HCO_3^- gradients are ever outwardly directed, mitochondrial bicarbonate efflux could be linked to phosphate influx, an extremely important step for continued production of ATP from ADP. If so, it represents another important selective pressure for the presence of CA within the mitochondrial matrix: by maintaining instantaneous equilibrium between CO_2 and HCO_3^-, the latter would be available for transport as well as metabolism. Very little is known about the inorganic ion transport capabilities of fish tissue mitochondria. But, because fish like all aquatic species have much lower total carbon dioxide contents than terrestrial mammals, their mitochondria would seem to be very interesting systems with which to study inorganic ion limitations to synthesis. As we saw above, metabolically generated HCO_3^- can be readily siphoned off for biosynthetic pathways, and consequently intramitochondrial bicarbonate concentrations are probably rather low in fish, minimizing the potential role for carbon dioxide efflux via bicarbonate transport. However, this possibility should be directly examined in fish mitochondrial preparations *in vitro*.

CO_2 can be directly generated in the cytoplasm, by the pentose phosphate shunt, in fish tissues where this activity is relatively high (e.g., liver[117], red blood cells[122], swim bladder gas gland[75]). Once in the cytoplasm, from whatever metabolic source, CO_2 could diffuse directly through the cytoplasm and the plasma membrane. However, recent models for mammalian muscle cells suggest that CO_2 is first rapidly converted to HCO_3^- by cytoplasmic CA, HCO_3^- then diffuses through the cytoplasm, and is reconverted to CO_2 near the plasma membrane by CA for diffusion through the membrane[39]. This evidence supports the concept that CO_2 transport is limited at the level of the membrane (discussed above), and that while cytoplasmic CA maintains intracellular equilibrium between chemical species, it is the membrane-associated CA that confers directionality on CO_2 movements[58]. Three major forms of CA (including two cytoplasmic ones) are known for mammalian muscle, and their kinetic constants, and observed rates of CO_2 transfer through muscle appear to fit this model[39]. Virtually nothing is known regarding intramuscular (or tissue–extracellular boundary layer) CA in aquatic species; studies of these enzymes would be of value in determining if these isozymes are a recent evolutionary acquisition necessitated by the increase in carbon dioxide content associated with air-breathing, or a general feature of tissue carbon dioxide elimination. Randall and Wright[86] proposed a CA-based CO_2-trapping mechanism in the boundary layer of the extracellular fluid/plasma membrane surface which would enhance diffusion, similar to the way it does at the gill/water interface (see below and Fig. 5). Experiments which test this hypothesis should be performed: immunochemical localization of CA and measurements of carbon dioxide excretion *in vitro* during selective inhibition of extracellular CA (by immobilized acetazolamide or antibodies) would be informative. Finally, although it would appear that the bulk of carbon dioxide elimination from fish tissues to the blood is via membrane diffusion of CO_2^{44}, it would also be interesting to determine if significant excretion can occur as HCO_3^-, by seeing if decreases in excretion rates occur *in vitro* when Cl^-/HCO_3^- exchange is blocked with the disulphonic-stilbene derivatives (SITS, DIDS, etc.). Such 'secondary' pathways might increase in importance during periods of elevated CO_2 production (e.g., exercise).

Once in the plasma, pathways of carbon dioxide excretion are well documented in fish (for recent reviews, see refs. 76, 78, 86). It appears that 90% of carbon dioxide excretion at the level of the gill is via CO_2 diffusion and at least two

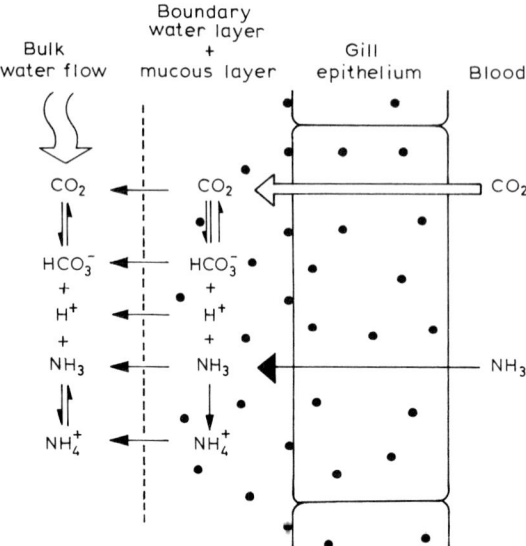

Fig. 5. A simplified cross-section through the gill epithelium showing the bulk water flower, the boundary water layer and associated mucus layer containing CA molecules (●). The catalyzed hydration of CO_2 in the boundary layer to HCO_3^- and H^+ facilitates carbon dioxide and ammonia excretion by promoting the removal of CO_2 and NH_3, thereby preventing reductions in the transepithelial CO_2 and NH_3 diffusion gradients due to accumulation of CO_2 and NH_3 (from ref. 132).

proteins (besides Hb) are vital to CO_2 transport from the tissues to the gills. According to current models (developed largely with active species, e.g., rainbow trout, striped bass, snapper, and dogfish), at the level of the tissues, plasma CO_2 diffuses into the red blood cell where it is converted to HCO_3^- via CA. HCO_3^- then exits the cell in exchange for extracellular Cl^- via the Band 3 (or capnophorin) Cl^-/HCO_3^- exchanger (the Hamburger shift). This process is reversed during red blood cell passage through the gills, with reconversion of plasma HCO_3^- to CO_2 (which can only be catalyzed by CA within the red blood cell) as the likely limiting step to excretion[46,78,102]. If intracellular branchial CA is involved in CO_2 excretion at all it is probably that fraction of CA that is present on the cytoplasmic face of the apical membrane of respiratory lamellae, and its putative function would be to maintain the pCO_2 gradient across the apical membrane/boundary layer via the localized catalysis of the dehydration of HCO_3^-[82]. Cameron[7] estimated that only 2–3% of the total CO_2 excreted from fish is in the form of HCO_3^-. Only two exceptions to this general scheme have been noted in fish. Hagfish erythrocytes have minimal Cl^- transport capacity and low titres of Band 3[31]. The exchanger is absent in lamprey erythrocytes[73,74,105], necessitating the carriage of carbon dioxide as intraerythrocytic HCO_3^-, at a lower overall capacity[105].

Once CO_2 reaches the respiratory surface, the role of CA is not complete. Recent evidence has demonstrated that extracellular CA, in the boundary-layer water/mucus traps CO_2 into HCO_3^- and maintains low boundary layer pCO_2 and an outwardly directed gradient[132]. Gill apical membrane Cl^-/HCO_3^- exchange may be of importance as a means of carbon dioxide excretion, but this role appears to be more related to ion regulation and acid–base balance *per se*.

Although the organismal and cellular level aspects of carbon dioxide excretion are well understood in fish, there are several avenues where biochemical and molecular approaches with fish systems should yield interesting information. Most of the functional study of the Cl^-/HCO_3^- exchanger has been with human erythrocyte ghosts, and although a great deal is known about the kinetic characteristics of this transporter, there are still aspects of the mechanism that are not clear. Most data are consistent with a ping-pong mechanism, i.e., one in which the transporter reversibly cycles through 'inward' and 'outward' directed conformations, but other mechanisms cannot yet be discarded[34,53,97]. Fish red blood cells appear to be potentially useful mechanistic models in at least two ways: (1) carbon dioxide contents of aquatic species are lower, and typically ectothermic metabolic rates and carbon dioxide transport requirements are lower, so fish erythrocytes may simply use a slower catalytic system, one more amenable to kinetic study; (2) the lamprey erythrocyte lacks Band 3[74], and thus lamprey erythrocytes could be engineered to contain controlled quantities and types of Band 3 from other species. Conversely, biochemical and molecular approaches used for mammalian cells can be used to further probe unanswered questions in fish carbon dioxide transport. (1) As pointed out above, CA is intimately involved in carbon dioxide transport, yet its precise intracellular localization in fish erythrocytes is unknown. Given its close functional linkage with Cl^-/HCO_3^- exchange, it is tempting to speculate that it is physically associated with Band 3. Even more intriguing is the possibility that CA reversibly associates with intracellular structures depending upon physiological conditions and its degree of association could markedly affect its activity, as for other metabolic enzymes[89,96]. Regardless of its exact intracellular location, much of the basic information on kinetic properties of the fish red cell CA(s), and how activity varies across species and acclimation regimens, is incomplete. (2) As will be pointed out below, Na^+/H^+ exchangers in mammalian cells are subject to covalent modification, by phosphorylation for example, during hormonal stimulation. Potential sites of phosphorylation on Band 3 and tyrosine kinases have been identified in human red cells[22,27], indicating that Cl^-/HCO_3^- exchanger functional properties may be subject to hormonal modification. Such modifications should be looked for in fish. (3) There are many specific inhibitors of Band 3 transport activity that can be used as labels to determine the number of exchangers per cell. Application of these molecules to fish red blood cells would allow species and acclimation comparisons to determine just how plastic these numbers are. The dynamics of channel recruitment could be studied. As noted above, models regarding carbon dioxide interconversion and transport by fish red blood cells have been developed using largely active species. It would be interesting to examine the amount of Band 3 in erythrocytes of less active species (e.g., toadfish, flatfish) where carbon dioxide excretion demands would be less.

2. Ammonia

Intracellular ammonia originates by several means in fish[14,15,67,108,110,123]. The first two processes occur largely in the liver and relate mostly to processing of dietary amino acids or nitrogenous compounds transported from other tissues. Transdeamination, the sum of transamination and subsequent deamination by glutamate dehydrogenase, leads to a net release of NH_4^+ within the liver mitochondrion. Hydrolysis of amide groups (Asn, Gln) also occurs within mitochondria, but as discussed above it is not clear if NH_3 or NH_4^+ is the released form. Little information is available on the modes of ammonia exit from mitochondria, probably owing to the fact that this is not the normal situation in mammals. It is likely that ammonia exits the mitochondria as NH_3 for several reasons: (1) mitochondria are the source of ammonia and the gradient should be directed outward; (2) the mitochondrial inner membrane is relatively impermeable to ions; (3) NH_4^+ and NH_3 are rapidly interconverted, and (4) the higher pH in the mitochondria should elevate pNH_3 relative to the cytoplasm.

During hypoxia and anoxia, two other pathways of ammonia production take place in fish muscle. During strenuous exercise, AMP deamination takes place in the muscle cytoplasm releasing NH_3, which then combines with a proton to yield NH_4^+; ammonia is later consumed as AMP is resynthesized *in situ*, a process which appears to begin immediately, but which takes up to 24 h to complete[67]. The muscle ammonia load for a 100 g rainbow trout (with approximately 70 g of muscle) is about 463 μmol at the end of strenuous exercise at 8°C[67]. By the end of 4 h recovery, the excess muscle ammonia load is still ca. 200 μmol, meaning that 263 μmol was removed by either excretion or metabolic means. During this 4 h period, recovering rainbow trout at 15°C excrete about 213 μmol more ammonia than control animals[63]. These data would indicate that much of the ammonia generated in the muscle can be excreted. Unfortunately the fate of intramuscular ammonia cannot be adequately addressed at present since data on muscle and whole body ammonia loads and excretion are not available in the same group of fish or from different studies performed at the same temperature.

Allowing that at least part of the ammonia produced in the muscle during exercise, and most of the ammonia produced in the liver, is excreted, how is it transported out of the tissue? Randall and Wright[85] concluded that pNH_3 gradients exist from the muscle to the plasma despite reductions in pH following exercise, allowing at least some of the transport to occur by NH_3 diffusion. The proposed CO_2 trapping mechanism[86] (Fig. 5) would have the benefit of enhancing NH_3 diffusion. However, Wright, Randall and Wood[131], Wright and Wood[130], and Wood et al.[126] have emphasized that tissue/blood ammonia distribution ratios in fully aquatic species appear to be set largely by the transmembrane potential, owing to a relatively low permeability ratio for pNH_3/pNH_4^+ (<20) compared to mammals (>300) (Fig. 6). Mammalian and amphibian skeletal muscle have Na^+/H^+ exchangers[1] and fish hepatocytes do possess Na^+/H^+ exchangers[114,116], so it is possible that exchange of NH_4^+ for Na^+ could take place in these tissues. Excretion of ammonia by this route would have the advantage of allowing NH_3 generated by AMP deaminase to first combine with a proton, thereby removing the acid from the tissue.

Similar accumulation of muscle IMP occurs during the initial hours of complete and extended anoxia in goldfish and carp[107]. During prolonged anoxia, ammonia

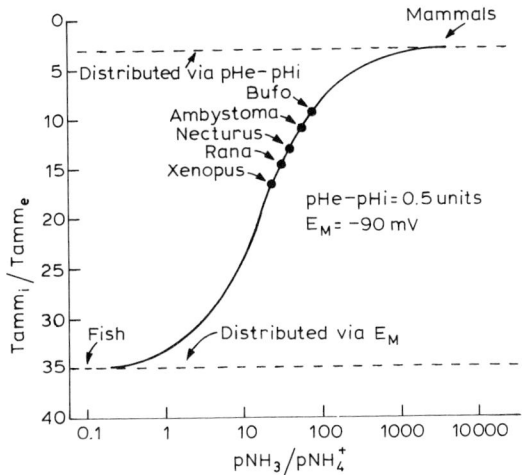

Fig. 6. The relationship between the distribution of ammonia at equilibrium between intracellular and extracellular compartments of muscle ($Tamm_i/Tamm_e$) and the relative permeability (pNH_3/pNH_4^+) of the cell membrane to NH_3 and NH_4^+ (from ref. 126).

excretion continues[108], ammonia does not accumulate appreciably in any tissues, muscle aspartate levels decrease and alanine and succinate levels increase[111]. These authors have interpreted these and other results to mean that aspartate, protein and AMP nitrogen is transferred to pyruvate to produce alanine. Total ammonia levels remain lower in blood than in tissues[111], but pH_i values for these time courses are not known. van Waarde et al.[111] assumed that ammonia transfer from the muscle continued, but alanine concentrations increase substantially in the blood and the liver, suggesting that it may also be a transported form of ammonia nitrogen as has been proposed for migrating salmon[70]. Information regarding mechanisms of alanine transport in fish is limited, but it appears that alanine uptake by hepatocytes and intestinal tissue takes place by Na^+ co-transport[4,98], the mechanisms most commonly responsible for amino acid transport in tissues (reviewed for fish in ref. 35).

Once in the blood, ammonia is transported to the gills where it can potentially be excreted by several routes[32]. Much evidence indicates that non-ionic diffusion constitutes the bulk of ammonia excretion, at least under basal conditions[11,33,85,86,132]. However, the role of Na^+/NH_4^+ exchange in ammonia excretion is not clear for several reasons: (1) uncertainties exist regarding exact gill tissue and boundary layer pNH_3 values and gradients; (2) whole organisms are difficult to manipulate and investigators disagree over the value and validity of in vitro preparations such as perfused heads and gills; and (3) real species differences may exist between, for example marine vs. freshwater fish. Biochemical studies that concentrate on the characteristics of the transporter may offer a fruitful avenue of research. Although it is widely quoted that substrate specificity for Na^+/H^+ is relatively flexible (see e.g., ref. 38), in only one case (rabbit renal microvillus vesicles) has it been directly demonstrated that Na^+/NH_4^+ exchange can occur[55].

Fish would appear to be interesting systems to examine because of the apparent lower relative permeability of their tissues to NH_3[126]. Transport studies should be performed with Na^+/H^+ exchangers purified from fish gills, liver, muscle, and red blood cells, and reconstituted into membrane lipids. Recently Kikeri et al.[54] have discovered a portion of the mouse kidney that is very impermeable to NH_3, and they speculated that this property can be attributed to a specialized lipid composition. This study indicates that measurements of ammonia (and carbon dioxide) permeabilities in bilayers of purified fish cell membrane lipids should be performed to insure that extrapolated values from other taxa are not in error. Furthermore, use of these different kidney lipids in reconstitution studies would allow study of fish $Na^+/H^+(NH_4^+)$ exchangers while factoring out effects due to NH_3 permeation, and allow the direct assessment of the kinetic capabilities and substrate preferences of the fish gill transporter. If NH_4^+ can be established as a true substrate for fish, other questions can be addressed. Does the transporter change structural and kinetic characteristics in times of increased ammonia excretion? Recently Sardet et al.[91] demonstrated that the Na^+/H^+ exchanger in human fibroblasts is phosphorylated during growth factor mediated stimulation. Are more transporters recruited during times of increased ammonia excretion? Dixon et al.[23] describe a method to count exchangers with [^{3}H]methyl isobutyl amiloride that may be useful in this regard.

One final aspect of ammonia excretion needs to be addressed from a biochemical perspective. As Mommsen[66] points out, the roles of glutamine in ammonia transport and glutaminase in generation of ammonia at the gill are not entirely clear. Detailed biochemical studies of enzymes and their compartmentation in the fish gill are called for. Studies of glutamine turnover with stable nitrogen isotopes would also clarify the role of this compound.

IV. Epilogue

Several suggestions for research directions were noted throughout this article, but two areas deserve emphasis. CA is truly a vital metabolic enzyme in addition to its role in carbon dioxide transport. It is very surprising that few studies of the distribution, compartmentation and kinetic characteristics of purified enzymes have been performed for fish. Furthermore precise substrate and product specificities, $K_m/[S]$ interrelationships, and novel allosteric/covalent modifications (e.g., with HCO_3^- and CO_2) should be explored for 'mainstream' metabolic enzymes. This molecular information will complement the extensive database on higher organizational levels and will be invaluable in understanding the complex processes governing the concentrations, fates and pathways of carbon dioxide and ammonia in fish and other aquatic species.

Acknowledgements. The authors' research is supported by NSF (DCB-88–01926 and RII-86–10669 to R.P.H. and DCB-8608728 to P.J.W.) and NIH (ES04361 to P.J.W.). We thank Drs. François Lallier and Chris Wood for critically reading the manuscript.

V. References

1. Aickin, C.C. Intracellular pH regulation by vertebrate muscle. *Annu. Rev. Physiol.* 48: 349–361, 1986.
2. Atkinson, D.E. and M.N. Camien. The role of urea synthesis in the removal of metabolic bicarbonate and the regulation of blood pH. *Curr. Top. Cell. Regul.* 21: 261–302, 1982.
3. Ballantyne, J.S., C.D. Moyes and T.W. Moon. Compatibility and counteracting solutes and the evolution of ion and osmoregulation. *Can. J. Zool.* 65: 1883–1888, 1987.
4. Ballatori, N. and J.L. Boyer. Characteristics of L-alanine uptake in freshly isolated hepatocytes of elasmobranch *Raja erinacea*. *Am. J. Physiol.* 254: R801–R808, 1988.
5. Boutilier, R.G., T.A. Heming and G.K. Iwama. Physicochemical parameters for use in fish respiratory physiology. In *Fish Physiology, Vol. XA*, edited by W. Hoar and D.J. Randall. New York: Academic Press, 1984, pp. 401–429.
6. Boyer, P.D. (ed.) *The Enzymes, Vols. I–XVIII* Orlando, FL: Academic Press, 3rd edn., 1970–1987.
7. Cameron, J.N. Branchial ion uptake in Arctic grayling: resting values and effects of acid–base disturbance. *J. Exp. Biol.* 64: 711–725, 1976.
8. Cameron, J.N. Excretion of CO_2 in water-breathing animals — a short review. *Mar. Biol. Lett.* 1: 3–13, 1979.
9. Cameron, J.N. *The Respiratory Physiology of Animals.* New York: Oxford University Press, 1989.
10. Cameron, J.N. and N. Heisler. Studies of ammonia in the rainbow trout: physico-chemical parameters, acid–base behavior and respiratory clearance. *J. Exp. Biol.* 105: 107–125, 1983.
11. Cameron, J.N. and N. Heisler. Ammonia transfer across fish gills: a review. In *Circulation, Respiration and Metabolism* edited by R. Gilles. Berlin: Springer-Verlag, 1985, pp. 91–100.
12. Cameron, J.N. and G.K. Iwama. Compensation of progressive hypercapnia in channel catfish and blue crabs. *J. Exp. Biol.* 133: 183–197, 1987.
13. Cameron, J.N. and G.K. Iwama. Compromises between ionic regulation and acid–base regulation in aquatic animals. *Can. J. Zool.* 67: 3078–3084, 1989.
14. Campbell, J.W. Nitrogen excretion. In *Comparative Animal Physiology.* edited by C.L. Prosser. Philadelphia: W.B. Saunders, 1973, 3rd edn., pp. 279–316.
15. Casey, C.A., D.F. Perlman, J.E. Vorhaben and J.W. Campbell. Hepatic ammoniagenesis in the channel catfish, *Ictalurus punctatus*. *Mol. Physiol.* 3: 107–126, 1983.
16. Cherniack, N.S., G.S. Longobardo and A.P. Fishman. The behavior of the carbon dioxide stores of the body during unsteady states. In *Carbon Dioxide and Metabolic Regulations*, edited by G. Nahas and K.E. Schaefer. New York: Springer-Verlag, 1974, pp. 324–338.
17. Cheung, C.-W., N.S. Cohen and L. Raijman. Channeling of urea cycle intermediates *in situ* in permeabilized hepatocytes. *J. Biol. Chem.* 264: 4038–4044, 1989.
18. Cohen, N.S., F.S. Kyan, C.-W. Cheung and L. Raijman. The apparent Km of ammonia for carbamoyl phosphate synthetase (ammonia) *in situ*. *Biochem. J.* 229: 205–211, 1985.
19. Cooper, T.G., T.T. Tchen, H.G. Wood and C.R. Benedict. The carboxylation of phosphoenolpyruvate and pyruvate. I. The active species of 'CO_2' utilized by phosphoenolpyruvate carboxykinase, carboxytransphosphorylase, and pyruvate carboxylase. *J. Biol. Chem.* 243: 3857–3863, 1968.
20. Coulson, R.A. and J.D. Herbert. A role for carbonic anhydrase in intermediary metabolism. *Ann. N.Y. Acad. Sci.* 429: 505–515, 1984.
21. Dejours, P., J. Armand and H. Beekenkamp. La toxicité de l'ammoniac est fonction du pH de l'eau. Étude chez la salamandre *Pleurodeles waltl*. *C.R. Acad. Sci. Paris*, 308, Série III: 55–60, 1989.
22. Dekowski, S.A., A. Rybicki and L.K. Drikamer. A tyrosine kinase associated with the red cell membrane phosphorylates band 3. *J. Biol. Chem.* 258: 2750–2753, 1983.
23. Dixon, S.J., S. Cohen, E.J. Cragoe and S. Grinstein. Estimation of the number and turnover rate of Na^+/H^+ exchangers in lymphocytes. *J. Biol. Chem.* 262: 3626–3632, 1987.
24. Dobson, G.P. and P.W. Hochachka. Role of glycolysis in adenylate depletion and repletion during work and recovery in teleost white muscle. *J. Exp. Biol.* 129: 125–140, 1987.
25. Dodgson, S.J. and R.E. Forster. Inhibition of CA V decreases glucose synthesis from pyruvate. *Arch. Biochem. Biophys.* 251: 198–204, 1986a.
26. Dodgson, S.J. and R.E. Forster. Carbonic anhydrase inhibition results in decreased urea production by hepatocytes. *J. Appl. Physiol.* 60: 646–652, 1986b.
27. Drikamer, L.K. Fragmentation of the band 3 polypeptide from the human erythrocyte membranes. Identification of regions likely to interact with the lipid bilayer. *J. Biol. Chem.* 252: 6909–6917, 1977.

28. Edsall, J.T. Carbon dioxide, carbonic acid and bicarbonate ion: physical properties and kinetics of interconversion. In CO_2: Chemical, Biochemical and Physiological Aspects, edited by R.E. Forster, J.T. Edsall, A.B. Otis and F.J.W. Roughton. Washington, DC: NASA Scientific Publication no. 188, 1969.
29. Effros, R.M., G. Mason and P. Silverman. Role of perfusion and diffusion in $^{14}CO_2$ exchange in the rabbit lung. *J. Appl. Physiol.* 51: 1136–1144, 1981.
30. Elder, J.A. and A.L. Lehninger. Respiration dependent transport of carbon dioxide into rat liver mitochondria. *Biochemistry* 12: 976–982, 1973.
31. Ellory, J.C., M.W. Wolowyk and J.D. Young. Hagfish (*Eptatretus stouti*) erythrocytes show minimal chloride transport activity. *J. Exp. Biol.* 129: 377–383, 1987.
32. Evans, D.H. and J.N. Cameron. Gill ammonia transport. *J. Exp. Zool.* 239: 17–23, 1986.
33. Evans, D.H., K.J. More and S.L. Robbins. Modes of ammonia transport across the gill epithelium of the marine teleost fish *Opsanus beta*. *J. Exp. Biol.* 144: 339–356, 1989.
34. Falke, J.J. and S.I. Chan. Molecular mechanisms of band 3 inhibitors. 3. Translocation inhibitors. *Biochemistry* 25: 7899–7906, 1986.
35. Ferraris, R.P. and G.A. Ahearn. Sugar and amino acid transport in fish intestine. *Comp. Biochem. Physiol.* 77A: 397–413, 1984.
36. Fritz, B.J.C. Massabuau and B. Ambroise. Physico-chemical characteristics of surface waters and hydrological behaviour of a small granitic basin (Vosges massif, France): annual and daily variations. In *Hydrochemical Balances of Freshwater Systems*, edited by E. Eriksson, IAHS-AISH Publication no. 150, 1984, pp. 249–261.
37. Gleeson, T.T. 1985. Glycogen synthesis from lactate in skeletal muscle of the lizard *Dipsosaurus dorsalis*. *J. Comp. Physiol.* 156: 277–283, 1985.
38. Grinstein, S. (ed.) Na^+/H^+ Exchange. Boca Raton, FL: CRC Press, 1988.
39. Gros, G. and S.J. Dodgson. Velocity of CO_2 exchange in muscle and liver. *Annu. Rev. Physiol.* 50: 669–694, 1988.
40. Gutknecht, J., M.A. Bisson and F.C. Tosteson. Diffusion of carbon dioxide through lipid bilayer membranes. Effects of carbonic anhydrase, bicarbonate, and unstirred layers. *J. Gen. Physiol.* 69: 779–794, 1977.
41. Häussinger, D., W. Gerok and H. Sies. Hepatic role in pH regulation: role of intracellular glutamine cycle. *Trends Biochem. Sci.* 9: 299–302,1984.
42. Häussinger, D. and W. Gerok. Hepatic urea synthesis and pH regulation: role of CO_2, HCO_3^-, and the activity of carbonic anhydrase. *Eur. J. Biochem.* 152: 381–386, 1985.
43. Heisler, N. Acid-base regulation in fishes. In *Acid-Base Regulation in Animals*, edited by N. Heisler. Amsterdam: Elsevier, 1986, pp. 309–356.
44. Heisler, N. Interactions between gas exchange, metabolism, and ion transport in animals: an overview. *Can. J. Zool.* 67: 2923–2935, 1989.
45. Heisler, N., G. Forcht, Ultsch, G.R. and J.F. Anderson. Acid-base regulation in response to environmental hypercapnia in two aquatic salamanders, *Siren lacertina* and *Amphiuma means*. *Resp. Physiol.* 49: 141–158, 1982.
46. Henry, R.P., N.J. Smatresk and J.N. Cameron. The distribution of branchial carbonic anhydrase and the effects of gill and erythrocyte carbonic anhydrase inhibition in the channel catfish, *Ictalurus punctatus*. *J. Exp. Biol.* 134: 201–218, 1988.
47. Herbert, J.D., R.A. Coulson, T. Hernandez and G. Ehrensvard. A carbonic anhydrase requirement for the synthesis of glutamine from pyruvate in the chameleon. *Biochem. Biophys. Res. Commun.* 65: 1054–1060, 1975.
48. Herbert, J.D., R.A. Coulson and T. Hernandez. Inhibition of pyruvate carboxylation in alligators and chameleons by carbonic anhydrase inhibitors. *Comp. Biochem. Physiol.* 75A: 185–192, 1983.
49. Herbert, J.D. and R.A. Coulson. A role for carbonic anhydrase in de novo fatty acid synthesis. *Ann. N.Y. Acad. Sci.* 429: 525–527, 1984.
50. Holland, H.D. *The chemical evolution of the atmosphere and oceans*. Princeton: Princeton University Press, 1984.
51. Holligan, P.M., W.M. Balch and C.M. Yentsch. The significance of subsurface chlorophyll, nitrite and ammonia maximum in relation to nitrogen for phytoplankton growth in stratified waters of the Gulf of Maine. *J. Mar. Res.* 42: 1051–1073, 1984.
52. Hyde, D.A., T.W. Moon and S.F. Perry. Physiologiocal consequences of prolonged aerial exposure in the American eel, *Anguilla rostrata*: blood respiratory and acid–base status. *J. Comp. Physiol.* 157: 635–642, 1987.
53. Jennings, M.L. Kinetics and mechanism of anion transport in red blood cells. *Annu. Rev. Physiol.* 47: 519–533, 1985.

54. Kikeri, D., A. Sun, M.L. Zeidel and S.C. Hebert. Cell membranes impermeable to NH_3. *Nature* 339: 478-480, 1989.
55. Kinsella, J.L. and P.S. Aronson. Interaction of NH_4^+ and Li^+ with the renal microvillus membrane Na^+/H^+ exchanger. *Am. J. Physiol.* 241: C220-C226, 1981.
56. Knox, D., M.J. Walton and C.B. Cowey. Distribution of enzymes of glycolysis and gluconeogenesis in fish tissues. *Mar. Biol.* 56: 7-10, 1980.
57. Lanaras, T.A.M. Hawthornthwaite and G.A. Codd. Localization of carbonic anhydrase in the cyanobacterium *Chlorogloeopsis fritschii*. *FEMS Microbiol. Lett.* 26: 285-288, 1985.
58. Maren, T.H. Current status of membrane-bound carbonic anhydrase. *Ann. N.Y. Acad. Sci.* 341: 246-254, 1980.
59. Massabuau, J.C. and B. Fritz. Respiratory gas content (O_2 and CO_2) and ionic composition of river water in the plain of Alsace (eastern France). In *Hydrochemical Balances of Freshwater Systems*, edited by E. Eriksson, IAHS-AISH Publication no. 150, 1984, pp. 107-115.
60. Meijer, A.J., W.H. Lamers and R.A.F.M. Chamuleau. Nitrogen metabolism and ornithine cycle function. *Physiol. Rev.* 70: 701-748, 1990.
61. Miller, S.L. and L.E. Orgel. *The Origins of Life on Earth*. Englewood Cliffs, NJ: Prentice-Hall, 1974.
62. Milligan, C.L. and C.M. Wood. Intracellular pH transients in rainbow trout tissues measured by dimethadione distribution. *Am. J. Physiol.* 248: R668-R673, 1985.
63. Milligan, C.L. and C.M. Wood. Intracellular and extracellular acid-base status and H^+ exchange with the environment after exhaustive exercise in rainbow trout. *J. Exp. Biol.* 123: 93-121,1986.
64. Milligan, C.L. and C.M. Wood. Tissue intracellular acid-base status and the fate of lactate after exhaustive exercise in the rainbow trout. *J. Exp. Biol.* 123: 123-144, 1986.
65. Milligan, C.L. and D.G. McDonald. In vivo lactate kinetics at rest and during recovery from exhaustive exercise in coho salmon (*Oncorhynchus kisutch*) and starry flounder (*Platichthys stellatus*). *J. Exp. Biol.* 135: 119-131, 1988.
66. Mommsen, T.P. Metabolism of the fish gill. In *Fish Physiology, Vol. XB*, edited by W.S. Hoar and D.J. Randall, New York: Academic Press, 1984, pp. 203-238.
67. Mommsen, T.P. and P.W. Hochachka. The purine nucleotide cycle as two temporally separated metabolic units-a study on trout muscle. *Metabolism* 36: 552-556, 1988.
68. Mommsen, T.P. and P.J. Walsh. Evolution of urea synthesis in vertebrates: the piscine connection. *Science* 243: 72-75, 1989.
69. Mommsen, T.P. and P.J. Walsh. Urea synthesis in fishes: evolutionary and biochemical perspectives. In *Biochemistry and Molecular Biology of Fish*, edited by P.W. Hochachka and T.P. Mommsen. Amsterdam: Elsevier, 1991, pp. 137-163.
70. Mommsen, T.P., C.J. French and P.W. Hochachka. Sites and patterns of protein and amino acid utilization during the spawning migration of salmon. *Can. J. Zool.* 58: 1785-1799, 1980.
71. Morrison, R.T. and R.N. Boyd. *Organic Chemistry, 2nd edition*. Boston: Allyn and Bacon, 1966.
72. Moyes, C.D., L.T. Buck and P.W. Hochachka. Temperature effects on pH of mitochondria isolated from carp red muscle. *Am. J. Physiol.* 254: R611-R615, 1988.
73. Nikinmaa, M. and E. Railo. Anion movements across lamprey (*Lampetra fluviatilis*) red cell membrane. *Biochim. Biophys. Acta* 899: 134-136, 1987.
74. Ohnishi, S.T. and H. Asai. Lamprey erythrocytes lack glycoproteins and anion transport. *Comp. Biochem. Physiol.* 81B: 405-407, 1985.
75. Pelster, B., H. Kobayashi and P. Scheid. Metabolism of the perfused swimbladder of the European eel: oxygen, carbon dioxide, glucose and lactate balance. *J. Exp. Biol.* 144: 495-506, 1989.
76. Perry, S.F. Carbon dioxide excretion in fishes. *Can. J. Zool.* 64: 565-572, 1986.
77. Perry, S.F., P.S. Davie, C. Daxboeck and D.J. Randall. A comparison of CO_2 excretion in a spontaneously ventilating blood-perfused trout preparation and saline-perfused gill preparations: contribution of the branchial epithelium and red blood cell. *J. Exp. Biol.* 101: 47-60, 1982.
78. Perry, S.F. and C.M. Wood. Control and coordination of gas transfer in fishes. *Can. J. Zool.* 67: 2961-2970, 1989.
79. Pörtner, H.-O. The importance of metabolism in acid-base regulation and acid-base methodology. *Can. J. Zool.* 67: 3005-3017, 1989.
80. Pörtner, H.-O. Determination of intracellular buffer values after metabolic inhibition by fluoride and nitrilotriacetic acid. *Respir. Physiol.*, 81: 275-288, 1990.
81. Pörtner, H.-O., R.G. Boutilier, Y. Tang and D.P. Toews. Determination of intracellular pH and pCO_2 after metabolic inhibition by fluoride and nitrilotriacetic acid. *Respir. Physiol.*, 81: 255-274, 1990.
82. Rahim, S.M., J.-P. Delaunoy and P. Laurent. Identification and immunocytochemical localization

of two different carbonic anhydrase isoenzymes in teleost fish erythrocytes and gill epithelia. *Histochem.* 89: 451–459, 1988.
83. Randall, D.J. and C. Daxboeck. Oxygen and carbon dioxide transfer across fish gills. In *Fish Physiology, Vol. XA*, edited by W.S. Hoar and D.J. Randall. New York: Academic Press, 1984, pp. 263–314.
84. Randall, D.J., C.M. Wood, S.F. Perry, H. Bergman, G.M.O. Maloiy, T.P. Mommsen and P.A. Wright. Urea excretion as a strategy for survival in a fish living in a very alkaline environment. *Nature* 337: 165–166, 1989.
85. Randall, D.J. and P.A. Wright. Ammonia distribution and excretion in fish. *Fish Physiol. Biochem.* 3: 107–120, 1987.
86. Randall, D.J. and P.A. Wright. The interaction between carbon dioxide and ammonia excretion and water pH in fish. *Can. J. Zool.* 67: 2936–2942, 1989.
87. Raven, J.A. *Energetics and Transport in Aquatic Plants.* New York: Alan R. Liss, 1984.
88. Riggs, A. Properties of fish hemoglobins. In *Fish Physiology, Vol. IV*, edited by W.S. Hoar and D.J. Randall, New York: Academic Press, 1970, pp. 209–252.
89. Roberts, S.J., M.S. Lowery and G.N. Somero. Regulation of binding of phosphofructokinase to myofibrils in the red and white muscle of the barred sand bass, *Paralabrax nebulifer* (Serranidae). *J. Exp. Biol.* 137: 13–27, 1988.
90. Robinson, B.H., J. Oei, S. Cheema-Dhadli and H.L. Halperin. Regulation of citrate transport and pyruvate dehydrogenase in rat kidney cortex mitochondria by bicarbonate. *J. Biol. Chem.* 252: 5661–5665, 1977.
91. Sardet, C., L. Counillon, A. Franchi and J. Pouysségur. Growth factors induce phosphorylation of the Na^+/H^+ antiporter, a glycoprotein of 110 kD. *Science* 247: 723–726, 1990.
92. Shankar, R.A. and P.M. Anderson. Purification and properties of glutamine synthetase from liver of *Squalus acanthias*. *Arch. Biochem. Biophys.* 239: 248–259, 1985.
93. Shiraiwa, Y. and S. Miyachi. Role of carbonic anhydrase in photosynthesis of blue-green alga (Cyanobacterium *Anabaena variabilis* ATCC 29413). *Plant Cell Physiol.* 26: 109–116, 1985.
94. Srivastava, D.K. and S.A. Bernhard. Metabolite transfer via enzyme-enzyme complexes. *Science* 234: 1081–1086, 1986.
95. Simon, S.A. and J. Gutknecht. Solubility of carbon dioxide in lipid bilayer membranes and organic solvents. *Biochem. Biophys. Acta* 596: 352–358, 1980.
96. Somero, G.N. and S.C. Hand. Protein assembly and metabolic regulation: physiological and evolutionary perspectives. *Physiol. Zool.* 63: 443–471, 1990.
97. Stein, W.D. *Transport and Diffusion Across Cell Membranes.* Orlando, FL: Academic Press, 1986.
98. Storelli, C., S. Vilella and G. Cassano. Na-dependent D-glucose and L-alanine transport in eel intestine brush border membrane vesicles. *Am. J. Physiol.* 251: R463–R469, 1986.
99. Stumm, W. and J.J. Morgan. *Aquatic Chemistry*, 2nd ed. New York: John Wiley, 1981.
100. Suarez, R.K. and P.W. Hochachka. Pyruvate carboxylase from rainbow trout liver. *J. Comp. Physiol.* 143: 281–288, 1981.
101. Suarez, R.K. and T.P. Mommsen. Gluconeogenesis in teleost fishes. *Can. J. Zool.* 65: 1869–1882, 1987.
102. Swenson, E.R. and T.H. Maren. Roles of gill and red cell carbonic anhydrase in elasmobranch HCO_3^- and CO_2 excretion. *Am. J. Physiol.* 253: R450-R458, 1987.
103. Truchot, J.-P. and A. Duhamel-Jouve. Oxygen and carbon dioxide in the marine intertidal environment: diurnal and tidal changes in rockpools. *Respir. Physiol.* 39: 241–254, 1980.
104. Tu, C., H. Spiller, G.C. Wynns and D.N. Silverman. Carbonic anhydrase and the uptake of inorganic carbon by *Synechococcus* sp. (UTEX 2380). *Plant Physiol.* 85: 72–77, 1987.
105. Tufts, B.L. and R.G. Boutilier. The absence of rapid chloride/bicarbonate exchange in lamprey erythrocytes: implications for CO_2 transport and ion distributions between plasma and erythrocytes in the blood of *Petromyzon marinus*. *J. Exp. Biol.* 144: 565–576, 1989.
106. Ultsch, G.R. The potential role of hypercarbia in the transition from water-breathing to air-breathing in vertebrates. *Evolution* 41: 442–445, 1987.
107. van den Thillart, G., A. van Waarde, H.J. Muller, C. Erkelens, A. Addink and J. Lugtenburg. Fish muscle energy metabolism measured by in vivo ^{31}P-NMR during anoxia and recovery. *Am. J. Physiol.* 256: R922–R929, 1989.
108. van Waarde, A. Aerobic and anaerobic ammonia production by fish. *Comp. Biochem. Physiol.* 74B: 675–684, 1983.
109. van Waarde, A. Biochemistry of non-protein nitrogenous compounds in fish including the use of amino acids for anaerobic energy production. *Comp. Biochem. Physiol.* 91B: 207–228, 1988.
110. van Waarde, A. and F. Kesbeke. Nitrogen metabolism in goldfish, *Carassius auratus* (L.).

Influence of added substrates and enzyme inhibitors on ammonia production of isolated hepatocytes. *Comp. Biochem. Physiol.* 70B: 499–507, 1981.
111. van Waarde, A., G. van den Thillart and F. Dobbe. Anaerobic metabolism of goldfish, *Carassius auratus* (L.). Influence of anoxia on mass-action ratios of transaminase reactions and levels of ammonia and succinate. *J. Comp. Physiol.* 147: 53–59, 1982.
112. van Waarde, A., G. van den Thillart and F. Kesbeke. Anaerobic energy metabolism of the European eel, *Anguilla anguilla* L. *J. Comp. Physiol.* 149: 469–475, 1983.
113. Walsh, C. *Enzymatic Reaction Mechanisms.* San Francisco, CA: W.H. Freeman, 1979.
114. Walsh, P.J. Ionic requirements for intracellular pH regulation in rainbow trout hepatocytes. *Am. J. Physiol.* 250: R24-R29, 1986.
115. Walsh, P.J. An *in vitro* model of post-exercise hepatic gluconeogenesis in the gulf toadfish *Opsanus beta*. *J. Exp. Biol.* 147: 393–406, 1989.
116. Walsh, P.J. Regulation of intracellular pH by toadfish (*Opsanus beta*) hepatocytes. *J. Exp. Biol.* 147: 407–419, 1989.
117. Walsh, P.J., T.W. Moon and T.P. Mommsen. Interactive effects of acute changes in temperature and pH on metabolism in hepatocytes from the sea raven *Hemitripterus americanus*. *Physiol. Zool.* 58: 727–735, 1985.
118. Walsh, P.J., T.P. Mommsen, T.W. Moon and S.F. Perry. Effects of acid–base variables on *in vitro* hepatic metabolism in rainbow trout. *J. Exp. Biol.* 135: 231–241, 1988.
119. Walsh, P.J., J.J. Parent and R.P. Henry. Carbonic anhydrase supplies bicarbonate for urea synthesis in toadfish (*Opsanus beta*) hepatocytes. *Physiol. Zool.* 62: 1257–1272, 1989.
120. Walsh, P.J. and C.L. Milligan. Coordination of metabolism and intracellular acid–base status: ionic regulation and metabolic consequences. *Can. J. Zool.* 67: 2994–3004, 1989.
121. Walsh, P.J., E. Danulat and T.P. Mommsen. Variation in urea excretion in the Gulf toadfish, *Opsanus beta*. *Mar. Biol.* 106:323–328, 1990.
122. Walsh, P.J., C.M. Wood, S. Thomas and S.F. Perry. Characterization of red blood cell metabolism in rainbow trout. *J. Exp. Biol.*, 154: 475–489, 1990.
123. Watts, R.L. and D.C. Watts. Nitrogen metabolism in fishes. In *Chemical Zoology, Vol. VIII*, edited by M. Florkin and B.T. Scheer, New York: Academic Press, 1974, pp. 369–446.
124. Weber, R.E. and F.B. Jensen. Functional adaptations in hemoglobin from ectothermic vertebrates. *Annu. Rev. Physiol.* 50: 161–179, 1988.
125. Wetzel, R.G. *Limnology*, 2nd edn. Philadelphia, PA: Saunders, 1983.
126. Wood, C.M., R.S. Munger and D.P. Toews. Ammonia, urea and H^+ distribution and the evolution of ureotelism in amphibians. *J. Exp. Biol.* 144: 215–233, 1989.
127. Wood, C.M., J.D. Turner, R.S. Munger, M.S. Graham. Control of ventilation in the hypercapnic skate *Raja ocellata*: II. Cerebrospinal fluid and intracellular pH in the brain and other tissues. *Respir. Physiol.*, 80: 279–298, 1990.
128. Wood, C.M., P.J. Walsh. S. Thomas and S.F. Perry. Control of red blood cell metabolism in rainbow trout after exhaustive exercise. *J. Exp. Biol.*, 154: 491–507, 1990.
129. Wood, C.M. and S.F. Perry. Respiratory, circulatory and metabolic adjustments to exercise in fish. In *Comparative Physiology and Biochemistry: Current Topics and Trends, Vol. A.* edited by R. Gilles, Berlin: Springer-Verlag, 1985, pp. 1–22.
130. Wright, P.A. and C.M. Wood. Muscle ammonia stores are not determined by pH gradients. *Fish Physiol. Biochem.* 5: 159–162, 1988.
131. Wright, P.A.D.J. Randall and C.M. Wood. The distribution of ammonia and H^+ between tissue compartments in lemon sole (*Parophrys vetulus*) at rest, during hypercapnia and following exercise. *J. Exp. Biol.* 136: 149–175, 1988.
132. Wright, P.A., D.J. Randall and S.F. Perry II. Fish gill water boundary layer: a site of linkage between carbon dioxide and ammonia excretion. *J. Comp. Physiol.* 158: 627–635, 1989.
133. Yagawa, Y., Y. Shiraiwa and S. Miyachi. Carbonic anhydrase from the blue-green alga (Cyanobacterium) *Anabaena variabilis*. *Plant and Cell Physiol.* 25: 775–783, 1984.

CHAPTER 9

Biochemical aspects of buoyancy in fishes

CHARLES F. PHLEGER

Department of Natural Science, San Diego State University, San Diego, CA 92182, U.S.A.

I. Introduction
II. Swimbladder
    1. Swimbladder gases and oxygen secretion
    2. Fat-invested swimbladders
    3. Fat-filled swimbladders
III. Skeleton
    1. Morphological consideration
    2. Triacylglycerol-filled bones
    3. Wax ester-filled bones
IV. Liver
    1. Squalene
    2. Glyceryl ethers
    3. Triacylglycerols and wax esters
V. Integument
VI. Muscle
VII. Digestive tract
VIII. Gills
IX. Blood
X. Eggs
XI. Hypotonicity
XII. Conclusions
XIII. References

I. Introduction

Fish living in an aqueous medium may be negatively buoyant, neutrally buoyant, or positively buoyant. Negatively buoyant benthic fishes, such as rays and other flatfish, must exert energy to swim off the bottom and remain afloat. Neutrally buoyant fishes, such as many midwater fishes, can hover motionless in sea water and exert very little energy to swim. Positive buoyancy is uncommon, but occurs in air-breathing fishes and certain other species such as swordfish when resting at the surface. To achieve neutral buoyancy, fishes have evolved a number of fascinating strategies. These include reduction of body components denser than seawater (seawater density is 1.024–1.026 g/ml), such as skeleton (2 g/ml or greater density for many fish) and swimming musculature (1.05–1.06 g/ml), storage of low density materials such as lipid (0.86–0.93 g/ml), use of a gas-filled swimbladder (which may be fat-invested or fat-filled), and maintenance of a tissue water content hypotonic to sea water.

The four most important lipids used for buoyancy purposes in marine teleosts include squalene, wax esters, alkyldiacylglycerol, and triacylglycerol (Fig. 1). Squa-

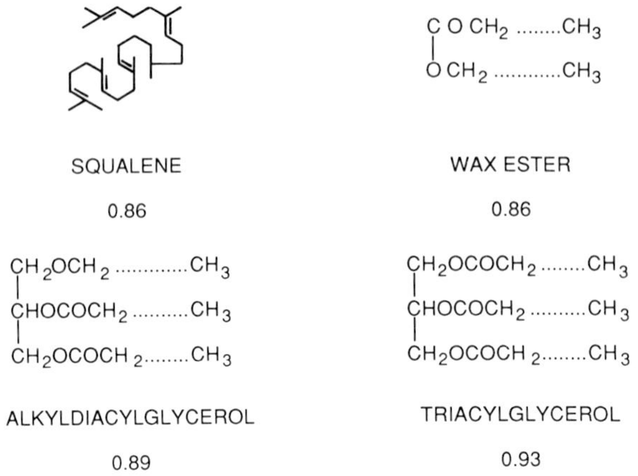

```
      C O CH₂ ........CH₃
      |
      O CH₂ ............CH₃
```

SQUALENE   WAX ESTER

0.86   0.86

```
CH₂OCH₂ ............CH₃      CH₂OCOCH₂ ........CH₃
|                            |
CHOCOCH₂ ..........CH₃       CHOCOCH₂ ..........CH₃
|                            |
CH₂OCOCH₂ ........CH₃        CH₂OCOCH₂ ........CH₃
```

ALKYLDIACYLGLYCEROL   TRIACYLGLYCEROL

0.89   0.93

Fig. 1. Principal lipids implicated in buoyancy of fishes with specific gravities. Specific gravities for wax esters, alkyldiacylglycerol, and triacylglycerol of similar chain length and unsaturation at 21°C[202] and squalene.

lene is the only one of these that is probably metabolically inert in the liver of certain sharks[157], and may therefore have buoyancy as its only function. Wax esters and alkyldiacylglycerols have a lower metabolic turnover rate than triacylglycerols, and thus may have a more important function in buoyancy than triacyglycerols[202]. The densities of wax esters and alkyldiacylglycerols are less than triacylglycerol (Fig. 1), and the upthrust they generate is also greater (Table 1). The storage location of lipids also relates to buoyancy. Extracellular lipid, although uncommon in marine fish, may function primarily in buoyancy, because the only known mechanism for mobilizing lipids is by a hormone-sensitive lipase releasing free fatty acids from intracellular lipid[68,69]. There are many other lipids in marine fish, such as monoglycerides, diglycerides, free fatty acids, phospholipids, plasmalogens, neutral plasmalogens and hydrocarbons[206], which although usually present in lesser quantities than the lipids in Fig. 1, are less dense than seawater and therefore could conceivably contribute to buoyancy under certain conditions. The hydrocarbon pristane, exogenously derived in marine teleosts, for example, is only present in minor amounts, yet has a density of 0.78 g/ml. It must also be recognized that temperature and pressure affect these compounds, and their

TABLE 1

Upthrust from lipids in seawater (density 1.025) [a]

Lipid	Density	g upthrust/g lipid	g upthrust/ml lipid
Triacylglycerol	0.93	0.103	0.096
Alkyldiacylglycerol	0.91	0.127	0.116
Wax ester	0.86	0.193	0.166
Squalene	0.86	0.193	0.166

[a] From ref. 204.

density at 2°C and 400 atm pressure (deep ocean) may be markedly different than at the ocean surface[243]. There are a number of reviews on buoyancy in fishes[17,24,112,141,229], and buoyancy is often discussed more briefly in marine lipid reviews[130,149,157,202,204]. This review will concentrate upon biochemical aspects of buoyancy from the point of view of different tissues of the fish such as swimbladder, skeleton, liver, integument, muscle, digestive tract, gills, blood, and eggs. Hypotonicity is discussed separately. References in the literature prior to 1970 are generally not included for the sake of brevity.

II. Swimbladder

The gas-filled swimbladder of fish functions primarily as a buoyancy organ. It has attracted much interest and is the subject of many reviews[18,142,216]. Most swimbladders to be discussed here are of the physoclist type; that is, there is no connection to the gut and gas secretion and absorption occurs within the swimbladder itself. Physostome swimbladders have a pneumatic duct to the gut of the fish. Therefore, these fishes can fill their swimbladders by swallowing air at the surface. This discussion will concentrate on swimbladder gases and oxygen secretion, fat-invested swimbladders, and fat-filled swimbladders.

1. Swimbladder gases and oxygen secretion

One of the most perplexing questions in animal physiology is the problem of gas secretion by fish into their swimbladders at great depths in the ocean. It is estimated that at least half of the benthic fishes living below a depth of 2000 m have normal functional swimbladders. For instance, a brotulid fish, *Bassogigas profundissimus,* trawled from a depth of more than 7000 m in the Sunda Trench, was found to have a well-developed functional swimbladder[163]. Most of the gas in deep fish swimbladders is oxygen while the remainder consists of nitrogen, argon, and carbon dioxide. In fishes living below a depth of 40–50 m in the ocean, oxygen comprises 90% or more of these gases. This means that the oxygen tension in the brotulid swimbladder at 7000 m can be 630 atm. At 10,000 m depth oxygen tensions in sea water are approximately four times the surface value. The fish is therefore capable of a seemingly impossible physiological feat, that of increasing the oxygen partial pressure from less than 0.8 atm in sea water to 630 atm in its swimbladder.

Considerable study has been devoted to the subject of gas secretion[18,50,216]. Oxygen is secreted by release from oxyhemoglobin caused by lactic acid pH lowering (the Bohr effect). Glucose is converted to lactic acid at a rapid rate in the gas secreting gland in the swimbladder[51]. The blood is enriched in lactic acid during its passage through the swimbladder epithelium. Gas gland cells do not show a Pasteur effect because glycolysis operates in a similar manner under both aerobic and anaerobic conditions. The activities of the glucose phosphorylating enzyme and the hexose monophosphate shunt enzymes are ten to twenty times higher in the richly vascularized gas gland of the cod, *Gadus morhua*, than in white muscle[25]. This suggests that a large amount of glucose-6-phosphate goes into the shunt, producing lactic acid and CO_2, for pH lowering and O_2 release from oxyhemoglobin. Also, the aerobic capacity of the gas gland is even lower than that of white muscle, as evidenced by low activities of the oxidative enzymes reducing levels of citrate and ATP[25,80]. The blood of marine fishes is far more sensitive to pH lowering than the blood of other animals. The term 'Root Effect' refers to the situation where acidity completely blocks the oxygen from part of the hemoglobin as a Root-off shift, and the opposite effect, when the pH is increased, as a Root-on shift. The single tetrameric hemoglobin of the spot, *Leiostomus xanthurus*, in equilibrium with air, will lose 30% of bound oxygen when the pH is decreased from 7 to 6[22]. The same pH drop will cause an 80% oxygen loss from hemoglobin in the presence of anions such as ATP or 2,3-bisphosphoglycerate. Evidence for organic phosphate regulation of oxygen release by hemoglobin has also been obtained[28]. Oxygen secreted into swimbladders is derived directly from molecular oxygen dissolved in the surrounding water. The cells of the swimbladder gas gland are apparently insensitive to the high oxygen pressures inside the swimbladder.

Primary concentration caused by the Bohr or Root effects and salting out are effectively maintained and multiplied by countercurrent exchange in the *rete mirabile*, a closely knit bundle of countless afferent and efferent capillaries which circulates blood through the swimbladder gas gland. Lengths of the *retia* range from 5 to 25 mm, attaining their greatest lengths in the deepest living species of fish. Despite multiplication by the countercurrent system, the primary Root effect for oxygen secretion in deep fishes is still limited by oxygen back pressure in the swimbladder. At pH 6 these effects are nullified at 50–100 atm oxygen pressure. The filling of the swimbladder of *Antimora rostrata* with gaseous oxygen at depths of a mile or more, was not explained[164]. The hemoglobin of *A. rostrata* displayed a typical Root effect, but the measured reduction in oxygen affinity was insufficient to fill the swimbladder. Therefore, fish with a functional swimbladder living at pressures in excess of 100 atm must utilize some additional stategy to secrete oxygen. It has been suggested that some diffusible x-substance which dissociates oxygen from oxyhemoglobin against much higher pressure than can lactic acid must be secreted by the gas gland cells[209].

An effective diffusion barrier is required to maintain oxygen tensions of hundreds of atmospheres in the swimbladder against tensions of a fraction of an atmosphere in the rest of the body of the fish and the surrounding sea water. The tough swimbladder wall performs this function admirably. The outer tunica externa of the swimbladder provides a remarkably effective barrier to gas diffusion. It consists of a silvery film of connective tissue impregnated with overlapping crystalline guanine[123,124] or collagen. An increase in guanine in the swimbladder wall of anacanthine fishes with greater depth of occurrence was found[200]. Guanine crystals are also found in the tapeta lucida structure in the retina of vertebrate eyes (including fish) which are also rich in very long chain polyenoic fatty acids (up to 36:6)[6,7]. In *Leiognathus equulus*, the guanine/hypoxanthine ratio in the swimbladder lining was responsible for reflection of bioluminescent light from the circumesophageal light organ[146]. Overlapping platelets were found in the submucosa layer of the swimbladder in atheriniform fishes[30] and in two other orders, Perciformes and Batrachoidiformes[29]. These papers suggested that the randomly disposed platelets with their highly ordered layering of membranes may function as a physical barrier to gas diffusion out of the swimbladder. The inner mucosa consists of a gas secretion and resportion part.

Ultrastructure of gas gland cells has been described[67,151]. The gas gland connects to the *rete mirabile*, which by countercurrent exchange effectively prevents swimbladder gases at high pressure from escape. The resorbent part (oval window) of the mucosa, consists of a single type of squamous epithelial cell in the toadfish, *Opsanus tau*[151]. It lacks a *rete mirabile* and a gas gland. The anterior chamber (for gas secretion) of the toadfish swimbladder is lined by columnar cells that contain lamellar bodies much like those in epithelial cells of vertebrate lungs[150]. Biochemically, however, the swimbladder differs considerably from lungs. Swimbladder surfactants in the Amazon obligate air breathing fishes, *Lepidosiren paradoxa* and *Arapaima gigas,* contained primarily phosphatidylcholine (39–46% of the phospholipids) with sphingomyelin (11–24%), phosphatidylethanolamine (7–19%) and lesser amounts of other phospholipids[185]. These phospholipid patterns are significantly different from those of mammals which have phosphatidyglycerol[88], not present in these air-breathing fish. Perhaps these swimbladder lipids are important in keeping water from leaking into the lung/swimbladder. The minimum surface tension of the lipid-lavages from the swimbladders were 20–25 dyn/cm[185], whereas mammalian lung lavages have surface tensions of 1–2 dyn/cm (see ref. 230 for a review of the biochemistry of the pulmonary surfactant system).

The effect of scattering by midwater fishes with swimbladders on acoustic propagation in the ocean has received much study[118,232]. The scattering seen between 1 and 20 kilohertz, caused by biological organisms, indicate that some of the fishes perform daily vertical migrations. Large myctophids apparently maintain constant swimbladder volume during these daily migrations including hundreds of meters, but at a level below that required by neutral buoyancy. The energetics of this process are not well understood, however. Inflating the swimbladder at depth after a downward vertical migration may be the most costly with respect to energetics because it requires 50 μl of oxygen/h/2 g myctophid fish[4]. This is complicated by the oxygen minimum layer in which the fish live during the day where the oxygen concentration is only 50 to 200 μl of oxygen/liter of seawater[232]. This might suggest an advantage to an oxygen-filled swimbladder as an oxygen reserve. Also, blood oxygen affinities are among the lowest ever reported and the hemoglobin is not fully functional. Buoyancy in these midwater fishes is aided by lipid which is essentially non-compressible, and present in a number of species[156,158].

2. Fat-invested swimbladders

Fat-investment of regressed swimbladders in midwater fishes is common in such fish as the lantern fish, *Lampanyctus leucopsarus,* and stromateoides[113,114]. Fat-investment involves accumulation of fat between the peritoneum and the tunica externa of the swimbladder, and is fairly common in deep water fishes. Wax esters are a major component of the invested fat. Regressed swimbladder lipid of myctophids probably consists of at least 90% wax esters[162]. In myctophids lipids take on a major buoyancy function as the swimbladder regresses with age and the swimbladder becomes invested with fat[32]. The adults make extensive diurnal vertical migrations whereas the juveniles do not. Regressed swimbladder lipid of the coelacanth, *Latimeria chalumnae,* consisted of 97% wax esters[160]. This wax ester-invested swimbladder undoubtedly plays a role in vertical migration[145]. The coelacanth is a nocturnal piscivorous drift hunter, and moves very slowly, often touching the bottom and taking advantage of up- or downwelling currents[78]. The fish is slightly negatively buoyant, and perches on reef platforms before swimming away to capture prey[145].

The swimbladder of the deep ocean orange roughy, *Hoplostethus atlanticus,* is also fat-invested. Over 90% of the lipid in the swimbladder in this fish is wax ester[86,87,173]. The swimbladder lipid appears to be extracellular and is contained within a three-dimensional network of collagen fibers[173]. The only known mechanism to mobilize lipids in animals is by a hormone-sensitive lipase that releases free fatty acids from intracellular lipids. Therefore, the finding that the swimbladder wax ester is apparently extracellular implies that this lipid is permanent since there is no way to mobilize it. Extracellular lipid storage in fish may well be a permanent source of buoyancy.

3. Fat-filled swimbladders

In contrast to near-surface fishes, with their gas-filled swimbladders and midwater fishes with fat-invested regressed swimbladders, many deep ocean benthopelagic fishes have fat-filled swimbladders (see ref. 149 for a recent review). These swimbladders are fully functional with respect to gas secretion, and contain hyperbaric oxygen at many hundreds of atmospheres pressure. The fat, or lipid, when observed in these fish after rapid ascent to the surface on baited free-vehicle long-lines[184] appears as a 'foam' since it filled with bubbles of oxygen gas which have expanded during ascent of the fish from the deep sea. Cholesterol and phospholipids in a ratio of 1:1, are the principal lipids in the macrourid fishes, *Coryphaenoides acrolepis* and *C. abyssorum*, as well as in the morid, *Antimora rostrata*[172]. Fatty acids of the phospholipids are mostly unsaturated (up to 82% in *C. acrolepis*)[172].

This cholesterol-phospholipid-rich foam has also been reported in the brotulid fishes *Dicrolene intronigra, Bassozetes* sp., and *Barathrodemus iris* from the Atlantic ocean where the fatty acids are also highly unsaturated (up to 89%)[189]. The high level of unsaturated fatty acids in the swimbladder is unusual. They resist auto-oxidative processes in the swimbladder which contains almost pure oxygen at many hundreds of atmospheres pressure. Recent studies[33] suggest the presence of a superoxide dismutase enzyme in swimbladders of *Salmo trutta*. The high concentration of superoxide dismutase in human lung tissue has been shown to increase in the presence of high oxygen pressures. There is also elevated activity of hexose monophosphate shunt enzymes in *S. trutta*, indicating high NADPH to maintain a supply of reduced-SH groups. Sulfhydryl compounds protect cells against damage due to superoxide anion production[33]. Anti-oxidant mechanisms such as these may also be important in deep-sea fish.

Sphingomyelin was found to be the principal phospholipid in *Coryphaenoides acrolepis*, whereas phosphatidylcholine was highest in *Antimora rostrata*[170]. A lipid to protein ratio of 1.5 to 2.1 was reported. Protein characterization yielded several proteins between 18,000 and 66,000 in molecular weight that were resolvable by dodecylsulphate PAGE, as well as a major unresolvable protein fraction of extremely large molecular weight ($\geqslant 40 \times 10^6$)[117,152]. Since naturally occurring membrane proteins of such high molecular weight are very unusual, it was postulated that rapid decompression of the membrane had caused irreversible aggregation of the membrane protein species. The swimbladder foams of *Coryphaenoides* (cf. *fernandezianus*) and a brotulid, *Parabassogigas* sp., collected from 3400–4575 m depth were characterized ultrastructurally as sheets of layered membranes having typical bilayers[186]. In mammalian plasma membrane the cholesterol/sphingomyelin ratios are highest and most closely approach a one to one ratio[169]. The gram quantities (20 g) of pure membrane available from one individual fish not only provide an unprecedented amount of material for the study of biological membranes, but must also relate to buoyancy control in the fish, since they are located inside the functional swimbladder. The swimbladder foam of *A. rostrata* also has a bilayer membrane configuration[117] as does *Bassozetus* sp. and two species of *Barathrodemus*[189]. Membrane biosynthesis in *C. acrolepis* and *A. rostrata* was studied[190] in the swimbladders. Excised tunica interna incorporated [1-^{14}C]acetate and [4-^{3}H]squalene into membrane lipids, cholesterol, phosphatidylethanolamine, phosphatidylcholine, cardiolipin, and to a lesser extent sphingomyelin. Ultrastructural study of the tunica interna of the swimbladder revealed large cytoplasmic vacuoles lined and filled with bilayer membrane. These membrane-lined vacuoles in the tunica interna cells may be an assembly point for the biosynthesis of the gram quantities of pure bilayer membrane found in the interior of the swimbladder. The tunica, with its numerous collagen fibers, provides a flexible, yet strong enclosure for the swimbladder with its cholesterol-rich membranous contents.

Bowne[26] examined the swimbladders of 44 species of deep-sea gadiform fishes and found a lipid deposit in the swimbladders of almost all specimens captured below 800 m, and in over half the specimens from shallow waters. She confirmed the membranous nature of the swimbladder lipid in *Coryphaenoides armatus*, finding that dorsally, it was attached to the oval sphincter, and ventrally it covered the retia. Phospholipids routinely found in the swimbladder lipid of these fishes included phosphatidylcholine (34%), sphingomyelin (29%), phosphatidylserine (23%) and phosphatidylethano-lamine (14%). The fatty acids were mostly unsaturated (76–90%) with high levels of mono- (43–61%) and polyunsaturates (17–44%).

The accumulation of cholesterol within the oxygen-filled swimbladders of these deep sea fishes suggests that certain steps in cholesterol biosynthesis are effected by hyperbaric oxygen. The lipid content of these swimbladders contained up to

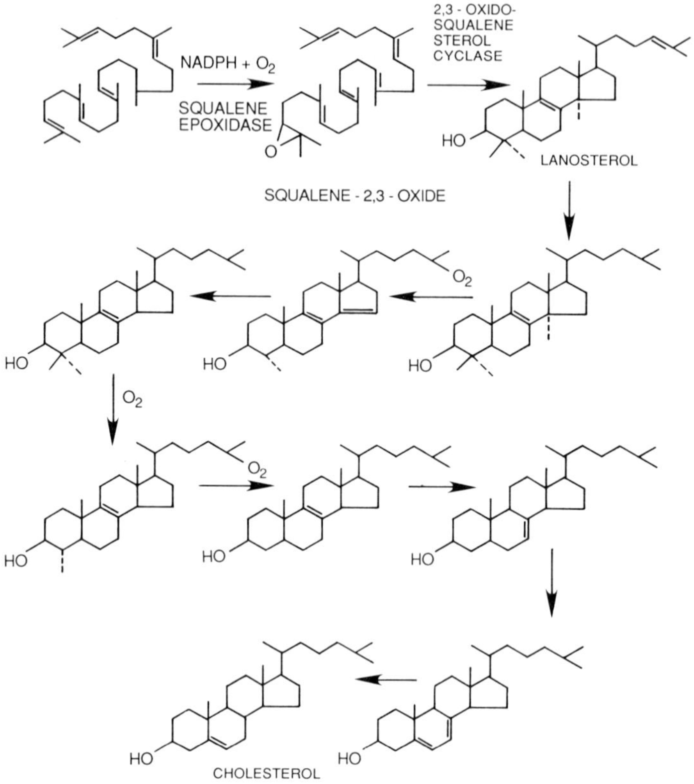

Fig. 2. The biochemical conversion of squalene to cholesterol[242,82].

49% (dry weight basis) of cholesterol[172]. The lavage-fluid from the goldfish (*Carassius auratus*) contained 17% to 30% cholesterol[55]. Deep ocean fishes, however, contain more total cholesterol in their swimbladders, as it is a major component of the gram quantities of pure membrane therein.

The swimbladder gas gland is a site of cholesterol synthesis in macrourid fishes (*Coryphaenoides* spp.) as well as various shallow water fishes with swimbladders (*Orthopristis forbesi, Seriola mazatlana*, and *Sphoeroides annulatus*)[181]. This means that the cholesterol is synthesized under hyperbaric oxygen conditions in the swimbladder. The pathway of cholesterol biosynthesis in fish probably does not differ from that in mammals[82] (Fig. 2), but it takes place in fish at a slower rate. Therefore, one might surmise that pressure and oxygen influence cholesterol synthesis in deep-sea fish.

During cyclization of squalene, a significant volume decrease occurs. The density of squalene-2,3-oxide is similar to squalene, 0.86 g/ml, whereas cholesterol has a density of 1.067 g/ml. Pressure (5000 and 10,000 psi) increased the rate of conversion of [4-^{3}H]squalene-2,3-oxide to cholesterol in the swimbladder gas gland of the grunt, *Anisotremus davidsoni*[183], and did not decrease cholesterol synthesis in the macrourid (*Coryphaenoides acrolepis*). Synthesis of cholesterol from [1-

[¹⁴C]acetate was enhanced by pressure in swimbladder gas gland and liver of an abyssal *Coryphaenoides* species, and a shallow water grunt, *Orthopristis forbesi*[181].

There are a number of steps in cholesterol biosynthesis which require molecular oxygen (Fig. 2). These include the oxygenation of squalene to squalene-2,3-oxide and three demethylation reactions occurring between lanosterol and cholesterol (Fig. 2). Oxygen enhancement of cholesterol synthesis has been demonstrated in an abyssal codling[179], and Kayama et al.[119] have demonstrated the importance of molecular oxygen in cholesterol synthesis in fish. Therefore, despite the density of cholesterol (1.067 g/ml) which is higher than that of sea water (1.024–1.026 g/ml), it is an integral component in an oxygen lipoprotein-filled buoyancy chamber, the swimbladder of fish.

The function of the fat-filled swimbladders of deep-sea fishes has never been explained. There are two major hypotheses to explain its function[26,149]. The first hypothesis is that the swimbladder wall is a lipid-rich barrier to oxygen diffusion[239]. In support of this hypothesis, the swimbladder wall has been observed to be brilliantly white and reflective in freshly captured specimens, which suggests that the lipids are in a highly ordered array. Rather than a 'foam' as it appears in fish observed at the surface after capture, the swimbladder lipid contents form a thick wall around the interior of the swimbladder in fish at depth in the ocean. Wittenberg et al.[239] suggest that the lipid has the same function as guanine crystals in silvery layers[200]. The argument is presented that these swimbladder membrane lipids are below their transition temperature, and thus, the diffusion coefficient of solutes and permeation rates are approximately one-thousand fold less than in fluid membranes[219]. Furthermore, addition of cholesterol to a fluid membrane formed from phospholipid reduces the diffusion constant approximately tenfold[75,239].

The second hypothesis[177] suggests that by dissolving gaseous oxygen in the swimbladder lipid the back pressure of oxygen on the gas gland is reduced allowing oxygen secretion by the Bohr and Root effects. The Bohr and Root effects are nullified at 50–100 atm oxygen pressure[164], and Scholander[209] suggested that some unknown substance could be secreted by the gas gland to dissociate oxygen from oxyhemoglobin at pressures higher than 50–100 atm. Fishes with functional oxygen-filled swimbladders are routinely captured at depths where the pressure is well in excess of 50–100 atm[26,186]. Since the fatty acids of the swimbladder membrane lipids are highly unsaturated[26,172,189], it is very unlikely that it acts as a solid-phase diffusion barrier. Furthermore, the oxygen permeability of phospholipid layers increases as the thickness of the layer increases, which suggests that the swimbladder multilamellar membrane system is not a more effective diffusion barrier than a single membrane[16,26]. The observation that the swimbladder wall is brilliantly white and reflective in freshly caught specimens[239] is not sufficient evidence to support the hypothesis that the lipid is a diffusion barrier. The lipids probably are highly ordered[239], but this does not mean they are below their phase transition temperature at depth in the ocean. Lower melting points of unsaturated fatty acids are known to help keep lipids in a liquid state and protoplasmic viscosity normal. It is well known that lower temperatures such as those deep sea fish experience, result in greater unsaturation of fatty acids[46,71,168].

Solubility of oxygen in these swimbladder lipid membranes has not been studied. Oxygen solubility in various marine triacylglycerols and wax esters has been measured[121]. From these data the volume percent of oxygen in a typical marine oil at (90 atm and 20°C) was calculated as 14.41 ml O_2/100 ml oil[16, 121]. This solubility of oxygen would not suffice to dissolve the whole gas phase (7% of the volume of the fish) in the 4–7 mls of swimbladder lipid normally present in *Coryphaenoides armatus*[26]. However, this calculation was performed for marine oil at 20°C and deep sea fish live at 2°C. Also, the marine oil used in the calculation is not the same as the swimbladder membrane lipid. Furthermore, the gas phase in the swimbladder of deep fishes may be much less than 7% of the volume of the fish. For example, Hemmingsen[104] has suggested that O_2 is present in fish swimbladders below 3500–4000 m depth as a clathrate hydrate of oxygen.

To clarify the function of lipid in swimbladders, two questions must be answered. First, is oxygen present as a gas phase in the swimbladder of the fish at the depth where it lives? Second, what is the solubility of oxygen in the swimbladder lipid at 2°C and the ambient pressure in the deep sea habitat of the fish?

To answer the first question, two hyperbaric traps were built which successfully brought the Pacific rattail, *Coryphaenoides acrolepis,* up from depth under pressure[192,238]. However, it was impossible to keep these fish alive, as no fish survived full decompression to atmospheric pressure. *Coryphaenoides acrolepis* was kept alive for 41 h in the hyperbaric trap[238]. If hyperbaric trap-aquarium maintenance of these fish becomes a reality, it may be possible to answer the first question; whether oxygen in the swimbladder is in gaseous phase, and if so, what is the volume? It could be possible to X-ray the fish, or remove gas from the swimbladder under pressure by a syringe.

To resolve the second question, very careful measurements of oxygen solubility must be made of the swimbladder lipid of freshly captured deep sea fishes at different temperatures and pressures. Also, thermal transitions of swimbladder membrane lipids must be determined directly. Physico-chemical studies of unsaturated fatty acids have been rather restricted[110]. The effects of high pressure on the structure and properties of lipid membranes have recently been reviewed[240]. High-precision gas chromatography[127–129] appears to be a useful technique for the measurement of oxygen solubility in swimbladder tissues at elevated pressure, including the effects of temperature and phase transitions. One would need only construct a high-pressure system and make use of swimbladder tissues (or models thereof) as stationary phases. The solubility of oxygen could then easily be carried out over the entire concentration range, infinite dilution to saturation.

III. Skeleton

1. Morphological considerations

Since fish are supported by the buoyancy of water, the skeleton does not support their weight, but rather provides support for the muscles to pull and propel the fish through the water[196]. Skeletons of larger terrestrial animals have an upper limit on maximum size; about 4×10^7g for larger dinosaurs[10], whereas the blue whale exceeds this by almost an order of magnitude. A study of whale skeletal weight allometry[197] confirmed their similarity to that of fishes.

The humeri and femora of a number of marine mammals exhibit depressed skeletal maturation and lack of a marrow cavity[233]. These include the California sea otter *Enhydra lutris* and the pinnipeds *Erignathus barbatus, Mirounga angustirostris, Leptonychotes weddelli, Halichoerus grypus, Phoca vitulina, Zalophus californianus, Eumetopias jubata, Arctocephalus townsendi, Callorhinus ursinus*, and *Odobenus rosmarus*. The skeletal neoteny possessed by these species and penguins, sirenia, cetacea, and marine reptiles may involve greater skeletal acid-based buffering capacity, due to greater rates of skeletal blood flow and bone–blood mineral exchange. This allows them to withstand acidosis from excess carbon dioxide during prolonged and repetitive diving.

The vertebrae of mammals have higher tissue densities and lower organ densities than fishes, where the bone tissue is the hard, solid part of the bone, and the bone organ includes all the fat marrow, loose water, and interstitial material including the bone itself[227]. Marrow containing vertebrae of baleen whales contains 24–68% lipid, the rib of the fin whale contains 27% lipid, whereas lipid contents of sea lion and sperm whale vertebrae are low[227]. Cetacean bones are generally characterized by extreme porosity. The relative skeletal weight (as percent total body weight) in adult delphinids, *Delphinus delphis*, is much less (3.7%) than that for larger cetaceans (10–17%[134]), and is the smallest for any vertebrate. A skeleton of a 20 m whale, recovered from 1240 m depth in the Santa Barbara basin off California, was found to be laden with oil smelling of sulphide[211]. Bacterial mats were supported by chemical reducing power from bone-oil seepage. These skeletons were estimated to occur once every 300 km^2 along the whale migration route with an average distance between nearest neighbor of 9 km[211].

The coelacanth, *Latimeria chalumnae,* has a spine which is a notochord made of tough, thick cartilage filled with liquid[213]. The name 'coelacanth' means 'hollow spine'. Perhaps this liquid is hypotonic and aids in buoyancy.

Reduction in bone ossification in deep-sea fishes is accompanied by protein reduction. This is an obvious advantage to the maintenance of neutral buoyancy, because skeletal tissue may have a density as high as 2–3 and protein 1.33 g/ml. The loss of protein occurs mostly in the trunk and caudal musculature of bathypelagic fishes, where it includes skeletal reduction[112].

Although the liver of sharks is probably the most important buoyancy organ since it may contain a good deal of low density squalene, density regulation also results from skeletal tissues as well as white muscle and the skin. Cartilage in sharks is secondarily derived from osseous endoskeletal tissue in ancestral forms. Calcification of cartilage occurs in living sharks in the vertebral column and the jaws, and is age dependant in some species[153]. In certain teleosts, such as the lumpsucker (*Cyclopterus lumpus*), low density is derived from a cartilagenous skeleton. The skull, vertebral column, and branchial skeleton are almost completely uncalcified, with the average density 1.04 g/ml[52]. Much persistent cartilage and weak skeletal ossification as well as incomplete mineralization of scales also characterizes the swimbladderless notothenioid (Antarctic toothfish) *Dissostichus mawsoni*, which is neutrally buoyant[60]. Cartilage (density of 1.1) is substituted for bone (density 2.0) in the skull, pectoral girdle and caudal skeleton of *D. mawsoni*[62]. Another Antarctic notothenioid, the silverfish *Pleurogramma antarcticum*, has gained neutral buoyancy by skeletal reduction and deposition of lipid[54,59]. The ashed skeleton weight in *Pleurogramma* is only 0.3% of body weight, and that of *Dissostichus* 0.6%[63]. The vertebrae in *Pleurogramma* are unconstricted, vertebral processes are reduced, and the centra are thin collars of bone. A persistent gelatinous notochord is found in hollow adult vertebral centra[61]. *Pagothenia borchgrevinki* and seven species of *Trematomus* (all Antarctic notothenioids) possess partially patent notochordal foramina which occupy 20–38% of the vertebral centra cross sectional area[58]. Large foramina are associated with the more buoyant species. It would be of interest to know if the skeletal components of these Antarctic fishes contained lipid, as an aid to attainment of neutral buoyancy.

2. Triacylglycerol-filled bones

Forty-eight species of fish from 30 families have been analyzed for skeletal lipid; thirteen of these families have representatives (15 species) with more than 24% lipid, as percent weight[173,182,193,227]. Thus, almost 30% of all fish analyzed to date have oil-filled bones. The mackerel, *Scomberomorus maculatus,* has 29% lipid in its vertebrae[227]. The lipids of the vertebrae of fall chum salmon *(Oncorhynchus keta)* varied between 26–32% (by weight) depending on the type of factory heat-softening process used[221]. Five fish species were analyzed for bone lipids and triacylglycerol was found to be the major lipid class[132]. Three marine fish, *Schedophilus medusophagus, Peprilus simillinus, and Anoplopoma fimbria* had oil-filled bones, with the neurocranium of *P. simillinus* and *A. fimbria* containing 68 and 60% lipid (by weight), respectively. The bones of two fish with fat-filled swimbladders, *Coryphaenoides acrolepis* and *Antimora rostrata,* contained only 0.2% lipid[132]. A role in neutral buoyancy was assumed. However, the wrasse, *Cheilinus rhodochrous,* has a swimbladder and has oil-filled bones (52–82% lipid[175]). *Sebastes ruberrimus,* a rockfish with a swimbladder, has oil-filled bones (24–32% lipid) and *Hydrolagus colliei,* a ratfish without a swimbladder, lacks oil-filled bones (1–3%)[174]. The swordfish, *Xiphias gladius,* has a swimbladder and has high lipid content with porous fatty bones[36]. Since the swimbladder may not be able to compensate for rapid vertical movements (swordfish may move from 100 m depth to the surface in 5 min), the bone lipid may help compensate by lowering the density of the fish. Swordfish, white marlin (*Tetraptures albidus*), and sailfish (*Istiophorus platypterus*) have a mass of white adipose tissue in the skull that insulates the brain and its brown adipose heating mechanism[35].

A possible role of skeletal lipid could be to maintain the posture of a fish. The negatively buoyant hawkfish, *Cirrhitus pinnulatus,* rests on the bottom in a head-upward position[175]. The enlarged head and lipid-filled cranial bones of these fish maintain the body in a head-upward position even in frozen specimens. The oil-filled skull of the giant hawkfish, *C. rivulatus*, also relates to its posture. It has a typical head-up inclination when perched at the entrance of a cave[176]. This could allow these fish to sight prey below when perched on coral or rock, enabling the eyes of the fish to protrude away from the perch. Other species of hawkfish have been photographed illustrating the head-free attitude[111]. *Ruvettus pretiosus* also has a head-upward position[23]. The blennid, *Exallias brevis*, stores more oil in its skull (27% lipid by weight) and has a head-up attitude when perched on coral reefs[174].

The oil globule in California grunion larvae (*Leuresthes tenuis*) provides lift to the head, the densest part of the body[65]. The buoyancy provided by this oil globule may allow these larval fishes to resume feeding after prolonged starvation. Starvation in other fish larvae (such as *Pleuronectes platessa* and *Clupea harengus*) may deplete cranial fat and change the center of gravity resulting in a head-down inclination making it impossible to resume feeding[66].

The fraction of the total body lipid held in the bones of fish may be very high. In the sheepshead wrasse *Pimelometopon* (*Semicossyphus*) *pulcher,* and the sablefish *Anoplopoma fimbria*, the bone lipid comprised 79–93% and 52–82% of the total body lipid, respectively[188]. In the doctorfish, *Acanthurus chirugus,* bone lipid comprised 81% of the total body lipid[193].

Triacylglycerol is the principal lipid in the skeleton of fishes with oil-filled bones and comprises 64–97% of the lipid[132,182,187,193]. Cholesterol and phospholipids are the only other major lipid classes present in fish bones rich in triacylglycerol. In contrast, orange roughy, *H. atlanticus,* and smooth oreo dory *Pseudocyttus maculatus*, both contain wax esters as the principal skeletal lipid[86,173]. The principal phospholipids of sablefish (*A. fimbria*) bones are phosphatidylcholine (31%), phosphatidylethanolamine (28%), phosphatidylserine (26%), and phosphatidylinositol (13%). In *Peprilus simillinus* bones, the primary phospholipids are phosphatidylcholine (19%), phosphatidylethanolamine (63%), and phosphatidylserine (15%)[132]. In both fish minor amounts of sphingomyelin, lysophosphatidylcholine, and cardiolipin were also noted. Bone phospholipids are similar to phospholipids of other fish tissues[225]. The major triacylglycerol fatty acids in *P. simillinus* and *A. fimbria* were 18:1, 16:0, and 16:1[132]. Phospholipid fatty acids of the bone in these fish were enriched in the polyunsaturated fatty acids 20:4, 20:5, 22:5, and 22:6. The percent polyunsaturation of the bone triacylglycerol fatty acids (12–34%)[175] in four species of Hawaiian fish was much higher than it is for bones of sablefish, *A. fimbria,* which had only 2% polyunsaturates[132]. Temperature influences bone oil fatty acid composition. The temperature in Hawaiian waters is 25–27°C. The percent unsaturation of fatty acids from Hawaiian fish bones ranges from 52–65% for the phospholipids and from 39–58% for the triacylglycerols[175]. The percent unsaturation of the bone fatty acids of *Sebastes ruberrimus,* collected in Canada in 10°C water, ranged from 69–72% for the phospholipids and 68% for the triacylglycerols[174]. This is approximately 10% greater unsaturation for colder water. An increase in skeletal polyunsaturated fatty acids was found in the deep cardinal snapper, *Pristopomoides macrophthalmus* (from 500 m depth; 10°C) and the rare arrowtail, *Melanonous zugmayeri* (from 1000 m depth; 3–4°C) as compared to two shallow water (5 m, 25–30°C) acanthurid species[187].

Skeletal fatty and composition is also controlled by diet. Polyunsaturated fatty acids in the bigeye, *Priacanthus arenatus* (25 m depth), comprise a significantly greater percentage of the fatty acids than in the acanthurids from 5 m depth, whereas the temperature at 25 m off Jamaica does not differ significantly from that at 5 m. *Priacanthus* is a nocturnal carnivore, whereas the acanthurids are diurnal herbivores which can be observed actively grazing on algae. *Priacanthus* is a zooplankton feeder, consuming primarily fish and fish larvae, shrimp, polychaetes, and crabs and crab larvae. Some of these larval forms are derived from adults found in colder, deeper water, and whose lipids are enriched in polyunsatured fatty acids. The more highly unsaturated fatty acids in *P. arenatus* may therefore be a consequence of diet rather than temperature.

Bone triacylglycerol is as metabolically active as other tissues such as viscera and flesh. The results of feeding 1-^{14}C-palmitic acid to the señorita, *Oxyjulus californica,* and the sablefish, *A. fimbria*, indicated a rapid uptake into triacylglycerol after 12 h, followed by a drop in radioactivity at later times (up to 96 h) in both these species[188]. There was an equal distribution of radioactivity between phospholipid and triacylglycerol in bone lipid of the sheepshead wrasse, *Pimelometopon* (*Semicossyphus*) *pulcher*. It was suggested that the sheepshead wrasse was just meeting the turnover requirements of its biomembranes with the administered dose, while the smaller sablefish, which received the same amount, was able to maintain its biomembranes easily and incorporate most of the radioactivity into neutral lipid. Neurocranial lipid (triacylglycerol) in the ocean surgeon, *Acanthurus bahianus,* is rapidly utilized during starvation stress, dropping from 9.1 and 12.9% in two individuals starved for 72 h[182,193]. ^{14}C-Radiolabelled palmitic acid fed to these fish appeared rapidly in skeletal lipid. The neurocranium in the tropical herbivorous ocean surgeon provides a readily available pool of energy-rich lipid for use during short term starvation stress (1–3 days) which might be incurred during intense tropical storms in the Caribbean[182,193].

Skeletal tissues in Pacific salmon are composed of significant amounts of lipid (which serves as an energy supply), that is then depleted during migration and freshwater spawning[187]. The vertebral centra and neurocrania of ocean prespawning pink salmon *Oncorhynchus gorbuscha,* although composed respectively of 2.9–13% and 0.5–6.3% lipid (each by weight) amount only to 0–1.6% and 0–1.4% in river postspawning individuals. Triacylglycerols of carbon numbers 47, 49 and 51 are selectively depleted during spawning, whereas there are increases in the relative amounts of those of carbon numbers 55, 57, 59 and 61. The patterns of variation of the skeletal lipids of coho salmon, *O. kisutch,* chum salmon, *O. keta,* and sockeye salmon, *O. nerka* are similar to those of *O. gorbuscha*[187]. The skull of sockeye salmon was found to have 20.7% lipid and to be richer in lipid than the flesh[241].

3. Wax ester-filled bones

Neutral buoyancy in the castor-oil fish, *Ruvettus pretiosus,* is attained by storage of large amounts of wax esters in the flesh and skeleton[130]. This fish stores 30% lipid in the frontal bone, and has its head pointing at an angle of 45%[23].

Several tissues of the orange roughy, *Hoplostethus atlanticus*, contain large amounts of lipid mainly as wax ester. These include the skin, a lipid-invested swimbladder, the skeleton and a fatty tissue located in the neurocranial cavity[31,86,207]. The function of the wax ester in these tissues remains to be elucidated. Metabolic studies[87] suggest that there may be turnover of wax esters thus, potentially, they may act as an energy reserve as triacylglycerols do in bones of other fish[182,193]. Alternatively, it has been suggested that they may have a role in determining the buoyancy of the fish which is normally caught at depths of 1000 m[173].

The intraneurocranial lipid accounts for 5% of the total lipid of the fish and the skeletal lipids comprise about 8% of the total body lipid. Between 94 and 96% of this lipid is wax ester[173]. The two major lipid containing fractions are the flesh which includes the muscle and the skin, the latter being the major wax ester source, and the viscera where the swimbladder probably accounts for at least 50% of the lipid in this fraction[86]. With the exception of the skin, all the major lipid containing tissues appear to be in the anterior region. The center of gravity at the surface is located in the same position as the swimbladder in these fish.

Although orange roughy do not live at the surface, several observations suggest that the lipids of the fish do modify the buoyancy of the fish at the surface. Firstly, the fish floats in a head up inclination consistent with the distribution of the major wax containing tissues. There is a difference of 8°C in the temperature of the water at the surface and at depth. Phase transition measurements of this wax ester-rich lipid made by differential scanning colorimetry revealed phase transitions occurring at $-23°C$, $-7°C$, and $+8°C$. The proportion of lipid solid at 6°C (the temperature at 1000 m depth where the fish occurs in reproductive aggregations) was 11–18% for neurocranium and 16–25% for the intraneurocranial lipid[173]. Swimbladder lipid at 6°C was 17% solid. Results were similar when samples were analyzed by pulsed nuclear magnetic resonance[137]. At 14°C all of the samples contained less than 1% lipid in the solid phase. The effect of temperature on the density of these lipids, revealed an increase in density of 0.013 going from 14°C to 6°C. The pressure at 1000 m depth is 100 atmospheres and this is likely to add further to the density of the wax esters found in *H. atlanticus* in its normal habitat. Studies with sperm whale oil have suggested that 100 atmospheres is likely to increase the density of the lipid by 0.005[40,43]. The effect of the increased pressure on the density of the surrounding seawater will also modify the net uplift produced by the lipid. The compressibility of natural oils and water are similar and thus the increased density of the lipid at high pressures should have no effect on the buoyancy of the fish. However, the wax ester-rich lipids of the copepod, *Calanus plumchrus,* have a marked change in density with pressure, and especially temperature[245]. Hydrostatic pressure is known to have major effects on the physical properties of lipids[37,215]. The intraneurocranial lipid (similar to the swimbladder lipid) is contained within a three-dimensional network of what appears to be collagen fibres[173]. The dimensions of the collagen network are considerably larger than those of normal cells and suggest that the lipid may well be extracellular. This implies that such lipid functions as a permanent source of buoyancy[68,69]. The

Antarctic notothenioid, *Pleuragramma antarcticum*, has been reported to contain large extracellular lipid stores that are proposed also to function in regulating buoyancy[61].

An interesting contrast to *H. atlanticus* is the oreo dory, *Pseudocyttus maculatus*. This fish is caught in the same trawls as *H. atlanticus* but contains neither the intraneurocranial lipid nor the fat-invested swimbladder[173]. Instead, there appears to be a fully functional swimbladder. Although the flesh and bones do contain wax esters, the total lipid of the fish is only one tenth that of *H. atlanticus* and hence not likely to provide a major contribution to determining buoyancy.

A possible phase change in the spermaceti oil of the sperm whale has been suggested as a buoyancy-control hypothesis[40,41,42,43]. When the whale dives (which may be in excess of 1000 m depth for almost one hour) the oil may turn partially solid, as the whale lowers the temperature of the oil from 33°C to 31°C by altering blood flow, becoming denser and therefore occupying less volume. This aids the whale in its descent to great depths. When ready to surface, blood flow could be resumed to the spermaceti oil (mostly wax esters, it may comprise up to four tons in males) causing it to become more liquid and less dense, making ascent a simpler task. Neutral buoyancy at depth would be advantageous for prey capture (large squid) and energy conservation.

The skeleton of the leatherback turtle, *Dermochelys coriacea*, is suffused with oil[194]. Although the skeletal oil in this turtle has not been described, subcutaneous depot and plastron fats have[2]. This turtle is warm blooded, and may have a body temperature of 18°C above sea water[77]. Deep adipose tissue in these turtles may be important in endothermy[83]. Since these turtles dive to 1000 m depth to eat siphonophores, it is possible that the oil may lessen decompression problems during rapid diving and resurfacing[194]. The turtle could shut off blood supply to the oily bones when it commences the dive (peripheral vasoconstriction), causing a partial phase transition to more dense solid. Prior to re-surfacing, it could re-establish peripheral blood supply warming up the bone oil to a more liquid and less dense state, making ascent easier as has been described for the sperm whale[43].

A study of wax ester metabolism in the orange roughy (*H. atlanticus*) revealed that both the skull and the intraneurocranial fat could synthesize lipid *de novo* from acetate as well as incorporate long chain fatty acids into more complex lipids[87]. Both the bone (skull) and the swimbladder were very active in lipid synthesis with the acetate incorporation rates approximately one sixth those of liver.

The rate of lipid synthesis in the skull was quite surprising given that bone is largely structural and metabolically inert. This would imply that the lipogenic cells associated with the bone are extremely active. The skull of *H. atlanticus*, with 15–20% lipid may function as an energy reserve during starvation stress as it does in other fish species[174,182,188,193], where radioactive palmitic acid was fed to whole live fish. Pure fish bone from orange roughy synthesized lipid from radioactive precursors as though it was a metabolically active adipose tissue[87]. It is not yet known whether these lipid reserves can be mobilized during starvation. To determine this, specimens of *H. atlanticus* would have to be captured and kept alive for some time.

There may be an inability of the tissues of *H. atlanticus* to elongate and/or desaturate fatty acids such as palmitate that is either synthesized *de novo* or taken up from the circulation. Evidence for this comes from effective exclusion of the palmitate and [^{14}C]acetate labels from the wax esters, in skull, intraneurocranial fat, and liver[87]. These tissues, however, have the ability to reduce oleate to oleyl alcohol which is also incorporated into wax esters. Oleate comprises over 50% of the wax ester fatty acids and over 60% of the fatty alcohols found in the wax esters are monounsaturated with chain lengths or 18, 20 and 22 carbons[31,86,207]. Clearly, the long chain fatty acids must come from dietary sources, but it is uncertain as to whether dietary fatty alcohols are required or whether they can be produced by reduction of the corresponding fatty acids. The diet of *H. atlanticus* is varied[201] but contains species rich in wax esters[130,158,201,205]. Serum lipoproteins of *H. atlanticus* contained wax esters suggesting that they are in transit from the gut to the peripheral tissues. The presence of long chain polyunsaturated fatty acids in different classes of the total serum lipids is also consistent with their originating from a dietary source[87].

Wax esters provide 70% more upthrust than the same volume of triacylglycerol (Table 1; refs. 130, 203). Therefore, these wax ester-filled bones provide more buoyancy to the fish than triacylglycerol-filled bones. However, some oily fish bones nevertheless end up on the sea floor. The deep sea limpet, *Cocculinella minutissima*, lives on fish bones, while *Osteapelta mirabilis*, was discovered on a whale skull at 1100 m depth[89].

IV. Liver

1. Squalene

Trace amounts of squalene occur commonly in fish tissues because it is an intermediate in cholesterol biosynthesis (Fig. 1). Squalene occurs in very high quantities in the liver oil of certain shark species[93]. These species belong to five families including Chlamydoselachidae, Odontaspidae, Lamnidae, Scyliorhinidae, and Squalidae[157], with the Squalidae having the most species with squalene-filled livers. Squalene may comprise up to 15.5% of the total body weight in *Centrophorus* spp., where the percent oil in the liver (as percent wet weight) may be 88%; squalene may comprise 90% of this liver oil[157]. Squalene is only found in appreciable amounts (17% in liver) in one species of teleost fish; the eulachon, *Thaleichthyes pacificus*[1].

Squalene cannot be catabolized to yield metabolic energy like triacylglycerols and diacylglyceryl ethers[208]. Squalene can only be metabolized to cholesterol and bile salts. Cholesterogenesis may be blocked just beyond squalene synthesis because of lack of oxygen or enzymes required for cyclization in livers of deep ocean sharks. Many sharks are almost neutrally buoyant because of the upthrust generated by squalene in the liver. Since squalene is essentially inert metabolically the only function advanced for it is buoyancy[157]. Liver oil was found to be the primary factor affecting the density of sharks[12], where specific gravities of sharks (69 individuals from 13 species in the eastern Gulf of Mexico) without the liver averaged 1.062–1.089 g ml^{-1} (sea water is 1.026 g ml^{-1}). Open-ocean, fast-swimming pelagic sharks were less dense than coastal sharks[12]. In the eulachon, squalene may be accumulated from hydrocarbons in their zooplankton diet[1]. The 17% squalene in the eulachon liver could result from depletion of metabolized depot fats during the spawning migration.

Shark starvation experiments support the buoyancy hypothesis for liver squalene. *Galeocerdo cuvieri*, starved 26 days in captivity, lost 10.1% of its weight; yet, its liver held 79.7% oil[11]. *Negaprion brevirostris* had 5.7 kg liver oil prior to starvation, and used 3.5 kg (61%) of oil during starvation[11]. The remaining oil was thought to function in buoyancy. Liver oil of sharks gives 5–6 times the buoyancy of an equal amount of solute-free water. As the shark ages the liver oil may increase from near zero to 25% of the weight of adult sharks, thereby decreasing their density[11,12]. Sharks lack a swimbladder and are usually negatively buoyant. While swimming, the pectoral fins and the heterocercal tail keep them off the bottom. Certainly, the low density of squalene (0.86 g/ml) is far superior to that of cod-liver oil (0.93).

2. Glyceryl ethers

Alkyldiacylglycerols are abundant components of shark liver oil[119,120], especially the family Squalidae. *Squalus acanthias* has been reported to have 38–45% alkyldiacylglycerols (as% of total lipid) and other species of *Squalus* may have up to 54%[202,208]. The liver of the deep shark, *Dalatias licha*, has 83.4% lipid, 27.8% as alkyldiacylglycerol and 40.1% as squalene[93]. The holocephalans *Chimaera barbouri*, *C. ogilbyi*, and *Hydrologus colliei*, have more than 50% alkyldiacylglycerols (as% total lipid in the liver)[202]. *Hydrolagus novaezealandiae* has 64.1% liver lipid, 65.8% of which is alkyldiacylglycerol, 12.8% neutral plasmalogens, and 10.4% triacylglycerol[99,100]. Neutral plasmalogens, or alkenyldiacylglycerols, are much less abundant than alkyldiacylglycerols in elasmobranch liver, and probably often occur in small amounts along with alkyldiacylglycerols[202]. Methoxy-glyceryl ethers, first isolated from greenland shark liver lipids, occur as 0.1–0.3% of liver lipid of some sharks (*S.acanthias*, *D. licha*, and *Scyliorhinus torazame*) and ratfish (*Hydrologus novazealandiae*, *H. barbouri*, and *Rhinochimaera pacifica*)[92]. These compounds have antibiotic activity and inhibit growth of tumors in mice. Glyceryl ethers are useful as surfacants in cosmetics and ointments, and diacylglyceryl ethers from ratfish liver are intermediates in synthesis of alkylacetoylglycerophosphocholine, a platelet activating factor[96,154]. Glyceryl ethers also occur in the liver lipid of two species of gonatid squids, *Berryteuthis magister* and *Gonatopsis makko*[96,97,101].

A convincing argument for buoyancy regulation as a function for alkyldiacylglycerols was made by Malins and Barone[139], who found the concentration ratios of diacylglyceryl ethers to triacylglycerols to significantly increase in livers of weighted *Squalus acanthias* when compared to an unweighted control group of dogfish. One gram of diacylglycerol ethers (specific gravity 0.908 at 21°C) gives 14% more lift than one gram of triacylglycerol (specific gravity 0.927 at 25°C) in sea water (specific gravity of 1.026). The change was not due to selective mobilization of triacylglycerols, but involved a net increase in alkyldiacylglycerols. Malins and Barone[139] suggested that *S. acanthias* regulates buoyancy by the selective metabolism of diacylglyceryl ethers and triacylglycerols during vertical migrations, and that this may substitute for the swimbladder in teleost fishes not found in sharks. Regulating alkyldiacylglycerol metabolism has also been carefully examined in a cell-free system from *S.acanthias* liver, where the data indicated regulation by enzymes controlling the fatty acid-fatty alcohol equilibrium[140]. Positional distribution of acyl groups on alkyldiacylglycerols has been described[138]. However, more experimental data are needed to conclusively prove the case for alkyldiacylglycerols in buoyancy regulation[202].

3. Triacylglycerols and wax esters

Much of the research on liver oils of sharks and teleost fishes has been stimulated by the presence of commercially valuable vitamins A and D[205]; particularly vitamin A in species such as the basking shark[157] and the soupfin shark[11].

A case for liver buoyancy has been made for the macrourid *Coryphaenoides armatus*[217]. In this species, the liver becomes larger in older fish who increase in girth, rather than length. Since the liver size increase occurs in both sexes and is independent of time of year or age stage, it may act as a buoyancy controlling device. Since the respiration rate decreases in larger fish, the glycogen and lipid in the liver could keep *C. armatus* alive for 186 days[212]. Also, the swimbladder does not increase in size as the fish increases in weight; therefore increased liver lipid could provide the necessary buoyancy in larger fish[217]. Presumably, triacylglycerols are the principal component of these macrourid liver lipids, as has been reported for *Coryphaenoides* sp. off Ecuador[181]. Liver consisted of $66 \pm 9\%$ lipid (as percent dry weight) for seven specimens of *Coryphaenoides* sp., compared to $28 \pm 16\%$ ($n = 7$) for shallow water fish from the same area (*Orthopristis forbesi*, a grunt, *Seriola mazatlana*, a jack, and *Sphoeroides annulatus*, a puffer[181]. The liver of *Bathysaurus* which increases in size in larger individuals, probably also contributes to buoyancy[143]. The inverse is true for the redlip blennny, *Ophioblennius atlanticus*. In this species, the liver makes up a much greater proportion of the larval body mass than it does in the adult and has more lipid in the larval fish[165]. Although, liver lipid buoyancy aids the pelagic larvae, it is secondary in importance to the function of energy for metabolism at metamorphosis[166].

Liver lipid (as percent dry weight) in the pink salmon, *Oncorhynchus gorbuscha*, decreased from 9.9% in pre-spawning ocean fish to 3.2% in river postspawning salmon[180]. The ability of the liver to synthesize triacylglycerols is lost after migration into fresh water and spawning[180]. Livers of chum salmon and coho salmon maintained in fresh water contained 3.9–9.5% lipid (as percent fresh weight), most of which was neutral lipid[223]. Lipid-rich livers are not common in fresh water fish under normal conditions[106].

Perisinusoidal (Ito) cells in the liver have been studied in nine species of Antarctic fishes[57,59]. In *Dissostichus mawsoni* these cells are common in the liver and contain numerous lipid droplets. These cells may be involved in lipid metabolism, integral to maintenance of peripheral lipid deposits important in neutral buoyancy of this fish[57,59].

Two species of Caribbean shark, *Centrophorus granulosa* and *Dalatias licha*, contained sufficient lipid in their livers to attain neutral buoyancy in sea water[231]. The liver comprised 77% and 61% respectively of body dry weight, and in both cases stored more than 99% of the total body lipids. In each species, wax esters were the major liver lipid class present with lesser amounts of triacylglycerol and hydrocarbon[231]. Seven other Caribbean shark species had 89% of their body lipids contained in the liver. The data for *Dalatias licha*, revealing wax esters as the major liver lipid[231] contrast with an earlier study for *D.licha* from the Pacific ocean, where alkyldiacylglycerol and squalene were reported to be the principal lipid classes[93].

Wax esters are not a major component of the liver of most teleost fishes[130], and are more important in roe and muscle. Some deep-sea cods (Moridae) have wax esters that may comprise up to 25–30% of the total liver lipid; these include *Lotella phycis*, *Laemonema morasum* and *Podonema longipes*[130]. An analysis of 40 specimens of *Laemonema (Podonema) longipes* revealed liver weights that accounted for about 10% of the total body weight and contained $74.0 \pm 6.3\%$ oil consisting of $73.6 \pm 11.6\%$ wax esters[95]. The wax esters consisted mostly of C_{38}, C_{40}, and C_{42} carbon chain lengths; the fatty alcohols were mainly 22:1, 20:1, and 16:0, and the fatty acids were rich in the monoenes 16:1, 18:1, 20:1 and 22:1 acids with some 20:5[91].

Cholesterol, being more dense than sea water (1.067 g ml^{-1}), does not contribute to buoyancy. Although not a major lipid class in the liver of fish, it accounts for 96% of the sterol which comprises 3.3 to 6.7% (as percent of the non-saponifiable liver fraction) in the channel catfish (*Ictalurus punctatus*), the vermillion snapper (*Rhomboplites aurorubens*), and the gray triggerfish (*Balistes capriscus*)[27]. A large, fatty liver characterizes *B. capriscus*, which may function in buoyancy, since this species has a small swimbladder. The liver of the blue shark (*Isurus oxyrinchus*), contained more cholesterol (1160 mg per 100g liver) than some teleost fish; the flatfish (*Microstomus achne*) contained 646 mg cholesterol per 100 g liver, the angler (*Lophius setigerus*) contained 646 mg cholesterol per 100 g liver, and the eel (*Anguilla japonica*) contained 304 mg cholesterol per 100 g liver[116].

V. Integument

Fish skin is a living tissue with its lipids in metabolic equilibrium with those of other body tissues, particularly the underlying muscle. In lipid-rich fish, the fat deposits under the skin are usually located in extracellular sinuses, between the muscle fibers[135,160]. Alkyldiacylglyceryls are important components of the skin of chimaera and dogfish and stromateoid fishes (*Stromateus maculatus*, *Centrolophus* sp., and *Cubiceps gracilis*), where chimyl and selachyl alcohols comprise 62–82% of the alkyldiacylglyceryl ethers[220]. Wax esters in the skin of the orange roughy, *Hoplostethus atlanticus*, which comprise $29 \pm 6\%$ (as percent wet weight) of the skin, include 18:1 as the major fatty acid and 16:0, 18:1, 20:1, and 22:1 as major fatty alcohols[86]. *Benthosema glaciale*, a mesopelagic teleost, has subcutaneous wax ester depots rich in 20:1 and 22:1[68], whereas *Maurolicus muelleri* has triacylglycerol-rich subcutaneous lipid deposits. The castor-oil fish, *Ruvettus pretiosus*, which lacks a swimbladder, also has high concentrations of wax ester in the skin[23]. Since the skin, skeleton, and swimbladder of the orange roughy contain most of its wax ester, buoyancy was suggested as a major function[86,173].

The Antarctic midwater fish, *Electrona antarctica* and *Gymnoscopelus nicholsi*, contain extensive lipid stores in subcutaneous lipid sacs, which include the majority of the body lipids[195]. In *E. antarctica*, 57.2% of the subcutaneous lipid is wax ester, whereas 19.7% includes triacylglycerols. The subcutaneous lipid of *Pleurogramma antarcticum* and *Protomyctophum bolini* does not contain the majority of the body lipid. In *P. antarcticum* the subcutaneous lipid is 67.8% triacylglycerols and 23.2% wax ester. Triacylglycerols in subcutaneous and intermuscular deposits may be utilized first for energy, leaving wax ester to function primarily in buoyancy[195]. This scenario is similar to the relative utilization of these lipids in copepods[14].

Antarctic notothenioid fishes store triacylglycerol rather than wax ester, and one good reason is that wax ester could be partially solid, and have less buoyant force at lower temperatures[62,244]. *Pleurogramma antarcticum* stores lipid in large subcutaneous and intermuscular sacs instead of adipose cells[58,61]. *Dissostichus mawsoni*, in contrast, stores lipid in adipose cells in subcutaneous deposits[63]. The lipid stored in these sacs may not be available for metabolism since there is no cellular metabolic machinery available to utilize it[61]. However, data is needed to confirm this hypothesis. The pelagic *D. mawsoni* skin and subcutaneous tissues have 74% lipid (as percent wet weight)[39], whereas the cryopelagic *Pagothenia borchgrevinki* has only 3.5% skin lipid. *Aethotoxis mitopteryx*, which is probably also neutrally buoyant, has a subcutaneous adipose layer (1.1–1.5 mm thick in a 157 mm specimen) almost completely encircling the body[62]. In the swimbladderless *D. mawsoni*, the 2–8 mm subcutaneous lipid layer, which occupies 5–6% of body volume and 4.7% of body weight, provides static lift[57,59,60]. Scales, in the neutrally buoyant *D. mawsoni* are incompletely mineralized at the posterior margin[62]. Deciduous scales are also found on *P. antarcticum* and *A. mitopteryx*[58].

Reduced skin mass in piscivorous fish improves both buoyancy and acceleration performance. Of eight centrarchid species analyzed, the more piscivorous species had lighter skin[237]. For example, skin mass was 6.1% of the body mass in the green sunfish, *Lepomis cyanellus*, whereas the skin mass of four cyprinid prey species was 11.6–12.1% of total body mass[237].

The blubber of whales is thick and rich in wax ester[130]. It insulates warm blooded whales against cold polar waters and is also used for energy storage[205] as well as buoyancy. The blubber of the sperm whale (*Physeter catadon*) may contain 66% wax esters (as percent lipid)[130]. Some whales, such as the marine little teeth whale (*Stenella caeruleo-alba*) contain more triacylglycerol than wax ester[148]. The triacylglycerols of this whale are rich in *ISO-5:0* acids which are related to the abundant short chain acids in the inner melon. In the wax ester, almost all the fatty acids are short-chain; long-chain fatty acids are present mostly in triacylglycerol[148].

VI. Muscle

Reduction of musculature, more dense than sea water, is fairly common in deep sea fishes and helps them maintain neutral buoyancy. Red muscle is reduced in some midwater fish that lack a swimbladder. Red muscle with aerobic fibers is used for sustained swimming, whereas white muscle, which respires anaerobically, is used for short burst swimming which characterizes some sluggish midwater fishes.

Oily muscles are also considered to be an important buoyancy mechanism in marine animals[52]. Wax esters are more commonly present in muscle of deep and midwater fishes than epipelagic fishes where they are usually associated with roe. Lee and Patton[130], list 7 families of teleost fishes which have more than 10% wax ester specifically in muscle (the percent wax ester as percent lipid is also given): Trachichthydae (*Hoplostethus islandicus* and *H. gilchristi*, 90–97%), Paralepididae (*Paralepsis rissoi*, 85%), Latimeridae (*Latimeria chalumnae*, 93–97%), Moridae (*Laemonema morosum*, 60%), Myctophidae (*Lampanyctus ritteri* 12–90%; *Stenobrachius leucopsarus*, 74%; *Triphoturus mexicanus*), Gempylidae (*Lepidocybium flavobrunneum*, 89%; *Ruvettus pretiosus*, 92% (see also 159), and Zeiformes (*Allocytus verrucosus*, 76%). In addition, wax esters are found in high levels in muscle of *Hoplostethus atlanticus*[86,87,173,207].

Hydrocarbons of fish muscle are a minor lipid class, being less than 1% in all fish species examined[157]. Squalene, which can be synthesized by fish, and pristane, which is mostly of dietary origin or converted by the fish from exogenous phytol, appear to be the two major components of fish muscle hydrocarbons.

Alkyldiacylglycerols, like wax esters, are more commonly found in deep and midwater fish muscle. Sargent[202] lists four shark and four teleost species with appreciable contents of muscle alkyldiacylglycerols (percent alkyldiacylglycerol is given for each species as percent total lipid): *Cetorhinus maximus* (29–52%), *Squalus acanthias* (21–35%), *Centroscymnus* sp.(10%), *Deania* sp. (22%), *Dexea solandri* (30%), *Centrolophus* sp. (64%), *Cubiceps gracilis* (58%), and *Stromateus maculatus* (47%).

Triacylglycerols are abundant lipids in the muscle of numerous fish species. Triolein is a major constituent of the muscle of the deep sea black cod, *Erilepis zonifer*[235]. Monoenic fatty acids comprised 71.1% of the triacylglycerols fatty acids of *E. zonifer* conpared to only 2.3% polyenes[236]. The capelin, *Mallotus villosus*, has 20% of its body weight as oil, mostly triacylglycerol. This may be 30% (as percent dry weight) in the muscle of males, which decreases significantly during sexual maturation and spawning[108,109]. Lumpsucker (*Cyclopterus lumpus*) muscle has a high lipid content (probably triacylglycerol). The high fat plus low osmolarity of the muscle give it a density of 1.01 g ml^{-1}. The dorsal muscle located above the abdomen may be a specialized buoyancy organ, not important in locomotion[52].

Pleurogramma antarcticum has large intermuscular sacs (0.5–3.0 mm diameter) filled with triacylglycerol-rich lipid. These lipid sac walls consist of several white adipocytes located circumferentially around a large lipid droplet[64]. While definitely functioning in buoyancy, sac lipid surrounded by adipocytes is also available for metabolism. The 23% triacylglycerol-rich lipid (as percent dry weight) of white muscle of *Dissostichus mawsoni* aids attainment of neutral buoyancy in this fish[58,63]. The red and white muscle of the Antarctic cryopelagic *Pagothenia borchgrevinki* is rich in triacylglycerol, whereas the muscle of the benthic *Trematomus bernacchi* contains less lipid[39].

Few fresh water fish have more than 10% muscle lipid (as percent wet weight of the fillet). These include common carp (*Cyprinus carpio*), chub (*Leucicthys reighardi*), lake trout (*Salvelinus namaycush*), whitefish (*Coregonus clupeaformis*), eel (*Anguilla japonica*), and Brazilian mandi (*Pimelodus claria*)[106].

VII. Digestive tract

Most instances of buoyancy control in the digestive tract involve air gulped at the surface by the fish. The catfishes, *Hoplosternum thoracatum* and *Brochis splendens*, use the intestine as an accessory respiratory organ. Air is gulped at the surface, passed through the gut, and expelled from the anus. The intestine provides approximately 75% of the buoyancy in these fishes; in contrast the minute swimbladder provides less than 5%[79]. The decrease in buoyancy between air breaths was 13.2% for *B. splendens* and 7.8% for *H. thoracatum*. *Hoplosternum* swims more actively in midwater, whereas *B. splendens* is primarily benthic, thus requiring a less precise buoyancy control[79]. Electric eels (*Electrophorus*) use the buccal chamber as an accessory respiratory organ, the volume of which decreases by about 11% between breaths[70]. It has been suggested that the New England common sand shark, *Odontaspis taurus*, gulps air at the surface, which provides it neutral buoyancy and allows it to rest 2 m above the bottom[12,13].

Although lipid may comprise 82–85% of adipose tissue in association with the gut and gonad mesentery, there is no evidence for a specific buoyancy function, and it functions primarily as an energy reserve[106]. Visceral adipose tissue in rainbow trout, *Salmo gairdneri* (now: *Oncorhynchus mykiss*), is adapted for the uptake and storage of preformed fatty acids. Liver triacylglycerol synthesis rates were higher than in adipose tissue. Viscera have been reported to have 13–35% lipid (as wet weight), composed mostly of triacylglycerol, in various trout species[106]. Adipose tissue in the coelacanth, *Latimeria chalumnae*, contained 97% wax ester (as percent lipid)[160]. Few analyses of fish intestine have been performed. Goldfish intestine contained 3.6 mg lipid/g mucosa, comprised of 58% phospholipid (48% phosphatidylcholine) and 42% neutral lipid (16% cholesterol)[147]. Diacylglyceryl ethers were reported to be the major lipid (mostly chimyl and batyl alcohols) in the intestine of three species of stromateids[221]. There is a special oil sac in juvenile

Lumpaneus maculatus (a common north Norwegian fjord bottom fish) located ventral to the digestive tract extending from the pectoral fins to the anus[68,69]. The pure triacylglycerols in the oil sac have 20:1 (n-9) and 22:1 (n-11) as the major fatty acids. No discrete adipocytes were observed in the oil sac, which is composed of large polygonal units; but it is not known whether the triacylglycerol is extracellular or not. If this oil sac functions primarily in buoyancy, its ventral location might cause inherent stability problems. Therefore, a primary role of the oil sac as an energy reserve was assumed[68,69].

The wax cetyl palmitate comprised 30% of total lipids which were 12% of the dry weight of stomach and contents of the corallivorous butterflyfish, *Chaetodon trifascialis*[122]. The wax, derived from coral polyps[15], was not detected in the head, ovaries, swimbladder, or intestine of this fish. It may be converted to triacylglycerol which comprises 66–80% of the skeleton (by weight) in *Chaetodon ornatissimus*[174].

VIII. Gills

Gills are composed mostly of biomembranes consisting of such molecules as protein, phospholipids, and cholesterol. Thus, they do not contribute to the overall buoyancy of the fish with respect to density. However, gill chloride secretory cells pump out monovalent ions (Cl^-, Na^+, K^+) to osmoregulate and maintain hypotonicity, which is important in buoyancy (see Section XI., this chapter).

There are few published studies on gill lipids of marine fish. Thomas and Patton[225] found gill phospholipids to be the dominant lipid class, ranging from 60 to 76% of the total gill lipid, in nine species of marine fish. Phosphatidylcholine (PC) and phosphatidylethanolamine (PE) were the principal gill filament phospholipids (PC/PE ratio of 2.8) with lesser amounts of phosphatidylserine, phosphatidylinositol, and sphingomyelin. No major differences in component phospholipids were observed in pre- and postspawning salmon (*Oncorhynchus gorbusha* and *O. keta*), fishes from shallow water (kelp bass *Paralabrax clathratus*, sand bass *P. nebulifer*, whitefish *Caulolatilus princeps*, and sheepshead *Semicossyphus pulcher*), and fishes from great depths (Pacific rattail, *Coryphaenoides acrolepis*, flatnose codling, *Antimora rostrata*, and sablefish, *Anoplopoma fimbria*). Similar results were obtained for four species of Amazon river fishes, including an obligate air-breather (*Arapaima gigas*), two facultative air-breathers (*Hoplerythrinus unitaeniatus* and *Erythrinus erythrinus*), and a non-air breather (*Hoplias malabaricius*)[178]. Environmental temperature does influence phospholipid composition[5,34,103]. Gill tissue from 5°C-acclimated rainbow trout (*Salmo gairdneri*) had more PE than gills from 20°C-acclimated trout[102]. Acidity also affects the PC/PE ratio in fish gills[21]. Nevenzel et al.[161] analyzed the plasmalogens in six species of fishes and found the highest values in catfish (*Ictalurus punctatus*) phospholipid fraction (286 mmol/g).

IX. Blood

Hagfish blood is isotonic with seawater. Sharks, rays, and the coelacanth have blood serum which is ureosmotic and high in trimethylamine oxide and therefore slightly hypertonic. Marine teleost and lamprey blood is hypotonic to sea water. Midwater teleosts are characterized by a very low hematocrit ($< 10\%$) and very low serum proteins (0.8 g/100 ml) which may relate to buoyancy as well as low metabolism[84]. The necessity for buoyancy regulation could have kept midwater teleosts from developing osmoregulatory strategies to accumulate extra solutes[84]. Relatively low hematocrits and hemoglobin values were found in cyclostomata, and elasmobranchs[125]. The benthopelagic macrourid, *Coryphaenoides rupestris*, had low plasma protein[125], possibly related to buoyancy and the low level of activity.

Midwater fishes that lack swimbladders have low hematocrits (5–9%), small hearts, and blood of low viscosity, whereas those fish with swimbladders have hematocrits of 14–35% and larger hearts[19]. Higher hematocrits characterize some active surface species (48–57%)[19].

Fish blood is hyperlipidemic, hypercholesterolemic, and hyperlipoproteinemic[8]. Up to 2000 mg of high density lipoproteins per 100 ml of plasma are found in fish. There appears to be an absence or low level of an albumin-type plasma fraction in fish, which results in low protein levels associated with lipoproteinemia, contrary to mammals[8]. The concentration of plasma apolipoproteins in some fish is 30–36%, but less than 10% in humans. In fish, plasma cholesterol is 2–6 times higher than mammalian blood, where it could be associated with heart disease. In fish, high cholesterol is normal[149]. The only fish reported to have coronary degeneration are spawning Pacific salmon and steelhead trout. Most of this plasma cholesterol is in the esterified form; in 22 species of marine fish from Sweden, 65–75% of plasma cholesterol was esterified[126]. The blood serum of *Pagothenia borchgrevinki* and *Trematomus bernacchi*, both Antarctic fishes, is rich in lipid (16–19.5 mg ml^{-1}) with high amounts of sterol ester, free sterol, and phospholipid, in contrast to triacylglycerol as the principal lipid class in other tissues of these fishes[39]. In the migrating lamprey, *Mordacia mordax*, both plasma and muscle cholesterol levels are very high[73]. There are lower levels of plasma cholesterol in elasmobranchs[126]. Since some elasmobranchs have high squalene in liver, their requirement of squalene for buoyancy may reduce liver cholesterol synthesis activity[149]. The serum of shark livers (*Deania* sp., *Centroscymnus* sp., *Squalus acanthias*, and *Prionace glauca*) contains cholesteryl esters (rich in oleic acid) and wax esters (rich in polyunsaturated fatty acids)[208]. Interestingly, in *S. acanthias*, wax esters transport essential polyunsaturated fatty acids, whereas in the Pacific sardine (*Sardinops caerula*) essential fatty acids are transported by cholesteryl esters in the serum[131].

X. Eggs

An oil droplet is very important in the buoyancy of freshwater fish eggs, whereas a dilute aqueous fluid is more important in the buoyancy of marine fish eggs. For example, the buoyant pelagic eggs of 12 species of marine teleosts had 92% water content and 10–17% lipid (as percent dry weight)[47,49]. The water accounted for 90% of the buoyancy of these eggs. Grenadier and ling had 27–35% lipid, but water still provided most of the buoyancy. The buoyant pelagic eggs of marine fish have the highest water content of any known vertebrate eggs; in eggs of freshwater fishes and other vertebrates water is usually 50–70%[47].

Neutral buoyancy of eggs and larvae of the sablefish, *Anoplopoma fimbria*, changes markedly during development[3]. During early development the egg appears to rise from 200 to 400 m to a depth of about 200 m. The larvae then sink to about 1200 m depth during the first seven days after hatching[3]. The vertical distribution of eggs of sprat (*Sprattus sprattus*) and pilchard (*Sardina pilchardus*) also reflects changes in buoyancy[45]. Sprat eggs increased in density to reach near neutral buoyancy by hatching. The recent osmotic environment of the pre spawn ing adult females may influence egg buoyancy[214]. Temperature and ocean mixing also may influence egg buoyancy[45].

In low-salinity water, eggs absorb the less-concentrated solution of ovary fluid, and increase in size. This effective osmoregulation, affecting egg density is important in the brackish landlocked fjords and river mouths. The spawning sciaenid fish, *Bairdiella icistia*, adjusted to a lower salinity (15‰), also produced larger, more buoyant eggs than those fish in 33‰ sea water[144]. However, the actual lower threshold for successful development might be determined by egg buoyancy, and *B. icistia* eggs spawned in salinities lower than 30‰ would probably sink[144]. Variations in the thickness of the egg chorion were responsible for the buoyancy of flatfish eggs in the low salinity Baltic Sea[136]. The vertical distribution of eggs of Atlantic halibut (*Hippoglossus hippoglossus*) closely correlates with ocean salinity which affects egg buoyancy[90]. Most eggs were suspended in intermediate layers of deep water, and the egg neutral buoyancy salinity corresponded well with the ocean salinities (34.2‰).

Biochemical changes which occur during fish egg ripening cause a massive influx of water after vitellogenesis but before ovulation[49]. There is a large influx of potassium and sodium into the eggs, and a four- to five-fold increase in amino acids and water content in cod (*Gadus morhua*) and plaice (*Pleuronectes platessa*). Protein phosphate becomes exhausted during egg maturation (in *P. platessa*, *G. morhua*, haddock *Melanogrammus aeglefinus*, and whiting *Merlangius merlangus*) which correlates with water uptake into the egg[47]. Some phospholipid phosphate also declines during egg maturation of plaice. Energy necessary for ion uptake (sodium and potassium) into eggs may be provided by the disappearing protein phospate, which can give rise to ATP as a source of energy for ion pumps. These changes in phosphoprotein are found to a lesser extent in species with demersal eggs, even if much water uptake occurs during ripening[48].

Tocher and Sargent[226] suggested that lipids in fish eggs do not play a primary role in buoyancy. In a study of seven species of northwest European marine fish, the eggs of capelin (*Mallotus villosus*) and sand eel (*Ammodytes lancea*) with highest lipid contents, were not planktonic and were spawned on the shore or on the bottom. Egg phospholipids probably function primarily as a source of biomembranes in the developing organism[226]. PC was the only lipid to decrease during embryogenesis of the cod, *Gadus morhua*[76]. One third of the 22:6 (n-3) fatty acid released from PC was incorporated into triacylglycerol and sterol ester; therefore, PC is an important source of long-chain (n-3) polyunsaturated fatty acids for developing eggs and larvae. Polyunsaturated fatty acids are more important components of triacylglycerol-rich gonad lipid in herring (*Clupea harengus*) than reserve body lipid[105].

Wax esters comprise 50% of neutral lipid in red drum (*Sciaenops ocellata*) eggs along with triacylglycerol[234]. During development, total egg lipid decreased by 30% and glycogen by 53%. Since the estuarine red drum produces eggs at low salinities where the egg may be hypertonic to the water, the wax ester with its lesser density than triacylglycerol, may play an important role in buoyancy[234]. High concentrations of wax ester have been found in eggs of the teleosts *Merluccius capensis* (codfish family), *Coryphaena hippurus* (dolphin family), *Lutjanus campechanus* (snapper family), *Mugil cephalus* and *M. japonicus* (mullet family), *Rachycentron canadum* (cobias family), *Cynoscion nebulosus* (croaker family), and *Euthynnus alletteratus* (tuna family)[130]. In contrast, the body lipids of these species have low levels of wax ester[130]. In addition high wax ester have been found in eggs of gouramis (*Trichogaster cosby*)[158], one of the few freshwater fishes that produces buoyant eggs. Although the muscle of orange roughy (*Hoplostethus atlanticus*) is rich in wax ester, the roe lipid includes triacylglycerol, phosphatidylcholine, and cholesterol as 85% of the total lipid[20]. The squaloid shark, *Dalatias licha* contains alkyldiacylglycerols in the ovaries (35% of total lipid). Alkyldiacylglycerols are also major components of neutral lipid of yolk sacs of newborn *Squalus acanthias*[202].

The eggs of the New Zealand hoki (*Macruronus novaezelandiae*) are positively buoyant with an oil droplet 30% the size of the egg[167]. Hoki eggs develop very rapidly (80 h at 12°C) to minimize their vulnerability to predation[167,224]. The yolk sac is absorbed after 5 days, and the larva also develops rapidly; ossification of the skull has begun in larvae less than 6 mm long. Meso- and bathypelagic fishes generally have positively buoyant eggs[141]. However, meso- and bathypelagic species from four fish families (Zoarcidae, Liparidae, Alepocephalidae, and Searsidae) have negatively buoyant eggs[199], which are generally smaller in size than positively buoyant eggs and have a sinking rate of 1.8–3.1 km/day. Negative buoyancy in eggs of halibut (*Hippoglossus hippoglossus*) is caused by a high content of organic matter in the egg proper[198].

XI. Hypotonicity

Marine teleosts are characteristically hypotonic to seawater, which helps them maintain neutral or near-neutral buoyancy resulting in conservation of energy. The maintenance of dilute body fluids in fish and their eggs, is accomplished by expenditure of energy (see sections on gills and eggs, this chapter), involving ion pumps and high energy phosphate compounds. One advantage of water over storage of high concentrations of other low density metabolites, is that water is non-toxic[49]. Furthermore, low density metabolites are scarce and they are normally present at much lower levels.

Deeper living mesopelagic fishes substitute dilute fluids for organic matter to achieve neutral buoyancy, conserve energy, and adapt to low food availability[9,38,218]. This allows for rapid growth with minimum energy input. Protein reduction and muscle mass decrease is associated with increase of water in deeper midwater fishes[19], as well as a lower metabolic rate[228].

Neighbors[155] found seven species of midwater fishes that lack gas-filled swimbladders and have low lipid to have buoyancy procuing high water contents ($\geq 87\%$). Six species without swimbladders had low lipid and low water ($\leq 85\%$) contents, meaning they might have to swim continuously to maintain position in the water column[155]. Eight species had inflated swimbladders. If a gas-filled swimbladder is used to achieve neutral buoyancy in mesopelagic fishes, the water level is lower (73.5%) than in fishes without swimbladders[38]. High water content and low lipid characterized neutrally buoyant midwater fishes (Myctophidae) whereas low water content and low lipid were characteristic of negatively buoyant fishes[156].

Four species of mesopelagic non-vertical migrating fishes lacking a swimbladder (*Bathylagus pacificus*, *B. milleri*, *Tactostoma macropus* and *Chauliodus macouni*) have gelatinous deposits rich in acidic glycosaminoglycans located between swimming muscle cells, subcutaneously, and along the dorsal midline and surrounding the spine[243]. High water (96%) and low protein (3%) characterize the buoyant subcutaneous gelatinous tissue of *B. pacificus*. The glycosaminoglycans bind large amounts of water with their negative charges[171].

Ureosmotic regulation characterizes sharks, their relatives, and the coelacanth. The coelacanth lacks a heavily ossified vertebral column, and the fluid-filled notochord is therefore the principal axial skeleton. Notochordial fluid contains lower levels of sodium, magnesium, calcium, bicarbonate, sulfate, carbohydrate glucose, lactate, cholesterol, bound phosphate and total proteins than serum[86]. Its osmolarity (1058 mOsm) is higher than serum (942 mOsm) due presumably to higher levels of potassium, chloride, urea, trimethylamineoxide, and total free amino acids. Nevertheless, this fluid-filled notochord is certainly less dense than an ossified vertebral column.

Acanthonus armatus (family Ophidiidae) is a slightly negatively buoyant benthopelagic fish without a swimbladder that has reduced tissues and a massive head. This species concentrates most of its mass around the huge buoyant cranium filled with a low density (1.008 g/ml) fluid which provides static lift for this fish whose weight in water ranged from 1.0 to 2.6 g[115]. This deep water fish has the smallest brain of any known teleost and possibly any vertebrate[74]. The osmotic concentration of this fluid (294 mOsm/l) was 45% lower than the concentrations of the plasma or perivisceral fluid. Sodium and potassium concentrations of the cranial fluid were also lower than those of plasma or perivisceral fluid[115].

The muscle of the lumpsucker, *Cyclopterus lumpus*, has low density (as low as 1.01 g/ml) resulting from watery, low osmolality tissue and high lipid[52]. The large watery, dorsal muscles located above the abdominal cavity appear to be specialized buoyancy organs with little locomotory significance.

A low density dilute aqueous ammonium-rich solution is used for buoyancy in the enormous coelom of certain pelagic squids. High concentrations of ammonia (up to 0.5 molar) replace sodium, and potassium in this acidified fluid. Since ammonia is an end product of nitrogen metabolism, the squid must only retain a fraction of this normal excretory product to attain buoyancy[81]. Sulphate ions of seawater may be replaced by chloride ions in some oceanic octopods and other pelagic animals with very gelatinous bodies and little muscle[44]. Twelve of 26 families of living teuthoids use ammonium chloride solutions for buoyancy. This solution, isotonic with seawater, is of lower density[44]. The soft bodied buoyant squids have replaced the extinct forms which had chambered gas-filled shells. Therefore, use for buoyancy has resulted in squids, colonizing much greater depths in the ocean. Nerves will not conduct in the presence of such high ammonium concentrations[44].

XII. Conclusions

There are a number of diverse strategies utilized by marine fish to remain buoyant in sea water. The gas-filled swimbladder has probably received the most attention by scientists, but fat-invested and fat-filled swimbladders have also been studied in great detail, and are most important in deep midwater and benthopelagic fish. Fishes store lipid in numerous body tissues including liver, integument, swimbladder, skeleton, muscle, digestive tract, and gonads. There are many cases where the major function of the lipid is buoyancy, such as metabolically inert squalene in shark liver and extracellular wax ester and triacylglycerol. However, energy reserve is the primary function of most stored lipid in fish. Fishes are hypotonic to sea water, and gills help to maintain hypotonicity by osmoregulation. Deep sea fishes have watery muscles in addition to reduced musculature and skeleton. The buoyant pelagic eggs of marine fishes have the highest water content of any known vertebrate eggs (up to 92% water).

Despite much effort, the mechanism by which deep ocean fish fill their swimbladders with oxygen gas has never been explained. Although the possibility of O_2 pumping mechanism other than the Bohr effect, the Root effect, and salting out should not be overlooked[164], the function of the cholesterol-rich membrane system in swimbladders of deep sea fishes should be clarified. There are two hypotheses to explain the function of this lipid membrane system. The first hypothesis is that the lipid (present in gram quantities of pure, bilayer membrane) functions as a diffusion barrier to oxygen out of the swimbladder[239]. The second hypothesis suggests that the oxygen under pressure in the deep sea dissolves in the swimbladder membrane lipid[177]. The oxygen back-pressure on the swimbladder gas gland is thereby reduced allowing oxygen secretion to occur by the Bohr and Root effects. Buoyancy regulation could be accomplished by a slight change in the density of the cholesterol-phospholipid membrane upon addition or deletion of oxygen[172]. Clearly, very careful gas solubility measurements in the swimbladder lipid must be made at ambient deep sea temperatures and pressures to substantiate or refute the second hypothesis. Since the validity of the first hypothesis depends upon the swimbladder lipid-rich membranes being below their phase transition temperature, very precise phase transition temperature measurements must be made on the swimbladder membranes. One final question that must be answered relates to the swimbladder in the fish at depth in the ocean; namely, is the oxygen gas present as a gas, or is it dissolved in the lipid membranes or present as a clathrate[104]?

Phase transition temperatures should be measured at different temperatures in fish lipids. Swimbladder invested wax ester in the orange roughy, for example, was found to be 17% solid at 6°C, the temperature at which the fish lives[173]. The densities of fish lipids implicated in possible buoyancy regulation should be carefully measured at different temperatures and pressures, such as has been done for copepods[244]. The properties of a lipid may change markedly during diurnal vertical migration due to density changes, altering the buoyancy of the fish or invertebrate.

More definitive experiments need to be performed quantitatively relating lipid composition changes to changes in upthrust[202], such as the alkyldiacylglycerol/triacylglycerol ratio in the liver of *Squalus acanthias* and its response to artificial weighing[139]. Since lipids are such an effective source of metabolic energy in fish, it is indeed difficult to make a strong case for their function in buoyancy regulation. Squalene, and possibly extracellular lipid, appear to represent exceptions to this. However, extracellular lipid may be metabolically active under certain circumstances; orange roughy intraneurocranial lipid and swimbladder wax ester appear to be extracellular yet both have lipid synthesis rates one-sixth that of liver[87]. Certainly, we must expect the existence of a diversity of biochemical strategies of buoyancy regulation to reflect the extreme diversity of fishes in Nature.

XIII. References

1. Ackman, R.G., R.F. Addison and C.A. Eaton. Unusual occurrence of squalene in a fish, the Eulachon *Thaleichthys pacificus*. *Nature* 220: 1033–1034, 1968.
2. Ackman, R.G., S.N. Hooper and J.C. Sipos. Distribution of trans-6-hexadecanoic and other fatty acids in tissues and organs of the Atlantic leatherback turtle *Dermochelys coriacea coriacea* L. *Int. J. Biochem.* 3: 171–179, 1972.
3. Alderdice, D.F., J.O.T. Jensen and F.P.J. Velsen. Preliminary trials on incubation of sablefish eggs (*Anoplopoma fimbria*). *Aquaculture.* 69: 271–290, 1988.
4. Alexander, R.McN. The energetics of vertical migration by fishes. In: *The Effects of Pressure in Organisms,* edited by M.A. Sleigh and A.G. MacDonald. London: Academic Press, pp. 273–294, 1972.
5. Anderson, T.R. Temperature adaptation and the phospholipids of membranes in goldfish (*Carassius auratus*). *Comp. Biochem. Physiol.* 33: 663–687, 1970.
6. Aveldaño, M.I. A novel group of very long chain polyenoic fatty acids in dipolyunsaturated phosphatidylcholines from verberate retina. *J. Biol. Chem.* 262: 1180–1186, 1987.
7. Aveldaño, M.I. and H. Sprecher. Very long chain (C_{24} to C_{26}) polyenoic fatty acids of the n-3 and n-6 series in dipolyunsaturated phosphatidylcholines from bovine retina. *J.Biol.Chem.* 262: 1180–1186, 1987.
8. Babin, P.J. and J.-M. Vernier. Plasma lipoproteins in fish. *J.Lipid Res.* 30: 467–489, 1989.

9. Bailey, T.G. and B.H. Robison. Food availability as a selective factor on the chemical compositions of midwater fishes in the eastern north Pacific. *Mar.Biol.* 91: 131–141, 1986.
10. Bakker, R.T. Dinosaur renaissance. *Sci. Am.* 232: 58–79, 1975.
11. Baldridge, H.D., Jr. Accumulation and function of liver oil in Florida sharks. *Copeia* 2: 306–325, 1972.
12. Baldridge, H.D., Jr. Sinking factors and average densities of Florida sharks as functions of liver buoyancy. *Copeia* 4: 744–754, 1970.
13. Bass, A.J. and J.A. Ballard. Buoyancy control in the shark *Odontaspis taurus* (Rafinesque). *Copeia* 3: 594–595, 1972.
14. Benson, A.A. and R.F. Lee. The role of wax in oceanic food chains. *Sci. Am.* 232: 77–86, 1975.
15. Benson, A.A. and L. Muscatine. Wax in coral mucus: energy transfer from corals to reef fishes. *Limnol. Oceanogr.* 19: 810–814, 1974.
16. Berezovskii, V.A., V.Yu. Gorchakov and B.S. Sushko. Kinetics of oxygen transport through precipitation phospholipid films. *Fiziol. Zh. (Kiev)* 23: 641–644, 1977 (English Abstr.)
17. Bilinski, E. Biochemical aspects of fish swimming. In: *Biochemical and Biophysical Perspectives in Marine Biology, Vol. 1*, edited by D.C. Malins and J.R. Sargent. London: Academic Press, 1974, pp.239–288.
18. Blaxter, J.H.S. and P. Tytler. Physiology and function of the swimbladder. *Adv. Comp. Physiol. Biochem.* 7: 311–367, 1978.
19. Blaxter, J.H.S., C.S. Wardle and B.L. Roberts. Aspects of the circulatory physiology and muscle systems of deep-sea fish. *J. Mar. Biol. Assoc. U.K.* 51: 991–1006, 1971.
20. Body, D.R. The composition of orange roughy (*Hoplostethus atlanticus*) roe lipids. *J. Sci Food Agric.* 36: 679–684, 1985.
21. Bolis, C.L., A. Cambria and A. Fama. Effects of acid stress on fish gills. In: *Toxins, Drugs, and Pollutants in Marine Animals*, edited by C.L. Bolis and R. Gilles, Berlin: Springer-Verlag, 1984.
22. Bonaventura, C., B. Sullivan, J. Bonaventura and M. Brunori. Spot hemoglobin: studies on the Root effect hemoglobin of a marine teleost. *J. Biol. Chem.* 251: 1871–76, 1976.
23. Bone, Q. Buoyancy and hydrodynamic functions of integument in the castor oil fish, *Ruvettus pretiosus* (Pisces: Gempylidae). *Copeia* 1: 78–87, 1972.
24. Bone, Q. and N.B. Marshall. *Biology of Fishes*. London: Blackie, 1982, 253 pp. (Chapter 4).
25. Boström, S.-L., R. Fänge and R.G. Johansson. Enzyme activity patterns in gas gland tissue of the swimbladder of the cod *(Gadus morhua)*. *Comp. Biochem. Physiol.* 43B: 473–478, 1972.
26. Bowne, P.S. Swimbladder lipid deposits: morphology and phospholipid composition in *Coryphaenoides armatus* and occurrence in related fishes. M.S. Thesis, University of Miami, 1978, 75 pp.
27. Bradford, V.S., E.J. Parish and J.M. Grizzles. A comparative study of fish liver sterol content in *Ictalurus punctatus, Rhomboplites aurorubens*, and *Balistes capriscus. Comp. Biochem. Physiol.* 93B: 589–590, 1989.
28. Brewer, G.J. and J.W.Eaton. Erythrocyte metabolism: interaction with oxygen transport. *Science* 171: 1205–1211, 1971.
29. Brown, D.S. and D.E. Copeland. Layered membranes: A diffusion barrier to gases in teleostean swimbladders. *Tissue Cell.* 10: 785–796, 1978.
30. Brown, D.S. and D.E. Copeland. Overlapping platelets: A diffusion barrier in a teleost swimbladder. *Science* 197: 383–384, 1977.
31. Buisson, D.H., D.R. Body, G.J. Dougherty, L. Eyres and F. Vlieg. Oil from deep water fish species as a substitute for sperm whale and jojoba oils. *J. Am. Oil Chem. Soc.* 59: 390–395, 1982.
32. Butler, J.L. and W.G. Pearcy. Swimbladder morphology and specific gravity of myctophids off Oregon. *J. Fish. Res. Bd. Can* 29: 1145–1150, 1972.
33. Calabrese, V., F. Guerrera, M.Avitabile, M. Fama and V. Rizza. Superoxide dismutase and reduced glutathione: possible defenses operating in hyperoxic swimbladder of fish. In: *Toxins, Drugs, and Pollutants in Marine Animals*, edited by L. Bolis, J. Zadunaisky and R. Gilles, Berlin: Springer-Verlag, 1984, pp. 130–136.
34. Caldwell, R.S. and F.J. Vernberg. The influence of acclimation temperature on the lipid composition of fish gill mitochondria. *Comp. Biochem. Physiol.* 34: 179–191, 1970.
35. Carey, F.G. A brain heater in swordfish. *Science* 216: 1327–1329, 1982.
36. Carey, R.G. and B.H. Robison. Daily patterns in the activities of swordfish, *Xiphias gladius*, observed by acoustic telemetry. *Fish. Bull.* 79: 277–292, 1981.
37. Ceuterick, F., J. Peeters, K. Heremans, H. De-Smedt and H. Olbrechts. Effects of high pressure, detergents and phospholipase on the break in the Arrhenius plot of *Azobacter nitrogenase. Eur. J. Biochem.* 87: 401–407, 1978.

38. Childress, J.J. and M. Nygaard. The chemical composition of midwater fishes as a function of depth of occurrence off southern California. *Deep-Sea Res.* 20: 1093–1109, 1973.
39. Clarke, A., N. Doherty, A.L. DeVries and J.T. Eastman. Lipid content and composition of three species of Antarctic fish in relation to buoyancy. *Polar Biol.* 3: 77–83, 1984.
40. Clarke, M.R. Buoyancy control as a function of the spermaceti organ in the sperm whale. *J. Mar. Biol. Assoc. U.K.* 58: 27–71, 1978b.
41. Clarke, M.R. Physical properties of spermaceti oil in the sperm whale. *J. Mar. Biol. Assoc. U.K.* 58: 19–26, 1978a.
42. Clarke, M.R. Structure and proportions of the spermaceti organ in the sperm whale. *J. Mar. Biol. Assoc. U.K.* 58: 1–17, 1978c.
43. Clarke, M.R. The head of the sperm whale. *Sci. Am.* 240: 128–141, 1979.
44. Clarke, M.R., E.J. Denton and J.B.Gilpin-Brown. On the use of ammonium for buoyancy in squids. *J. Mar. Biol. Assoc. U.K.* 59: 259–276, 1979.
45. Coombs, S.H., C.A. Fosh and M.A. Keen. The buoyancy and vertical distribution of eggs of sprat (*Sprattus sprattus*) and pilchard (*Sardina pilchardus*). *J. Mar. Biol. Assoc. U.K.* 65: 461–474, 1985.
46. Cossins, A.R. and A.G. MacDonald. Homeoviscous adaptation under pressure. III. The fatty acid composition of liver mitochondrial phospholipids of deep-sea fish. *Biochim. Biophys. Acta* 860: 325–335, 1986.
47. Craik, J.C.A. and S.M. Harvey. Biochemical changes occurring during final maturation of eggs of some marine and freshwater teleosts. *J. Fish. Biol.* 24: 599–610, 1984.
48. Craik, J.C.A. and S.M. Harvey. Phosphorus metabolism and water uptake during final maturation of ovaries of teleosts with pelagic and demersal eggs. *Mar. Biol.* 90: 285–289, 1986.
49. Craik, J.C.A. and S.M. Harvey. The causes of buoyancy in eggs of marine teleosts. *J. Mar. Biol. Assoc. U.K.* 67: 169–182, 1987.
50. D'Aoust, B.G. Experiments in the physiology of the swimbladder. *Exp. Physiol. Biochem.* 6: 33–46, 1973.
51. D'Aoust, B.G. The role of lactic acid in gas secretion in the telost swimbladder. *Comp. Biochem. Physiol.* 32: 637–668, 1970.
52. Davenport, J. and E. Kjørsvik. Buoyancy in the lumpsucker *Cyclopterus lumpus*. *J. Mar. Biol. Assoc. U.K.* 66: 159–174, 1986.
53. DeBuffrénil, V., A. Collet and M. Pascal. Ontogenetic development of skeletal weight in a small delphinid, *Delphinus delphis* (Cetacea, Odontoceti). *Zoomorphology* 105: 336–344, 1985.
54. DeVries, A.L. and J.T. Eastman. Lipid sacs as a buoyancy adaptation in an Anatarctic fish. *Nature (Lond.)* 271: 352–353, 1978.
55. Doneen, B.A. and D.H. Gutman. Lipid compositon and *in vitro* biosynthetic rates of neutral lipids and phosphatidylcholine in anterior and posterior chambers of the goldfish swimbladder. *Comp. Biochem. Physiol.* 69A: 291–295, 1981.
56. Douglas, E.L., W.A. Friedl and G.V. Pickwell. Fishes in oxygen minimum zones: blood oxygenation characteristics. *Science* 191: 957–959, 1976.
57. Eastman, J.T. and A.L. DeVries. Buoyancy adaptations in a swim-bladderless Antarctic fish. *J. Morphol.* 167: 91–102, 1981.
58. Eastman, J.T. and A.L. DeVries. Buoyancy studies of Notothenioid fishes in McMurdo Sound, Antarctica. *Copeia* 2: 385–393, 1982.
59. Eastman, J.T. and A.L. DeVries. Hepatic ultrastructural specialization in Antarctic fishes. *Cell. Tissue Res.* 219: 489–496, 1981.
60. Eastman, J.T. Buoyancy adaptations in a swim-bladderless Antarctic fish. *Am. Zool.* 19: 911, 1979.
61. Eastman, J.T. Lipid storage systems and the biology of two neutrally buoyant Antarctic Notothenioid fishes. *Comp. Biochem. Physiol.* 90B: 529–537, 1988.
62. Eastman, J.T. The evolution of neutrally buoyant Notothenioid fishes: their specializatons and potential interactions in the Antarctic marine food web. In: *Antarctic Nutrient Cycles and Food Webs*, edited by W.R. Siegfried, P.R. Condy and R.M. Laws, Berlin: Springer-Verlag, 1985, pp. 430–436.
63. Eastman, J.T. and A.L. DeVries. Antarctic fishes. *Sci. Am.* 254: 106–114, 1986.
64. Eastman, J.T. and A.L. DeVries. Untrastructure of the lipid sac wall in the Antarctic Notothenioid fish *Pleurogramma antarcticum*. *Polar Biol.* 9: 333–335, 1989.
65. Ehrlich, K.F. and G. Muszynski. Effects of temperature on interactions of physiological and behavioral capacities of larval California grunion: adaptations to the planktonic environment. *J. Exp. Mar. Biol. Ecol.* 60: 223–244, 1982.
66. Ehrlich, K.F., J.H.S. Blaxter and R. Pemberton. Morphological and histological changes during the growth and starvation of herring and plaice larvae. *Mar. Biol.* 35: 105–118, 1976.

67. Fahlén, G. The gas bladder of *Argentina silus* L. with special reference to the ultrastructure of the gas gland cells and the countercurrent vascular bundles. *Z. Zellforsch.* 110: 350–372. 1970.
68. Falk-Petersen, I-B., S. Falk-Petersen and J.R. Sargent. Nature, origin and possible role of lipid deposits in *Maurolicus muelleri* (Gmelin) and *Benthosema glaciale* (Reinhart) from Ullsifjorden, northern Norway. *Polar. Biol.* 5: 235–240, 1986.
69. Falk-Petersen, S., I.-B. Falk-Petersen and J.R. Sargent. Structure and function of an unusual lipid storage organ in the Arctic fish *Lumpenus maculatus* Fries. *Sarsia* 71: 1–6, 1986.
70. Farber, J. and H. Rahn. Gas exchange between air and water and the ventilation pattern in the electric eel. *Resp. Physiol.* 9: 151–161, 1970.
71. Farkas, T. and R. Roy. Temperature mediated restructuring of phosphatidylethanolamines in livers of freshwater fishes. *Comp. Biochem. Physiol.* 93B: 217–222, 1989.
72. Fänge, R., Å. Larsson and U. Lidman. Fluids and jellies of the acusticolateralis system in relation to body fluids in *Coryphenoides rupestris* and other fishes. *Mar. Biol.* 17: 180–185, 1972.
73. Fellows, F.C.I. and R.M. McLean. A study of the plasma lipoproteins and the tissue lipids of the migrating lamprey, *Mordacia mordax*. *Lipids*, 17: 741–747, 1982.
74. Fine, M.L., M.H. Horn and B. Cox. *Acanthonus armatus*, a deep-sea teleost fish with a minute brain and large ears. *Proc. R. Soc. Lond.* B230: 257–265, 1987.
75. Finkelstein, A. Water and nonelectrolyte permeability of lipid bilayer membranes. *J. Gen. Physiol.* 68: 127–165, 1976.
76. Fraser, A.J., J.C. Gamble and J.R. Sargent. Changes in lipid content, lipid class composition, and fatty acid composition of developing eggs and unfed larvae of cod (*Gadus morhua*). *Mar. Biol.* 101: 307–313, 1989.
77. Friar, W., R.G. Ackman and N. Mrosovsky. Body temperatures of *Dermochelys coriacea*: warm turtles from cold water. *Science* 177: 791–793, 1972.
78. Fricke, H., O.Reineke, H.Hofer and W. Nachtigall. Locomotion of the coelacanth *Latimeria chalumnae* in its natural environment. *Nature* 329: 331–333, 1987.
79. Gee, J.H. and J.B. Graham. Respiratory and hydrostatic functions of the intestine of the catfishes *Hoplosternum thoracatum* and *Bronchis spendens* (Callichthydae) *J. Exp. Biol.* 74: 1–16, 1978.
80. Gesser, H. and R. Fänge. Lactate dehydrogenase and cytochrome oxidase in the swimbladder of fish. *Int. J. Biochem.* 2: 163–166, 1971.
81. Gilpin-Brown, J.B. Buoyancy mechanisms of cephalopods in relation to pressure. *Symp. Soc. Exp. Biol.* 26: 251–259, 1972.
82. Goad, L.J. Sterol biosynthesis. In: *Natural Substances Formed Naturally from Mevalonic Acid*, edited by T.W. Goodwin. London: Academic Press, 1970, pp. 45–77.
83. Goff, G.P. and G.B. Stenson. Brown adipose tissue in leatherback sea turtles: a thermogenic organ in an endothermic reptile. *Copeia* 4: 1071–1075, 1988.
84. Griffith, R.W. Composition of the blood serum of deep-sea fishes. *Biol. Bull.* 160: 250–264, 1981.
85. Griffith, R.W., M.B. Mathews, B.L. Umminger, B.F. Grant, P.K.T. Pang, K.S. Thomson and G.E. Pickford. Composition of fluid from the notochordal canal of the coelacanth, *Latimeria chalumnae*. *J. Exp. Zool.* 192: 165–172, 1975.
86. Grigor, M.R., C.R. Thomas, P.D. Jones and D.H. Buisson. Occurrence of wax esters in the tissues of the orange roughy (*Hoplostethus atlanticus*). *Lipids.* 18: 585–588, 1983.
87. Grigor, M.R., W.H. Sutherland and C.F.Phleger. Wax ester metabolism in the orange roughy (*Hoplostethus atlanticus*) (Bericiformes: Trachichthydae). *Mar. Biol.* 105: 223–227, 1990.
88. Hallman, M. and L.Gluck. Phophatidylglycerol in lung surfactant. III. Possible modifier of surfactant function. *J. Lipid Res.*17: 257–262, 1976.
89. Haszprunar, G. Anatomy and relationships of the bone-feeding limpets, *Cocculinella minutissima* (Smith) and *Osteopelta mirabilis* Marshall (Archeogastropoda). *Moll. Stud.* 54: 1–20, 1988.
90. Haug, T., E. Kjorsvik and P. Solemdal. Vertical distribution of Atlantic halibut (*Hippoglossus hippoglossus*) eggs. *Can. J. Fish. Aquat. Sci.* 41: 798–804, 1984.
91. Hayashi, K. and I. Kashiki. Level and composition of wax esters in the different tissues of deep-sea teleost fish *Laemonema longipes*. *Bull. Jpn. Soc. Sci. Fish.* 54: 135–140, 1988.
92. Hayashi, K. and T. Takagi. Characteristics of methoxyglyceryl ethers from some cartilaginous fish liver lipids. *Bull. Jpn. Soc. Sci. Fish.* 48: 1345–1351, 1982.
93. Hayashi, K. and T. Takagi. Distribution of squalene and diacyl glyceryl ethers in the different tissues of deep sea shark, *Dalatias licha*. *Bull. Jpn. Soc. Sci. Fish.* 47: 281–288, 1981.
94. Hayashi, K. Isolation of alkyl glyceryl ethers from liver oil unsaponifiables by recrystallization. *Bull. Jpn. Soc. Sci. Fish.* 52: 1475, 1986.
95. Hayashi, K. Liquid wax esters in liver oils of the deep sea teleost fish *Laemonema longipes*. *Bull. Jpn. Soc. Sci. Fish.*53: 2263–2267, 1987.

96. Hayashi, K. and K. Kawasaki. Unusual occurrence of diacyl glyceryl ethers in liver lipids from two species of Gonatid squids. *Bull. Jpn. Soc. Sci. Fish.* 51 (4): 593–597, 1985.
97. Hayashi, K. and S. Yamamoto. Content and composition of alkyl glyceryl ethers in liver of Gonatid squid *Berryteuthis magister* from the northwestern Pacific. *Bull. Jpn. Soc. Sci. Fish.* 53 (1): 137–140, 1987.
98. Hayashi, K. and S. Yamamoto. Distribution of diacylglyceryl ethers in the different tissues and stomach contents of Gonatid squid *Berryteuthis magister. Bull. Jpn. Soc. Sci. Fish.* 53 (6): 1057–1063, 1987.
99. Hayashi, K. and T. Takagi. Composition of diacylglyceryl ethers in the liver lipids of ratfish, *Hydrolagus novaezealandiae. Bull. Jpn. Soc. Sci. Fish.* 46 (7): 855–861, 1980a.
100. Hayashi, K. and T. Takagi. Occurrence of neutral plasmalogens in the liver lipids of rat fish, *Hydrolagus novaezealandiae. Bull. Jpn. Soc. Sci. Fish.* 46 (8): 1043–1049, 1980b.
101. Hayashi, K., Y. Okawa and K. Kawasaki. Liver lipids of Gonatid squid *Berryteuthis magister:* a rich source of alkyl glyceryl ethers. *Bull. Jpn. Soc. Sci. Fish.* 51 (9): 1523–1526, 1985.
102. Hazel, J.R. Determination of the phospholipid composition of trout gill by iatroscan TLC/FID: effect of thermal acclimation. *Lipids* 20: 516–520, 1985.
103. Hazel, J.R. Effects of temperature on the structure and metabolism of cell membranes in fish. *Am. J. Physiol.* 246: R460–R470, 1984.
104. Hemmingsen, E.A. Clathrate hydrate of oxygen: does it occur in deep sea fish? *Deep-Sea Res.* 22: 145–149, 1975.
105. Henderson, R.J. and S.M. Almatar. Seasonal changes in the lipid composition of herring (*Clupea harengus*) in relation to gonad maturation. *J. Mar. Biol. Assoc. U.K.* 69: 323–334, 1989.
106. Henderson, R.J. and D.R. Tocher. The lipid composition and biochemistry of freshwater fish. *Progr. Lipid Res.* 26: 281–347, 1987.
107. Henderson, R.J. and J.R. Sargent. Lipid biosynthesis in rainbow trout, *Salmo gairdnerii,* fed diets of differing lipid content. *Comp. Biochem. Physiol.* 69C: 31–37, 1981.
108. Henderson, R.J., J.R. Sargent and B.J.S. Pirie. Fatty acid catabolism in the capelin, *Mallotus villosus* (Müller), during sexual maturation. *Mar. Biol. Lett.* 5: 115–126, 1984b.
109. Henderson, R.J., J.R. Sargent and C.C.E. Hopkins. Changes in the content and fatty acid composition of lipid in an isolated population of the capelin *Mallotus villosus* during sexual maturation and spawning. *Mar. Biol.* 78: 255–263, 1984a.
110. Hiramatsu, N., T. Inoue, M. Suzuki and K. Sato. Pressure study on thermal transistions of oleic acid polymorphs by high-pressure differential thermal analysis. *Chem. Phys. Lipids.* 51: 47–53, 1989.
111. Hobson, E. and E.H. Chave. *Hawaiian Reef Animals.* Hawaii: University Press, 1972, p. 135.
112. Hochachka, P.W. and G.N. Somero. *Strategies of Biochemical Adaptation.* London: W.B. Saunders, 358 pp. (Chapter 9), 1973.
113. Horn, M.H. Swimbladder state and structure in relation to behavior and mode of life in stromateoid fishes. *Fish. Bull.* 73: 95–109, 1975.
114. Horn, M.H. The swimbladder as a juvenile organ in stromateoid fishes. *Breviora* 359: 1–8, 1970.
115. Horn, M.H., P.W. Grimes, C.F. Phleger and L.L. McClanahan. Buoyancy function of the enlarged fluid-filled cranium in the deep-sea ophidiid fish, *Acanthonus armatus. Mar. Biol.* 46: 335–339, 1978.
116. Iwasaki, M. and R. Harada. Cholesterol content of fish gonads and livers. *Bull. Jpn. Soc. Sci. Fish.* 50: 1623, 1984.
117. Josephson, R.V., R.B. Holtz, J.P. Misock and C.F. Phleger. Composition and partial protein characterization of swimbladder foam from deep sea fish *Coryphaenoides acrolepis and Antimora rostrata. Comp. Biochem. Physiol.* 52B: 91–95, 1975.
118. Kalish, J.M.C.F. Greenlaw, W.G. Pearcy and D. Van Holliday. The biological and acoustical structure of sound scattering layers off Oregon. *Deep-Sea Res.* 33: 631–653, 1986.
119. Kayama, M., S. Zafar, W.Rizvi and S. Asakawa. Biosynthesis of squalene and cholesterol in the fish. I. *In vitro* studies in acetate incorporation. *J. Fac. Fish Anim. Husb. Hiroshima Univ.* 10: 1, 1971.
120. Kayama, M., Y. Tsuchiya and J.C. Nevenzel. The glyceryl ethers of some shark liver oils. *Bull. Jpn. Soc. Sci. Fish.* 37: 111–118, 1971.
121. Ke, P.J. and R.G. Ackman. Bunsen coefficient for oxygen in marine oils at various temperatures determined by an exponential dilution method with a polarographic oxygen electrode. *J. Am. Oil Chem. Soc.* 50: 429–435, 1972.
122. Kung, S.S. and L.S. Ciereszko. Occurrence of the wax cetyl palmitate in stomachs of the corallivorous butterflyfish *Chaetodon trifascialis. Coral Reefs* 4: 45–46, 1985.

123. Kutchai, H. and J.B. Steen. The permeability of the swimbladder. *Comp. Biochem. Physiol.* 39A: 119–123, 1971.
124. Lapennas, G.N. and K. Schmidt-Nielsen. Swimbladder permeability to oxygen. *J. Exp. Biol.* 67: 175–196, 1977.
125. Larsson, Å, M.-L. Johansson-Sjöbeck and R. Fänge. Comparative study of some haematological and biochemical blood parameters in fishes from the Skagerrak. *J. Fish Biol.* 9: 425–440, 1976.
126. Larsson, Å. and R. Fänge. Cholesterol and free fatty acids (FFA) in the blood of marine fish. *Comp. Biochem. Physiol.* 57B: 191–196, 1977.
127. Laub, R.J. Probe — solute study of mesomorphic polysiloxane (MEPSIL) solvents. Discontinuities in family retention plots that traverse phase transitions. *Mol. Cryst. Liq. Cryst.* 157: 369–385, 1988.
128. Laub, R.J., J.H. Purnell, P.S. Williams, M.W.P. Harbison and D.E. Martire. Meaningful error analysis of thermodynamic measurements by gas-liquid chromatography. *J. Chromatogr.* 155: 233–240, 1978.
129. Laub, R.J. and R.L Pecsok. Physicochemical Applications of Gas Chromatography. New York: John Wiley, 1978.
130. Lee, R.F. and J.S. Patton. Alcohol and waxes. In: *Marine Biogenic Lipids, Fats and Oils, Vol. 1*, edited by R.G. Ackman, 1989, pp. 73–102.
131. Lee, R.F. and D.L. Puppione. Serum lipoproteins of the Pacific sardine (*Sardinops caerula* Girard). *Biochim. Biophys. Acta* 270: 272–278, 1972.
132. Lee, R.F., C.F. Phleger and M.H. Horn. Composition of oil in fish bones: possible function in neutral buoyancy. *Comp. Biochem. Physiol.* 50B: 13–16, 1975.
133. Lewis, R.W. The densities of three classes of marine lipids in relation to their possible role as hydrostatic agents. *Lipids* 5: 151–153, 1970.
134. Lockyer, C. Body weights of some species of large whales. *Cons. Int. Explor. Mer.* 36 (3): 259–273, 1976.
135. Love, R.M. *The Chemical Biology of Fishes.* London: Academic Press, 1970, p. 36.
136. Lönning, S. and P. Solemdal. The relation between thickness of chorion and specific gravity of eggs from Norwegian and Baltic flatfish populations. *Fiskdir. Skr. (Serie Havunders.)* 16: 77–88, 1972.
137. MacGibbon, A.H.K. and W.D. McLennan. Hardness of New Zealand patted butter: seasonal and regional variations. *N.Z.J. Dairy Sci. Tech.* 22: 143–156, 1987.
138. Malins, D.C. and U. Varanasi. The ether bond in marine lipids. In: *Ether Lipids Chemistry and Biology*, edited by F. Snyder. New York: Academic Press, 1972, 297 pp.
139. Malins, D.C. and A. Barone. Glyceryl ether metabolism: regulation of buoyancy in dogfish, *Squalus acanthias. Science* 167: 79–80, 1970.
140. Malins, D.C. and J.R. Sargent. Biosynthesis of alkyl diacylglycerols and triacylglycerols in a cell-free system fom the liver of dogfish (*Squalus acanthias*). *Biochemistry* 10: 1107–1110, 1970.
141. Marshall, N.B. *Deep Sea Biology: Developments and Perspectives.* London: Garland STPM Press, 1979, 566 pp. (Chapter 9).
142. Marshall, N.B. Swimbladder organization and depth ranges of deep-sea teleosts. *Symp. Soc. Exp. Biol.* 26: 261–272, 1972.
143. Marshall, N.B. and N.R. Merrett. The existence of a benthopelagic fauna in the deep-sea. In: *A Voyage of Discovery*, edited by M. Angel, Oxford: Pergamon Press, 1977, pp. 483–498.
144. May, R.C. Factors affecting buoyancy in the eggs of *Bairdiella icistia* (Pisces: Sciaenidae). *Mar. Biol.* 28: 55–59, 1974.
145. McCosker, J.E. The biology and physiology of the living coelacanth. *Occas. Papers Cal. Acad. Sci.* 134: 17–23, 1979.
146. McFall-Ngai, M.J. Adaptions for reflection of bioluminescent light in the gas bladder of *Leiognathus equulus* (Perciformes: Leiognathidae). *J. Exp. Zool.* 227: 23–33, 1983.
147. Miller, N.G., M.W. Hill and M.W. Smith. Positional and species analysis of membrane phospholipids extracted from goldfish adapted to different environmental temperatures. *Biochim. Biophys. Acta* 445: 644–654, 1976.
148. Morii, H. Fatty acid and fatty alcohols of triglycerides and wax esters in subcutaneous tissues of marine little teeth whales *Stenella caeruleo*-alba. *Bull. Jpn. Soc. Sci. Fish.* 48: 227–235, 1982.
149. Morris, R.J. and F. Culkin. Fish. In: *Marine Biogenic Lipids, Fats and Oils, Vol. II,* edited by R.G. Ackman, Florida: CRC Press, 1989, pp. 145–178.
150. Morris, S.M. and J.T.Albright. Cytochemical study of the lamellar bodies in the swimbladder of the toadfish, *Opsanus tau* L. *Cell. Tissue Res.* 185: 77–87, 1977.
151. Morris, S.M. and J.T.Albright. The ultrastructure of the swimbladder of the toadfish, *Opsanus tau* L. *Cell Tissue Res.* 164: 85–104, 1975.

152. Mosholder, R.S., R.V. Josephson and C.F. Phleger. Swimbladder membrane protein of an abyssal fish, *Coryphaenoides acrolepis. Physiol. Chem. Physics* 11: 37–47, 1979.
153. Moss, M.L. Skeletal tissues in sharks. *Am. Zool.*17: 335–342, 1977.
154. Muramatsu, T., N. Totani and H.K. Mangold. A facile method for the preperation of 1-*O*-alkyl-2-*O*-acetoyl-sn-glycero-3-phosphocholines (platelet activating factor). *Chem. Phys. Lipids.* 29: 121–127, 1981.
155. Neighbors, M.A. Triacylglycerols and wax esters in the lipids of deep midwater teleost fishes of the southern California Bight. *Mar. Biol.* 98: 15–22, 1988.
156. Neighbors, M.A. and B.G. Nafpaktitis. Lipid compositions, water contents, swimbladder morphologies and buoyancies of nineteen species of midwater fishes (18 myctophids and 1 neoscopelid). *Mar. Biol.* 66: 207–215, 1982.
157. Nevenzel, J.C. Biogenic hydrocarbons of marine organisms. In: *Marine Biogenic Lipids, Fats, and Oils, Vol. 1*, edited by R.G. Ackman, 1989, pp. 3–71.
158. Nevenzel, J.C. Occurrence, function and biosynthesis of wax esters in marine organisms. *Lipids* 5: 308–319, 1970.
159. Nevenzel, J.C., W. Rodegker and J.F. Mead. The lipids of *Ruvettus pretiosus* muscle and liver. *Biochemistry* 4 (8): 1589–1594, 1965.
160. Nevenzel, J.C., W. Rodegker, J.F.Mead and M.S. Gordon. Lipids of the living coelacanth, *Latimeria chalumnae. Science* 152: 1753–1755, 1966.
161. Nevenzel, J.C., A. Gibbs and A.A. Benson. Plasmalogens in the gill lipids of aquatic animals. *Comp. Biochem. Physiol.* 82B: 293–297, 1985.
162. Nevenzel, J.C., W.Rodegker, J.S. Robinson and M.Kayama. The lipids of some lantern fishes (family myctophidae). *Comp. Biochem. Physiol.* 31: 25–36, 1969.
163. Nielsen, J.G. and O. Munk. A hadal fish (*Bassogigas profundissimus*) with a functional swimbladder. *Nature* 204: 494–495, 1964.
164. Noble, R.W., R.R. Pennelly and A. Riggs. Studies of the functional properties of the hemoglobin from the benthic fish, *Antimora rostrata. Comp. Biochem. Physiol.* 52B: 75–81, 1975.
165. Nursall, J.R. Buoyancy is provided by lipids of larval redlip blennies, *Ophioblennius atlanticus* (Teleostei: Blennidae) *Copeia* 3: 614–621, 1989.
166. Nursall, J.R. and L.J. Turner. The liver supports metamorphosis in the Caribbean reef blennid *Ophioblennius atlanticus.* In: *Fifth International Coral Reef Congress, Vol. 5,* edited by M. Harmelin Vivien and B. Balvat. Tahiti: Antenne du Museum National d' Histoire Naturelle et de l'Ecole Pratique des Hautes Etudes en Polynesie Francaise, Antenne Museum — EPHE, Moorea, French Polynesia, 1985, pp. 457–462.
167. Patchell, G.J., M.S. Allen and D.J. Dreadon. Egg and larval development of the New Zealand hoki, *Macruronus novaezelandiae. N.Z.J. Mar. F.W. Res.* 21: 301–313, 1987.
168. Patton, J.S. The effect of pressure and temperature on phospholipid and triglyceride fatty acids of fish white muscle: a comparison of deepwater and surface marine species. *Comp. Biochem. Physiol.* 52B: 105–110, 1975.
169. Patton, S. Correlative relationship of cholesterol and sphingomyelin in cell membranes. *J. Theor. Biol.* 29: 489–491, 1970.
170. Patton, S. and A.J. Thomas. Composition of lipid foams from swimbladder of two deep ocean fish species. *J. Lipid Res.*12: 331–335, 1971.
171. Pfeiler, E. Towards an explanation of the developmental strategy in leptocephalus larvae of marine teleost fishes. *Environ. Biol. Fish.* 15: 3–13, 1986.
172. Phleger, C.F. and A.A. Benson. Cholesterol and hyperbaric oxygen in swimbladders of deep-sea fishes. *Nature* 230: 122, 1971.
173. Phleger, C.F. and M.R. Grigor, Role of wax esters in determining buoyancy in *Hoplostethus atlanticus* (Beryciformes: Trachichthydae). *Mar. Biol.* 105, 229–233, 1990.
174. Phleger, C.F. and P.W. Grimes. Bone lipids of marine fishes. *Physiol. Chem. Phys.* 8: 447–456, 1976.
175. Phleger, C.F. Bone lipids of Kona coast reef fish: skull buoyancy in the hawkfish, *Cirrhitus pinnulatus. Comp. Biochem. Physiol.* 52B: 101–104, 1975.
176. Phleger, C.F. Bone lipids of tropical reef fishes. *Comp. Biochem. Physiol.* 86B: 509–512, 1987.
177. Phleger, C.F. *Cholesterol and Hyperbaric Oxygen in Swimbladders of Deep Sea Fishes.* Ph. D. Thesis, University of California, San Diego, 1972, 113 pp.
178. Phleger, C.F. Gill phospholipids of Amazon fishes. *Can. J. Zoology.* 56: 793–794, 1978.
179. Phleger, C.F. Lipid synthesis by *Antimora rostrata*, an abyssal codling from the Kona coast. *Comp. Biochem. Physiol.* 52B: 97–99, 1975.

180. Phleger, C.F. Liver triglyceride synthesis failure in post-spawning salmon. *Lipids* 6: 347–349, 1971.
181. Phleger, C.F. Pressure effects on cholesterol and lipid synthesis by the swimbladder of an abyssal *Coryphaenoides* species. *Am. Zool.* 11: 559–570, 1971.
182. Phleger, C.F. The importance of skull lipid as an energy reserve during starvation in the ocean surgeon *Acanthurus bahianus*. *Comp. Biochem. Physiol.* 91A: 97–100, 1988.
183. Phleger, C.F., A.A. Benson and A.A. Yayanos. Pressure effect on squalene-2,3-oxide cyclization in fish. *Comp. Biochem. Physiol.* 45B: 241–247, 1973.
184. Phleger, C.F. and A. Soutar. Free-vehicles and deep-sea biology. *Am. Zool.* 11: 409–418, 1971.
185. Phleger, C.F. and B.S. Saunders. Swimbladder surfactants of Amazon air-breathing fishes. *Can. J. Zool.* 56: 946–952, 1978.
186. Phleger, C.F. and R.B. Holtz. The membranous lining of the swimbladder in deep sea fishes — I. Morphology and chemical composition. *Comp. Biochem. Physiol.* 45B: 867–873, 1973.
187. Phleger, C.F. and R.J. Laub. Skeletal fatty acids in fish from different depths off Jamaica. *Comp. Biochem. Physiol.* 94B: 329–334, 1989.
188. Phleger, C.F., J. Patton, P. Grimes and R.F. Lee. Fish-bone oil: percent total body lipid and carbon-14 uptake following feeding of 1-C^{14} palmitic acid. *Mar. Biol.* 35: 85–90, 1976.
189. Phleger, C.F., P.W. Grimes, A. Pesely and M.H. Horn. Swimbladder lipids of five species of deep benthopelagic Atlantic ocean fishes. *Bull. Mar. Sci.* 28: 198–202, 1978.
190. Phleger, C.F., R.B. Holtz and P.W. Grimes. Membrane biosynthesis in swimbladders of deep sea fishes *Coryphaenoides acrolepis* and *Antimora rostrata*. *Comp. Biochem. Physiol.* 56B: 25–30, 1977.
191. Phleger, C.F., R.J. Laub and A.A. Benson. Skeletal lipid depletion in spawning salmon. *Lipids* 24: 286–289, 1989.
192. Phleger, C.F., R.R. McConnaughey and P. Crill. Hyperbaric fish trap: operation and deployment in the deep sea. *Deep-Sea Res.* 26A: 1405–1409, 1979.
193. Phleger. C.F. Bone lipids of Jamaican reef fishes. *Comp. Biochem. Physiol.* 90B: 279–283, 1988.
194. Pritchard, P.C.H. *Encyclopedia of turtles*. New Jersey: T.F.H. Publications, 1979, pp. 722–727.
195. Reinhardt, S.B. and E.S. Van Vleet. Lipid composition of twenty-two species of Antarctic midwater zooplankton and fish. *Mar. Biol.* 91: 149–159, 1986.
196. Reynolds, W.W. and W.J. Karlotski. The allometric relationship of skeleton weight to body weight in teleost fishes: a preliminary comparison with birds and mammals. *Copeia.* 1: 160–163, 1977.
197. Reynolds, W.W. Skeleton weight allometry in aquatic and terrestrial vertebrates. *Hydrobiologia* 56: 35–37, 1977.
198. Riis-Vestergaard, J. Water and salt balance of halibut eggs and larvae (*Hippoglossus hippoglossus*). *Mar. Biol.* 70: 135–139, 1982.
199. Robison, B.H. and T.M. Lancraft. An upward transport mechanism from the benthos. *Naturwissenschaften* 71: 322–324, 1984.
200. Rose, L.G. and J.D.M. Gordon. Guanine and permeability in swimbladders of slope-dwelling fish. In: *Twelfth European Symposium on Marine Biology — Physiology and Behavior of Marine Organisms*, edited by P.S. McLusky and A.J. Berry. London: Pergamon Press, 1978, pp. 113–121.
201. Rosecchi, E., D.M. Tracey and W.R. Weber. Diet of orange roughy, *Hoplostethus atlanticus* (Pisces: Trachichthyidae) on the Challenger Plateau, New Zealand, *Mar. Biol.* 99: 293–306, 1988.
202. Sargent, J.R. Ether-linked glycerides in marine animals. In: *Marine Biogenic Lipids, Fats, and Oils, Vol. 1*, edited by R.G. Ackman, 1989, pp. 175–197.
203. Sargent, J.R. Marine wax esters. *Sci. Prog. Oxford* 65: 437–458, 1978.
204. Sargent, J.R. The structure, metabolism, and function of lipids in marine organisms. In: *Biochemical and Biophysical Perspectives in Marine Biology, Vol. III*, edited by D.C. Malins and J.R. Sargent, London: Academic Press, 1976, pp. 149–212.
205. Sargent, J.R., R.F. Lee and J.C. Nevenzel. Marine waxes. In: *Chemistry and Biochemistry of Natural Waxes*, edited by P.D. Kolattukudy, Amsterdam: Elsevier, 1976, pp. 49–91.
206. Sargent, J.R. and K.J. Whittle. Lipids and hyrocarbons in the marine food web. In: *Analysis of Marine Ecosystems*, edited by A.R. Longhurst, London: Academic Press, 1981, pp. 491–533.
207. Sargent, J.R., R.R. Gatten and N.R. Merrett. Lipids of *Hoplostethus atlanticus* and *H. mediterraneus* (Beryciformes: Trachichthydae) from deep water west of Britain. *Mar. Biol.* 74: 281–286, 1983.
208. Sargent, J.R., R.R. Gatten and R. McIntosh. The distribution of neutral lipids in shark tissues. *J. Mar. Biol. Assoc. U.K.* 53: 649–656, 1973.
209. Scholander, P.F. Secretion of gases against high pressures in the swimbladder of deep-sea fishes. II. The rete mirabile. *Biol. Bull.* 107: 260–277, 1954.
210. Shimma, H. and H. Shimma. Studies on liver oil of a frill shark. *Bull. Jpn. Soc. Sci. Fish.* 36: 1157–1162, 1970.

211. Smith, C.R. and H. Kukert. Vent fauna on whale remains. *Nature* 341: 27–28, 1989.
212. Smith, K.L., Jr. Metabolism of the abyssopelagic rattail *Coryphaenoides armatus* measured *in situ*. *Nature* 274: 362–364, 1978.
213. Smith, M.M. and D.C. Heemstra. *Smith's Sea Fishes*. Berlin: Springer-Verlag, 1986, pp. 152–153.
214. Solemdal, P. Transfer of Baltic flatfish to a marine environment and the long term effects on reproduction. *Oikos* Suppl. 15: 268–276, 1973.
215. Somero, G.N., J.F. Siebenaller and P.W. Hochachka. Biochemical and physiological adaptations of deep-sea animals. In: *The Sea, Vol. 8, Deep Sea Biology*, edited by G.R. Rowe, 1983, pp. 261–330.
216. Steen, J.B. The swimbladder as a hydrostatic organ. In: *Fish Physiology, Vol. 4*, edited by W.S. Hoar and D.J. Randall. New York: Academic Press, 1970, pp. 413–443.
217. Stein, D.L. and W.G. Pearcy. Aspects of reproduction, early life history, and biology of macrourid fishes off Oregon, U.S.A. *Deep-Sea Res.* 29: 1313–1329, 1982.
218. Stickney, D.G. and J.J. Torres. Proximate composition and energy content of mesopelagic fishes from the eastern Gulf of Mexico. *Mar. Biol.* 103: 13–24, 1989.
219. Szabo, G., G. Eisenman, R. Laprade, S.M. Ciani and S. Krasne. Experimentally observed effects of carriers on the electrical properties of bilayer membranes-equilibrium domain. In: *Membranes, a Series of Advances, Vol. 2. Lipid Bilayers and Antibiotics*, edited by G. Eisenman, New York: Marcel Dekker, 1973, Chapter 3, pp. 265–269.
220. Takada, K., H. Kamiya and Y. Hashimoto. Studies on lipids of some stromateidei fishes. *Bull. Jpn. Soc. Sci. Fish.* 45 (5): 605–610, 1979.
221. Omitted.
222. Takama, K., T. Furui and K. Nishimoto. Bone lipids of chum salmon migrated littorally for spawning. *Bull. Fac. Fish Hokkaido Univ.* 36: 157–162, 1985.
223. Takeuchi, T. and T. Watanabe. Effects of various polyunsaturated fatty acids on growth and fatty acid compositions of rainbow trout *Salmo gairdneri*, coho salmon, *Oncorhynchus kisutch*, and chum salmon, *Oncorhynchus keta*. *Bull. Jpn. Soc. Sci. Fish.* 48: 1745–1752, 1982.
224. Theilacker, G.H. and R. Lasker. Laboratory studies of predation by euphausid shrimps on fish larvae. In: *The Early Life History of Fishes*. edited by J.H.S. Blaxter, New York: Springer-Verlag, 1974, pp. 287–300.
225. Thomas, A.J. and S. Patton. Phospholipids of fish gills. *Lipids* 7: 76–78, 1972.
226. Tocher, D.R. and J.R. Sargent. Analysis of lipids and fatty acids in ripe roes of some northwest European marine fish. *Lipids* 19: 492–499, 1984.
227. Tont, S.A., W.G. Pearcy and J.S. Arnold. Bone structure of some marine vertebrates. *Mar. Biol.* 39: 191–196, 1977.
228. Torres, J.J., B.W. Belman and J.J. Childress. Oxygen consumption rates of midwater fishes as a function of depth of occurrence. *Deep-Sea Res.* 26A: 185–197, 1979.
229. Tytler, P. Buoyancy. In: *Environmental Physiology of Animals*, edited by J. Bligh, J.L. Cloudsley-Thompson and A.G. MacDonald. New York: John Wiley, 1976, Chapter 18, pp. 369–388.
230. Van Golde, L.M.G., J.J. Batenberg and B.Robertson. The pulmonary surfactant system: biochemical aspects and functional significance. *Physiol. Rev.* 68: 374–455, 1988.
231. Van Vleet, E.S., S. Candileri, J. McNeillie, S.B. Reinhardt, M.E. Conkright and A. Zwissler. Neutral lipid components of eleven species of Caribbean sharks. *Comp. Biochem. Physiol.* 79B: 549–554, 1984.
232. Vent, R.J. and G.V. Pickwell. Acoustic volume scattering measurements with related biological and chemical observations in the northeastern tropical Pacific. In: *Oceanic Sound Scattering Prediction*, edited by N.R. Anderson and B.J. Zahuranec. New York: Plenum Press, 1977, pp. 697–716.
233. Versaggi, C.S. Studies on the structure and function of bone in marine mammals. *Sci. Biol. J.* p.305–306, March-April, 1977.
234. Vetter, R.D. and R.E. Hodson. Energy metabolism in a rapidly developing marine fish egg, the red drum (*Sciaenops ocellata*). *Can. J.Fish Aquat. Sci.* 40: 627–634, 1983.
235. Wada, S., C. Koizumi, A. Takiguchi and J. Nonaka. Recognition of triolein in black cod lipid. *Bull. Jpn. Soc. Sci. Fish.* 44: 1167, 1978.
236. Wada, S., C. Koizumi, A. Takiguchi and J. Nonaka. Triglyceride composition of black cod lipid — I. Triglyceride composition based on the acyl carbon number and degree of unsaturation. *Bull. Jpn. Soc. Sci. Fish.* 45: 611–614, 1979.
237. Webb, P.W. and J.M. Skadsen. Reduced skin mass: an adaptation for acceleration in some teleost fishes. *Can. J. Zool.* 57: 1570–1575, 1979.
238. Wilson, R.R. and K.L. Smith, Jr. Live capture, maintenance and partial decompression of a deep-sea grenadier fish (*Coryphaenoides acrolepis*) in a hyperbaric trap-aquarium. *Deep-Sea Res.* 32: 1571–1582, 1985.

239. Wittenberg, J.B., D.E. Copeland, R.L. Haedrich and J.S. Child. The swimbladder of deep-sea fish: the swimbladder wall is a lipid-rich barrier to oxygen diffusion. *J. Mar. Biol. Assoc. U.K.* 60: 263–276, 1980.
240. Wong, P.T.T., D.J. Siminovitch and H.H. Mantsch. Structure and properties of model membranes: new knowledge from high-pressure vibrational spectroscopy. *Biochim. Biophys. Acta* 947: 139–171, 1988.
241. Yamada, M. and K. Hayashi. Fatty acid composition of lipids from 22 species of fish and mollusk. *Bull. Jpn. Soc. Sci. Fish.* 41: 1143–1152, 1975.
242. Yamamoto, S. and K. Bloch. Enzymatic studies on the oxidative cyclizations of squalene. In: *Natural Substances formed Biologically from Mevalonic Acid*, edited by T.W. Goodwin. London: Academic Press, 1970, pp. 35–43.
243. Yancey, P.H., R. Lawrence-Berrey and M D. Douglas. Adaptations in mesopelagic fishes. I. Buoyant glycosaminoglycan layers in species without diel vertical migrations. *Mar. Biol.* 103: 453–459, 1989.
244. Yayanos, A.A., A.A. Benson and J.C. Nevenzel. The pressure-volume-temperature (PVT) properties of a lipid mixture from a marine copepod, *Calanus plumchrus:* implications for buoyancy and sound scattering. *Deep-Sea Res.* 25: 257–268, 1978.
245. Yin, M.C. and J.H.S. Blaxter. Temperature, salinity tolerance and buoyancy during early development and starvation of Clyde and North Sea herring, cod, and flounder larvae. *J. Exp. Mar. Biol. Ecol.* 107: 279–290, 1987.

CHAPTER 10

Movement in water: constraints and adaptations

IAN A. JOHNSTON * AND JOHN D. ALTRINGHAM **

*Gatty Marine Laboratory, Department of Biology and Preclinical Medicine, University of St. Andrews, St. Andrews, Fife, Scotland KY16 8LB, U.K., and ** Department of Pure and Applied Biology, The University, Leeds LS2 9JT, W. Yorkshire, U.K.*

I. Introduction
II. The anatomy of the neuromuscular system
   1. The organisation of myotomal muscles
   2. Innervation patterns
III. The biomechanics of sub-carangiform swimming
IV. Muscle function under 'locomotory' conditions
V. Matching muscle function to locomotory requirements
   1. Optimising twitch duration
   2. Optimising stress and strain
      2.1. Myofibrillar packing
      2.2. Sarcomere structure
      2.3. Contractile protein isoforms
VI. Metabolism — matching supply with demand
   1. Aerobic metabolism
   2. Anaerobic metabolism
   3. Scaling effects
VII. The energetics of muscle contraction
VIII. References

I. Introduction

Our brief was to review the biochemical consequences of movement in water. Evolution leads towards an optimisation of all processes underlying movement, bound by the constraints of life in water. The biochemical consequences of this evolutionary process can only be understood within the context of the hydrodynamic and biomechanical constraints operating on a swimming fish. These constraints influence the structural design of fish, and the properties of their neuromuscular system. The hydrodynamic constraints are determined by the physical properties of water and the speed, size and shape of the fish moving through the water. The power required to swim increases as a cubed function of speed[13]. High speed swimming thus demands large numbers of muscle fibres which can deliver a high power output, with the metabolic support to match. The high energy demands can be met only transiently by anaerobic metabolism. Slow swimming can be powered by sustainable aerobic pathways. Wardle[71] has provided a complete description of swimming performance in adult Atlantic mackerel (*Scomber scom-*

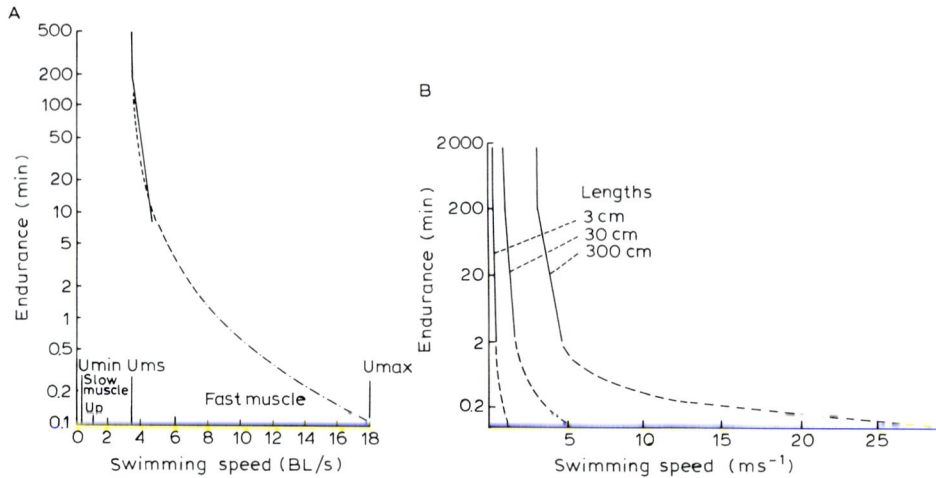

Fig. 1. A: swimming endurance of a 30 cm mackerel, *Scomber scombrus*. B: the effects of body size on endurance in mackerel. From Wardle[71].

brus L.) (Fig. 1A). Mackerel lack a swim bladder and must swim continuously at > 0.4 bodylengths/second (L/s) to generate lift and control their swimming depth. Speeds of up to 3.5 L/s can be maintained for long periods, and in the wild, tagged mackerel can cover 1,088 km in 13 days. Once 3.5 L/s is exceeded endurance declines rapidly, and the maximum speed of 18 L/s can only be sustained for 0.1 min (Fig. 1A). Scaling effects add a new dimension to the problem: the effects of size on swimming speed and endurance are summarized in Fig 1B. An understanding of the biochemical basis of the relationship between endurance and speed, and of scaling effects, demands a consideration of many interacting factors. Fig. 2 shows the inter-relationships of some of the elements which influence muscle biochemistry, and ultimately locomotion, and highlights a number of features either of particular importance to swimming fish, or of uncertain significance. A consideration of all of these elements is beyond the scope of the present review, and attention will be given primarily to aspects of muscle structure and function.

II. The anatomy of the neuromuscular system

1. The organisation of myotomal muscles

The axial muscles in fish are segmentally arranged into myotomes. Individual muscle fibres insert via short tendons into sheets of connective tissue (myosepta). The myosepta form a series of overlapping cones which are stacked along axes parallel to the midline, giving rise to the characteristic W-shape appearance of myotomes in longitudinal section. Each cone makes an acute angle to the body axis that is in the opposite direction to the acute angle made by the fibre to the body

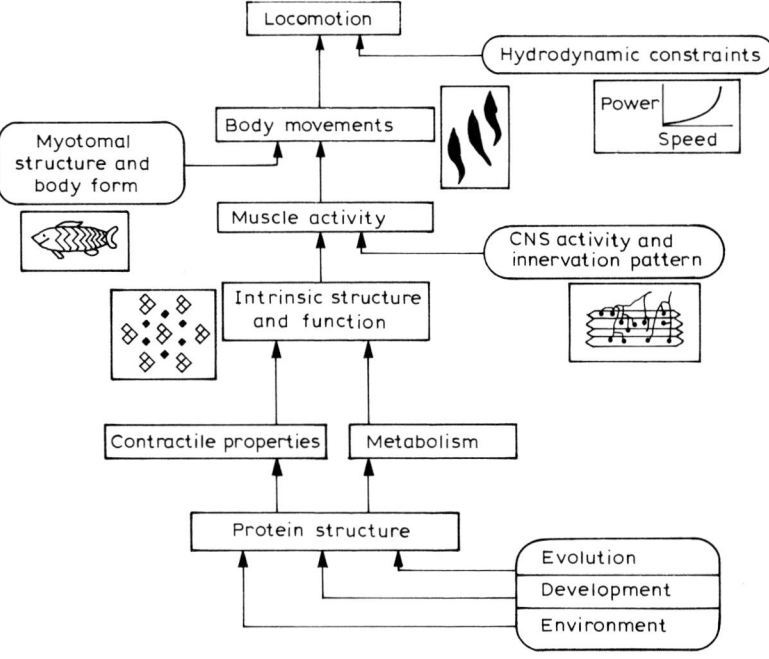

Fig. 2. Factors influencing muscle structure and function, and locomotion in fish. A number of elements are highlighted, which are either of particular importance to swimming fish (the arrangement of the body musculature into myotomes, undulatory body movements, and the high power demands of fast swimming), or of uncertain significance (the simple lattice structure of myosin and the presence of polyneuronal innervation in the fast fibres of many teleosts).

axis. The shape of successive myotomes changes along the length of the body, the angles between the fibres and myosepts becoming more acute towards the tail. Fig. 3A shows the complex arrangement of a typical myosept in the atlantic cod (*Gadus morhua* L.). The majority of superficial fibres run parallel to the longitudinal axis of the body, whilst deeper fibres make angles of up to $40°$[3]. In teleosts, the muscle fibre trajectories form a series of helices arranged in coaxial bundles (Fig. 3B), which may allow fibres at different distances from the median plane to contract by a similar amount as the body bends[3]. The analysis of Alexander[3] also predicts that deep muscle fibres would need to shorten at only 0.25 the speed of superficial fibres for a given change in curvature of the body. The geometric arrangement of myotomes and muscle fibres is thought to provide a significant mechanical advantage during flexion of the fish, helping to generate large forces with a small percentage shortening of muscle blocks[24]. The last vertebra has appendages which connect the caudal peduncle and tailfin. Whereas the effective transfer of forces to the tailfin is best achieved by rigid non-elastic structures (tendinous ligaments), the ability to adjust the position of the finray heads requires some flexibility of movement. Not suprisingly the anatomy of the skeletal elements and muscles in the caudal part of the fish varies enormously with swimming style, particularly with the power needed to be transferred to the tailfin[70].

A

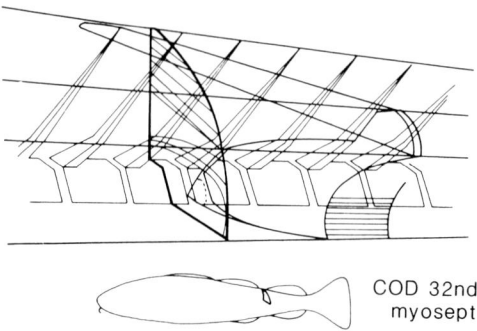

B

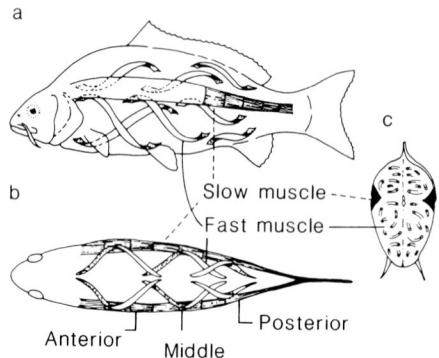

Fig. 3. A: oblique projection of the 32nd dorsal myosept in the atlantic cod (*Gadus morhua* L.). From Videler[70]. B: trajectories of muscle fibres in the carp. From Rome et al.[61]. Reprinted by permission from *Nature* (Vol. 335, p. 824); copyright © 1988, Macmillan Magazines Ltd.

Histochemical staining techniques reveal a number of different muscle fibre types in the myotomes, which are usually separated into anatomically discrete regions (Fig. 4). These can be divided into two main types, tonic and twitch. Tonic muscle fibres occur as a thin superficial layer in some bottom living sharks[16], and in certain teleosts[48], and are thought to have a postural role in bending the trunk. The bulk of the remaining muscle is composed of fast contracting (white) twitch fibres, which utilise anaerobic metabolism (see refs. 15, 37). Slowly contracting twitch fibres form a thin superficial layer of aerobic red muscle (Fig. 4), which increases as a proportion of total muscle cross sectional area from anterior to posterior myotomes. Typically, there is also a range of fibre types with intermedi-

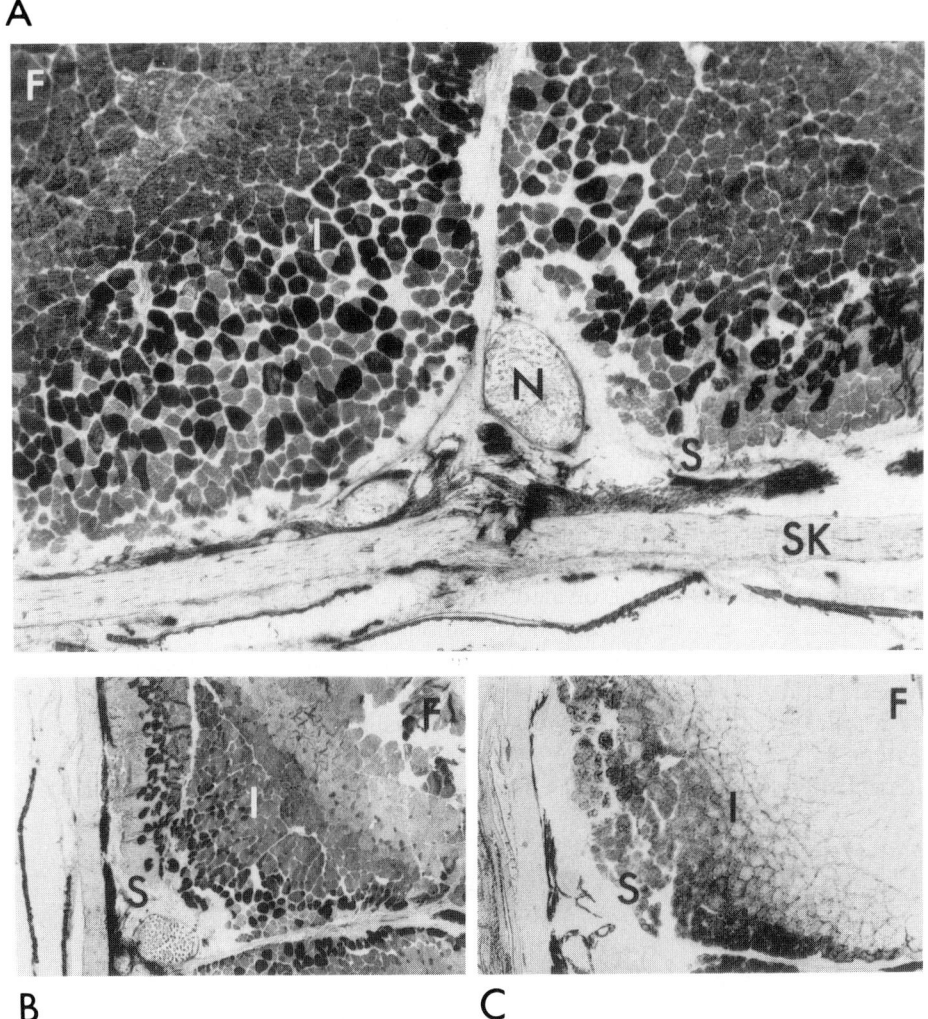

Fig. 4. A: transverse frozen section (10 μm) of myotomal muscle from the cod (*Gadus morhua*) stained for myosin ATPase activity at pH 9.4. N = lateral line nerve; S = slow muscle fibres; I = intermediate muscle fibres; F = fast muscle fibres. B: similar section stained for myosin ATPase following 5 min preincubation at pH 10.4 (see ref. 45 for methodological details). Note the wide range of fibre diameters and staining patterns in the intermediate muscle zone[1]. C: similar section stained for the mitochondrial enzyme, succinic dehydrogenase.

ate contractile and metabolic properties situated between the slow and fast muscle zones (Fig. 4).

2. Innervation patterns

Because of the complex myotomal arrangement, several overlapping myotomes need to shorten in order to contract a complete transverse section. Co-ordination is presumably accomplished through the CNS motor program, and the pattern of peripheral innervation. The latter is related to the electrophysiological and mechanical properties of the muscle fibres, and exhibits unique features in fish. Slow twitch fibres in all fish are multiply innervated[14], and may be activated by both junction potentials and action potentials[5]. The fast muscle fibres in elasmobranchs and primitive teleosts are focally innervated[14,18]. In elasmobranchs, certain sarcopterygians, and some actinopterygians, the endplates on these fibres are innervated by two motor axons each with a distinct vesicular morphology[57]. What is the functional significance of this dual innervation and are other transmitter substances than acetylcholine involved? As yet we do not know. Fast muscle fibres in the vast majority of bony fishes are polyneuronally innervated[14,18]. The normal response of white fibres to spinal nerve stimulation is an overshooting action potential and a fast mechanical twitch[5]. We know little about the peripheral organisation of the motor innervation. In the sculpin (*Myoxocephalus scorpius*), the majority of fast fibres in a given myotome are innervated by both adjacent spinal nerves, and in addition, up to 15% receive input from sub-adjacent myotomes[7]. Typically each fast fibre in the sculpin has 8–20 endplates derived from 4–6 axons, suggesting a modest amount of preterminal branching[7]. Fast muscle motorneurones in the Zebrafish can be divided into two classes, primary and secondary, on the basis of their cell body sizes and positions. Individual muscle fibres receive inputs from a single primary motorneurone, and as many as 3 secondary motorneurones[72]. The three primary neurones on each side of each segment all innervate a specific sub-set of muscle fibres in mutually exclusive regions of the body segment[72]. There is evidence that this complex pattern of innervation has evolved independently on several occasions during the adaptive radiation of the teleosts[18].

III. The biomechanics of sub-carangiform swimming

How does this complex neuromuscular system operate during swimming? In the interests of simplicity and brevity, we will confine our discussion to steady-state sub-carangiform swimming. This involves the sequential stimulation of muscle fibres on alternate sides of the body to produce a caudally running wave of lateral bending. A kinematic analysis of swimming in the saithe, *Pollachius virens*, has shown that the resulting bending moment runs faster than the wave of lateral bending[32]. There is also evidence that the mechanical wave proceeds faster than the electromyographical wave, resulting in systematic phase differences between force and length cycles along the length of the body[28,73]. This implies that muscle fibres are active at different strain rates along the body. Van Leeuwen, Lankheet, Akster and Osse[69] investigated recruitment patterns and strain fluctuations for the slow myotomal muscles of the common carp (*Cyprinus carpio*) using synchronized electromyography and cinematography (Fig. 5A). The lengths of the actin and myosin filaments and the bare zone on the myosin filaments were also measured by electron microscopy. This enabled muscle power output to be modelled on the basis of twitch kinetics, the length–tension and force–velocity relationships, and

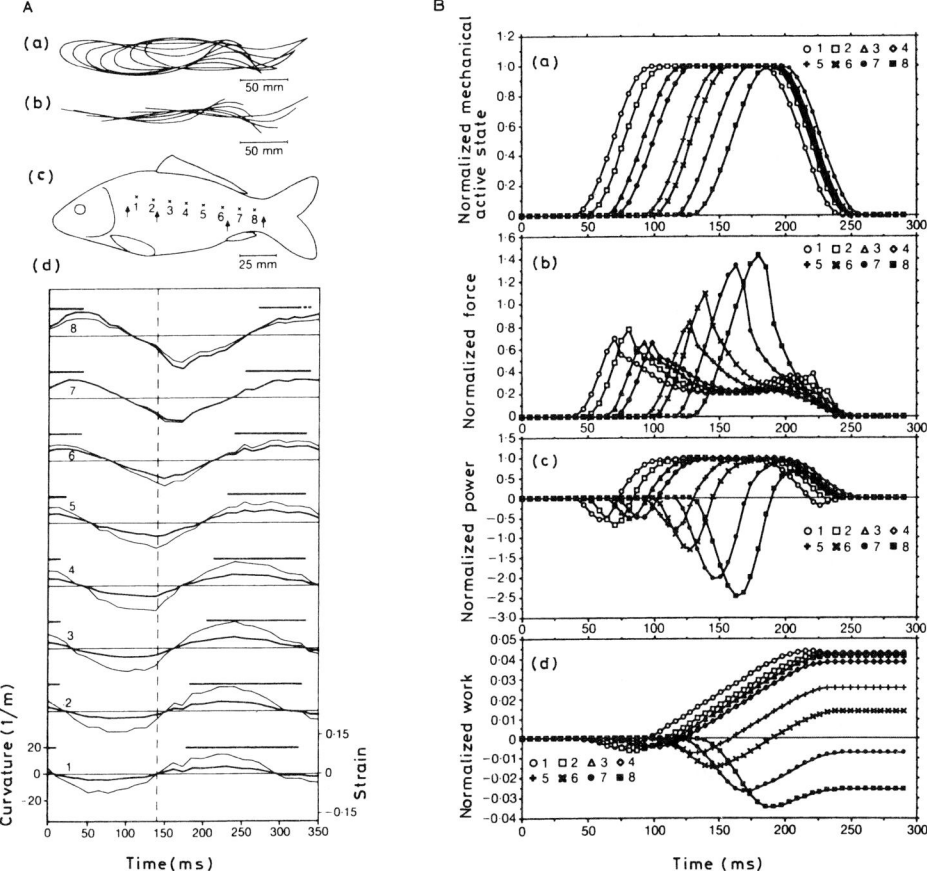

Fig. 5. A cinematographic and electromyographic analysis of near steady-state swimming in the carp, *Cyprinus carpio*. A: (a) body outlines during swimming, (b) calculated central axes, (c) positions 1–8 of the electrode pairs in the slow muscle on the left side of the body. (d) Curvature of body axis (heavy curves), calculated slow muscle strain (light curves), and periods of slow muscle electrical activity (horizontal bars), during a single swimming cycle, at the different positions along the body. Positive bending is to the right, negative to the left. The vertical dashed line represents $t = 0$ ms for the figure on the right. B: calculated, normalised values of mechanical active state (a), force (b), power (c) and (d) work during a single tail-beat cycle for slow muscle fibres at points 1–8 along the body. From van Leeuwen et al.[69]. Reprinted by the permission of the Zoological Society of London.

the stimulation pattern of fibres. The modelling results confirm that during swimming muscle fibres in successive segments operate in different strain ranges and at different contraction speeds. The slow fibres in anterior myotomes were mainly active during shortening, producing net positive work over the entire tail-beat cycle (Fig. 5B). Towards the anus muscle fibres were active during shortening and lengthening, such that the amounts of positive and negative work almost balanced each other (Fig. 5B). In the region of the caudal peduncle active fibres were mainly being stretched so that they produced net negative work (Fig. 5B). These results suggest that most work is done by the anterior myotomes resulting in the stretching of muscle and collagen fibres in and near the caudal peduncle.

The evolution of polyneuronal innervation in teleosts presumably has some functional significance to fibre recruitment during swimming. Its role remains obscure, but clues exist. The focal pattern of innervation is associated with a sharp transition between sustained and burst swimming. Pacific herring swim at 3–4 L/s almost indefinitely, but with the recruitment of focally innervated fast fibres at > 4.5 L/s, fatigue within 2 min[17]. In contrast, many teleosts recruit their polyneuronally innervated fast fibres at sustainable swimming speeds involving either steady or flick-glide swimming[35,42]. Capillary and mitochondrial volume densities are several-fold higher in fast muscles with polyneuronal, relative to focal, innervation[44].

IV. Muscle function under 'locomotory' conditions

One approach to understanding muscle function during swimming is to measure the forces generated by isolated muscle fibres under conditions which simulate their *in vivo* activity. Kinematic analysis of fish locomotion suggests that during steady swimming the length changes of the myotomal muscle fibres approximate a sine wave[32,69]. Altringham and Johnston[8] isolated slow and fast fibres from the anterior myotomes of the sculpin (*Myoxocephalus scorpius* L.) (0.65 bodylengths from the head) and subjected them to sinusoidal length changes, symmetrical about *in situ* rest length. Fibres were stimulated at selected phases in each cycle and work loops generated by plotting force against muscle length for each cycle (Fig. 6). The power output of the muscle fibres, which is the work per cycle multiplied by the cycle (tail-beat) frequency, is a complex function of the strain amplitude, cycle frequency and the timing of stimulation. The effects of these various parameters on power output can be determined by experiment[8,9]. Power output was found to be maximal at a strain of $\pm 5\%$ of resting fibre length, which corresponds to muscle length changes *in vivo*[3,32,69]. For anterior myotomes the force is maximal during shortening and power output is positive throughout most of the cycle[32,69]. The above parameters were adjusted to obtain the maximum positive work loops, to approximate these conditions. A small stretch in the active state prior to shortening, increased force above the isometric level, increasing the net work performed[8,9]. A small initial stretch prior to shortening has also been observed in kinematic studies of anterior myotomal fibres in the common carp[69].

Grillner and Kashin[28] found that in rainbow trout the duration of the EMG occupied an almost constant proportion of each cycle as swimming speed increased. Thus the number of nervous stimuli delivered per tail-beat cycle is likely to decrease with increasing swimming speed *in vivo*. Oscillatory work experiments on isolated fibres are consistent with this view. In the fast muscle fibres of the sculpin, a single stimulus was sufficient to produce maximum power output at high cycle frequencies (Fig. 7). As the frequency decreased more stimuli were needed, since the shortening phase, during which positive work could be performed,

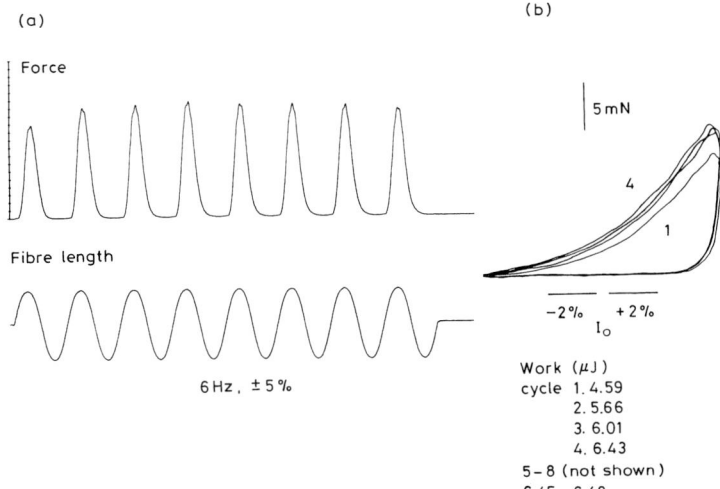

Fig. 6. The measurement of oscillatory work in muscle fibre bundles from *Myoxocephalus scorpius* L. A fast fibre preparation subjected to 8 sinusoidal length change cycles, of ±5% resting fibre length at 6 Hz. A single supramaximal stimulus was given 20° after the start of each cycle from rest length. (a) Force and fibre length plotted against time. (b) Force plotted against fibre length to produce a loop for each cycle. From Altringham and Johnston[8].

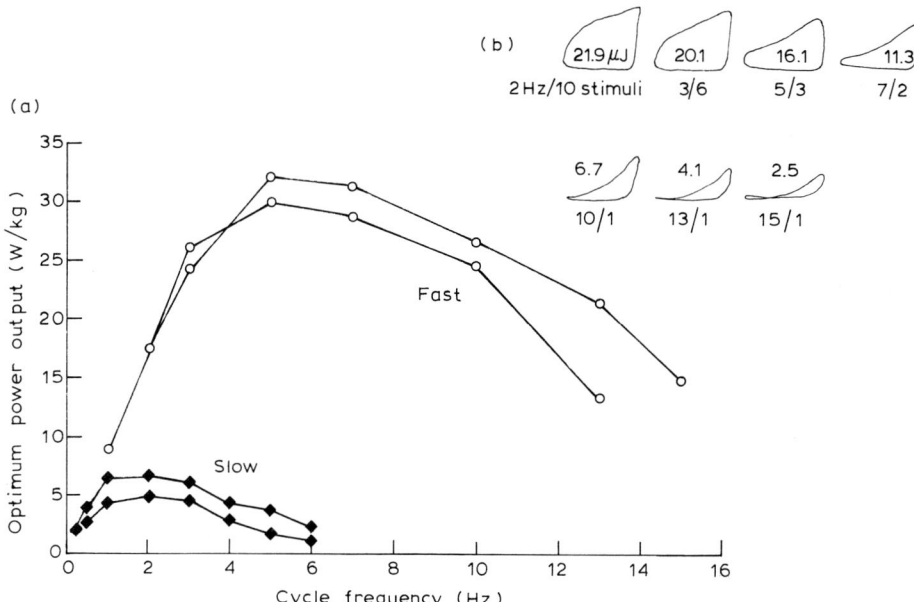

Fig. 7. (a) Optimum power output plotted against cycle ('tail-beat') frequency for representative fast and slow fibre preparations from *Myoxocephalus scorpius* L. (b) Optimum work loops for each point on the upper fast fibre curve of (a). Cycle frequency, number of stimuli and net work are given for each. From Altringham and Johnston[8].

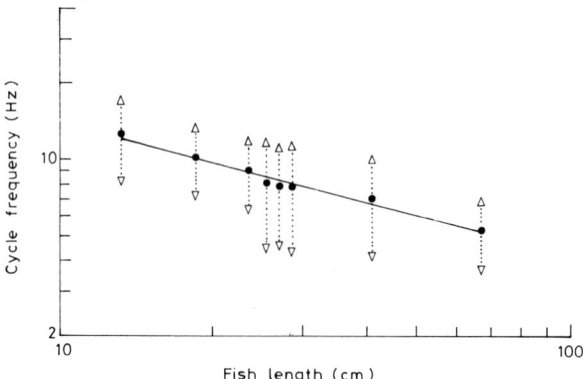

Fig. 8. Log·log plot of the cycle frequency range (vertical bars) over which muscle fibre preparations, from different sizes of cod (*Gadus morhua*), generate more than 90% of maximum power output. The midpoints of each bar (f_{opt}) can be fitted to the equation: $f_{opt} = 1.67 L^{-0.52}$, $r^2 = 0.97$, L = body length. From Altringham and Johnston[9].

became longer[8,9]. Electromyography of mackerel swimming at 1–14 L/s also indicates that the number of bursts of muscle activity decreases with increasing swimming speed (C.S. Wardle, personal communication).

The maximum power output of slow and fast fibres for 23–29 cm sculpin was measured at different cycle frequencies by optimising all parameters (Fig. 7). Slow fibres developed maximum power at around 2 Hz, a tail-beat frequency that can be maintained almost indefinitely. In contrast, fast fibres produce around five times more power at an optimal frequency of 5–7 Hz (at 5°C) (Fig. 7). It should be noted that the power output of slow fibres is negligible relative to that of fast fibres at high tail-beat frequencies, and that no positive work is generated above 8 Hz. This provides one explanation of why different muscle fibre types are needed for locomotion (see also ref. 62). As swimming speed increases there is a sequential recruitment of muscle fibre types which produce higher maximum stresses, shorter twitch durations and higher contraction speeds[17,42,61].

In cod, isometric twitch contraction time for fast fibres decreased as fish length decreased, scaling to $L^{0.29}$ (ref. 10). As a result of the change in twitch duration the number of stimuli needed at a given cycle frequency decreases with increasing fish length[9]. We also found that as fish length and twitch duration increased, the cycle frequency for maximum power output (f_{opt}) decreased from 12.5 Hz (13 cm fish) to 5 Hz (67 cm fish) (Fig. 8). However, twitch contraction time alone could not fully explain the scaling of f_{opt} ($\alpha L^{-0.52}$). This decline in cycle frequency for optimum power output with increasing size is to be expected, given the known decrease in tail-beat frequency.

V. Matching muscle function to locomotory requirements

Kinematic and electromyographical studies, and measurements of work during cyclical contractions, provide an insight into the parameters which need to be optimised to maximize power output at a given tail-beat frequency. These include properties intrinsic to the muscle, such as maximum tension (or stress), twitch duration, length–force and force–velocity curves, and external factors comprising muscle strain, strain rates and stimulation patterns. These studies also emphasise that there are likely to be significant differences in the biochemical properties of muscle fibres between anterior and posterior myotomes.

The optimisation of external parameters has been dealt with in the last section. In this section we will discuss the ways in which variations in intrinsic contractile performance can be accomplished. Modifications to contractile properties can be achieved by changes in the organisation and proportions of myofibrils and other cell components, by changes in sarcomere structure and changes in the composition of the muscle proteins themselves.

1. Optimising twitch duration

Twitch duration is dependent on a balance between the rate and amount of calcium released from the sarcoplasmic reticulum (SR) and its re-sequestration. The half-times for relaxation during isometric twitches are around 3–6 times shorter for teleost fast than slow fibres[2, 6]. These large differences in relaxation time are not accompanied by corresponding differences in the extent of the SR[2,58]; instead they are probably brought about by biochemical differences. The rate of Ca^{2+} uptake by frozen sections and purified SR is 2 times faster for fast than slow myotomal muscles in the rainbow trout[53]. There are also differences in the protein composition of the SR in these muscles[53]. In addition, fish fast muscle contains 0.5–1.5 mM parvalbumin compared with only around 50 μM in slow muscle[25]. Parvalbumins are cytoplasmic Ca^{2+}-binding proteins with molecular masses in the range 11–12 kDa. They are thought to have an important role as a temporary calcium sink at the onset of relaxation and as a shuttle protein between troponin and the sarcoplasmic reticulum as relaxation continues[31].

The rate of force development is also dependent on myofibril diameter (i.e., diffusion distances), the pCa–force characteristics of the muscle fibres, and their cross bridge cycle times. Demembranated muscle fibres have been used to investigate the pCa–force relationship of fast and slow muscles[4]. In the cod (*Gadus morhua* L.) the threshold for force generation is around 0.06 μM calcium and maximum force is produced at 7 μM calcium. The calcium concentration required for half maximal force in cod is significantly higher for fast (0.83 μM) than slow (0.38 μM) fibres[4]. These studies suggest that there is a higher degree of co-operativity between the binding sites on troponin C in the fast muscle[4]. Fast and slow fibres contain distinct isoforms of troponin C, I and T (unpublished results). Two troponin T isoforms with similar isoelectric points, but slightly different molecular weights, have been reported in the fast axial muscles of the carp[19]. Cross-bridge cycle times will depend on the isozymes present of the various muscle proteins, as discussed below.

2. Optimising stress and strain

2.1. Myofibrillar packing

The maximum stress produced is proportional to the number of myosin filaments per unit cross-sectional area (assuming constant filament and bare zone length). Tonic fibres are densely packed (76%) with myofibrils which are large diameter (2–4 μm) and irregular in cross-section[16,48]. Twitch fibres contain smaller diameter myofibrils (0.5–1.0 μm) which occupy from 40 to 65% of fibre volume in slow fibres and from 70 to 90% in fast fibres[16,40].

2.2. Sarcomere structure

The length of actin filaments, the *in vivo* sarcomere length, and hence the degree of overlap possible between actin and myosin filaments, all show considerable variation between muscle types[1]. During locomotion muscle strain occurs over a limited range of sarcomere lengths, over which they produce near maximum force[62].

The structure of the A-band in teleost muscle is different from that of other vertebrates, including other fish[30,65]. Electron microscopy of myosin filaments in the bare region of the A-band in teleosts reveals a simple lattice structure[59]. Low-angle X-ray diffraction studies have shown that the myosin filament array in bony fish has a crystalline structure with good three-dimensional regularity that is coherent across each myofibril[30,65]. In contrast, the myosin filaments in other vertebrates can have one of two different orientations 180° apart introducing statistical disorder into the lattice (superlattice) (Fig. 9). Understanding the functional significance of the simple lattice in teleosts, like polyneuronal innervation, remains a major challenge.

2.3. Contractile protein isoforms

The majority, if not all, of the myofibrillar proteins can exist as isoforms, each with slightly different functional properties, which are expressed in a developmental stage- and tissue-specific manner. They occur either as multiple gene families, such as myosin heavy chains, or are produced via an alternate RNA-splicing mechanism from a common primary transcript, as in the case of troponin T. The types of myosin isoforms expressed are major determinants of both contraction speed and maximum tension generation. Myosin has a common sub-unit structure in all vertebrate muscles, consisting of two heavy chains (HC) and four light chains (LC). Cloning studies indicate that common carp possess at least 28 different myosin heavy chain genes[26], about twice the number reported in mammals[51]. By means of electrophoresis in non-denaturing conditions three isoforms of myosin can be identified in fast myotomal muscle and two in slow muscle in the roach (*Rutilus rutilus* L.)[47]. The myosin isolated from fast muscle contains 1 mol of alkali light chains and 1 mol of P-light chain (LC2). There are two isoforms of alkali light chain (LC1$_f$ and LC3$_f$) which differ in sequence at their NH$_2$ terminus, and in other vertebrates these are encoded by a single gene with two transcriptional sites[55]. The molecular mass of the light chains varies between species but is usually

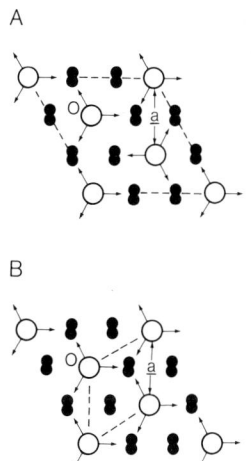

Fig. 9. A: the filament lattice arrangement found in amphibians, reptiles, birds, cartilagionous fish and others. The myosin filament superlattice is shown by the dashed outline. The open circles represent the three-stranded vertebrate myosin filaments with the origins of the myosin heads indicated by arrows (143 Å spacing). The double filled circles represent the actin filaments at the trigonal points of the myosin filament lattice (a = 410 Å). B: the filament lattice arrangement in most bony fish in which all myosin filaments have the same orientation. The dashed lines show the true simple unit cell structure (a = 470 Å). Adapted from Squire[65].

within the range 16–27 kDa[22,47,63]. For example, the fast muscle of rainbow trout (*Salmo gairdneri*) contains three light chains of molecular mass 26.3 kDa ($LC1_f$), 20.5 kDa ($LC2_f$) and 17.6 kDa $LC3_f^{63}$. $LC2_f$ is phosphorylated and dephosphorylated *in vitro* by myosin light chain kinase and phosphatase respectively. Fast muscle from anaesthetised trout contains 0.16 mol P/mol $LC2_f$. In five trout subject to strenuous exercise, two fish showed a slight increase in myosin phosphorylation (0.32 mol P/mol $LC2_f$), and the other three no significant change[75]. The alkali light chains are thought to modulate the myosin heavy chain ATPase activity. In roach the three fast muscle isomyosins most probably correspond to the $LC3_f$ homodimer, $LC1_f$-$LC3_f$ heterodimer and $LC1_f$ homodimer.

In some fish three main layers of muscle fibres can be distinguished histochemically on the basis of the pH-lability of the myosin ATPase reaction: slow (red), intermediate (pink) and fast (white)[45]. In the common carp (*Cyprinus carpio* L.) peptide mapping studies have shown that each of these 3 fibre types contains a distinct isoform of myosin heavy chain (HC_s, HC_i, HC_f)[64]. Slow muscle fibres contain two slow light chain isoforms ($LC1_s$ and $LC2_s$)[22,63], whereas the intermediate fibres contain $LC1_f$, $LC2_f$ and $LC3_f^{42}$. Single muscle fibres in other vertebrates have been shown to express variable proportions of myosin heavy chain and light chain isoforms[60], giving rise to a large number of theoretical isomyosins. The ratio $LC3_f : LC1_f$ in fast muscle is usually > 1, and shows considerable interspecific variation[36].

In the common carp, cold acclimation (8°C) results in modest increases in the maximum contraction velocity (V_{max}) of slow muscle fibres from the pectoral fin adductor muscle. Whereas slow muscles from warm-acclimated (20°C) carp contain almost exclusively slow myosin light chains ($LC1_s$ and $LC2_s$), similar fibres from 8°C-acclimated fish also contain a significant proportion of fast myosin light chains ($LC1_f$ and $LC2_f$)[49]. Cold acclimation results in much larger increases in V_{max} ($\geqslant 2$) for fast myotomal muscle fibres at low temperature[46]. Crockford and Johnston[19] found evidence for the expression of a super-fast isoform of myosin light chain ($LC3_{sf}$) in the fast muscle of cold-acclimated carp, together with a significant decrease in the $LC3_f : LC1_f$ ratio. The myosin heavy chain composition is generally thought to be the major determinant of V_{max} in other vertebrate muscles[50] and temperature acclimation in carp may also involve changes in myosin HC expression[26]. The existence of such a large number of isomyosins provides an explanation for the plasticity of contractile properties between different muscle fibres, with increasing body size, and in response to environmental stress.

VI. Metabolism — matching supply with demand

So far we have discussed only the contractile aspects of muscle function. The metabolic pathways supplying ATP to the cross bridges must also be appropriate to the demands made by each muscle type.

1. Aerobic metabolism

The main fuels for aerobic metabolism are fatty acids from triacylglycerol in the adipose tissue and glucose from liver and muscle glycogen stores. These substrates are present in large amounts and have a high ATP yield/mol substrate[33]. This makes them ideal for continuous work. In general, the rate at which aerobic pathways can supply ATP is limited by the abilities of the respiratory and cardiovascular systems to supply oxygen and substrates. Aerobic metabolism is able to support very high frequency contractions, but only at the expense of a low mechanical power output. A good example of this is the sound producing swim bladder muscle of the toadfish, *Opsanus tau* which operates at 300 Hz, and can complete a contraction in a little over 10 ms[66] The complex sarcoplasmic reticulum present in the fibres comprises 30% of muscle volume such that the ratio of SR : mitochondria : myofibrils approaches $1:1:1$[23].

The volume density of mitochondria is usually 10–30 times higher in slow than fast myotomal fibres (see Table II in ref. 37). Expansion of aerobic capacity is achieved both by increasing the proportions of aerobic fibre types in the myotomes, and by increasing the mitochondrial content of individual fibres. The posterior myotomes of continuously active swimmers such as anchovies contain 29–38% of slow twitch fibres per cross-sectional area[38]. The individual fibres are flattened or plate-like in transverse section to increase their surface : volume ratio[27]. Mitochondria with dense cristae comprise almost half (45.3%) of the fibre volume and the density of myofibrils is correspondingly reduced[38]. The net result is that more than 95% of myofibrils are adjacent to mitochondria. However, the volume density of mitochondria is not always a good descriptor of aerobic capacity, due to variations in cristae density and aerobic enzyme activities[11]. For example, slow muscles in certain sluggish antarctic fish living at $-1°C$ contain almost 60% mitochondria by volume[41]. In this case the amplification of mitochondria is thought to represent an adaptation to overcome catalytic and diffusional limitations at low temperature[39]. A similar proliferation of mitochondria occurs with seasonal temperature acclimation in teleosts[43]. For example, in striped bass (*Morone saxatilis*) the fraction of cell volume occupied by mitochondria increases from 0.29 to 0.45 between 25°C and 5°C[21]. The mean intermitochondrial spacing decreased from 2.6 to 1.4 μm over this temperature range.

2. Anaerobic metabolism

The immediate energy supply for contraction comes from the hydrolysis of phosphocreatine which is catalysed by a single near-equilibrium enzyme, creatine phosphokinase[33]. Following some delay this is supplemented by the activation of anaerobic glycogenolysis resulting in the accumulation of lactic acid. In fatigued muscle ATP is also provided by the adenylate kinase pathway (2ADP to ATP + AMP). Since this reaction lies close to equilibrium the end products must be removed for it to proceed. This is achieved by the activation of AMP deaminase which converts AMP to IMP + NH_3. The relative importance of these energy supplying pathways is clearly related to swimming behaviour. 'Sit-and-wait' predators generally have low endurance at high speed since predatory manoeuvres occur over short distances. Typically, the fast muscle in species using this feeding strategy, such as the antarctic fish *Notothenia neglecta*, have high activities of creatine phosphokinase, adenylate kinase and AMP deaminase and low activities of glycolytic enzymes and modest glycogen stores. *N. neglecta* swum to exhaustion (1–3 min) show a respiratory but not a metabolic acidosis with no significant accumulation of lactate in the blood or muscles[20]. In contrast, species with greater endurance at maximum speed, such as the skipjack tuna, have muscle glycogen concentrations of 100 μmol glucosyl unit/g, and have extremely high activities of glycolytic enzymes[29]. Muscle lactate concentrations can reach 40 mM following 15 min strenuous activity in salmonids[12]. Thus the relative proportions of the different anaerobic energy supplying pathways are matched to behaviour.

3. Scaling effects

The metabolic profile will, like other muscle properties, be influenced by scaling effects. In fish, as far as we know, they have been looked at only in terms of the activities of metabolic enzymes. In common with all other animals the mass-specific activity of aerobic enzymes decreases with increasing body size[67,68]. In mammals, this is associated with a decline in the mitochondrial fractional volume of the muscle[34]. A simple-minded interpretation is that small fish have higher tail-beat frequencies, and sustained muscle activity will require a higher aerobic supply of ATP. The mass-specific activities of enzymes of anaerobic metabolism, at least in pelagic fish, increase with increasing body size. Somero and Childress[68] suggest that this is to provide increased power during burst swimming in larger fish.

VII. The energetics of muscle contraction

The efficiency of vertebrate muscle during the initial contractile events, defined as the ratio of the work done to the free energy change of the driving reactions, is known to vary from at least 30–70%[74]. Woledge[74] has suggested that there is a trade-off between power output and muscle efficiency. A high muscle power output is required for acceleration, and in overcoming the viscous resistance of the water during fast swimming. The maximum power available during burst swimming is not dependent on efficiency, but only on the rate at which work can be obtained from the phosphocreatine pools within the muscles fibres. In contrast, continuously available mechanical power is limited by the supply of oxygen to the muscles. Woledge[74] has pointed out that 'if the efficiency is higher, each mole of oxygen can give more work, and so the available power for a given oxygen supply will be higher'. Evolution has produced a whole range of muscle fibre types with properties that are matched to the power/efficiency requirements of each speed.

Recently, we measured the energy cost of contraction for fast muscle fibres in the cod (*Gadus morhua*) under conditions for optimum oscillatory power output at 4°C[54]. Fibre bundles were incubated in Ringer solution containing 0.5 mM iodoacetic acid and bubbled with nitrogen to block glycolysis and aerobic metabolism respectively. Following up to 64 cycles of work the fibre bundles were frozen, freeze-dried and the concentration of adenylates and creatine compounds determined by high performance liquid chromatography and compared with values for unstimulated fibres. The concentration of creatine increased and creatine phosphate decreased with increasing numbers of cycles (Fig. 10). Adenylate levels were not significantly different from control values, although there was a significant increase in IMP concentrations at 64 cycles, indicating the activation of the AMP deaminase reaction. We calculated economy as the ratio of work done per gram dry weight of muscle to the breakdown of phosphocreatine per gram dry weight of muscle. The economy at 5 Hz cycle frequency increased from around 7 mJ/μmol creatine phosphate after 8 cycles to 11–13 mJ/μmol creatine phosphate following 16–64 cycles. The Gibbs' force free energy change for PCr splitting is 55 kJ/mol[74], which gives efficiency values of 12–23%. This compares with a maximum efficiency of 30% for fast fibres from the dogfish (*Scyliorhinus canicula*) determined from heat + work measurements during isotonic shortening[74]. The cod values were determined during oscillatory work, when shortening velocity changed continuously, and the conditions were those yielding maximum power output. Maximum efficiency will almost certainly be found under different conditions.

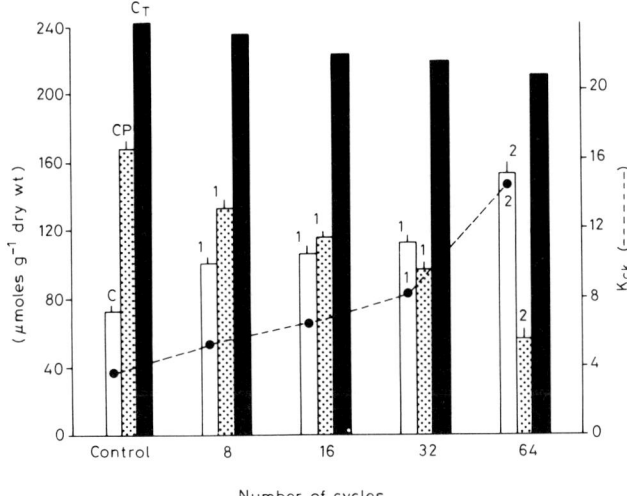

Fig. 10. Concentrations of creatine (C), creatine phosphate (CP) and total creatine (C_T) (mean ± S.E.M.) in control and muscle fibre bundles subjected to 8, 16, 32 and 64 cycles of oscillatory work. K_{CK} is the apparent creatine kinase equilibrium constant. [1]Significantly different from control ($P < 0.01$); [2]significantly different from control and 32 cycles ($P < 0.01$). From Moon, Altringham and Johnston[54].

VIII. References

1. Akster, H.A., H.L.M. Granzier and H.E.D.J. Ter Keurs. Force-sarcomere length relations vary with thin filament length in muscle fibres of the perch (*Perca fluviatilis*). *J. Physiol. (Lond.).* 353: 61P, 1984.
2. Akster, H.A., H.L.M. Granzier and H.E.D.J. Ter Keurs. A comparison of quantitative ultrastructural and contractile characteristics of muscle fibre types of the perch, *Perca fluviatilis* L. *J. Comp. Physiol. B.* 155: 685–691, 1985.
3. Alexander, R. McN. Orientation of muscle fibres in the myomeres of fishes. *J. Mar. Biol. Assoc. U.K.* 49: 263–290, 1969.
4. Altringham, J.D. and I.A. Johnston. The pCa-tension and force–velocity characteristics of skinned fibres isolated from fish fast and slow muscles. *J. Physiol. (Lond.).* 333: 421–449, 1982.
5. Altringham, J.D. and I.A. Johnston. Activation of multiply innervated fast and slow myotomal muscle fibres of the teleost *Myoxocephalus scorpius*. *J. Exp Biol.* 140: 313–324, 1988.
6. Altringham, J.D. and I.A. Johnston. The mechanical properties of polyneuronally innervated, myotomal muscle fibres isolated from a teleost fish (*Myoxocephalus scorpius*). *Pflügers Arch.* 412: 524–529. 1988.
7. Altringham, J.D. and I.A. Johnston. The innervation pattern of fast myotomal muscle in the teleost *Myoxocephalus scorpius*: a reappraisal. *Fish Physiol. Biochem.* 6: 309–314. 1989.
8. Altringham, J.D. and I.A. Johnston. Modelling muscle power output in a swimming fish. *J. Exp. Biol.* 148: 395–402, 1990.
9. Altringham, J.D. and I.A. Johnston. Scaling effects on muscle function: power output of isolated fish muscle fibres performing oscillatory work. *J. Exp. Biol.* 151: 453–467, 1990.
10. Archer, S.D., Altringham, J.D. and I.A. Johnston. Scaling effects on the neuromuscular system, twitch kinetics and morphometrics of the cod, *Gadus morhua*. *Mar. Behav. Physiol.* 17: 137–146, 1990.
11. Archer, S.D and I.A. Johnston. Cristae density and distribution of mitochondria in the slow muscle fibres of Antarctic fish. *Physiol. Zool.* 64: 242–258, 1991.
12. Black, E.C., A.C. Robertson and R.R. Parker. Some aspects of carbohydrate metabolism in fish. In *Comparative physiology of carbohydrate metabolism in heterothermic animals*, edited by A.W. Martin. Seattle, WA: University of Washington Press, 1961, pp. 89–124.

13. Blake, R.W. *Fish Locomotion*. Cambridge: Cambridge University Press, 1983, p. 208.
14. Bone, Q. Patterns of muscular innervation in the lower chordates. *Int. Rev. Neurobiol.* 6: 99–147, 1964.
15. Bone, Q. Locomotor muscle. In *Fish Physiology, Vol. 7*, edited by W.S. Hoar and D.J. Randall. Academic Press, New York and London, 1978, pp. 361–424.
16. Bone, Q., I.A. Johnston, A. Pulsford and K.P. Ryan. Contractile properties and ultrastructure of three types of muscle fibre in the dogfish myotome. *J. Muscle Res. Cell Motil.* 7: 46–56, 1986.
17. Bone, Q., J. Kiceniuk and D.R. Jones. On the role of the different fibre types in fish myotomes at intermediate swimming speeds. *Fish. Bull. (Washington)* 76: 691–699, 1978.
18. Bone, Q. and R.D. Ono. Systematic implications of innervation patterns in teleost myotomes. *Brevoria* 470: 1–23. 1982.
19. Crockford, T. and I.A. Johnston. Temperature acclimation and the expression of contractile protein isoforms in the skeletal muscles of the common carp (*Cyprinus carpio* L.). *J. Comp. Physiol. B* 160: 23–30, 1990.
20. Dunn, J.F. and I.A. Johnston. Metabolic constraints on burst-swimming in the Antarctic teleost *Notothenia neglecta*. *Marine Biology* 91: 433–440, 1986.
21. Egginton, S. and B.D. Sidell. Thermal acclimation induces adaptive changes in subcellular structure of fish skeletal muscle. *Am. J. Physiol.* 256: R1–R9, 1989.
22. Focant, B., F. Huriaux and I.A. Johnston. Subunit composition of fish myofibrils: the light chains of myosin. *Int. J. Biochem.* 7: 129–133, 1976.
23. Franzini-Armstrong, C. and G. Nunzi. Junctional feet and particles in the triads of a fast twitch muscle fibre. *J. Muscle Res. Cell Motil.*, 4: 233–252, 1983.
24. Gardner-Medwin, A.R. and N.A. Curtin. Implication of the septal structure in fish muscle. *J. Physiol.* 426: 36P. 1990.
25. Gerday, C. Soluble calcium-binding proteins from fish and invertebrate muscle. *Mol Physiol.* 2: 63–87, 1982.
26. Gerlach, G.-F., L. Turay, K.T.A. Malik, J. Lida, A. Scutt and G. Goldspink. The mechanism of temperature acclimation in the carp: A molecular biology approach. *Am. J. Physiol.* in press.
27. Greer-Walker, M.G., J. Horwood and L. Emerson. On the morphology and function of red and white skeletal muscle in the anchovies *Engaulis encrasicolus* L. and *E. mordax* Girard. *J. mar. biol. Assoc. U.K.*, 60: 31–37, 1980.
28. Grillner, S. and S. Kashin. On the generation and performance of swimming in fish. In *Neural Control of Locomotion*, edited by R.M. Herman, S. Grillner, P.S.G. Stein and D.G. Stuart. New York: Plenum Press, 1976, pp. 181–201.
29. Guppy, M., W.C. Hulbert and P.W. Hochachka. Metabolic sources of heat and power in tuna muscles. II. Enzyme and metabolite profiles. *J. Exp. Biol.* 82: 303–320. 1979.
30. Harford, J. and J. Squire. 'Crystalline' myosin cross-bridge array in relaxed bony fish muscle. Low-angle X-ray diffraction from plaice fin muscle and its interpretation. *Biophys. J.* 50: 145–155, 1986.
31. Heizmann, C.W. Parvalbumin, an intracellular calcium-binding protein; distribution, properties and possible roles in mammalian cells. *Experentia* 40: 910–921, 1984.
32. Hess, F. and J.J. Videler. Fast continuous swimming of saithe (*Pollachius virens*): a dynamic analysis of bending moments and muscle power. *J. Exp. Biol.* 109: 229–251, 1984.
33. Hochachka, P.W. Fuels and pathways as designed systems for support of muscle work. *J. Exp. Biol.* 115: 149–164, 1985.
34. Hochachka, P.W., B. Emmett and R.K. Suarez. Limits and constraints in the scaling of oxidative and glycolytic enzymes in homeotherms. *Can. J. Zool.* 66: 1128–1138, 1988.
35. Hudson, R.C.L. On the function of the white muscles in teleosts at intermediate swimming speeds. *J. Exp. Biol.* 58: 509–522, 1973.
36. Huriaux, F and B. Focant. Electrophoretic and immunological study of myosin light chains from freshwater teleost fishes. *Comp. Biochem. Physiol.* 82B: 737–743, 1985.
37. Johnston, I.A. Structure and function of fish muscles. In: *Vertebrate Locomotion, Symposium of the Zoological Society of London, Vol. 48,*, edited by M.H. Day. London: Zoological Society of London, 1981, pp. 71–113.
38. Johnston, I.A. Quantitative analyses of ultrastructure and vascularization of the slow muscle fibres of the anchovy. *Tissue Cell* 14: 319–328, 1982.
39. Johnston, I.A. Antarctic fish muscles – structure, function and physiology. *Antarct. Sci.* 1:97–108, 1989.
40. Johnston I.A. and J.-P. Camm. Muscle structure and differentiation in pelagic and demersal stages of the Antarctic teleost, *Notothenia neglecta*. *Mar. Biol.* 94: 183–190, 1987.

41. Johnston I.A., J.-P. Camm and M.G. White. Specialisations of swimming muscles in the pelagic antarctic fish *Pleuragramma antarcticum*. *Mar. Biol.* 100: 3–12. 1988.
42. Johnston, I.A., W. Davison and G. Goldspink. Energy metabolism of carp swimming muscles. *J. Comp. Physiol.* 114: 203–216, 1977.
43. Johnston I.A. and J. Dunn. Temperature acclimation and metabolism with particular reference to teleost fish. In *Temperature and Animal Cells*, edited by K. Bowler and B.J. Fuller. Soc. Exp. Biol. Symp., 41: 67–93, 1987.
44. Johnston, I.A. and T.W. Moon. Fine structure and metabolism of multiply innervated fast muscle fibres in teleost fish. *Cell Tissue Res.* 219: 93–109, 1981.
45. Johnston I.A., S. Patterson, P.S. Ward and G. Goldspink. The histochemical demonstration of myofibrillar adenosine triphosphatase activity in fish muscle. *Can. J. Zool.* 52: 871–877, 1974.
46. Johnston, I.A., B.D. Sidell and W. Dreidzic. Force-velocity characteristics and metabolism of carp muscle fibres following temperature acclimation. *J. Exp. Biol.* 119: 239–250, 1985.
47. Karasinski, J. and W. Kilarski. Polymorphism of myosin isoenzymes and myosin heavy chains in histochemically typed skeletal muscles of the roach (*Rutilus rutilus* L., Cyprinidae, fish). *Comp. Biochem. Physiol.* 92B: 727–731, 1989.
48. Kilarski, W. and M. Kozlowska. Comparison of ultrastructural and morphometric analysis of tonic, white and red muscle fibres in the myotome of teleost fish (*Noemacheilus barbatulus* L.). *Z. mikrosk.-anat. Forsch. (Leipzig)* 101: 636–648, 1987.
49. Langfeld, K., T.C. Crockford and I.A. Johnston. Temperature acclimation in the common carp: force-velocity characteristics and myosin sub-unit composition of slow muscle fibres. *J. Exp. Biol.* 155: 291–304, 1991.
50. Lannergren, J. Contractile properties and myosin isoenzymes of various kinds of *Xenopus* twitch muscle fibres. *J. Muscle Res. Cell Motil.* 8: 260–273, 1987.
51. Leinwald, L.A., L. Saez, E. McNally and B. Nadal-Ginard. Isolation and characterisation of human myosin heavy chain genes. *Proc. Natl. Acad. Sci. U.S.A.* 80: 3716–3722, 1983.
52. Luther, P.K., P.M.G. Munro and J.M. Squire. Three dimensional structure of the vertebrate muscle A-band III: M-region structure and myosin filament symmetry. *J. Mol. Biol.* 151: 703–730, 1981.
53. McArdle, H.J. and I.A. Johnston. Ca^{2+}-uptake by tissue sections and biochemical characteristics of sarcoplasmic reticulum isolated from fish fast and slow muscles. *Eur. J. Cell Biol.* 25: 103–107, 1981.
54. Moon, T.W., J.D. Altringham and I.A. Johnston. Muscle energetics and power output of isolated fish fast muscle fibres performing oscillatory work. *J. Exp. Biol.*, submitted.
55. Nabeshima, Y., Y. Fujii-Kuriyama, M. Muramatsu and K. Ogata. Alternative transcription and two modes of splicing result in two myosin light chains from one gene. *Nature* 308: 333–338, 1984.
56. Nag, A. Ultrastructure and adenosine triphosphatase activity of red and white muscle fibres of the caudal region of a fish, *Salmo gairdneri*. *J. Cell Biol.* 55: 42–57, 1972.
57. Ono, R.D. Dual motor innervation in the axial musculature of fishes. *J. Fish Biol.* 22: 395–408, 1983.
58. Patterson, S. and G. Goldspink. The fine structure of red and white myotomal muscle fibres of the coalfish (*Gadus virens*). *Z. Zellforsch. mikrosk. Anat.* 146: 463–474. 1972.
59. Pepe, F.A. Structure of the myosin filament of striated muscle. In *Progress in Biophysics and Molecular Biology, Vol. 22*, edited by J.A.V. Butler and D. Noble. New York: Pergamon Press, 1971, pp. 77–96.
60. Reiser, P.J., R.L. Moss, G.G. Giulan and M.L. Greaser. Shortening velocity in single fibres from rabbit soleus muscles is correlated with myosin heavy chain composition. *J. Biol. Chem.* 260: 9077–9080, 1985.
61. Rome, L.C., P.T. Loughna and G. Goldspink. Muscle fibre activity in carp as a function of swimming speed and temperature. *Am. J. Physiol.* 247: R272–279, 1984.
62. Rome, L.C., R.P. Funke, R.McN. Alexander, G. Lutz, H.D.J.N. Aldridge, F. Scott and M. Freadman. Why animals have different muscle fibre types. *Nature* 355: 824–827, 1988.
63. Rowlerson, A., P.A. Scapolo, F. Mascarello, E. Carpene and A. Veggetti. Comparative study of myosins present in the lateral muscle of some fish: species variations in myosin isoforms and their distribution in red, pink and white muscle. *J. Muscle Res. Cell Motil.* 6: 601–640, 1985.
64. Scapolo, P.A. and A. Rowlerson. Pink lateral muscle in the carp (*Cyprinus carpio* L.): histochemical properties and myosin composition. *Experientia* 43: 384–386, 1987.
65. Squire, J.M., K.P. Luther and E.P. Morris. Organisation and properties of the striated muscle sarcomere. In *Molecular Mechanisms in Muscular Contraction*, edited by John M. Squire. London: MacMillan Press, 1990.
66. Skoglund, C.R. Functional analysis of swimbladder muscles engaged in sound production of the toadfish. *J. Biophys. Biochem. Cytol.* 10: 187–200, 1961.

67. Somero, G.N. and J.J. Childress. A violation of the metabolism-size scaling paradigm: Activities of glycolytic enzymes in muscle increase in larger-size fish. *Physiol. Zool.* 53: 322–337, 1980.
68. Somero G.N. and J.J. Childress. Scaling of ATP-supplying enzymes, myofibrillar proteins and buffering capacity in fish muscle: relationship to locomotory habit. *J. Exp. Biol.* 149: 319–333, 1990.
69. Van Leeuwen, J.L., M.J.M. Lankheet, H.A. Akster and J.W.M. Osse. Function of red axial muscles of carp (*Cyprinus carpio*): recruitment and normalized power output during swimming in different modes. *J. Zool.* 220: 123–145, 1990.
70. Videler, J.J. Fish swimming movements: A study of one element of behaviour. *Neth. J. Zool.* 35: 170–185, 1985.
71. Wardle, C.S. Understanding fish behaviour can lead to more selective fishing gears. In *Proceedings of the World Symposium Fishing Gear and Fishing Vessel Design*. Newfoundland, Canada: Marine Institute, 1988, pp. 12–18.
72. Westerfield, M., J.V. McMurray and J.S. Eisen. Identified motoneurons and their innervation of axial muscles in the zebrafish. *J. Neurosci.* 6: 2267–2277, 1986.
73. Williams, T.L., S. Grillner, V.V., Smojaninov, P. Wallen, S. Kashin and S. Rossignol. Locomotion in lamprey and trout: the relative timing of activation and movement. *J. Exp. Biol.* 143: 559–566, 1989.
74. Woledge, R.C. Energy transformation in living muscle. In: *Energy Transformations in Cells and Organisms*, edited by W. Wieser and E. Gnaiger. Stuttgart, New York: Georg Thieme Verlag, 1989, pp. 36–45.
75. Yancey, P.H. and I.A. Johnston. Effect of electrical stimulation and exercise on the phosphorylation state of myosin light chains from fish skeletal muscle. *Pflügers Arch.* 393: 334–339, 1982.

CHAPTER 11

Endothermy in fish: thermogenesis, ecology and evolution

BARBARA A. BLOCK

Department of Organismal Biology and The Committee on Evolutionary Biology, The University of Chicago, 1025 East 57th Street, Chicago, IL 60637, U.S.A.

I. Introduction
   1. Relationships among scombroid fish
II. Brain heaters in billfish
   1. Telemetry of brain temperature in swordfish
   2. Structure of the heater organ
   3. How much heat is generated in the heater organ?
   4. How fish have built a furnace out of muscle
   5. Excitation–thermogenic coupling in heater cells
   6. Cycling calcium for heat production
   7. Ecological implications of heater organs
   8. Heater organs in scombrid fish
III. Regional endothermy *versus* whole body endothermy
IV. Behavioral thermoregulation: the importance of thermal inertia and large body size
V. Endothermy in tunas and lamnid sharks: what's new?
   1. Biomechanics of tunas and endothermy
VI. Why be warm?
   1. Increased power output and warm temperatures
   2. High metabolic rates and endothermy
   3. Oxygen and metabolite flux rates
   4. Niche expansion as the driving force for endothermy
   5. Endothermy in fish and the evolution of endothermy in birds and mammals
   6. Conclusions
Acknowledgements
VII. References

I. Introduction

Most of the 25,000 species of teleost fishes are ectotherms with body temperatures within 1–2°C of ambient water temperature[25,82]. The temperatures of core tissues of ectotherms are determined by the surrounding environment, not by metabolically derived heat production. Most fish are the same temperature as the water in which they swim because they breathe with a gill. The oxygen content of water is 1/40th that of air while the heat capacity is 3000 times greater resulting in large heat losses when respiring via a gill (thermal diffusion is 50 times more rapid than molecular diffusion ensuring that the blood comes to water temperature in the gill). Fish lose almost all of the metabolic heat generated aerobically in the tissues during passage of the blood through the gill. Heat loss by conduction to the body surface can be an important factor in small fish (< 1 kg) but is secondary to convective heat transfer by the blood in larger fish[16,87].

Endothermy, the ability to maintain elevated body temperatures by metabolic means, is unusual in teleostean fish and has only been documented within one major assemblage of large oceanic fishes, the Scombroidei (mackerels, tunas, bonitos and billfishes). Certain sharks (of the family Lamnidae which includes the mako, porbeagle and white sharks) are also warm and share features convergent with the scombroid fishes[20,25,30]. There are 27 species of fish and sharks that use some form of endothermy or thermogenesis to elevate tissue temperatures above water temperature. Several requirements are necessary for endothermy, foremost is a large body size, a heat source and a mechanism (heat exchangers) for conserving the heat. Two distinct strategies for elevating tissue temperatures have evolved and both utilize skeletal muscle as the major source of heat. This chapter focuses on recent developments in our understanding of how fish stay warm, with emphasis on the recent discovery of regional endothermy in the billfishes, involving a thermogenic organ beneath the brain[2,19]. A major goal of the present chapter is to define the strategies that fish use to stay warm and to provide a historical perspective on the evolution of endothermy in scombroid fishes.

Among scombroid fishes endothermy primarily has been associated with the 13 species of tunas[22,25,82]. Tunas and 5 species of lamnid sharks maintain body temperatures up to 21°C above that of the surrounding water by conserving metabolic heat in the muscle, viscera, and brain[7,20,22-24,27,29,70,87]. Fishermen of various cultures have known for centuries that tunas are warm-bodies. Their high temperatures were noted in 1835 by the British physician Davy[38] who observed that skipjack tunas were up to 10°C warmer than the water. Endothermy in the so-called 'warm-fish' is associated with the elevation of metabolic rate[15,51,98] and the presence of vascular heat exchange systems that are essential for limiting heat loss and heat gain to the environment[22,30]. Eschrict and Muller[43] provided the first anatomical descriptions of the heat exchangers surrounding the red muscle in tunas and elegant details of the elaborate vasculature of tunas appeared in a classical monograph by Kishinouye[66]. Anatomists early on also had examined and detailed some of the 'peculiarities' of the circulation in the lamnid sharks[18,74]. Carey and Teal[22-24] provided the initial physiological evidence detailing how tunas and lamnid sharks are able to conserve metabolic heat and elevate body temperature. The major source of metabolic heat is thought to be derived from the slow oxidative swimming musculature that powers sustained cruising[55]. Endothermy in tunas is associated with two major anatomical changes: (1) the internal location of the red oxidative muscle mass, and (2) a complex counter-current heat exchange system in the circulation to the oxidative muscle mass which retains heat in the deep musculature. Endothermic fish and sharks have numerous well documented anatomical, physiological and biochemical specializations associated with the maintenance of elevated body temperatures[15,20,30,52,57,58,83,85,88,89].

While the elaborate anatomical and physiological specializations of tunas have long fascinated physiologists and morphologists, the billfishes have received far less study. Recently it was discovered that a relative of tunas, the swordfish, *Xiphias gladius*, is capable of raising its brain temperature up to 13°C above ambient[19]. In billfishes, the body temperature remains close to water temperature and only the cranial cavity is warm. The basis for this 'regional endothermy' in the brain and eyes is the modification of muscle into a novel thermogenic organ[2,3,19]. Research in the past decade on the physiology, behavior, and biomechanics of billfishes indicate that they are highly specialized for life in the pelagic ocean and use different strategies than tunas to optimize performance[5,6,19,21,26,37,41,56,90]. Regional endothermy in billfishes allows this clade to extend their thermal environment in a similar fashion to many endothermic scombrids and lamnid sharks but presumably at a lower energetic cost. A goal of the present chapter is to examine how billfish have built a furnace out of muscle and to understand the ecological advantages afforded by regional endothermy.

1. Relationships among scombroid fish

Cladograms indicating relationships among scombroid fishes (Istiophoridae, Xiphiidae and Scombridae, Trichiuridae and Gempylidae) and the acquisition of endothermy are shown in Fig. 1. The scombroid fishes are a suborder of Perciformes and relationships among families within this group, which includes many of the largest living teleosts, have received considerable study[36,60]. Endothermy has evolved independently three times within the scombroid fishes (Fig. 1). Whole body endothermy is characteristic of the true tunas (Scombridae: tribe Thunnini) while regional endothermy involving a thermogenic organ has evolved in the billfishes (Xiphiidae and Istiophoridae) and independently in one scombrid fish thought to be primitive, the butterfly mackerel *Gasterochisma melampus*. As discussed below, the relationships between *Gasterochisma* and other scombroid fish remain controversial; likewise the relationships of the billfishes (Xiphiidae plus Istiophoridae) to the Scombridae are uncertain. When considering the historical aspects of the evolution of endothermy in scombroid fishes, the exact nature of the relationships must be ascertained and new molecular data presented below address some of the uncertain relationships.

Kishinouye[66] used the elaborate vascular anatomy of tunas to separate them into a distinct order separate from both scombrids and all other Teleostei. More recent examination of the internal characteristics of the Scombridae has led to a rearrangement of the phylogenetic relationships[34,35]. The four genera of endothermic bonitos and tunas, members of the tribe Thunnini, are unique among teleost fishes (yet related to mackerels and other bonitos and thus placed within this family) in having rete mirabile forming countercurrent heat exchange systems in the vasculature associated with the muscle, viscera and brain[21]. Additionally, this tribe is distinguished in having internal placement of the red oxidative muscle that powers sustained swimming.

As mentioned above, the interrelationships between billfishes and other scombroids remain controversial and the relationships of the two families of fish comprising this assemblage, Xiphiidae (swordfish) and Istiophoridae (marlins, sailfish, spearfish), have been the subject of some debate[35,60,75]. A rigorous phylo-

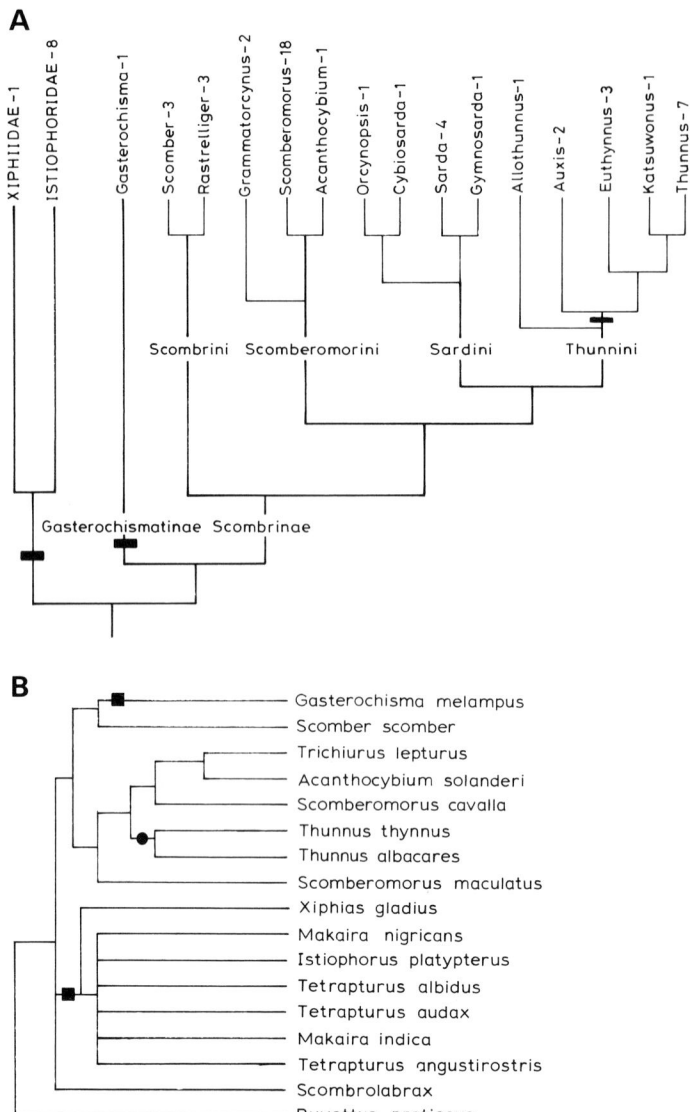

Fig. 1. A: relationships among several families, tribes, and genera (species indicated by number) of scombroid fishes indicating the evolution of endothermy three separate times (black bar). This tree is modified from Collette[34]. The Scombridae are composed of two tribes the Scombrinae (mackerels, bonitos and tunas) and Gasterochimatinae (butterfly mackerel). The four genera of tunas are unique among bony fishes in having heat exchangers surrounding the red muscle. Three genera (*Auxis, Euthynnus* and *Katsuwonus*) have central heat exhangers whereas *Thunnus* have more well developed lateral heat exchangers. *Gasterochisma* is considered a more primitive evolutionary lineage of scombrids and has heater organs associated with the lateral rectus muscles. Heater organs are present in all billfishes (Istiophoridae and Xiphiidae) associated with the superior rectus eye muscles. B: relationships of scombroid fishes based on mitochondrial DNA. 612 base pairs of sequence from cytochrome *b* were compared. With the primitive Gempylid, *Ruvettus pretiosus,* or *Scombrolabrax* as the outgroup, *Gasterochisma* falls within the Scombridae. As above regional endothermy has evolved at least two different times.

genetic analysis of the billfishes based on morphological data has not been undertaken in a manner comparable to Collette et al.'s[34,35] contributions to understanding the relationships among the Scombridae. In the most recent systematic review of the billfishes, Nakamura[75] differed from previous workers and concluded that the similarities between Xiphiidae (swordfish) and Istiophoridae (marlins, sailfish and spearfishes) are convergent and that the two families do not share a common ancestor within the Scombroidei. Collette and others have argued that several synapamorphies unite the two families and have established this monophyletic group (Istiophoridae and Xiphiidae) as the sister group to Scombridae. Block[2] and Carey[19] have demonstrated that all billfishes have a novel type of muscle tissue capable of thermogenesis, associated with the superior rectus eye muscles. The presence of the modified muscles is associated with other morphological features of the cranial cavity and blood supply to the brain and eye that are shared between the two families[2,5,6]. Thus the complex of characteristics associated with modification of the eye muscle into a heat generating tissue is a shared derived characteristic of the billfishes that strongly supports a monophyletic lineage for the two families[2,5]. This set of characteristics was recently used by Johnston[60] who also agreed that the billfishes shared a common ancestor.

Heater organs are also present in the distinctive scombrid fish *Gasterochisma melampus*[2,6,19]. The heater organ in *Gasterochisma* is associated with the lateral rectus eye muscle rather than the superior rectus. This extraocular muscle is innervated by the abducens nerve, while the superior rectus is associated with the oculomotor nerve. The difference in cranial innervation indicates that the embryonic source of the heater tissue (the preotic myotomes) is distinctly different in *Gasterochisma* from that in the billfishes[3,5] and supports an independent evolution of the heater in *Gasterochisma*. The taxonomic position of *Gasterochisma* has been uncertain. Collette et al.[35] and Kohno[67] referred to it as a primitive member of the Scombridae while Johnston[60], not having enough information on this rare fish, found it necessary to leave it out of his analysis of the Scombroidei. However Johnston did indicate that if he included *Gasterochisma* in his parsimony analysis it was consistently placed near the mackerels (Scombrini), a position that is not inconsistent with Collette's[35] placement of the fish as sister group to all other Scombridae.

In an attempt to resolve some of the ambiguities cited above, the relationships between the Xiphiidae, Istiophoridae and Scombridae have recently been examined with molecular techniques[11]. A 656 bp region of the cytochrome b gene from the 21 taxa of scombroid fishes was amplified by the polymerase chain reaction and directly sequenced. Phylogenetic comparisons of the sequences support the conclusions stated above, specifically the monophyly of the billfishes and the placement of *G. melampus* outside of this monophyletic group[11]. Thus the presence of heater tissue in the head of *Gasterochisma* is not indicative of any association with the billfishes and represents independent acquisition of regional endothermy. *Gasterochisma* has large cycloid scales and has previously been thought to be a more primitive scombrid fish[34,67]. This is supported by the mtDNA data which indicates that *Gasterochisma* is most certainly not a billfish and shares the most sequence similarities with the scombrini (mackerels). The molecular results, when combined with morphology, confirm that heater organs evolved independently two times within the scombroid lineage. The molecular data also suggest that both the swordfish and *Gasterochisma* are relatively primitive when compared to other scombroid fishes.

In summary, whole-body endothermy is only found in the Scombridae whereas regional endothermy is associated with the Xiphiidae, Istiophoridae and one of the most primitive members of the Scombridae. The two strategies are distinctly different, the tunas are warm in a manner that is analogous to mammals and birds; elevated metabolism[15,51,98] combined with elaborate physiological control of heat loss, while the billfishes and *Gasterochisma*, utilize a strategy that is unique among vertebrates—a furnace beneath the brain. As discussed toward the end of the chapter, consideration of the historical relationships amongst scombroid fishes *suggests* that acquisition of brain heaters and regional endothermy is both independently derived and more primitive than the whole-body endothermy characteristic of tunas.

II. Brain heaters in billfish

1. Telemetry of brain temperature in swordfish

In contrast to tunas and lamnid sharks, billfishes maintain only a limited region of endothermy that primarily involves warming of the brain and eyes. The best evidence for physiological regulation of brain temperatures has come from experiments with free-swimming swordfish[19,21] (Fig. 2). Carey and Robison[26] have shown that swordfish swim to depths as deep as 600 m during the day but spend most of the night at the surface. During these vertical excursions, which primarily occur at dusk and dawn, the fish encounter rapid water temperature changes of up to 19°C in a short period. Experiments with free-swimming swordfish[19,21] in which a thermistor was positioned close to the brain provide direct evidence that these fish are able to warm the cranial cavity (Figs. 3 and 4). While at depth, the swordfish was able to maintain a relatively constant cranial temperature. In the first experiment (Fig. 3), temperature in the cranial cavity remained at 28°C ± 1 for 36 h while the fish encountered water temperatures that ranged from 13 to 17°C. Fig. 4 demonstrates unequivocally that swordfish are able to *regulate* the brain temperature, maintaining a larger thermal gradient when in deeper, cooler waters than at the surface. In this experiment, in contrast to the one presented in Fig. 3, the swordfish dove below the thermocline, through a 10°C decrease in water temperature and spent 8 h in 6°C water. During this long period in cold water, temperature in the cranial cavity decreased only 5°C and reached an equilibrium temperature that was well above the ambient water temperature. The ability to control brain temperature implies that the swordfish heater organ is under neural or hormonal regulation. This has recently been confirmed with morphological studies of the heater organ which demonstrate neural control of the heat generating cells (Block and Kim, unpublished).

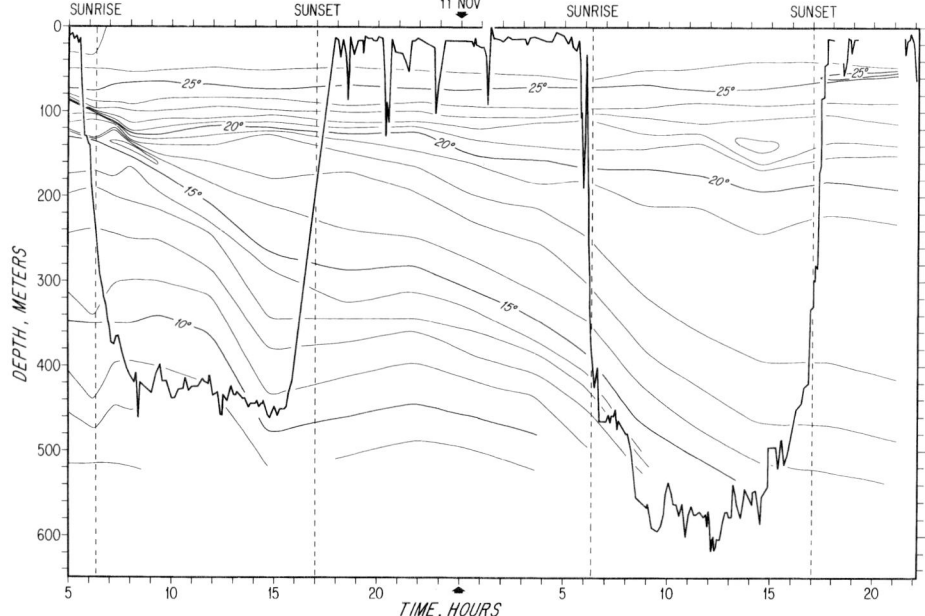

Fig. 2. Telemetry of the depth movements of the swordfish, *Xiphias gladius*, indicating the diurnal vertical migration pattern typical of these fish. At dawn and dusk the fish have large vertical excursions that take them through water temperatures of great extremes (from ref. 21).

Direct telemetry of brain temperature is difficult in large oceanic fishes and there are at present no measurements available for free-swimming istiophorids. Temperatures have been taken from the head region of billfishes captured on various types of fishing gear[3,6,19]. The results from such measurements, which provide some indication of thermogenesis in several other billfish species, show significant elevations of 2 to 6 °C above *surface* water temperatures (Fig. 5).

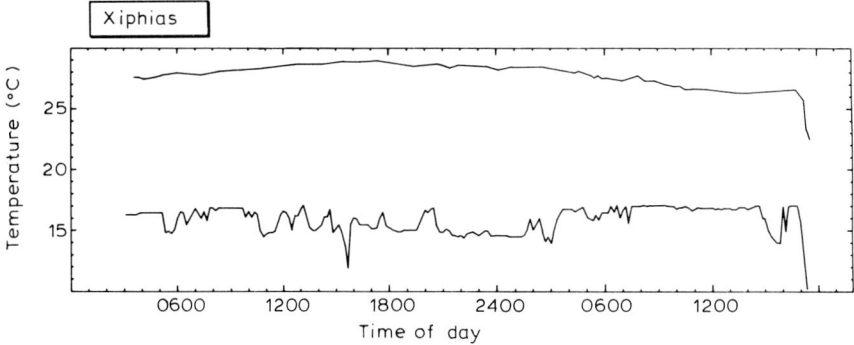

Fig. 3. Temperature telemetry record from a swordfish. Upper line is cranial temperature, lower line is water temperature (from ref. 21).

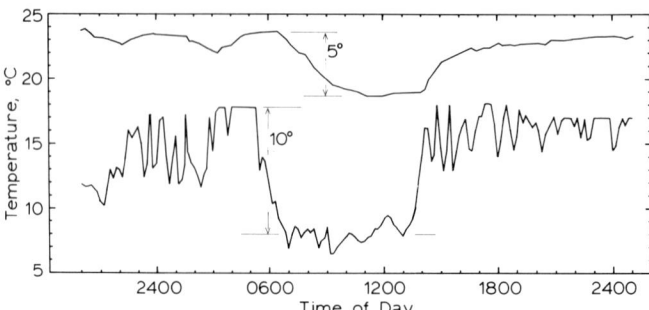

Fig. 4. Telemetry record of swordfish cranial temperature, upper line, and water temperature, lower line. When the swordfish entered 8°C water the cranial temperature cooled 5°C but reached a new equilibrium within 4 h. The 10°C change in water temperature produced only a 5°C change in cranial temperature. This is a strong indication of thermoregulation in swordfish (from ref. 21)

2. Structure of the heater organ

Situated at the base of the braincase in all billfishes is the heater, a mass of thermogenic tissue that is deep red in color and has the appearance and texture of liver (Fig. 6). The thermogenic organ is associated with the superior rectus eye muscle. Heaters from both sides of the head converge at the base of the braincase. In the swordfish, the brain is embedded in the dorsal surface of the heater, as the basisphenoid bone, which forms the base of the braincase, has been reduced to a membranous sheath. Large amounts of adipose tissue surround the brain and insulate the brain, eyes, and heater organ from large conductive heat losses.

The superior rectus muscle of billfishes has an unusual structure when compared to the same muscle in other vertebrates. A unique type of muscle fiber that

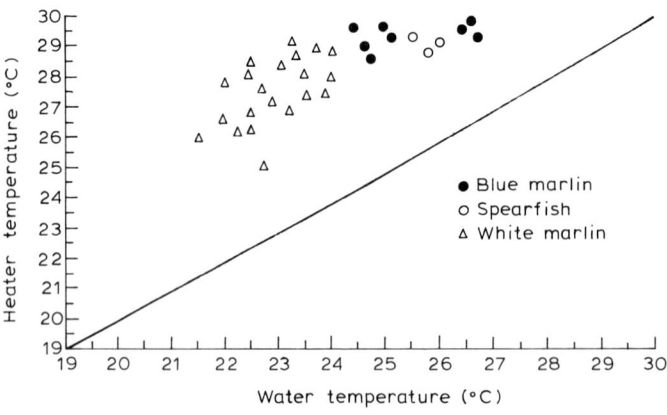

Fig. 5. The relationship between surface water temperature and heater tissue temperature in blue marlin (*Makaira nigricans*), white marlin (*Tetrapturus albidus*), shortbill spearfish, (*Tetrapturus angustirostris*). Measurements for white marlin are taken with the fish in the water while others are taken within three to five minutes of the fish being boated (data from refs. 3, 5). Isothermal line is where tissue temperatures for most other fish would appear.

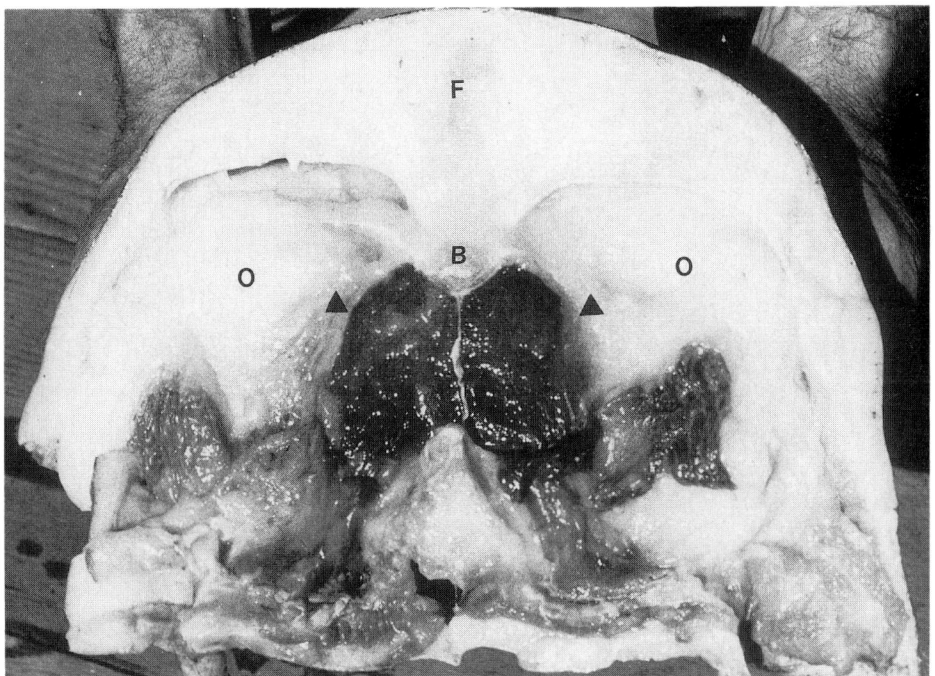

Fig. 6. Transverse section through the head of a swordfish showing position of the heater tissue (arrowhead) beneath the brain (B) and within the orbit (O). The head is packed with fat (F) to reduce conductive heat losses.

lacks the myofibrillar lattice typical of most muscle cells is found in the eye muscle and is expressed in the region of the muscle closest to the braincase[2,3,8]. This fiber type, called a 'heater cell', is an extraordinarily oxidative cell that is specialized for generating heat and is no longer capable of force generation[3,4,8]. A similar fiber type is expressed in the lateral rectus muscles of the scombrid fish, *G. melampus*[6]. The major difference between heater organs in the billfishes is one of size, the heater organ being largest in *Xiphias*. The size differences appear to be associated with quantitative differences in expression of the heater phenotype in the head region of various billfishes and, as discussed below, are correlated with the thermal ecology of the species[5,6]. The larger the thermal gradients experienced by the billfish, the more skeletal muscle appears to have been converted to heater phenotype[6]. In other words, all billfishes have furnaces beneath the brain, but expression of this trait is variable between species and correlates most with temperature.

A striking alteration in blood supply to the head accompanies the presence of the heat-generating muscle tissue. A major branch of the carotid passes through the heater and continues on supplying warm blood to the retina and brain[2]. At the base of the heater, a counter-current heat exchanger formed from the carotid artery and the venous return from the heater, prevents dissipation of heat from the

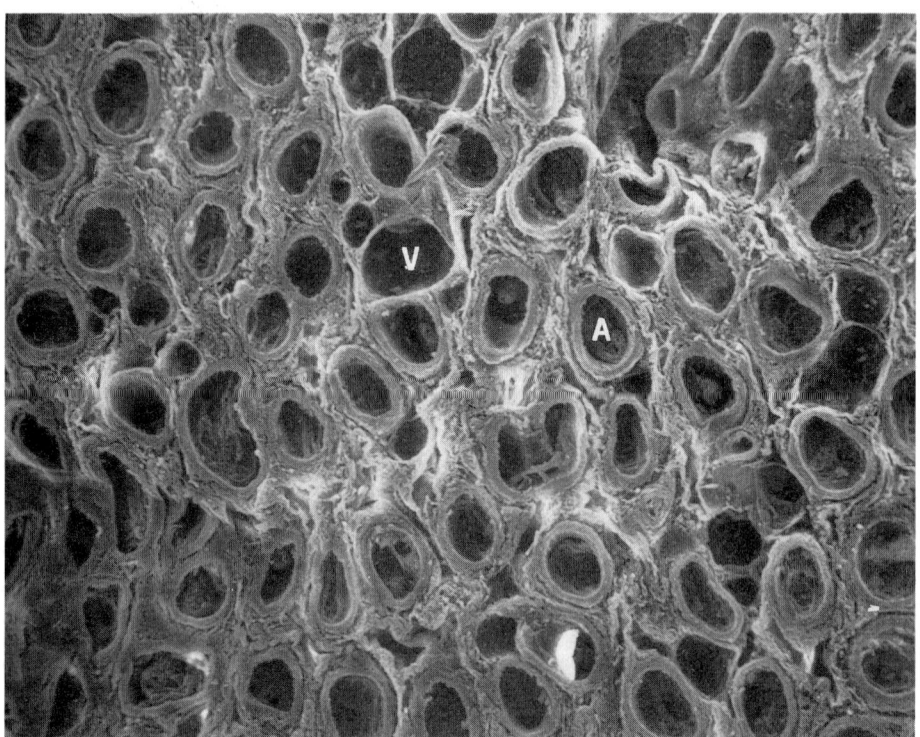

Fig. 7. Scanning electron micrograph of the heat exchanger situated at the base of the heater organ. The heat exchanger prevents dissipation of heat from the head region to the gills. A = artery; V = vein.

head region. The carotid heat exchanger is well developed in all of the billfishes and the size of the heat exchanger is correlated with the size of the heater organ in the various species of billfishes[2]. Diameters of arteries and veins comprising the heat exchanger range in size from 0.08 to 0.30 mm and the dimensions and absolute number of vessels involved varies between species. A cross-section through the heat exchanger in a blue marlin, *Makaira nigricans*, is shown in Fig. 7.

After passage through the arterial side of the heat exchanger, the carotid circulation breaks up into numerous capillaries that bathe the oxidative heater cells with blood. At the dorsal surface of the heater there is a large portal artery that is formed from numerous branches within the heater tissue and courses to the back of the eye. This is the major outlet for blood leaving the heater organ and provides strong evidence of its physiological role in warming both the brain and eyes. The warm blood supply entering the eye breaks up into a fine network of arteries that covers the entire ventral surface of the retina in a manner that is similar to the choroidal circulation of other vertebrates. In addition to the warm blood supply to the retina, the billfishes, like all teleost fish, receive a separate oxygenated source of blood to the retina by way of the choroid rete mirabilia[99]. The choroid retina is extraordinarily enlarged in the istiophorid billfishes and is supplied with blood via a separate pathway from the heater organ[5]. This insures a well-oxygenated blood supply to the eye, given that the blood coursing through the heater organ is presumably depleted of oxygen by the time it reaches the retina.

3. How much heat is generated in the heater organ?

The morphology of the heater system combined with the measured 15°C difference between brain and water temperature in free-ranging swordfish[21], allows one to make a first approximation of how much heat is being generated in the heads of swordfish. Blood flow to the head of a teleost fish is by way of the first gill arch and one can consider the head region to be separated from the body mass due to the insertion of the gills. Using a spherical model for heat conduction to calculate heat transfer in the head of the billfish, one can consider the bilaterally situated heater organs as the thermogenic core in this model. The one-dimensional equation for the Fourier law of conduction for a sphere used is[93]:

$$q = \frac{4\pi r_o r_i k (T_i - T_o)}{r_o - r_i}$$

where T_o equals the water temperature, T_i the brain temperature, r_o and r_i are the outer and inner shell radii (m^2) and k is the thermal conductivity (W/m/k). Assuming that the conductivity term of the outer core lies close to that of fat (0.205 W/m/°C) and that, for modelling purposes, the carotid heat exchanger acts as a perfect insulator (100% efficiency), then to account for the 15°C elevation of brain temperature in 14°C water, the swordfish heater organ must produce about 3 W of heat. The 40 kg fish involved in this experiment would have approximately 60 g of heater tissue and thus must have a heat production capacity equivalent to 50 W/kg which is 1.5-fold higher than the measured heat production of the human heart at rest (33.6 W/kg, ref. 80). If the assumption that the counter-current heat exchanger is 100% efficient is removed from this calculation, or a thermal conductivity slightly larger than that used for fat is applied, then far more heat (5–7 W) must be generated to account for the 15°C elevation in brain temperature.

4. How fish have built a furnace out of muscle

The heater cells have been most extensively studied in the istiophorid billfishes[4,6,8,9] and most of the discussion that follows applies primarily to marlins, sailfishes and spearfishes. Swordfish heater cells share many similarities to those of istiophorids and the details of the molecular mechanism of thermogenesis are presently being studied. One important question to be determined is if all three families of fishes with heater organs utilize the same molecular pathway for heat generation.

The heater cell is a modification of an oxidative fiber type within the extraocular muscle[8,9]. The most conspicuous feature of the heater cell is the absence of contractile filaments. Two morphological features stand out in all electron micro-

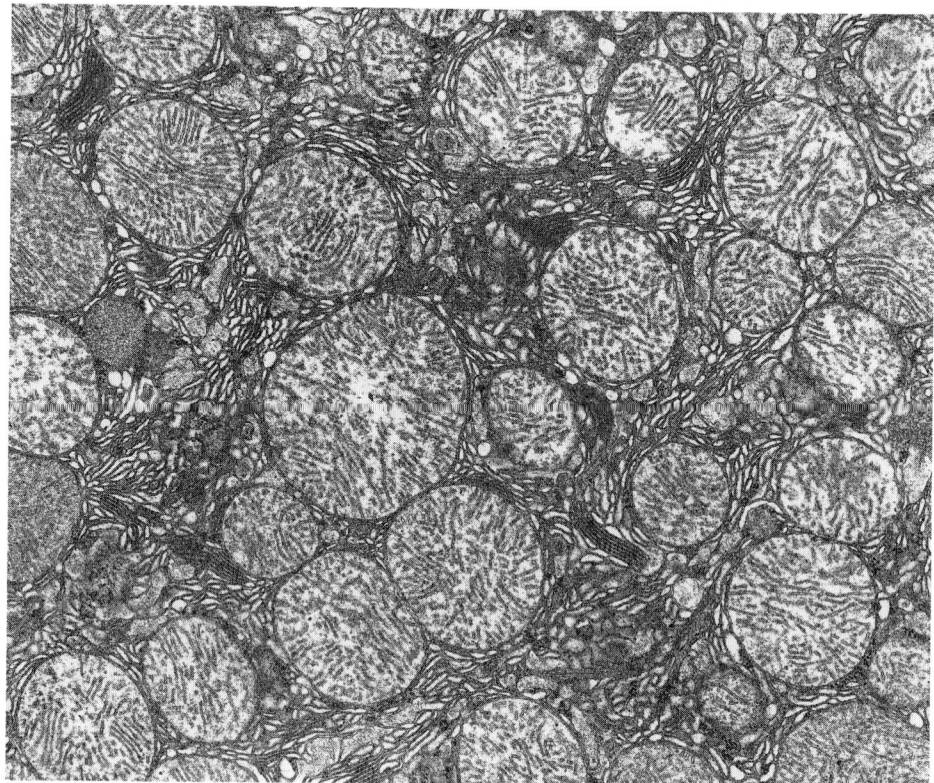

Fig. 8. Electron micrograph of a heater cell from a blue marlin, *Makaira nigricans*. The modified muscle cell lacks contractile filaments but is packed with mitochondria and extensive smooth membranes that regulate calcium release and reuptake (SR and transverse T tubules).

graphs of heater cells: (1) they are tightly packed with mitochondria and (2) they have an extensive smooth membrane system comprised of tubules, cisternae, and membranous stacks located throughout the cytoplasm in between the mitochondria (Fig. 8). The mitochondria of the heater cell occupy 63% of the cell volume in istiophorids (3,5; Table 1). The high oxidative capacity of the tissue is also reflected by the high myoglobin concentration present in the heater organ (Table 2). The smooth membranes are homologous to the sarcoplasmic reticulum (SR) and transverse tubule (T) system of normal skeletal muscle cells and occupy approximately 30% of the heater cell volume[8].

Billfish heater cells have the highest volume of mitochondria of any cell in the animal kingdom. As expected, oxidative enzyme activities of these cells are also extraordinarily high (Fig. 9). The oxidation of fuel molecules in mitochondria is potentially a major thermogenic process in animal cells[57]. The rate of mitochondrial respiration determines the rate at which substrate energy is released. The amount of heat given off at the mitochondria is dependent on the extent to which electron transport is coupled to oxidative phosphorylation (ATP synthesis). The

TABLE 1

Mitochondrial volume of oxidative muscle cells

Cell type	Mitochondrial volume as % of cell volume	Reference
Pigeon pectoralis muscle	29%	59
Anarctic fish oxidative muscle	30%	71
Trout red muscle	31%	61
Cicada tymbal muscle	33%	65
Finch cardiac muscle	34%	14
Mackerel red muscle	36%	13
Insect flight muscle	44%	43
Anchovy red muscle	45%	61
Billfish heater cells	63%	5

TABLE 2

Myoglobin concentration in muscle cells

Cell type	Myoglobin concentration (μmol/kg wet weight)	Reference
Cat soleus	127	100
Bat ventricle	230	100
Pigeon breast muscle	285	100
Dog gastrocnemius	311	100
Beef ventricle	317	100
Blue marlin heater organ	400	Block & Wittenberg, unpublished
Tuna red muscle	650–2100	Block & Wittenberg, unpublished

ability to utilize the proton gradient generated during electron transport for heat production has been well documented in mammals[77] but has not been demonstrated in ectothermic animals. Mitochondria from the billfish heater tissue possess

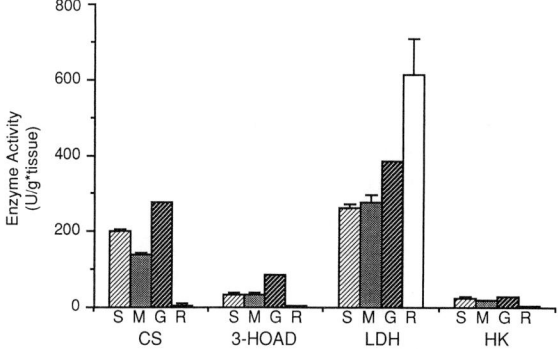

Fig. 9. Metabolic profiles of heater organs in three families of scombroid fishes; swordfish (S), blue marlin (M), and *Gasterochisma* (G). Red muscle (R) from extraocular muscle and epaxial muscle (LDH) was used as a control. Citrate synthase (CS), 3-hydroxyacyl CoA dehydrogenase, (3-HOAD), hexokinase (HK) and lactate dehydrogenase (LDH). (From Tullis, Block and Sidell unpublished data.)

TABLE 3

Respiration rates and ADP/O ratios in isolated mitochondria from the heater tissue of blue marlin, *Makaira nigricans*, and sailfish, *Istiophorus platypterus*

Fish	Substrate [a]	Oxygen uptake [b] (nmol O_2/mg protein/min)	ADP/O
Blue marlin	Glutamate/malate	68– 90	2.2
	Pyruvate/malate	91–125	2.3
Sailfish	Glutamate/malate	70–100	2.3
	Pyruvate/malate	48– 60	2.3

[a] Substrate additions were 10 mM glutamate, 10 mM malate or 10 mM pyruvate.
[b] Oxygen uptake was measured after the addition of 120 μM ADP. The values reported are the range (low and high values) of oxygen consumption measurements. Respiration without ADP (state 4) was usually under 25 nmol O_2/mg protein/min. Respiration in the presence of uncouplers such as FCCP [carbonyl cyanide *p*-(trifluoromethoxy)phenylhydrazone] or DNP (2,4 dinitrophenol) usually was two to three times higher than the ADP-stimulated respiration rates.

a functionally intact respiratory chain and are able to couple oxidation of fuel molecules to the production of ATP[3]. The respiratory and phosphorylating activities of the isolated mitochondria from heater tissue are shown in Table 3. Heater mitochondria exhibit a reasonable capacity for carrying out oxidative phosphorylation. ADP/O ratios, which measure the dependence of the rate of substrate oxidation on the presence of a phosphate acceptor such as ADP, approach values expected of coupled mitochondria. These results indicate that under normal physiological states, electron transport is probably coupled to phosphorylation and substrates are oxidized when a phosphate acceptor such as ADP is available for phosphorylation to ATP. The heater mitochondria seem to have a normal capacity to establish an electrochemical gradient across the inner mitochondrial membrane.

Heater cells share many morphological similarities to the heat generating cells of brown adipose tissue in mammals[77]. Mitochondria dominate electron micrographs of both tissues but only the heater has extensive SR and T tubule membranes between the mitochondria. Lipid droplets are conspicuous features of brown adipose cells, but in the billfishes, lipid is only found in large quantities in adult Pacific swordfish (but not juvenile) heater cells. The similarity in function between these two furnaces provides some reason to suspect similar mechanisms for generating heat. However, the evolutionary origin from different tissues, muscle in the case of the billfish heater and adipose for the mammalian version, suggests caution when looking for biochemical similarity[4]. Thermogenesis in the mammalian brown fat is based on a 32 kDa uncoupling protein unique to brown fat mitochondria[77]. Polyclonal antibodies raised against hamster uncoupling protein do not recognize any epitopes in the billfish mitochondrial fraction[3,4]. Although the morphological similarities between the two furnaces are significant, the biochemical strategies for generating heat are most certainly different.

5. Excitation–thermogenic coupling in heater cells

To understand how billfish have converted muscle into a furnace it is necessary to examine which molecular properties of a muscle fiber are best suited for generating heat. The loss of the contractile filament fraction of the heater cell indicates that thermogenesis is not based on the heat released during cross-bridge cycling[4]. Instead, hypertrophy of the membrane volume of the heater cell in istiophorid billfishes focuses attention on the energy dependent process of calcium release and re-uptake. Central to this process are two molecules, the SR calcium ATPase and the SR calcium release channel.

In vertebrate skeletal muscle, contraction and relaxation are regulated by the cytoplasmic calcium concentration which, in turn, is controlled by the SR and T tubule membranes[44,45]. The SR of skeletal muscle is a highly organized intracellular membrane system that plays a crucial role in calcium uptake, release and storage. Several key proteins found in the SR membrane are required for cycling of calcium in and out of the SR, and the particular isoform or quantity of these proteins expressed in a given muscle cell type determines the speed at which calcium can be removed or released from the sarcoplasm. Calcium is pumped into the SR by the calcium ATPase (calcium pump protein) which is the major protein constituent of the SR bilayer[44]. This enzyme uses energy derived from ATP hydrolysis to generate significant calcium ion gradients between the interior of the SR and the muscle cell cytoplasm. Release of calcium from the SR follows depolarization of the transverse T tubules. This coupling of the depolarization and calcium release events is termed excitation–contraction (EC) coupling and has been studied for many years[81]. Only in recent years has a molecular basis for nervous control of muscle contraction been elucidated in great detail[10,44,45,91,92,103]. Calcium release is now known to occur at morphological sites where the T tubules and SR form junctions. A direct interaction between T and SR occurs at these specialized junctional regions, called triads, which contain the proteins mediating the process. At the triad, a molecular bridge (Figs. 10 and 11) between the T and SR membranes[10,45] is composed of two molecules that together regulate the release of calcium; the T tubule voltage sensor (dihyropyridine receptor) and the SR calcium release channel (ryanodine receptor).

The SR calcium release channel is a large tetrameric protein (composed of 5037 amino acid residues) that has an SR membrane domain which forms the actual calcium pore, and a cytoplasmic domain which stretches across the 10 nm gap between SR and T tubules[10,45,91,103]. In skeletal muscle, the signal for Ca^{2+} release is initiated at the neuromuscular junction[45,79,81]. The action potential then spreads down the T tubules and is presumed to initiate a calcium release event from SR by direct mechanical coupling of the molecules spanning the triad junction[10,44,79,92]. The exact mechanism of signalling between the T tubule voltage sensor and the SR calcium release remains to be determined. The leading hypothesis suggests that depolarization of the T results in a conformational change in the voltage sensing molecule (the dihydropyridine receptor of T tubules) which directly gates calcium release via the SR calcium release channel[81,91,92]. Such direct molecular coupling

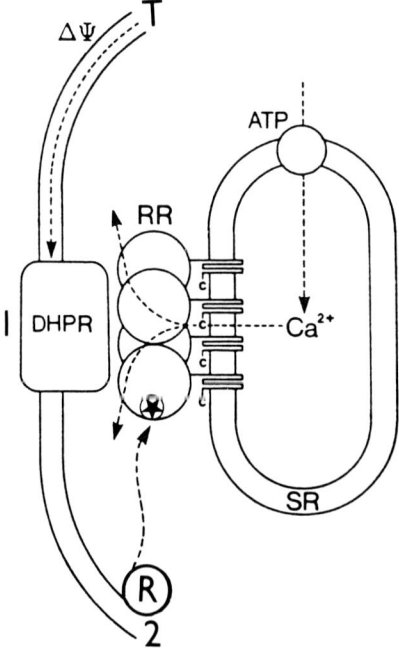

Fig. 10. Diagram of the transmembrane topology and molecular architecture of the two molecules involved in calcium release and reuptake at the T tubule/SR junction based on molecular structure revealed from morphological and sequence data. The large foot portion of the SR calcium release channel or ryanodine receptor (RR), spans the gap between SR and T tubule and presumably receives a signal from the T tubule voltage sensor molecule (1; DHPR) that results in opening and closing of the channel spanning the SR bilayer. An alternative hypothesis (2) is that depolarization results in release of inositol triphosphate from the T tubule membrane which then binds and opens the SR calcium release channel.

between T and SR provides a precise means for transducing the depolarization event into a tightly regulated calcium release signal and explains the tight coupling between membrane potential and tension in skeletal muscle. Alternative hypotheses invoke slower mechanisms involving a possible chemical transmitter[94]. The transmitter substance is believed to be released from T tubular membranes upon depolarization (Fig. 10). Binding of the transmitter to the cytoplasmic domain of the SR calcium release channel results in calcium release.

5. Cycling calcium for heat production

In heater cells, thermogenesis appears to be regulated by the uptake and release of calcium. To examine the molecular properties of the SR in heater cells for comparison with normal muscle a relatively simple procedure is used (Fig. 12). Biochemical analysis of the SR membrane fraction of the istiophorid heater cells indicates that the most abundant protein is the Ca^{2+} ATPase[8] (Fig. 13). Calcium-stimulated ATPase activity and phosphoenzyme formation are indicators of the

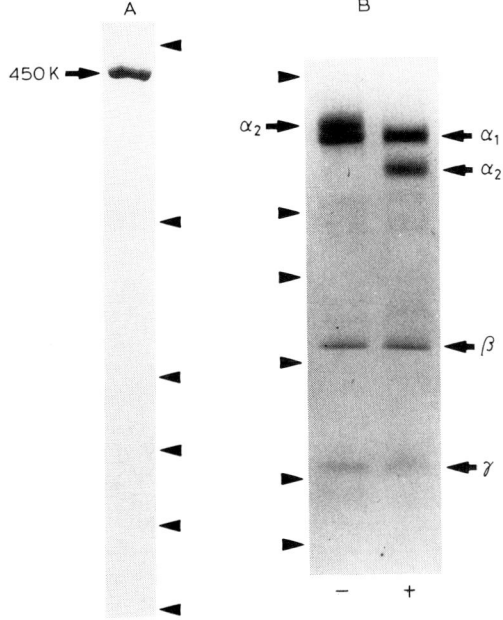

Fig. 11. Purified SR calcium relase channel (A) and purified DHP receptor (B) from rabbits were subjected to SDS-PAGE and stained with Coomassie Blue. The arrow (450 K) indicates the position of the 450 kDa release channel. B: 15 μg of purified DHP receptor separated on a 5–16% gradient gel under non-reducing (−) or reducing (+) conditions. Subunits of the DHP receptor are indicated by arrows; Molecular mass standards (nebulin, 500 kDa [A only], myosin, 200 kDa, phosphorylase b, 97 kDa; BSA, 68 kDa; ovalbumin, 45 kDa; chymotrypsinogen, 25 kDa; are indicated by arrowheads in A and B (from ref. 8).

activity of this enzyme and have been found to be similar to skeletal muscle[9]. Additionally, the calcium coupling ratio (calcium ions pumped per ATP) is low when compared to normal muscle, indicating a less efficient (higher ATP utilizing) isoform within the heater. Thus, more heat will be generated per Ca^{2+} ion translocated across the SR membrane. Although activity of the pump is not extraordinary the concentration of pumps within the heater cells is high[8]. Heater tissue in blue marlin has approximately 34,000 Ca^{2+}ATPases per μm^2 and up to 25% of the cell occupied by the SR membrane system[8]. The large cytoplasmic head of the molecule protrudes above the SR bilayer and can easily be visualized by freeze-fracture of the heater cells. Fig. 14 reveals the dense covering of 12 nm particles on the cytoplasmic leaflet of the heater SR which correspond to the tightly packed Ca^{2+} ATPase molecules. The presence of numerous mitochondria capable of generating ATP, and a membrane system loaded with a calcium activated ATPase, has led to the hypothesis that 'non-shivering thermogenesis' in the heater cell occurs via a calcium-mediated thermogenic cycle.

A model of the cellular basis for thermogenesis in the heater puts the emphasis of the heat producing cycle on the ATP dependent transport of calcium ions across

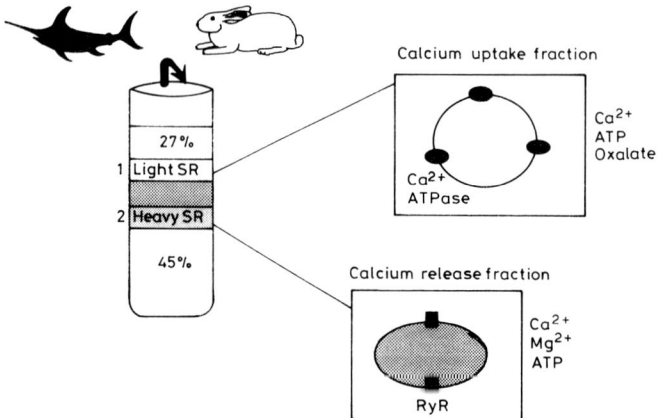

Fig. 12. The protocol for examining the calcium transport capabilities of the fish SR are similar to those developed for examining rabbit SR fractions[9]. Biochemical properties of SR membrane are examined by gentle homogenization of muscle followed by differential and sucrose gradient fractionation of the crude membrane fraction. Light SR contains only the calcium ATPase while heavy SR contains the terminal cisternae rich in calsequestrin and the SR calcium release channel. Physiological properties of the release and reuptake mechanisms can be studied in isolated vesicles in the presence of calcium, ATP and other compounds.

the SR bilayer by the calcium pump (Fig. 15). The stimulus for ATP hydrolysis and hence heat production is the presence of calcium in the cytoplasm. Essential to this model for non-shivering thermogenesis is the ability to regulate the release of calcium. 'Excitation–thermogenic coupling' may occur whereby 1) a nervous impulse results in calcium release and 2) as a result of this calcium release event, heat is produced via a calcium stimulated process (either ATP utilization by the pump or proton-calcium exchange by the mitochondria). The hypothesis that such a mechanism would exist in the heater cell is based on the notion that within a normal skeletal muscle fiber a precise mechanism for translating a nerve depolarization into a calcium release event is present. In normal muscle, firing of the motor nerve supplying the muscle fiber leads to T tubule depolarization, and a possible conformational change in the T tubule voltage sensing molecule (DHP receptor), which then gates the opening of the SR calcium release channel[92]. The subsequent rise in cytoplasmic calcium initiates cross-bridge cycling and force production. Pumping of calcium back into the SR via the Ca^{2+} ATPase, and the lowering of the cytoplasmic calcium concentration, inhibits the interaction between the contractile filaments.

In heater cells the network required for the physiological transmission of a nervous impulse into a calcium release event is also present. An extensive T tubule membrane network and neuromuscular junctions (Fig. 16A,B), principal components for *nervous regulation* are present in the heater cell, indicating that thermogenesis is mediated by the same neuromuscular pathway found in normal skeletal muscle. Stimulation of the oculomotor nerve should lead to T tubule depolarization and presumably calcium release, much as it does in normal muscle. The

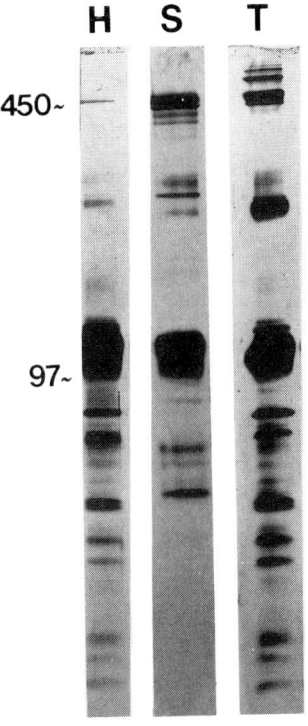

Fig. 13. SDS-PAGE 5–15% gradient gel of the crude membrane fraction of a fast-contracting vertebrate muscle (T, toadfish swimbladder muscle), white muscle (S, shark) and heater tissue (H) revealing the proteins involved in calcium release and reuptake. Heater and toadfish contain the most prominent protein bands at 100 kDa corresponding to the calcium ATPase (phosphorylase b standard marks 97 kDa). The high molecular weight protein at the position above the 450-kD arrow corresponds to the SR calcium release channel. Toadfish muscle which contracts at 400 Hz is enriched in the pump and the release channel while heater tissue has much more of the pump than the high conductance calcium release channel.

presence of calcium in the cytoplasm would stimulate hydrolysis of ATP, pumping of calcium back into the SR, and heat (and aerobic metabolism) as the byproduct of this activity. Heat production would continue in the heater cell as long as the motorneuron fired and an ensuing calcium release event occurred. The major point of difficulty with this model is understanding whether or not the heat is a result of the translocation of calcium ions into the SR by the ATP dependent pump, or is a result of calcium ions short circuiting the mitochondrial proton gradient[73]. In this latter scenario, the SR would take on its more customary role as the storehouse of calcium and release would stimulate utilization of the proton gradient across the inner mitochondrial membrane and subsequent oxidation of substrates. Calcium cycling at the inner mitochondrial membrane is a well documented energy-dependent process that can take precedence over oxidative phosphorylation. The role of calcium ions in regulation of oxidative processes within fish mitochondria is poorly understood. However, it should be emphasized that the

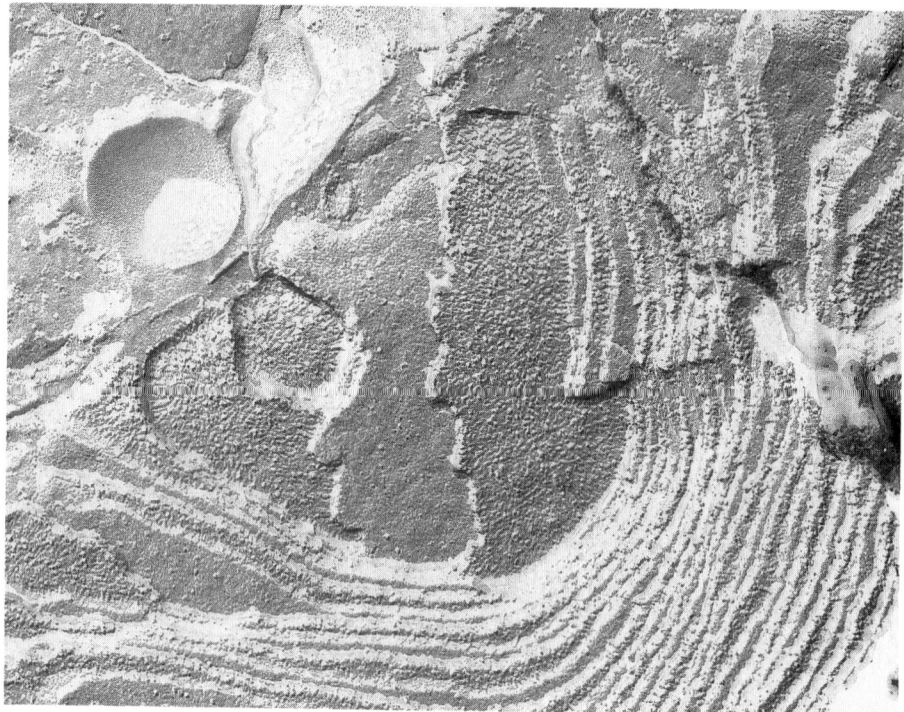

Fig. 14. Freeze fracture through a pancake-like stack of SR membranes in the heater cell reveals the numerous intramemranous particles on the cytoplasmic leaflet which represent the Ca^{2+}-ATPase molecules. Pumps appear like pebbles on the cytoplasmic leaflet and are absent on the luminal leaflet.

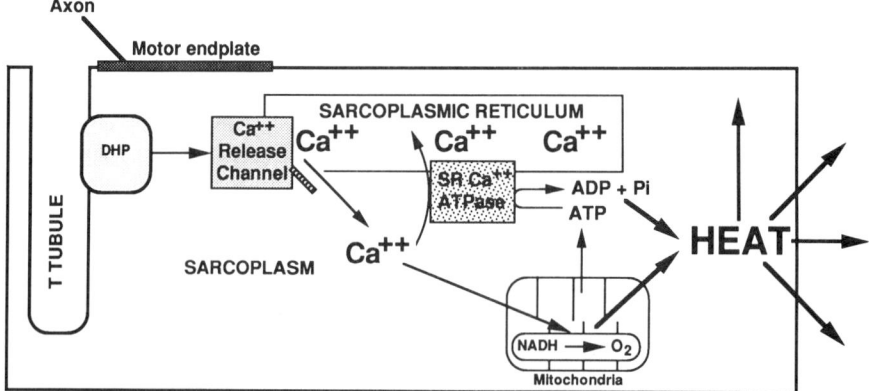

Fig. 15. Excitation–thermogenic coupling in the heater cell. This hypothesis suggests that a nervous impulse stimulates thermogenesis via the same molecular components found in the EC coupling pathway. Heat would be produced in this scheme during prolonged calcium release events due to cycling of calicum at the SR. The high mitochondria and myoglobin content provides ample ATP and oxygen for the calcium mediated thermogenic cycle.

two modes of heat generation are not mutually exclusive and the main point of the model for thermogenesis is that neurally controlled release of calcium ions regulates the thermogenic cycle.

As in normal muscle, a controlled release and reuptake of calcium requires that the molecules (the SR calcium release channel and T tubule voltage sensor) involved in excitation–contraction coupling be present in the heater cell. Heater cells have a low concentration of the SR calcium release channel and no information is yet available on the T tubule voltage sensor[9]. In normal muscles, the calcium release channel is found in abundance in fast-contracting fibers where speed is important. Ample sites for calcium release naturally increase the ability to saturate the cytoplasm with calcium quickly. Slower contracting fibers (oxidative or tonic fibers) have fewer sites for calcium release and thus activation times are longer[45]. The low amount of the SR calcium release channel in the heater cell emphasizes the difference between exciting a muscle for heat production rather than force generation. Initiating a synchronous contraction event throughout a muscle fiber selects for speed while turning up a furnace only requires the initial stimulation and presumably prolonged duration of the response. In redesigning a muscle fiber for heat production purposes, it appears that the ability to turn on the furnace via the same nerve-to-muscle pathway found in all skeletal muscles has been retained. The most important point is that this allows for controlled thermogenesis based on the same ATP dependent calcium release and reuptake pathway inherent in skeletal muscle.

The above results suggest that major 'rewiring' was not necessary for the transition from a force generating to heat generating cell. Instead, one can postulate that functional differences are associated with molecular regulation of either the T tubule voltage sensor, the SR calcium release channel or the SR calcium ATPase. One hypothesis is that there are molecular differences between the SR calcium release channel in the heater and normal muscle and that nervous stimulation leads to locking open of the heater calcium channel. A channel with prolonged open states would result in a continuous leak of calcium and constant stimulation of the energy dependent calcium translocation processes. An intriguing parallel exists in mammalian muscle. The site of a mutation in malignant hyperthermia, a fatal heat producing muscle disease, has recently been mapped to a chromosomal location close to the SR calcium release channel indicating that leaky SR membranes may also be responsible for intense heat production in mammalian muscles[72].

In summary, structural and biochemical results indicate that a heater cell is a modified muscle fiber with hypertrophy of the SR, T and mitochondrial volumes along with a loss of the contractile filaments. Neural control of the heater has been retained as evidenced by the extensive T system and the presence of neuromuscular junctions. The components for excitation–thermogenic coupling are present in the heater cell and represent a controllable neural pathway for initiating calcium release and reuptake, resulting in heat production. Release of calcium in the modified muscle cell does not result in contraction due to the absence of actin, myosin and troponin, however it should stimulate the pumping of calcium back into the SR at the expense of ATP. ATP would be supplied continually by the high oxidative potential of these cells. It is also possible that the presence of calcium activates the calcium/proton transporter of the mitochondria. As long as calcium is leaking into the cytoplasm, the heater cell could generate heat as a byproduct of both of these processes. It is difficult to know at this point which process should be emphasized: the control afforded by a thermogenic cycle dependent upon calcium release and reuptake (hence the mitochondrion supplies ATP to the pump), or the oxidative potential of the rich mitochondrial pool (the pump is present for regulating calcium release and reuptake and supplying calcium to the mitochondria). The enormous concentration of SR, and concomitantly high pump and calsequestrin concentrations favors, emphasis on the calcium pump. However, it is hard to believe that the oxidative potential of the mitochondrial pool would not be harnessed directly. The high mitochondrial volume and myoglobin content are indicative of high oxidative capacity in this tissue, indicating that heat production could continue for long periods. Sustained performance of the heater is clearly present in swordfish which dive to great depths yet maintain an elevated brain temperature for many hours (see Figs. 1–4).

6. Ecological implications of heater organs

The major morphological difference between heater organs of various billfishes is their size. The variation in size of these organs may correlate with thermal ecology of the species[6]. Fish spending more time in cold waters have more of the superior rectus muscle converted to the heater phenotype[5,6]. The heater is largest in *Xiphias*, where only a small portion of the superior rectus muscle remains intact. Additionally, in the swordfish there is the loss of the basisphenoid bone which forms the base of the braincase[19]. This allows for direct contact between the brain and the two lobes of the heater organ from the muscles on either side of the head. The large size of the swordfish heater organ coupled with loss of the basisphenoid bone at the base of the braincase are features which maximize warming of the brain. Among the istiophorid billfishes there is a gradient of expression of the heater phenotype within the extraocular muscle with more of the heat-generating cell-type expressed in *Makaira* and less in *Istiophorous*[3,5]. When comparing the behavior of the swordfish observed with acoustic telemetry versus that obtained from *Makaria* or *Tetrapturus*, it becomes evident why the furnaces are built with different overall dimensions. Swordfish spend 12 h on a diurnal cycle in water as cool as 5°C at depths of 600 meters. On the other hand, blue marlin only occasionally dive through the thermocline and spend most of their time in the top 210 meters of the water column (Fig. 17). The blue marlin ranges through water temperatures of 17–27°C while the swordfish encounters temperatures from 5 to 26°C. Thus, it comes as no surprise that the swordfish requires a larger furnace to keep the brain and eyes warm.

One striking difference between the swordfish and marlin heater organs is the rich supply of potential fuel in the former. Lipid droplets are evident in the adult swordfish heater cells but are less commonly seen in the istiophorid heater. The 12

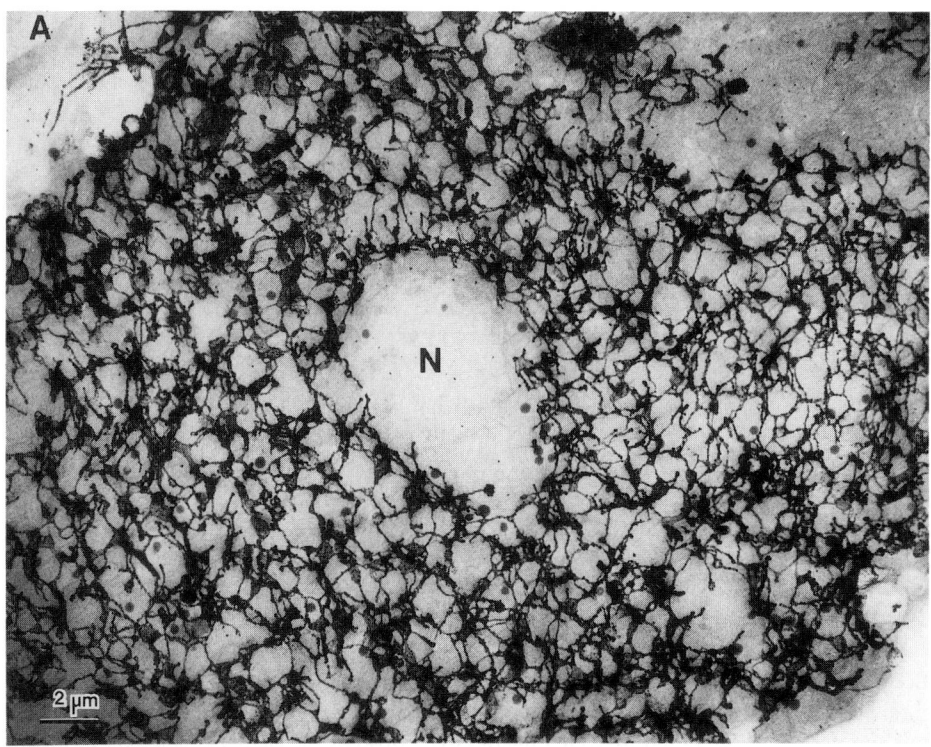

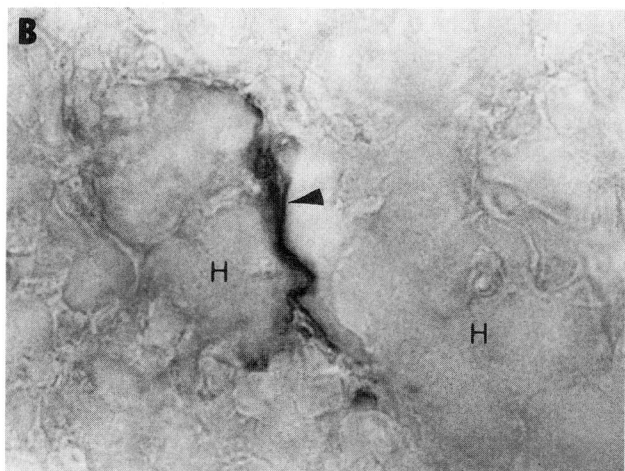

Fig. 16. A: Golgi-stained thick sections of heater (a and b) reveal the disposition of the T tubule network in both of these cells. In muscle the tubules course transversely between the myofibrils whereas in the heater cell the T tubules take a circular route around the numerous mitochondria in the cytoplasm. In both cells the T network acts as an electrical cable system propagating the action potential throughout the cell (from ref. 8). B: light micrograph of a neuromuscular junction in the heater cell. Biotin conjugated α-bungarotoxin labelled with a chromagen tag was used to localize the junction in heater. Arrows indicate junctional area. H = heater cell.

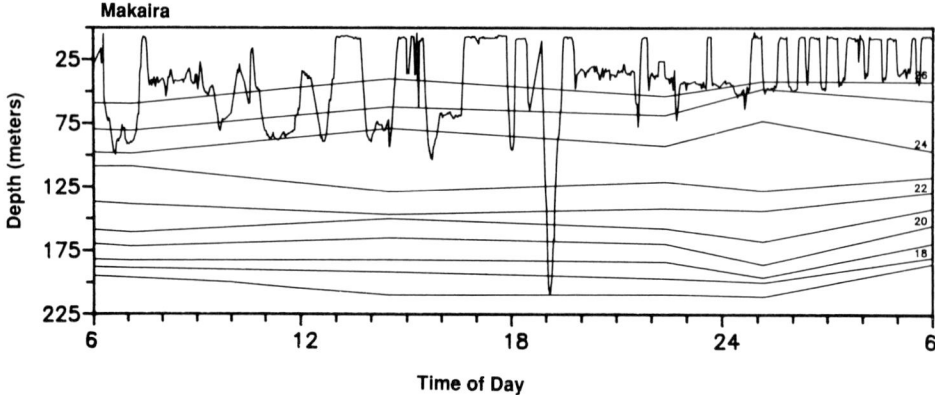

Fig. 17. Depth record from telemetry of a blue marlin off the coast of Hawaii. Marlin have a warmer water preference (17–27°C) when compared to swordfish (Fig. 2) and tend to stay above the thermocline.

h period that the swordfish spends daily in cold water requires continuous heat generation and the large stores of lipid in the cells presumably fuel this process. Juvenile swordfish (1–2 kg) have less intracellular lipid stores which may indicate that these small swordfish spend less time at depth and rely more on fuels in the blood. Surprisingly, the juvenile swordfish heater cells have a striking similarity to blue marlin, perhaps due to the similarity in ecological behavior between marlin and the young swordfish which are more surface dwelling as evidenced by stomach contents and recent telemetry results from both species (ref. 12 and Block, personal observation). The istiophorid heater of marlins and spearfishes has little in the way of an obvious fuel source within the cells (either glycogen or lipid), however, the ability of the heater cells to metabolize lipids and carbohydrates is very high (Tullis, Block and Sidell, unpublished data) indicating an opportunistic utilization of available fuels from the blood. Oxidative enzyme profiles and fuel preference in the heater tissue of *Gasterochisma*, *Xiphias*, *Makaira* and *Tetrapturus* are presently being examined and preliminary results suggest correlations of enzymatic activity with the thermal ecology of the species.

7. Heater organs in scombrid fish

The Scombridae are the most closely related outgroup to the monophyletic assemblage consisting of the Istiophoridae and Xiphiidae. Although tunas have the capacity to warm the brain and eye region, this capacity is not related to a specific heat-generating tissue[70,86]. As previously mentioned, a heater tissue modified from an eye muscle has been found in a primitive scombrid fish, the butterfly mackerel, *Gasterochisma melampus*. Understanding the appearance of the heater complex in this species is critical to the formation of a hypothesis for the evolution of endothermy among scombroid fishes. This unusual fish, which was once called the scaly tuna due to the large cycloid scales covering the tunniform body, is an active,

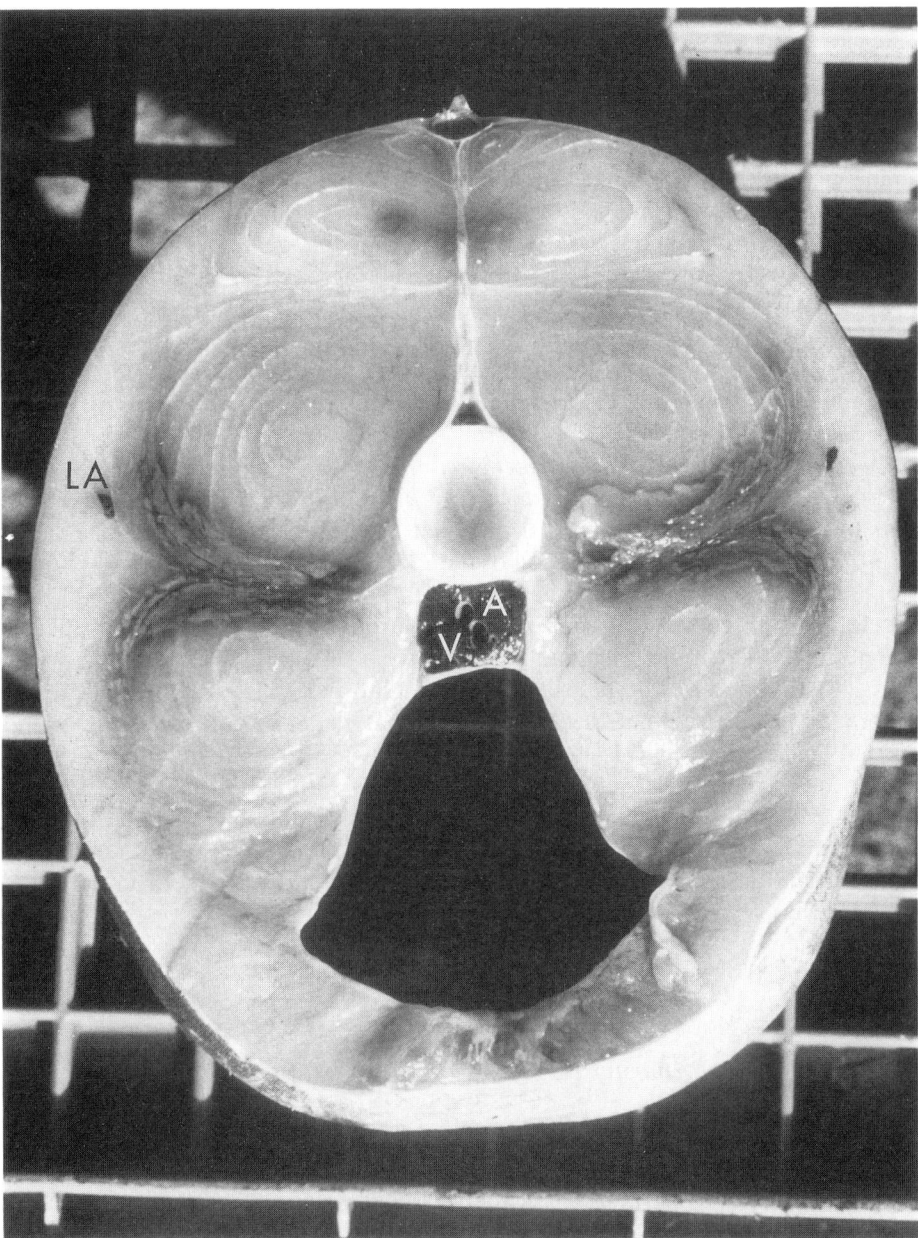

Fig. 18. Cross-section of swordfish taken near the posterior end of the body cavity. The red muscle is located well in from the surface. It has two arterial blood supplies: one through a rete (not visible in this photo) from the dorsal aorta and the second through a black pigmented artery with an extraordinarily thick, muscular wall. The thick walled artery can be seen lateral to the wings of the red muscle. A = aorta; V = postcardinal veing; LA = thick-walled lateral artery (from ref. 21).

predacious fish found in the cold waters of the Southern Ocean. The taxonomic status of *Gasterochisma* has been troublesome but, whether considered a member of the Scombridae or a monotypic family within the scombroid assemblage, it is clearly a primitive member of this clade.

The heater in *Gasterochisma* is large and is notably different from the billfishes, complementing the phylogenetic placement of this fish outside the xiphioid and istiophorid lineages. Expression of the heat-generating fiber type within the lateral rectus muscle brings about positional changes in the heater organ beneath the brain and a different association with the circulatory system. The counter-current heat exchanger is supplied with blood from a vessel off the lateral dorsal aorta which is more caudal in origin than the carotid supply of blood to the billfish heat exchanger (Finnerty, personal communication). As in the billfishes, the bilaterally situated heater organs converge along the midline but in *Gasterochisma* the furnace is situated behind the braincase.

Electron micrographs of *Gasterochisma* heater cells look similar to the istiophorid heater with mitochondria dominating the picture. Transitional cells with remnants of myofibrils scattered throughout the cytoplasm are more frequently found in the heat generating portion of the *Gasterochisma* lateral rectus. As in the istiophorids this is a direct indication that there is differentiation of skeletal muscle fiber precursor cell such as a myoblast into a heater cell. SDS-gels of heater tissue membrane fractions prepared from *Gasterochisma,* using the same protocol developed for the billfishes, indicate that the SR Ca^{2+} pump is present in the protein profile of the heater organ. Preliminary quantification of the amount of calcium ATPase in the *Gasterochisma* heater suggests a lower amount of pump than in other billfishes. Association of the heater with extraocular muscle, presence of numerous mitochondria within the heater cells and localization of calcium ATPase in the *Gasterochisma* heater membrane fraction suggest that there may be similarities in the molecular mechanisms for building a heater out of muscle in all three families of scombroid fishes. The profile of oxidative and glycolytic enzymes of the scombrid heater organ indicates that *Gasterochisma* has the highest oxidative capacity of all fish examined (*Xiphias, Makaira, Tetrapturus*).

III. Regional endothermy versus whole body endothermy

Istiophorids are the largest teleost fish in the ocean and forage for food in the epipelagic zone. These fish travel long distances (8,000 km or more) as they cross the ocean in yearly migrations. Telemetry studies on free-swimming blue marlin indicate that they prefer to remain in the top 210 m of the water column[12,102]. The blue marlin spends more time above the thermocline than swordfish and only occasionally do they dive into cooler waters. Blue marlin feed most often on endothermic scombrids and one of their major predators is the warm-bodied mako shark.

To examine why selection favored regional endothermy in the billfishes rather than whole body endothermy as in the tunas and lamnid sharks, it is necessary to consider the differences between these three groups. Most importantly, whole body endothermy in modern billfishes is constrained by their body plan. Numerous features of billfishes appear convergent with tunas such as the streamlined body design, the rigid, lunate tails, the presence of caudal keels and the ability to retract fins into grooves. The overall similarity in external design has lead to the assumption that billfishes and tunas locomote in a similar manner and are often assumed to be tunniform swimmers with a propulsive wave of low amplitude and high frequency[95]. More recent direct observations of swimming blue marlin combined with biomechanical observations indicate that they swim with a sinuous motion with lateral bending occurring throughout the length of the fish[56]. This difference in swimming performance is reflected in the distribution of the aerobic red muscle mass which powers sustained cruising.

Red muscle fibers comprise 5–6% of the total muscle mass in striped marlin[37] and these fibers are widely dispersed throughout the myomeres. This is similar to the proportion of red fibers found in other sub-carangiform swimmers but significantly lower than in most of the endothermic tunas (4–13%[52]). Additionally, the istiophorids lack the internal positioning of red muscle found in tunas. Although billfishes have a smaller proportion of their propulsive muscle mass associated with red aerobic fiber types, the activity of oxidative and glycolytic enzymes within the muscle fibers is similar to warm scombroid fishes[90]. This enzymatic study concluded that blue marlin red and white muscles have an oxidative potential that is similar to the tunas yet remains unexploited for endothermic purposes.

Oxidative capacity of red muscles in many fishes is similar to that of the endothermic tunas[83,90] and surprisingly close to that of birds and insect flight muscle. When comparing red fibers used for prolonged swimming in fish with red fibers from terrestrial animals traditionally considered to be highly aerobic, it can be seen that the mitochondrial volumes are similar (Table 1). Similarly, spectrophotometric determination of myoglobin concentration in tuna and marlin red muscle indicates a two- to five-fold higher myoglobin concentration than in flight muscle from birds or slow oxidative muscles from terrestrial mammals (Table 2).

Endothermy requires a high capacity for metabolic heat production as well as a system for conserving metabolic heat. The muscles that power active fishes through water, a medium of relatively high viscosity and density, have an enormous oxidative potential, a necessary precursor to the evolution of endothermy in scombroid fishes. The use of an alternative strategy in the billfishes emphasizes one of the most important specializations required for whole body endothermy in teleosts. Although many fish have a high aerobic metabolism associated with the red muscle mass (in turn reflecting the requirements for sustained power from a small proportion of the total muscle mass), the major requirement for whole body endothermy is the reorganization of the epaxial musculature. Few studies have examined the biomechanical importance of the movement of the red muscles into a central position[78] and it would be interesting to find out if there are ergometric advantages.

Although the swordfish appears externally similar to the istiophorid billfish and much of the epaxial musculature is composed of white fibers, there are important

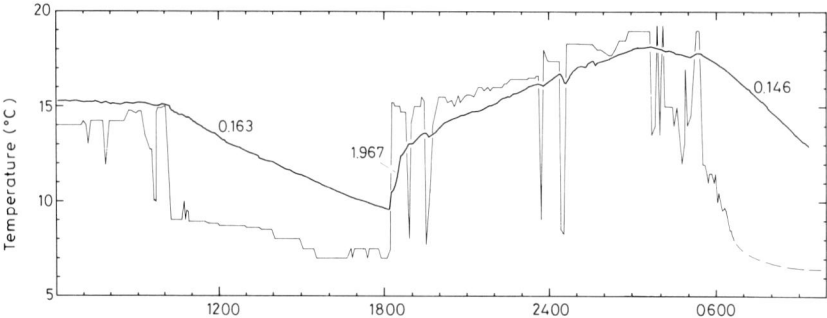

Fig. 19. Muscle temperature of a swordfish, heavy line and water temperature light line. When this first went below the thermocline at 10.00 h the first day, it cooled slowly. Muscle temperature gradually declined from 6 to 3°C above water temperature over an 8 h period so that the fish stayed significantly warmer than the cold bottom water. The muscle rewarmed very rapidly when the fish came back up thorugh the thermocline at 18.00 h. The rate of warming was more than 10 times as fast as the rate of cooling. When the fish entered cold water at 06.00 h on the second day, it again cooled very slowly (from ref. 21)

differences involving the positioning of the red muscle. The swordfish is unique among billfishes in having a small wedge of internally located red muscle[21] (Fig. 19). A simple rete mirabilia with only a few layers of arteries and veins is found in the red muscle of swordfish. This internal positioning of the red muscle mass and heat exchanger coupled with the large body size of the swordfish enables these fish to benefit maximally from its thermal inertia when diving below the thermocline[21,76]. Carey[21] has recently shown that during vertical excursions of the swordfish, the muscle cools at a much slower rate than it warms (Fig. 20). Thus the swordfish is not only able to keep brain temperature elevated, but is also able to reduce the rate of heat loss from the swimming musculature as it forages in cold waters.

IV. Behavioral thermoregulation: the importance of thermal inertia and large body size

Endothermy in teleost fishes remains the domain of scombroid fishes and most of the members of this clade are large in body size. In the evolution of endothermy, large body size and the ability to use thermal inertia to reduce thermal conductance was an essential factor contributing to the initial retention of heat. In large fish the only important route of heat transfer is convection by the blood. The development of heat exchangers to reduce convective heat loss through the circulatory system was an essential factor enabling these fish to retain heat. Recent telemetry experiments with swordfish and blue sharks[21,31] demonstrate the remarkable ability for behavioral thermoregulation in these fish (Figs. 20 and 21). As in terrestrial ectotherms, behavioral exploitation of the thermal environment results in elevation of body temperatures above ambient. These fish and most certainly others, use heat acquired at the surface in the mixed layer to stay significantly warmer than the cold water below the thermocline. Basking in the warm surface

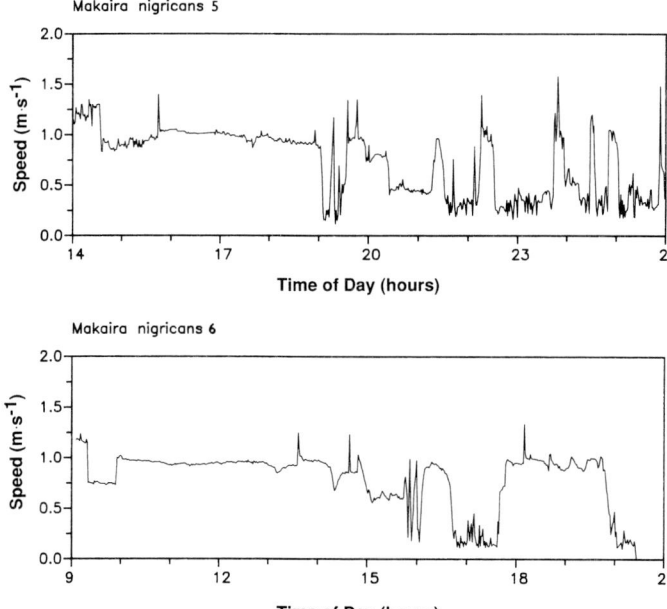

Fig. 20. Direct telemetry of swimming speed from blue marlin, Makaira nigricans, indicates cruising speeds of 1 m/s.

waters permits heat gain from the environment and retention of this heat is facilitated by reduced thermal conductance due to large body size and by reduced convective heat transfer by the presence of heat exchangers[21,47].

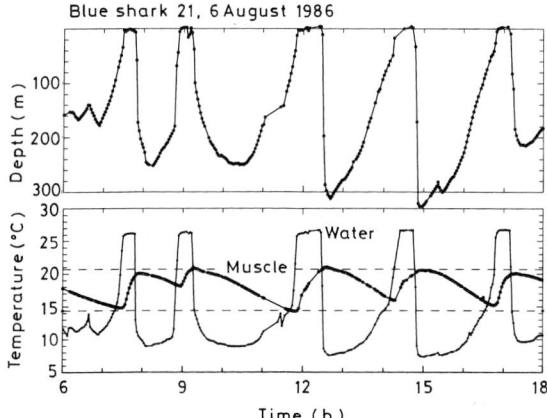

Fig. 21. Depth and temperature records for the blue shark, *Prionace glauca*. The muscle warmed faster than it cooled. As a result, average muscle temperature was 4°C above average water temperature during the period illustrated (from ref. 32).

Both swordfish and blue sharks make long vertical excursions to depths over 600 m. The swordfish regularly remains at depth during daylight hours for up to 12 h but the blue sharks swim up and down every few hours[26,32]. In the swordfish, circulatory specializations associated with the red muscle appear to play a role in allowing the muscle to cool at a rate up to 10 times more slowly than warming. The ability to regulate circulation slows down the rate of cooling of the musculature when the fish is below the thermocline, but promotes rapid rewarming upon return to the surface[26]. Behavioral thermoregulation in the swordfish, when combined with physiological mechanisms for retarding heat loss, provides a greater degree of freedom from ambient water temperature than in the blue shark. Internal localization of the red muscle and the presence of a small heat exchanger in its blood supply, along with the brain heater, are specializations that permit the swordfish to stay down in cold waters and forage for more prolonged periods.

Blue sharks appear to acquire heat at the surface, retaining it as they dive and coming back to the surface when the muscle temperature becomes substantially cooler[32] (Fig. 21). The shark data indicate a much more repetitive pattern of ascent and descent which may be due to the lack of a heat exchange system within the red muscle. Perhaps most intriguing is the fixed pattern of coming back to the surface before the muscle temperature drops below a critical point (15°C). Muscle temperature and water temperature in blue sharks are tightly coupled; these are ectothermic sharks. However, results of Carey and Scharold[32] demonstrate that muscle cools slowly, but warms rapidly during the repetitive dives. As in the swordfish, the slow cooling and rapid warming imply some circulatory adjustments[31]. This thermal hysteresis provides the shark with a significantly warmer muscle operating temperature for a few hours while in the cool waters at depth. Like the swordfish, sharks rewarm quickly when they return to the mixed layer. As a result of this behavioral thermoregulation temperature averaged 18°C, during a series of dives where the average water temperature was 13.7°C (Fig. 21). This 4°C elevation in muscle temperature relative to deep water is maintained by surface heat carried to depth and is achieved without any of the heat exchangers or elevated metabolic rates found in warm sharks or tunas.

Behavioral thermoregulation has been widely demonstrated among diverse terrestial ectotherms as the major strategy for regulating body temperatures. Therefore it is not too surprising to find that large fish use this strategy to regulate body temperature. Interestingly, demonstration of the importance of behavioral thermoregulation among the large pelagic fish provides a major clue in understanding the evolution of endothermic strategies amongst scombroid fishes and puts an emphasis on the importance of large body size for dampening temperature oscillations.

V. Endothermy in tunas and lamnid sharks: what's new?

Convergence between the specializations of lamnid sharks and the scombrid tunas for maintaining endothermy is remarkable. Lamnid sharks utilize the same strategies as tunas to elevate muscle, viscera and brain temperatures. Separate heat exchange systems are associated with each of these tissues and tissue temperatures range between 5 and 21°C above ambient water temperature depending upon the species involved[7,30,101]. Whereas it is obvious how the two groups are warming their muscles, the generation of heat in the head region remains more of a problem. The design and function of heat conservation systems consisting of counter-current exchangers in the head region have been described in both tunas and sharks[7,70,86]. In the lamnid sharks a flow of warm venous blood from the posteriorly located red muscle appears to contribute to warming in brain and eye. It is significant to note that all endothermic fish have a mechanism for warming the brain.

The purpose of the present chapter is not to review endothermic mechanisms in tunas, especially since this topic has received substantial discussion[20,25,39,49]. Instead the focus is on two questions: (1) Are tunas endotherms in the same sense as mammals and birds? and (2) Why are tunas warm?

The most significant results of the past decade of work on warm tunas are the papers demonstrating the elevated metabolic rates of tunas[15,48,50,51,53,54]. Many studies have attempted to ascertain the metabolic potential of tuna red and white muscles to ascertain whether the warmth of tunas was directly associated with a high metabolic potential of these muscles. The metabolic sources of heat and power in tuna muscles have been extensively examined using ultrastructural, biochemical and physiological approaches. Results of these studies indicate that, while tunas have extraordinary heat generating capacities, the highly oxidative red muscles are not exceptional when compared to other large ectothermic fishes such as the marlins. Thus, it remains a mystery how tuna achieve metabolic rates significantly higher than all other fishes.

1. Biomechanics of tunas and endothermy

One of the most important factors associated with endothermy in tunas is the positioning of the red muscle. What remains unusual and unexplained about the body plan of tunas is the placement of the deep red muscle along the vertebral column. This deep red muscle is the major source of heat in the tuna body plan and this is emphasized by the vascular counter-current heat exchanger in the blood supply to this internal furnace. Given that this muscle mass is highly oxidative and clearly the heat source for the temperature elevations apparent in tuna muscle, perhaps it is time to address why the red muscle is situated in this position. As discussed above, the combination of behavioral thermoregulation combined with a small amount of red muscle in an internal location retards heat loss from the epaxial musculature of the swordfish. Was the first step in the evolution of endothermy in tunas the movement of a small wedge of red muscle to an internal location? What advantage is there in moving the muscle internally? Is the 'motivation' biomechanical or is it thermogenic? The ability of swordfish to retard heat loss for up to 12 h with only a small amount of red muscle and a primitive heat exchange system indicates that perhaps the ancestor of modern tunas might have evolved along a similar path.

A major unanswered question concerning the red muscle in tuna is whether power output of the red muscle mass is affected by the new positioning. Is the frequency of contraction higher with the more internal localization of the red muscle mass? Perhaps this location results in a higher rate of firing and a more continual rate of heat production during sustained swimming. Graham et al.[52] recently examined the comparative distribution of red muscle in ectothermic and endothermic scombrids. They found a significant difference in the placement of the deep red muscle in the endothermic tunas to a more anterior as well as internal position. A fruitful set of experiments would be to record EMG's from the external red muscle, the deep red muscle and white muscle simultaneously while monitoring metabolic rate. Rayner and Keenan[78] pioneered this approach early on, yet few have followed up on the results.

VI. Why be warm?

The ability to regulate tissue temperatures above ambient is confined to the scombroid fishes and lamnid sharks. Many aspects of the behavior and lifestyles of these two groups are similar and convergence is apparent in external as well as internal features. Why is it that endothermy is confined to these large, predacious species and what are the selective factors influencing the evolution of endothermy in fish? Both of these questions have permeated the warm fish literature and several reviews on the subject are available[20,25,39,49]. In light of the recent acquisition of new data on the metabolism of tunas[15,48,50,51,53,54], the discovery of brain heaters in billfishes and a primitive scombrid fish, and the demonstration that behavioral thermoregulation plays a major role in the thermal biology of large fish, it is possible to re-evaluate the question of 'why be warm'?

1. Increased power output and warm temperatures

The question of why scombroid fishes and lamnid sharks are warm brings to bear many of the same sorts of speculation that arise in any discussion of the evolution of endothermy in the animal kingdom. The evolution of endothermy in birds and mammals has been the subject of considerable debate[1] and centers on whether selection is for a high body temperature and the adaptive benefits of being warm (independence from environmental temperature) or a result of an increase in aerobic capacities to support sustained activity. The evolution of endothermic mechanisms among the 27 species of pelagic fishes offers an additional perspective on the evolution of endothermy in vertebrates. The questions raised in discussions of the evolution of endothermy in higher vertebrates are no different than those raised in any discussions of why tunas and lamnid sharks are warm bodied.

Originally, Carey et al.[25] suggested that warm muscle temperatures enable tunas and lamnid sharks to swim at greater speeds. Thus warmer implied higher performance capabilities. The initial hypothesis was based on reports in the literature which indicated tuna speeds in excess of 70 km/h[96] combined with experimental results demonstrating the power available from a given muscle mass increases with temperature[83]. A substantial literature exists documenting swimming performance of fish at different temperatures. Both power output and sustained and burst swimming performance are affected adversely by cold temperature or acute temperature changes[83]. Most of the studies of swimming performance are laboratory experiments in which the fish are pushed to perform at their maximum capacity. Although warmer muscle generates more power, one important question to address is how relevant is this to fish if they are not performing at the maximum levels observed under experimental conditions?

Brill and Dizon[17] and Johnston and Brill[62] directly tested the effect of temperature on isotonic twitch in tuna muscles by examining the thermal dependence of contractile properties of red and white muscle from warm tunas. Results from both of these studies indicate that, while it is possible to gain some advantage in terms of contraction frequency, the calculated increase in maximum power output and swimming speed was not substantial enough to warrant a great deal of confidence in the hypothesis that being warm results in greatly improved swimming performance.

The similarity in external features apparent in many of the warm fishes such as a powerful lunate tail, caudal keels, pectoral fins designed for lift generation and fins that are retractable into grooves have been used to suggest that the streamlined designs were optimized for 'high-speed' swimming. Many ectothermic fish such as the wahoo (*Acanthocybium solandri*) and marlins also have these features but do not maintain warm muscles. At the same time, 'high-speed' swimming is a term often attributed to tunas despite the fact that there are few measurements providing detailed information about their sustained and burst swimming performance. The only measurements of actual high-speed swimming are the brief (20 s) recordings by Walters and Fierstine[96] who reported a maximum burst speed of 70 km/h for an ectothermic scombrid, the wahoo. The burst speed of a yellowfin tuna was not significantly different even though the tuna had a higher body temperature[17,88]. Thus, selection for an epipelagic existence, where migratory movements involve travel over long distances results in convergence in external morphologies that are known to reduce drag and increase efficiency in swimming.

One way to test the validity of the warmer-is-faster hypothesis would be to obtain simultaneous telemetry recordings of muscle core temperature and swimming speed. This would permit comparison of sustained and burst swimming speeds in endothermic and ectothermic fishes. Recent telemetry of swimming speed in free-swimming blue marlin, *Makaira nigricans*, indicate sustained swimming speeds (8–14 h) of 3–5 km/h[12]. Maximum burst speeds during these tracks ranged up to 8 km/h. Telemetry of body temperature indicates that blue marlin have muscle temperatures that are no more than 2°C above water temperature. Although sport fishermen are acutely aware that these fish can catch a bait trolled behind a boat at speeds in excess of 30 km/h, it is interesting that the free-ranging fish cruise at much slower speeds (Fig. 21).

Similar swimming speeds for blue marlin were observed by Yuen[102]. Tag and recapture data show good correspondence to these measurements of cruising speed. A black marlin, *Makaira indica*, was tagged in California and recaptured off Peru (5778 km distance) after only three months[37]. This gives a straight line speed of 61 km/day or 2.5 km/h. These reliable data are in sharp contrast to the suggested maximum reported speeds which permeate the literature. Estimates in the literature suggest that marlin can exceed 130 km/h[37,95]. Although it is certain that these fish can exceed 30 km/h the relevance of the extraordinarily high estimates of swimming speed to the actual daily swimming speeds of marlin remain questionable.

There is little in the way of direct measurement of sustained swimming speed available for free-swimming warm tunas. However, these fish are referred to as high performance animals capable of extraordinary feats. Until direct measurements are obtained and tunas proved to be faster than the sustained speeds of ectothermic billfish or the burst speed of ectothermic scombrids (such as the recordings from a wahoo) it is fruitless to suggest that warm tunas are faster than other fishes because they are warm. A record of a bluefin tuna tagged in the Gulf of Mexico and captured 50 days later in Norway suggests a swimming speed of 3 km/h. This speed is similar to the marlin cruising speed. Skipjack tunas have been observed travelling at speeds substantially greater than this for up to 1 h[15] and telemetry experiments also indicate higher sustained speeds for short periods. Estimates of maximal swim speed in scombrid fishes suggest that being warm enhances burst swimming performance. It is entirely possible that selection pressures in these fish were for faster burst speeds to escape predation or catch prey. The development of multiplex acoustic transmitters that permit recording of body temperature and speed simultaneously bodes well for obtaining new data on this question in the coming years.

2. High metabolic rates and endothermy

In the past decade physiological studies with albacore and skipjack tunas have increased in sophistication, enabling measurement of metabolic rate and blood gas parameters in actively swimming fish[15,41,53,64,98]. Such physiological studies have unequivocally demonstrated that endothermic tunas have metabolic capabilities close to those of mammals. This necessitates numerous specializations such as increased gill surface area[15], large blood volume, unusual hemoglobin and myoglobin properties[33,47,88], and high blood pressure and cardiac output[68,98]. Documentation of these specializations contradicts the notion that warm tuna have elevated body temperatures simply due to 'intense muscular activity'[1]. As in mammals and birds, tunas are endothermic because they have intrinsically high metabolic rates (linked in part to their large volume of aerobic musculature) and are able to regulate heat exchange with the environment.

High metabolic rate is a requirement for endothermy and future studies should address its mechanistic basis in tunas. Although there is an inextricable link between swimming performance and metabolic rate, the notion that high speed swimming or sustained swimming requires endothermy and an elevated body temperature is clearly not the case[39,40]. One possibility for further exploration is to resolve the endothermic and mechanical consequences of moving the aerobic mass to an interior position along the midline. One consequence that results from the interior positioning of the red muscle of tunas is the necessity for a higher frequency of contraction than external red muscle to generate the same power output as laterally placed red muscle. Perhaps the red muscle is in a biomechanically less efficient position (with ultimately higher work rate) and therefore produces more heat.

3. Oxygen and metabolite flux rates

An alternative idea on the adaptive advantage of being warm is that the warm temperatures increase metabolite and oxygen flux rates elevating metabolic rates in endothermic tunas to those more similar to mammals than fish[48,88,97]. Such a circular argument suggests that the increase in body temperature necessarily causes an increase in the metabolic scope and an increased potential for power output and hence performance. The increase in temperature has been postulated to accelerate the rate at which myoglobin delivers oxygen to mitochondria. Several components of myoglobin flux are temperature sensitive although an understanding at the molecular level of how temperature increases facilitate diffusion are not completely understood. What is clear, is that tuna muscle has extraordinarily high myoglobin and mitochondrial concentrations, and the ability to deliver oxygen and other metabolites such as ATP may be enhanced by the increase of temperature[83,84]. More recently, Weber et al.[97] demonstrated that skipjack tuna can support lactate and glucose turnover rates similar to those reported for mammalian species. The higher turnover rate of these metabolites most likely is directly associated with the increase in operating temperature of tuna red and white muscles. Brill[15] offers several ecological advantages related to foraging efficiency, linked to high metabolic rates and the ability to metabolize lactate quickly.

While it is clear that being warm is inextricably linked to the higher metabolism and that to fuel this high metabolism increases in flux of substrates and oxygen are necessary and that these attributes increase the potential for greater muscular performance, this explanation does not provide a strong selective argument for the appearance of endothermy in scombroids other than the tunas. Tunas and lamnid sharks are not only warming their muscles but keeping their brain, eyes, and viscera at elevated temperatures. The recent discovery that swordfish and other billfishes primarily elevate only their brain and eye temperatures, suggests that selection for increased tissue temperatures in the billfish centered upon warming of the central nervous system. As outlined below, the most unifying theory for understanding the evolution of endothermic mechanisms is based on the most important advantage of independence in movement afforded by endothermy.

4. Niche expansion as the driving force for endothermy

Carey et al.[25] and Graham[49] suggested warm bodied fish probably achieve a marked independence from environmental temperature permitting them to make rapid vertical and latitudinal migrations without the necessity of thermal acclimation. It is well established that acute temperature changes have detrimental affects on both central nervous system processes and contractile performance in muscle of fish[7,83]. The ability of many of the warm fishes to maintain constant deep body or cranial temperatures while moving rapidly through the thermocline is remarkable. Extension of the ranges of the Scombridae, Lamnidae, and Xiphiidae into cold temperate and polar seas where surface water temperatures are less than 15°C and at depth temperatures of 4°C are common is perhaps the most notable feature of these lineages.

Most fish are limited in their distribution and short term movements by temperature[83]. Acute temperature changes are known to have lethal effects on teleosts with the most sensitive organ system being the central nervous system. The aquatic environment is stratified into discrete thermal zones that are exploited by teleosts which have, in most instances, acclimated to their thermal niche. For many marine fishes, barriers to movement in the sea are most often defined by discrete masses of water with differing temperatures. Thermal acclimation is a slow process that requires compensatory changes at the molecular level that takes days or weeks to occur.

The recent discovery of brain heaters in the swordfish, billfishes and *Gasterochisma melampus* provides a new historical perspective recognizing that warming of the nervous system and ultimately niche expansion are the driving selective forces in the evolution of endothermy among scombroid fishes. Phylogeny suggests that brain heaters are more primitive than whole body endothermy in fishes. This is especially apparent in *G. melampus*. The morphology and the mtDNA data on this scaly scombrid corroborates its phylogenetic position as a primitive *scombroid* fish. The appearance of heater organs in this species is the strongest indication that regional endothermy was a first approach at exploiting colder habitats.

Regional endothermy, combined with behavioral thermoregulation is most certainly an attempt to gain one of the main advantages of being warm (independence from ambient temperature) at a lower energetic cost. Brain heaters buffer the central nervous system from deleterious effects of cold and allow fish such as the swordfish and *Gasterochisma* to forage for food in very cold waters. These fish are constrained from using whole body endothermy because they lack the vascular heat exchanger and the large volumes of red muscle. The relationships of scombroid fishes (Fig. 1), and in particular the position of *Gasterochisma*, indicate that this form of endothermy *has to be more primitive* than whole body endothermy of tunas. Thus, regional endothermy must be considered *the first attempt* in the scombroid lineage to obtain the benefits of being warm. With only brain heaters as the physiological specialization it is easier to accept that selection must have been for buffering the CNS from the effects of acute temperature changes.

The swordfish might be considered the most advanced thermoregulator of the regional endotherms. Telemetry data indicate that brain temperature remains constant during large temperature changes and muscle temperature remains significantly above water temperature due to a combination of physiological and behavioral mechanisms. Swordfish utilize a unique strategy to achieve something closer to whole body endothermy than the rest of the billfishes, and in some ways, may represent a strategy that was used by the ancestor of the modern tunas. Red muscle is located internally in these fish and a simplified heat exchange system has developed with this migration of the red muscle along the midline.

5. Endothermy in fish and the evolution of endothermy in birds and mammals

As mentioned above, the question of why has endothermy evolved in scombroid fishes ingenders many of the same conflicts as discussion of the evolution of endothermy in mammals. Selective factors influencing the evolution of endothermy in birds and mammals have been the subject of numerous articles[1,36] which center on the issue of whether endothermy evolved for warmth and environmental versatility or as a consequence of selection for aerobic performance (speed and endurance)? Crompton et al.[36] argued for niche expansion as the selective pressure driving the evolution of endothermy in mammals. Endothermy in scombroid fishes notably supports their hypothesis.

Many cetaceans migrate to polar seas during the summer months to feed and travel back to the tropics for reproductive purposes. Bluefin tuna and salmon sharks (*Lamna ditropis*), the warmest of the endothermic fishes, also extend their ranges far to the North. They feed on baitfish that are plentiful in the plankton rich seas of temperate and polar latitudes. Recent telemetry of the diving behavior of elephant seals has revealed that much like the swordfish (Fig. 22), these groups are often specialists on vertically migrating species such as squid. The depth

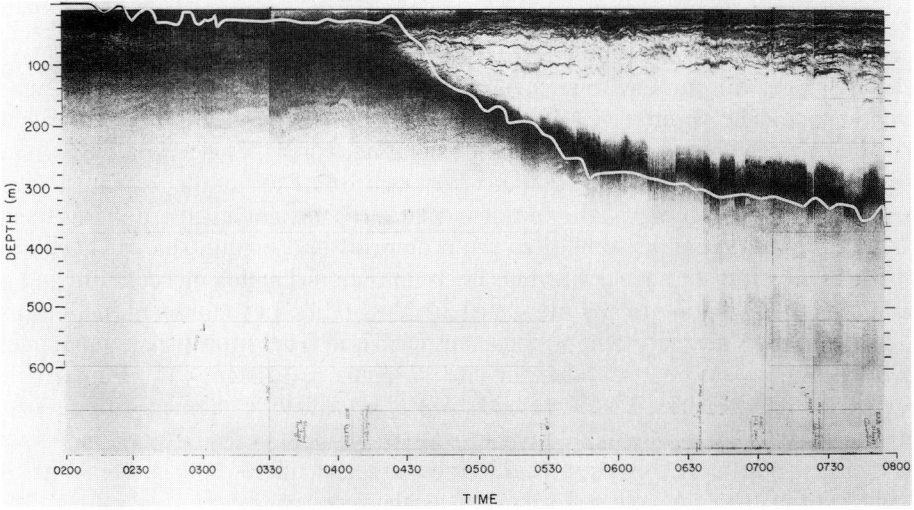

Fig. 22. Depth record for a swordfish superimposed on 50 kHz echogram. The fish was out over deep water and a heavy layer of sound scattering organisms that had been near the surface at night started to descend about 45 min before dawn. The swordfish followed and remained in the scattering layer. Diurnal movement of swordfish is most likely associated with the behavior associated with preying on organisms in the deep scattering layer (from ref. 21).

movements of swordfish and elephant seals are similar to one another (except that the mammal has to come to the surface to breathe). Is a swordfish warm for increased power output, or is it warm in order to take advantage of foraging beneath the thermocline? Being warm and the avoidance of the deleterious effects of acute temperature changes on the CNS appear to allow access to resources on a daily basis that most fish, because of their limited ability to endure rapid temperature change, cannot take advantage of.

6. Conclusions

As discussed above, brain- and eye-warming strategies found in marine fishes fall into two categories: whole body endothermy and regional or selective endothermy. Behavioral thermoregulation was most certainly a factor used in the evolution of both of these strategies. Whole body endothermy (as exhibited by the bluefin tuna, *Thunnus thynnus*) may represent a more advanced state of endothermy in teleost fishes, while regional endothermy as represented by the phylogenetically more primitive scombrid, *Gasterochisma melampus*, is most-likely an early attempt to gain the advantages of endothermy in the sea without paying a high metabolic price. The evolution of endothermy in scombroid fishes is a combination of the large body size coupled with the presence of aerobic red muscle both in the head and in the epaxial musculature along with the heat exchange systems.

In the warm-bodied tunas and sharks, vascular heat exchangers are found in the swimming muscle, cranial region and viscera but a specialized thermogenic tissue has not been identified. A high metabolic rate is associated with the elevated body temperatures in warm fish along with an elaborate means for limiting thermal convection. Thus endothermy in tunas occurs in a manner that is analogous to birds and mammals (elevated metabolism combined with the ability to regulate heat exchange with the environment), and is not entirely due to intense muscular activity alone. The absence of a thermogenic organ in these fish is due to the fact that their muscle generates sufficient metabolic heat and they need only conserve this heat to maintain warm tissue temperatures. The marlins, sailfish and spearfishes, and presumably the butterfly mackerel, do not maintain warm body temperatures in a manner similar to the scombrids and lamnid sharks. They are constrained by the abundance of white fibers in their swimming musculature and a more diffuse disposition of red muscle. A highly modified eye muscle functions as a brain and eye heater in these ectothermic fishes. The swordfish is unique in having a large brain heater and small rete mirabilia associated with an internally located red muscle mass. The large body size of swordfish, combined with circulatory specializations that regulate heat loss at the gill and body surfaces, increases the thermal inertia of this large ectothermic fish and allows it to remain significantly warmer than the cold water below the thermocline.

In billfishes, the correlation of heater size with thermal ecology indicates that there is plasticity in building this organ efficiently. Conversion of muscle cells into heater cells appears to be related to how much heat needs to be generated in the head. Although there are no metabolic measurements on billfishes, the cost of running a brain heater while letting body temperature fluctuate or, in the case of swordfish, using behavioral thermoregulation, must be a less costly metabolic strategy than whole body endothermy. The brain heaters in swordfish permit extended periods of foraging beneath the thermocline, opening the vast resources of the mesopelagic zone to this species. In essence, both regional and whole body endothermy at their highest level (as represented by the bluefin tuna and swordfish) provide the same end result: the exploitation of the marine environment in a manner similar to many cetaceans.

Acknowledgements. The preparation of this chapter was enhanced by discussions with Drs. Bruce Sidell, Frank Carey, and David Booth. I thank Alexa Tullis, Shyril O'Steen, Soo Young Kim, John Finnerty, Karen Mazurkiewicz and David Wang for useful discussions and for help in preparation of the manuscript. Support during preparation of the manuscript was from NSF DCB 895-8225 and NIH AR40246-01.

VI. References

1. Bennett, A.F. and J.A. Ruben. Endothermy and activity in vertebrates. *Science* 206: 649-673, 1979.
2. Block, B.A. Structure of the brain and eye heater tissue in marlins, sailfish, and spearfishes. *J. Morphol.* 190: 169-189, 1986.
3. Block, B.A. *Brain and Eye Warming in Billfishes (Istiophoridae): The Modification of Muscle Into a Thermogenic Tissue.* Ph.D. Thesis. Durham, NC: Duke University, 1986.
4. Block, B.A. Billfish brain and eye heater: A new look at nonshivering heat production. *NIPS* 2: 208-213, 1987.
5. Block, B.A. Physiology and ecology of brain and eye heaters in billfish. In *Planning the future of Billfishes*, edited by R.H. Stroud. Savanah, GA: National Coalition Marine Conservation, 1990, pp. 123-136.
6. Block, B.A. Evolutionary novelties: how fish have built a heater out of muscle. *Am. Zool.* in press.
7. Block, B.A. and F.G. Carey. Warm brain and eye temperatures in sharks. *J. Comp. Physiol. B* 156: 229-236, 1985.
8. Block, B.A. and C. Franzini-Armstrong. The structure of the membrane systems in a novel muscle cell modified for heat production. *J. Cell Biol.* 107: 1099-1112, 1988.
9. Block, B.A. and G. Meissner. Heater Tissue contains a functional sarcoplasmic reticulum. *Am. J. Physiol.* submitted.
10. Block, B.A., T. Imagawa, K.P. Campbell and C. Franzini-Armstrong. Structural evidence for direct interaction between the molecular components of the transverse tubule/sarcoplasmic reticulum junction in skeletal muscle. *J. Cell Biol.* 107: 2587-2600, 1988.
11. Block, B.A. and A.F.R. Stewart. Evolution of endothermy in scombroid fishes: systematic relationships examined with direct sequencing of mitochondrial DNA. *Am. Zool.* 30: 1990.
12. Block, B.A., F.G. Carey and D. Booth. Depth, swimming speed, and body temperature of the blue marlin, *Makaira nigricans*, observed by acoustic telemetry. *Physiologist* 33: 90, 1990.
13. Bone, Q. Myotomal muscle fiber typers in *Scomber* and *Katsuwonus*. In: *The Physiological ecology of tunas*, edited by G.D. Sharp and A.E. Dizon. New York: Academic Press, 1978, pp. 183-204.
14. Bossen, E.H., J.R. Sommer and R.A. Waugh. Comparative stereology of the mouse and finch left ventricle. *Tissue Cell* 10: 773-784, 1978.
15. Brill, R.W. On the standard metabolic rates of tropical tunas, including the effect of body size and acute temperature change. *Fish. Bull.* 85: 25-35, 1987.
16. Brill, R.W., D.L. Guernsey and E.D. Stevens. Body surface and gill heat loss rates in restrained skipjack tuna. In: *The Physiological Ecology of Tunas*, edited by G.D. Sharp and A.E. Dizon. New York: Academic Press, 1978, pp. 261-276.

17. Brill, R.W. and A.E. Dizon. Effect of temperature on isotonic twitch of white muscle and predicted maximum swimming speeds of skipjack tuna, Katsuwonus pelamis. *Environ. Biol. Fish.* 4: 199–205, 1979.
18. Burne, R.H. Some peculiarities of the blood vascular system of the porbeagle shark (*Lamna cornubica*). *Phil. Trans. Roy. Soc. Lond.* 212B: 1923–1924, 1923.
19. Carey, F.G. A brain heater in the swordfish. *Science* 216: 1327–1329, 1982.
20. Carey, F.G. Warm fish. In: *A Companion to Animal Physiology*, edited by C.R. Taylor, K. Johansen and L. Bolis. Cambridge: Cambridge University Press, 1982, pp. 216–233.
21. Carey, F.G. Further observations on the biology of the swordfish. In: *Planning the Future of Billfishes*. edited by R.H. Stroud. Savannah, GA: National Coalition for Marine Conservation, 1990.
22. Carey, F.G. and J.M. Teal. Heat conservation in tuna fish muscle. *Proc. Natl. Acad. Sci. U.S.A.* 56: 1464–1469, 1966.
23. Carey, F.G. and J.M. Teal. Mako and porbeagle: Warm-bodied sharks. *Comp. Biochem. Physiol.* 28: 199–204, 1969.
24. Carey, F.G. and J.M. Teal. Regulation of body temperature by the bluefin tuna. *Comp. Biochem. Physiol.* 28: 205–213, 1969.
25. Carey, F.G., J.M. Teal, J.W. Kanwisher and K.D. Lawson. Warm-bodied fish. *Am. Zool.* 11: 135–145, 1971.
26. Carey, F.G. and B.H. Robison. Daily patterns in the activities of swordfish, *Xiphias gladius*, observed by acoustic telemetry. *Fish. Bull.* 79: 277–291, 1981.
27. Carey, F.G., J.W. Kanwisher, O. Brazier, G. Gabrielson, J.G. Casey and H.L.J. Pratt. Temperature and activities of a white shark, *Carcharodon carcharias*. *Copeia*. 2: 254–260, 1982.
28. Carey, F.G. and R.J. Olson. Sonic tracking experiments with tunas. *Collective Volumes of Scientific Papers, International Commission for the Conservation of Atlantic Tunas (SCRS-1981)*. 17: 458–466, 1982.
29. Carey, F.G., J.W. Kanwisher and E.D. Stevens. Bluefin tuna warm their viscera during digestion. *J. Exp. Biol.* 109: 1–20, 1984.
30. Carey, F.G., J.G. Casey, H.L. Pratt, D. Urkuhart and J.E. McCosker. Temperature, heat production and heat exchange in lamnid sharks. *Mem. Southern Calif. Acad. Sci.* 9: 92–108, 1985.
31. Carey, F.G. and Gibson. Blood flow in the muscle of free-swimming fish. *Physiol. Zool.* 60: 138–148, 1987.
32. Carey, F.G. and J.V. Scharold. Movements of blue sharks in course and depth. *Mar. Biol.*: in press.
33. Cech, J.J., R.M. Laurs and J.B. Graham. Temperature-induced changes in blood gas equilibria in the albacore, *Thunnus alalunga*, a warm-bodied tuna. *J. Exp. Biol.* 109: 210–34, 1984.
34. Collette, B.B. Adaptations and systematics of the mackerels and tunas. In: *The Physiological Ecology of Tunas*, edited by G.D. Sharp and A.E. Dizon. New York: Academic Press, 1978, pp. 7–40.
35. Collette, B.B., T. Pothoff, W.J. Richards, S. Ueyanagi, J.L. Russo and Y. Nishikawa. Scombroidei: Development and relationships. In: *Ontogeny and Systematics of Fishes*, edited by H.G. Moser. Lawrence, KS: Allen Press, 1984, pp. 591–620.
36. Crompton, A.W., C.R. Taylor and J.A. Jagger. Evolution of homeothermy in mammals. *Nature* 272: 333–336, 1978.
37. Davie, P.S. *Pacific Marlins: Anatomy and Physiology*. Palmerston North, New Zealand: Simon Print, 1990.
38. Davy, J. On the temperature of some fishes of the *Genus Thunnus*. *Proc. Roy. Soc. Lond.* 3: 327–328, 1835.
39. Dizon, A.E. and R.W. Brill. Thermoregulation in tunas. *Am. Zool.* 19: 249–265, 1979.
40. Dizon, A.E. and R.W. Brill. Thermoregulation in yellowfin tuna, *Thunnus albacares*. *Physiol. Zool.* 52: 581–593, 1979.
41. Dobson, G.P., S.C. Wood, C. Daxboeck and S.F. Perry. Intracellular buffering and oxygen transport in the Pacific blue marlin (*Makiara nigricans*): adaptations to high-speed swimming. *Physiol. Zool.* 59: 150–156, 1986.
42. Elder, H.Y. Muscle structure. In: *Insect Muscle*, edited by P.R. Usherwood. London: Academic Press, 1975, pp. 1–74.
43. Eschricht, D.F. and J. Muller. Nachtrag zur Abhandlung der Hr. Eschricht und Muller uber die Vulpes. *Abhandl. d. koniglichen Akad. Wiss. Berlin*, 1835–1837: 325–328, 1837.
44. Fleischer, S. and M. Inui. Biochemistry and biophysics of excitation–contraction coupling. *Annu. Rev. Biophys. Chem.* 18: 333–364, 1989.

45. Franzini-Armstrong, C., B.A. Block and D.G. Ferguson. Molecular architecture of T-SR junctions: evidence for a junctional complex that directly connects the two membrane systems. In: *Transduction in Biological Membranes*. New York: Plenum Press, 1990, pp. 339–361.
46. George, J.C. and E.D. Stevens. Fine structure and metabolic adaptation of red and white muscles in tuna. *Environ. Biol. Fish.* 3: 185–191, 1978.
47. Gibson, Q.H. and F.G. Carey. The function of high hemoglobin in large fish. In: *Structure and Function Relationships in Biochemical Systems*, edited by C. Bossa and Finazzi-Argro and Strom. New York: Plenum Press, 1982, pp. 49–55.
48. Gooding, R.M., W.H. Neill and A.E. Dizon. Respiration rates and low-oxygen tolerance limits in skipjack tuna, *Katsuwonus pelamis. Fish. Bull.* 79: 31–48, 1981.
49. Graham, J.B. Heat exchange in the yellowfin tuna, *Thunnus albacares*, and skipjack tuna, *Katsuwonus pelamis*, and the adaptive significance of elevated body temperatures in scombrid fishes. *Fish. Bull.* 73: 219–229, 1975.
50. Graham, J.B. and K.A. Dickson. Physiological thermoregulation in the albacore *Thunnus alalunga. Physiol. Zool.* 54: 470–486, 1981.
51. Graham, J.B. and R.M. Laurs. Metabolic rate of the albacore tuna, *Thunnus alalunga. Mar. Biol.* 72: 1–6, 1982.
52. Graham, J.B., F.J. Koehrn and K.A. Dickson. Distribution and relative proportions of red muscle in scombrid fishes: consequences of body size and relationships to locomotion and endothermy. *Can. J. Zool.* 61: 2087–2096, 1983.
53. Graham, J.B., W.R. Lowell, N.C. Lai and R.M. Laurs. O_2 tension, swimming-velocity, and thermal effects on the metabolic rate of the Pacific albacore, *Thunnus alalunga. Exp. Biol.* 48: 89–94, 1989.
54. Graham, J.B., H. Dewar, N.C. Lai, W.R. Lowell and S.M. Arce. Aspects of shark swimming performance determined using a large water tunnel. *J. Exp. Biol.* 151: 175–192, 1990.
55. Guppy, M., W.C. Hulbert and P.W. Hochachka. Metabolic sources of heat and power in tuna muscles. II. Enzyme and metabolite profiles. *J. Exp. Biol.* 82: 303–320, 1979.
56. Hebrank, J.H., M.R. Hebrank, J.H.J. Long, B.A. Block and S.A. Wainwright. Backbone mechanics of the blue marlin *Makiara nigricans* (Pisces, Istiophoridae). *J. Exp. Biol.* 148: 449–459, 1990.
57. Hochachka, P.W. Regulation of heat production at the cellular level. *Fed. Proc.* 33: 2162–2169, 1974.
58. Hulbert, W.C., M. Guppy, B. Murphy and P.W. Hochachka. Metabolic sources of heat and power in tuna muscles. I. Muscle fine structure. *J. Exp. Biol.* 82: 289–301, 1979.
59. James, N.T. and G.A. Meek. Stereological analyses of the structure of mitochondria in pigeon skeletal muscle. *Cell Tissue Res.* 202: 493–503, 1979.
60. Johnson, G.D. Scombroid phylogeny: an alternative hypothesis. *Bull. Mar. Sci.* 39: 1–41, 1986.
61. Johnston, I.A. and T.W. Moon. Fine structure and metabolism of multiple innervated fast muscle fibers in teleost fish. *Cell Tissue Res.* 219: 93–109, 1981.
62. Johnston, I.A. and R. Brill. Thermal dependence of contractile properties of single skinned muscle fibres from Antartic and various warm water marine fishes including skipjack tuna (*Katsuwonus pelamis*) and kawakawa (*Euthynnus affinis*). *J. Comp. Physiol. B.* 155: 63–70, 1984.
63. Johnston, I.A. and J. Salamonski. Power output and force-velocity relationship of red and white muscle fibres from the Pacific blue marlin (*Makiara nigricans*). *J. Exp. Biol.* 111: 171–177, 1984.
64. Jones, D.R., R.W. Brill and D.C. Mense. The influence of blood gas properties on gas tensions and pH of ventral and dorsal aortic blood in free-swimming tuna, *Euthynnus affinis. J. Exp. Biol.* 120: 201–213, 1986.
65. Josephson, R.K. and D. Young. A synchronous insect muscle with an operating frequency greater than 500 Hz. *J. Exp. Biol.* 118: 185–208, 1985.
66. Kishinouye, K. Contributions to the comparative study of the so-called scombroid fishes. *J. Coll. Agric. Imper. Univ. Tokyo* 8: 293–475, 1923.
67. Kohno, H. Osteology and systematic position of the butterfly mackerel, *Gasterochisma melampus. Jpn. J. Ichthyol.* 31: 268–286, 1984.
68. Lai, N.C., J.B. Graham, W.R. Lowell and R.M. Laurs. Pericardial and vascular pressures and blood flow in the albacore tuna, *Thunnus alalunga. Exp. Biol.* 46: 187–192, 1987.
69. Laurs, R.M. and R.J. Lynn. Seasonal migration of North Pacific albacore, *Thunnus alalunga*, into North American coastal waters: distribution, relative abundance, and association with Transition Zone waters. *Fish. Bull.* 75: 795–822, 1977.
70. Linthicum, D.S. and F.G. Carey. Regulation of brain and eye temperatures by the bluefin tuna. *Comp. Biochem. Physiol.* 43A: 425–433, 1972.
71. Londraville, R.L. and B.D. Sidell. Ultrastructure of aerobic muscle in Antartic fishes may contribute to maintenance of diffusive fluxes. *J. Exp. Biol.* 150: 205–220, 1990.

72. MacLennan, D.H., C. Duff, F. Zorzato, J. Fujii, M. Philips, R.G. Komeluk, W. Frodis, B.A. Britt and R.G. Worton. Ryanodine receptor gene is a candidate for predisposition to malignant hyperthermia. *Nature* 343: 559–561, 1990.
73. McCormack, J.G., A.P. Halestrap and R.M. Denton. Role of calcium ions in regulation of mammalian intramitochondrial metabolism. *Physiol. Rev.* 70: 391–425, 1990.
74. Muller, J. Vergleichende Anatomie der Myxinoiden. *Abhandl. königl. Akad. Wiss. Berlin* 1839: 175–303, 1839.
75. Nakamura, I. Systematics of the billfishes (*Xiphiidae* and *Istiphoridae*). *Publ. Seto Mar. Biol. Lab.* 28: 255–396, 1983.
76. Neil, W.H., R.K. Chang and A.E. Dizon. Magnitude and ecological importance of thermal inertia in skipjack tuna, *Kasuwonnus pelamis* (Linneaus). *Environ. Biol. Fish* 6: 61–80, 1976.
77. Nicholls, D.G. and R.M. Locke. Thermogenic mechanisms in brown fat. *Physiol. Rev.* 64: 1–65, 1984.
78. Rayner, M.D. and M.J. Keenan. Role of red and white muscles in the swimming of the skipjack tuna. *Nature* 214: 392–393, 1967.
79. Rios, E. and G. Brum. Involvement of dihydropyridine receptors in excitation–contraction coupling in skeletal muscle. *Nature* 325: 717 720, 1987.
80. Schmidt-Nielsen, K. *Animal Physiology*. Cambridge: Cambridge University Press, 1986.
81. Schneider, M.F. and W.K. Chandler. Voltage dependent charge movement in skeletal muscle: A possible step in excitation–contraction coupling. *Nature* 242: 244–246, 1973.
82. Sharp, G.D. and A.E. Dizon. *The Physiological Ecology of Tunas*. 260–276, 1978.
83. Sidell, B.D. and T.S. Moerland. Effects of temperature on muscular function and locomotory perfomance in teleost fish. In: *Advances in Comparative and Environmental Physiology, Vol. 5*. Berlin: Springer-Verlag, 1989, pp. 116–152.
84. Stevens, E.D. The effect of temperature on facilitated oxygen diffusion and its relation to warm tuna muscle. *Can. J. Zool.* 60: 1148–1152, 1982.
85. Stevens, E.D. Energetics of locomotion in warm-bodied fish. *Annu. Rev. Physiol.* 44: 121–131, 1982.
86. Stevens, E.D. and F.E. Fry. Brain and muscle temperatures in ocean caught and captive skipjack tuna. *Comp. Biochem. Physiol.* 1971: 203–211, 1971.
87. Stevens, E.D. and A.M. Sutterlin. Heat transfer between fish and ambient water. *J. Exp. Biol.* 65: 131–145, 1976.
88. Stevens, E.D. and F.G. Carey. One why of the warmth of warm-bodied fish. *Am. J. Physiol.* 240: R151–R155, 1981.
89. Stroud, R.H. Planning the future of billfishes: Research and management in the 90's and beyond. In: *Proceedings of the 2nd International Billfish Symposium*. National Coalition for Marine Conservation, 1989.
90. Suarez, R.K., M.D. Mallet, C. Daxboeck and P.W. Hochachka. Enzymes of energy metabolism and gluconeogenesis in the Pacific blue marlin, *Makiara nigricans*. *Can. J. Zool.* 64: 694–697, 1986.
91. Takeshima, H., S. Nishimura, T. Matsumoto, H. Ishida, K. Kangawa, N. Minamino, H. Matsuo, M. Ueda, M. Hanoaka, T. Hirose and S. Numa. Primary structure and expression from complementary DNA of skeletal muscle ryanodine receptor. *Nature* 339: 439–445, 1989.
92. Tanabe, T., K.G. Beam, J.A. Powell and S. Numa. Restoration of excitation–contraction coupling and slow calcium current in dysgenic muscle by dihydropyridine receptor cDNA. *Nature* 336: 134–139, 1988.
93. Turner, J.S. Body size and thermal energetics: How should thermal conductance scale? *J. Therm. Biol.* 13: 103–117, 1988.
94. Volpe, P., G. Salviati, F. diVirgilio and T. Pozzan. Inositol 1,4,5-trisphosphate induces calcium release from sarcoplasmic reticulum of skeletal muscle. *Nature* 316: 347–349, 1985
95. Walters, V. Body form and swimming performance in the scombrid fishes. *Am. Zool.* 2: 143–149, 1962.
96. Walters, V. and H. Fierstine. Measurements of swimming speeds of yellowfin tuna and wahoo. *Nature* 202: 208–209, 1964.
97. Weber, J.-M., R.W. Brill and P.W. Hochachka. Mammalian metabolite flux rates in a teleost: lactate and glucose turnover in tuna. *Am. J. Physiol.* 250: R452–R458, 1986.
98. White, F.C., R. Kelly, S. Kemper, P.T. Schumacker, K.R. Gallagher and R.M. Laurs. Organ blood flow haemodynamics and metabolism of the albacore tuna *Thunnus alalunga* (Bonnaterre). *Exp. Biol.* 47: 161–169, 1988.
99. Wittenberg, J.B. and B.A. Wittenberg. The choroid rete mirabile of the fish eye I. Oxygen secretion and structure: comparison with the swimbladder rete mirabile. *Biol. Bull.* 146: 116–136, 1974.

100. Wittenberg, J.B. and B.A. Wittenberg. Mechanisms of cytoplasmic hemoglobin and myoglobin function. *Annu. Rev. Biophys. Chem.* 19: 217–241, 1990.
101. Wolf, N.G., P.R. Swift and F.G. Carey. Swimming muscle helps warm the brain of lamnid sharks. *J. Comp. Physiol. B.* 157: 709–715, 1988.
102. Yuen, H.S., A.E. Dizon and J.H. Uchiyama. Notes on the tracking of the Pacific blue marlin, *Makaira nigricans*. In: *Proceedings of the International Billfish Symposium, Kailua-Kona, Hawaii, 9–12 August 1972*. U.S. Dept. Commerce, NOAA TR NMFS SSRF-675. pp. 265–f268, 1972.
103. Zorzato, F., J. Fujii, K. Otsu, M. Phillips, N.M. Green, F.A. Lai, G. Meissner and D.H. MacLennan. Molecular cloning of cDNA encoding human and rabbit forms of the Ca^{2+} release channel of skeletal muscle sarcoplasmic reticulum. *J. Biol. Chem.* 265: 2244–2256, 1990.

CHAPTER 12

Temperature: the ectothermy option

P.W. HOCHACHKA

Department of Zoology, University of British Columbia, Vancouver, B.C., Canada V6T 2A9

I. Introduction
II. Factors determining temperature sensitivity
III. Catalytic efficiency and the Q_{10} values for enzyme reactions
IV. Enzyme–substrate affinities can influence reaction rates
V. Enzyme–modulator interactions can also affect thermal sensitivity
VI. Regulatory enzymes can influence the Q_{10} of their target enzymes
VII. Semistability of macromolecules can affect thermal sensitivity
VIII. Homeoviscous adaptation can influence Q_{10} values
IX. Differential temperature sensitivities: megaproblem of ectothermy
X. Q_{10} as an evolved property
XI. Ectothermy *versus* endothermy: what are the trade offs?
XII. References

I. Introduction

Although in principle the options of endothermy (body temperature determined by internal metabolic and physiological mechanisms) vs ectothermy (body temperature determined by external temperature) are available to fishes[2,12], most species are in fact ectothermic. Cell temperatures (and therefore the temperatures at which biochemical processes occur) are either the same as, or only modestly higher than, ambient. If biochemical and biophysical processes in living systems were unregulated and were dominated by thermodynamics, then two categories of temperature effects would lie at the base of most of the thermal relationships of fishes. These are temperature effects (1) on reaction rates and (2) on reaction equilibria, especially those equilibria involving the formation or rupture of noncovalent (weak) chemical bonds.

The expression Q_{10} is used to refer to the effect of temperature change on reaction rates, where Q_{10} is defined as the ratio of reaction velocities at temperatures (T) 10°C or °K apart. That is,

$$Q_{10} = \frac{\text{Velocity } (T° + 10°)}{\text{Velocity } (T°)}$$

A more general expression for Q_{10} which allows a Q_{10} to be computed for any temperature interval is:

$$Q_{10} = \left(\frac{k_2}{k_1}\right)^{10/(T_2° - T_1°)}$$

where k_1 and k_2 are the velocity constants at temperatures $T_1°$ and $T_2°$, respectively.

In any given system, only a small fraction of the total population of molecules possesses enough energy to react, and changes in temperature lead to marked increases in the size of this reactive population. The reactive molecules are referred to as having energies equal to or greater than the minimal *activation energy* required for the reaction. In the present context involving temperature effects on reaction velocities, the activation energy referred to is the *activation enthalpy* ($\Delta H^{\ddagger}$) expressed in kcal/mol. The formal expression for the temperature dependence of reaction velocity includes the $\Delta H^{\ddagger}$ term:

$$\frac{\delta \ln k}{\delta T} \simeq \frac{\Delta H^{\ddagger}}{RT}$$

Activation enthalpy is related to the free energy of activation ($\Delta G^{\ddagger}$) and activation entropy ($\Delta S^{\ddagger}$) by the well known equation

$$\Delta G^{\ddagger} = \Delta H^{\ddagger} - T\Delta S^{\ddagger}$$

Temperature change also influences the equilibrium constants of biochemical reactions, especially those involving the reversible formation of weak chemical bonds (Table 1). Covalent bonds linking amino acid residues of a protein or the nucleotide bases of DNA and RNA are relatively strong, and in the absence of enzymes catalyzing the hydrolysis of these bonds, covalent bonds are relatively invulnerable to thermal perturbation at biological temperatures. In contrast, the hydrogen bonds, charge-charge interactions, van der Waals interactions, and hydrophobic interactions responsible for a broad suit of biological functions are readily disrupted by temperature changes[12].

TABLE 1

The approximate enthalpy changes associated with the formation of weak bonds

Kind of weak bond	Enthalpy of formation (kcal/mol)
Van der Waals interactions	−1
Hydrogen bonds	−[3–7]
Electrostatic (ionic) bonds	−5
Hydrophobic interactions	+[1–3]

Note: the net change in enthalpy during weak bond formation between groups on a protein (or between nucleic acids, proteins and lipids, etc.) also usually require rupturing of weak bonds with the surrounding water. Thus, the net change in enthalpy is likely to be quite small since the heat released during hydrogen bond formation between two protein groups, for example, in effect helps to 'pay back' the heat expended to break the water-protein group hydrogen bonds that previously existed. Modified from Hochachka and Somero[12].

The magnitudes of temperature effects on weak bond based structures or functions cannot be as easily generalized as can the effects on reaction rates (where Q_{10} values of 2–3 are so common they are almost dogma). This is because these effects of temperature vary both in magnitude and direction depending upon the kinds of weak bonds that are involved. Thus, hydrogen bonds, charge-charge interactions, and van der Waals interactions form with variable but always negative enthalpy change, whereas hydrophobic interactions form with positive enthalpy change (Table 1). Thus, increasing temperature destabilizes the first three types of weak bonds, while stabilizing hydrophobic interactions. Due to such differing directions of temperature effect on the different types of weak bonds, the effects of a temperature change on weak-bond-dependent systems depend in part on the class of weak bond playing dominant roles in stabilization (see refs. 5, 12, 21 for references in this area).

II. Factors determining temperature sensitivity

Given the above, some interesting generalizations arise concerning the origins of temperature effects on ectothermic systems. Although the principles will be general, it is easiest (and perhaps best) to illustrate the situation with some specific process; so I will base our analysis on enzyme catalyzed reactions which *in vivo* are closely regulated biochemical processes.

III. Catalytic efficiency and the Q_{10} values for enzyme reactions

Since catalysts such as enzymes reduce the activation energy required for reaction while temperature influences the fraction of molecules with enough energy to react, a primary determinant of the inherent temperature sensitivity of any given reaction is the enzyme catalytic efficiency (usually defined as turnover number). Enzymes which have very high turnover numbers (highly efficient catalysts) typically display very low Q_{10} values. Acetylcholinesterases and catalases are two such examples, so efficient as catalysts that their Q_{10} values under substrate saturating conditions are in the 1.2–1.5 range[10]. The Q_{10} for diffusion is in this range or is only marginally lower! In contrast, enzymes which are catalytically relatively inefficient display higher Q_{10} values under saturating conditions. Na^+K^+ATPase is one such example, Q_{10} values usually being well over three[15]. Pyruvate kinases also may display quite high Q_{10} values under saturating substrate concentrations[16].

IV. Enzyme–substrate affinities can influence reaction rates

In vivo enzymes are rarely saturated with respect to their substrate concentrations; in fact, for many 'mainline' metabolic enzymes, such as those of glycolysis, the concentration of active sites for any given enzyme may well equal or even exceed the concentrations of their substrates[17]. Under these conditions, as has been appreciated for many years now, enzyme affinities are strong determinants of reaction rates[12]. Since enzyme–substrate interactions usually are dominated by weak bonds, they are exquisitely sensitive to temperature. Hence, saying that enzyme–substrate affinities can influence reaction rates is tantamount to saying that they can also influence the Q_{10} of reactions under subsaturating concentrations of substrates[12]. For many enzymes, falling temperatures lead to increasing enzyme–substrate affinities; this reduces Q_{10} values and the lower the substrate concentrations, the lower the Q_{10}. For such enzymes, the Q_{10} can sometimes be less than one[12], and these effects can be exaggerated if pH is allowed to change as well[12,21]. In contrast, enzyme substrate interactions that are stabilized mainly by hydrophobic bonds display reduced enzyme–substrate affinities at low temperatures. The net effect is to increase temperature sensitivity because as temperature drops enzyme affinities for substrate decrease as does thermal kinetic energy. Over critical (near 0 °C) temperatures, the Q_{10} for such enzymes can get quite high[8,12]. Acetylcholinesterase is again an interesting case in point. At saturating substrate levels, as mentioned above, the enzyme displays minimal temperature sensitivity (Q_{10} close to 1), but at subsaturating substrate levels, the Q_{10} can become quite high because of the disrupting effects of low temperature on the hydrophobic contributions to enzyme–substrate binding[8].

V. Enzyme–modulator interactions can also affect thermal sensitivity

In addition to catalytic efficiency and enzyme–substrate affinity having effects on thermal sensitivity of enzyme function, it is clear that so can enzyme–modulator equilibria (which again are based on reversible weak bond formations). Whether thermal sensitivity rises or falls depends upon the enzyme and modulator in question. For example, AMP is a physiologically relevant inhibitor of fructose bisphosphatase in all tissues where this enzyme plays a role in gluconeogenesis. Whereas the inherent Q_{10} value for this enzyme under saturating conditons is usually about 2, under subsaturating conditions and in the presence of AMP as negative modulator, the Q_{10} is nearly 20 (ref. 20)! This effect arises from a very large increase in enzyme–AMP affinity as temperature drops; in effect thermal kinetic energy is falling while AMP inhibition is rising, leading to an unusually high thermal sensitivity. In contrast to AMP effects on FBPases, AMP acts as a positive modulator of phosphofructokinases (PFKs) operating in these same tissues; for the same reasons as before, enzyme–AMP binding is favored as temperature drops, but in this case the reaction rate is accelerated. In this case, the enzyme–modulator interaction would serve to reduce the temperature sensitivity, assuming all other conditions were somehow held constant. For an enzyme such as PFK, this is a crucial proviso, since it is a closely regulatory allosteric enzyme and all of its array of negative and positive modulators show analagous temperature dependent binding[12]. Be that as it may, it is well appreciated that in physiologically relevant milieu (with substrate and modulators all present at appropriate concentrations), enzyme–modulator interactions can powerfully influence the apparent Q_{10} of the reaction.

VI. Regulatory enzymes can influence the Q_{10} of their target enzymes

In addition to enzyme–substrate and enzyme–modulator effects on reaction rates in living cells, some enzymes are regulated by covalent modifications catalyzed by controlling enzymes. The most common such process is phosphorylation and there are literally thousands of proteins that are now known to be substrates for endogenous protein kinases (presumably because they are regulated by such phosphorylation–dephosphorylation reactions[14]). Glycogen phosphorylase is a well known example of an enzyme that is regulated by such a kinase reaction (by phosphorylase b kinase), and by a phosphatase. Pyruvate dehydrogenases and oxoglutarate dehydrogenases are two additional enzymes under kinase and phosphatase control. Although I am not aware of a careful study of the temperature sensitivity *in vivo* of any one of these reactions, it is clear that the kinase and phosphatase affinities for their target enzymes are strongly perturbed by temperature changes (because they, like subunit–subunit interactions in many proteins[6], largely or solely involve weak bonding interactions between the kinases or phosphatases and their protein substrates). Hence the apparent Q_{10} values for enzymes such as glycogen phosphorylases *in vivo* are necessarily influenced by the effects of temperature on their interactions with the kinases and phosphatases that determine the amount of the target enzyme in active form.

VII. Semistability of macromolecules can affect thermal sensitivity

All cellular structures stabilized by weak bonds such as proteins, nucleic acids, and membranes share a common property: they are structures that are not rigid and invariant but rather are flexible and 'semistable'[1] since they must undergo conformational changes during the performance of their normal biological activities (for an especially interesting case, see ref. 6). Enzymes, for example, undergo conformational changes during binding of substrates and modulators and during catalysis *per se* (for proper alignment of reactive centers in the binding or catalytic sites with reactive groups on substrates and modulators). Thus the three-dimensional (3-D) structures of macromolecules such as enzymes are semistable and determined by a large series of internal weak bonding interactions (within a given polypeptide subunit or between different subunits). Since these are strongly influenced by temperature, the catalytic and regulatory properties of enzymes may display some thermal sensitivities that arise solely from these effects. Pyruvate kinase (PK) may supply us with an example of such a case, since in some species the enzyme shows a sharp break (a transition temperature) in Arrhenius plots[16]. It is not unreasonable to assume that the basis for this thermal behaviour is to be found in the thermal dependence of the enzyme's semistable 3-D structure[20]. At temperatures above and below the transition zone the enzyme adopts different 'semistable' conformations and thus displays different Q_{10} values.

VIII. Homeoviscous adaptation can influence Q_{10} values

Such transition temperatures are perhaps best known for membrane based functions because of phase transitions of the lipid bilayer of cell membranes, a phenomenon termed homeoviscous adaptation[21]. Enzymes, channels, exchangers, active transporters, receptor proteins, and signal transduction proteins identify the kinds of functions whose thermal sensitivities in ectothermic fishes may in part be set by the lipid bilayer in which they reside[10]. Although not all of these kind of membrane-based systems have been analyzed, when they have been studied above and below the phase transition temperatures, the Q_{10} values are characteristically different, as in fact theory would predict they usually would be[12,21].

IX. Differential temperature sensitivities: megaproblem of ectothermy

I have intentionally developed the topic of temperature sensitivity of biochemical functions in a manner emphasizing different categories of mechanisms contributing to overall thermal behaviour to bring attention to a frequently overlooked problem for ectotherms; namely, the problem of differential effects of temperature on different links in integrated functions.

Some idea of the magnitude of this problem may be obtained from a brief analysis of a specific, well understood set of physiological functions (for example, events at the motor end plate). Starting with presynaptic Ca^{2+} activation, the first two Q_{10} values of concern are for Ca^{2+} diffusion (Q_{10} close to 1) and for Ca^{2+} channel regulation (Q_{10} may be as high as 3–4). Acetylcholine (ACh) release is a complex process involving exocytotic release of ACh-filled vesicles; although the thermal characteristics of this complex process have not been extensively studied, it is reasonable to assume that they would be changed by temperature dependent membrane phase transitions, and hence would not be constant over a wide biological thermal range. Acetylcholine diffusion across the synaptic gap can be assumed to be nearly temperature independent (like all diffusion processes). The Q_{10} for ACh binding to the end plate receptor (AChR) is also unknown, but clearly is subject to both AChR affinity changes with temperature and to phase transitions of the cell membrane through which the receptor traverses several times. Acetylcholine binding to acetylcholinesterase (AChE) displays inverse temperature dependence at near $0°C$ but is largely temperature independent at elevated temperatures; the Q_{10} for catalysis *per se* varies with substrate concentration but is typically close to 1. The Q_{10} for diffusion of Na^+ and K^+ through the AChR channel is as expected close to unity; finally, the Q_{10} for regulation of AChR channel function (for example, the Q_{10} for inactivation) may be much higher than 1 (see refs. 5, 8, 10 for literature in this area). Thus a change in temperature may well differentially perturb nine or more of the currently known links in what *in vivo* must remain a simple integrated function: the transfer of an activating signal from nerve to muscle. Obviously this problem at synapses is multiplied over and over *in vivo*, in principle by as many times as there are physical, chemical, and biochemical reactions making up living ectothermic systems. That is why integrating temperature effects on individual physiological steps to achieve overall regulated *in vivo* function is not a simple problem for ectotherms: it is a megaproblem (well illustrated in a recent symposium series in *Am. J. Physiol.* 259: R189–R265, 1990).

If left to thermodynamic forces alone, it would be a miracle if the myriad of *in vivo* biochemical and physiological functions ever displayed similar Q_{10} values and it would be similarly miraculous if they displayed Q_{10} values near 2, the temperature dependency of simple uncatalyzed reactions. Yet it is surprising how often these kinds of Q_{10} values are reported in the literature on thermal relations of ectothermic organisms, including fishes. They are reported so commonly that an implicit assumption — that observed Q_{10} values are a matter of simple thermodynamics fully analogous to test tube chemical reactions — pervades the biological literature and the thinking of many (probably most) biologists and comparative physiologists.

X. Q_{10} as an evolved property

To properly appreciate the inadequacy of this common interpretation we must understand why, if Q_{10} values for biological functions can fall anywhere from 1 to very high values, the 2–3 range is so frequently observed. The answer as to why this is so, I would argue, is because *in vivo* Q_{10} values are *evolved* thermal properties. Although biochemical processes *in vivo* are influenced by thermodynamic constraints, they are not determined by them. That is, nature should be able to set Q_{10} for given biological functions at pretty well any value, and, as it turns out, sometimes does. For example, some biosynthetic pathways in insects show Q_{10} values at about 1 over part of the biological temperature range, but these values exceed 3 at temperatures above some critical transition zone[18]. Similarly, the metabolic rates of aquatic turtles display relatively 'normal' temperature dependence above 10–12°C;, however, below this critical temperature Q_{10} values for oxygen consumption rates rise to over 10 (ref. 4)! Data along these lines (temperature ranges over which Q_{10} values change drastically) are also available for fish metabolic, growth, and swimming rates (for example, see ref. 3), and are clearly consistent with the concept of *evolved* (versus simply thermodynamically determined) temperature sensitivities of biological processes in ectotherms. Nevertheless, such observations still remain a distinct minority and it is my impression that most studies of temperature effects on biological functions in ectotherms report Q_{10} values in the range of 2–3[10,12]. So something is still missing in the analysis. I suggest that the missing element is selective advantage; i.e., there must be some advantage or advantages for ectotherms to regulate or hold Q_{10} values for linked and integrated functions at (or close to) this value of 2.

One possibility is that setting Q_{10} values near 2 automatically integrates the temperature sensitivities of regulated biochemical processes and unregulated chemical and physical processes which are crucial to overall physiological function. The point of departure for this interpretation is the assumption that *in vivo* biological processes frequently involve coupling between (1) physical, (2) uncatalyzed (potentially unregulated) chemical, and (3) catalyzed (almost always closely regulated) biochemical reactions to achieve integrated function. Since the Q_{10} values for purely physical processes (e.g., diffusion) or for uncatalyzed chemical reactions (e.g., CO_2 hydration) are set by thermodynamics, their Q_{10} values are not adaptable. Thus, evolutionary fine tuning of the temperature dependence of biochemical processes is the only option available for integrating the effects of temperature change on *in vivo* physiological functions in ectotherms. In other words, biochemical processes of ectotherms must be temperature adapted to appropriately match the Q_{10} values of uncatalyzed chemical and physical steps to which they are coupled *in vivo*; otherwise, it would be difficult to achieve overall integrated physiological functions at variable temperatures. Although exceptions may arise with further work, at this time it appears that if ectotherms use temperature change to advantage, then evolved Q_{10} values may deviate widely from the values of 2-3. For example, overwintering ectotherms often 'use' falling temperatures as a means for suppressing energy demand and energy supply processes; evolved Q_{10} values in such cases may be very high, 10–12 for example[4]. The other extreme of not taking advantage of temperature change presumably implies reliance on a simpler adaptational strategy: evolving temperature sensitivities of biochemical processes to Q_{10} values of about 2 to match those of uncatalyzed physical-chemical reactions upon which living systems also rely. In either event, it appears that the temperature coefficients of ectothermic systems are the outcome of evolution and adaptation, constrained but not determined by thermodynamics.

XI. Ectothermy versus endothermy: what are the trade-offs?

From the above discussion it is evident that at a biochemical level ectotherms are really no more 'at the mercy' of the thermal environment than are endotherms. In evolutionary terms, organisms were seemingly faced with two options: either to regulate temperature effects on biochemical processes by regulating body temperature (endothermy) or to regulate the temperature coefficients of biochemical processes and tolerate varying body temperatures (ectothermy). Many workers have tried to evaluate the advantages and disadvantages of the two strategies of temperature adaptation. Aside from emphasizing obvious ecological consequences (for example, endothermy requiring higher energy turnover and foraging rates), the evaluations usually center on the costs and effects of thermal regulation, since endotherms control body temperature by controlling heat production and loss.

Heat by definition is a waste product of metabolism and thus endothermy almost by definition is based on regulated biochemical inefficiencies. There are various ways of generating heat in a regulated way (shivering and various non-shivering thermogenic mechanisms) all of which are essentially wasteful of energy. According to recent analyses, rather drastic up-regulation of thyroid hormone activity is considered the pivotal evolutionary event allowing the transition from ectothermy to endothermy[13]; amongst other effects, this led to lowered efficiencies in the conversion of metabolic energy to mechanical work (see refs. 11, 13, 19) as well as ionic pumping work[13]. From the best evidence currently available, both of these biochemical inefficiencies (or heat generation mechanisms) are based on 'leakier' membranes than found in ectotherms[13,19] and thus to the release of more heat in the ATP dependent pumping of ions in order to maintain electrochemical gradients and ionic homeostasis. What is more, this kind of apparent adaptation for endothermy makes cell function highly stenothermic; most tissues from endotherms are not able to sustain significant temperature change because of serious mismatching between temperature coefficients of downhill diffusion across membranes and the Q_{10} values for ATP-linked ion pumps[9,10]. By implication, none of these problems characterize ectothermic tissues, in which activities regulated by Ca^{2+} ATPases, Na^+K^+-ATPases, actomyosin ATPases, and ion channels should be operating at close to their thermodynamically allowable maximum efficiencies.

XII. References

1. Alexandrov, Y.Ya. *Cells, Molecules, and Temperature*. Berlin: Springer-Verlag, 1977.
2. Block, B.A. Endothermy in Fishes: thermogenesis, ecology and evolution. This volume, Chapter 11, 1991, pp. 269–311.
3. Evans, D.O. Temperature independence of the annual cycle of standard metabolism in the pumpkinseed. *Trans. Am. Fish. Soc.* 113: 494–512, 1984.
4. Herbert, C.V. and D.C. Jackson. Temperature effects on the responses to submerges of the turtle *Chrysemys picta bellii*. II. Metabolic rate, blood acid-base, and ionic changes, and cardiovascular functions in aerated and anoxic water. *Physiol. Zool.* 58: 670–681, 1985.
5. Hille, B. *Ionic Channels in Excitable Tissues*. Sunderland, MA: Sinauer, 1984.
6. Himes, R.H. and H.W. Detrich III. Dynamics of Antarctic fish microtubules at low temperature. *Biochemistry* 28: 5089–5095, 1989.
7. Hochachka, P.W. Regulation of heat production at the cellular level. *Fed. Proc.* 33: 2162–2169, 1974.
8. Hochachka, P.W. Acetylcholinesterase: temperature and pressure adaptation of the anionic binding site. *Biochem. J.* 143: 535–539, 1975.
9. Hochachka, P.W. Defense mechanisms against hypoxia and hypothermia. *Science* 231:234–241, 1986.
10. Hochachka, P.W. Metabolic-, channel-, and pump-coupled functions: Constraints and compromises of coadaptation. *Can. J. Zool.* 66: 1015–1027, 1988.
11. Hochachka, P.W., C. Stanley, G.O. Matheson, D.C. Mckenzie, P.S. Allen, and W.S. Parkhouse. Metabolic and work efficiencies during exercise in Andean native. *J. Appl. Physiol.* 70: 1720–1730, 1991.
12. Hochachka, P.W. and G.N. Somero. *Biochemical Adaptation*. Princeton, NJ: Princeton University Press, 1984.
13. Hulbert, A.J. Thyroid hormones, membranes and evolution of endothermy. In: *Advances in Physiol Research*, edited by H. McLennan, J.R. Ledsome, C.H.S. McIntoch and D.R. Jones. New York: Plenum Press, 1987, pp. 305–319.
14. Krebs, E. Role of cyclic AMP dependent protein kinases in signal transduction. *JAMA* 262: 1815–1818, 1989.

15. Matsuda, T. and H. Iwata. Species differences in temperature dependence of cardiac Na^+K^+ATPase activity. *Biochem. Pharmacol.* 34: 2343–2346, 1985.
16. Somero, G.N. and P.W. Hochachka. The effect of temperature on catalytic and regulatory properties of pyruvate kinases from rainbow trout and the Antarctic fish *Trematomus bernacchii. Biochem. J.* 110: 395–400, 1968.
17. Srivastava, D.K. and S.A. Bernard. Metabolite transfer via enzyme-enzyme complexes. *Science* 234: 1081–1086, 1986.
18. van Handel, E. The thermal dependence of the rates of glycogen and triglyceride synthesis in the mosquito. *J. Exp. Biol.* 44: 523–528, 1966.
19. van Hardeveld, C. Effects of thyroid hormone on oxygen consumption, heat production, and energy economy. In: *Thyroid Hormone Metabolism*, edited by G. Hennemann. New York: Marcel Dekker, 1989, pp. 579–608.
20. van Tol, A. On the occurrence of a temperature coefficient (Q_{10}) of 18 and a discontinuous Arrhenius plot for homogenous rabbit muscle fructose diphosphatase. *Biochem. Biophys, Res. Commun.* 62: 750–756, 1975.
21. White, F.N. and G.N. Somero. Acid-base regulation and P-lipid adaptation to temperature: time courses and physiological significance of modifying the milieu for protein function. *Physiol. Rev.* 62: 40–90, 1982.

Hochachka and Mommsen (eds.). Biochemistry and molecular biology of fishes, vol. 1
© 1991 Elsevier Science Publishers B.V. (Academic Publishing Division)

CHAPTER 13

Pressure as an environmental variable: magnitude and mechanisms of perturbation

JOSEPH F. SIEBENALLER

Department of Zoology and Physiology, Louisiana State University, Baton Rouge, LA 70803, U.S.A.

I. Introduction
II. Bases of pressure effects
III. Perturbation by pressure
IV. Proteins: subunit assembly and stability
V. Proteins: catalytic rate
VI. Proteins: ligand binding
VII. Membranes: homeoviscous adaptation
VIII. Membrane-associated processes
IX. Biosynthesis and gene regulation
X. Coping with pressure variation
Acknowledgements
XI. References

I. Introduction

One inevitable consequence of life in the aquatic environment is an increase in hydrostatic pressure with increasing depth. This physical parameter can have profound consequences for the biology of fishes. Pressure shapes the molecular evolution of deep-living fishes[82,87], may be important in the speciation process in the marine environment[80], and may play a role in determining the bathymetric limits of a species' distribution[78,80,82,88]. A variety of data indicate that pressure also influences the behavior and morphology of deep-sea organisms, as well biochemical and physiological processes (e.g., refs. 9–11, 44, 50).

There is a gradient of pressure from 1 atm (= 101,325 Pascals) at the water surface to approximately 1100 atm in the deepest region of the ocean, the Challenger Deep in the Mariannas Trench. Pressure increases approximately one atmosphere for every 10 m depth increase in the marine water column[70]. At the average depth of the ocean, the pressure is approximately 400 atm.

There are two basic problems posed by pressure for fishes. Species, because of their depth distributions, may experience (1) high hydrostatic pressure and also (2) variation in pressure. In a species occurring over a broad bathymetric range, populations inhabiting different depths will experience different pressure regimes. Some deep-living demersal fish species have depth ranges of 1000 m or more (e.g., ref. 12). Thus, conspecific populations may experience pressure regimes differing by 100 atm. Individuals may experience a range of pressures because of their life history characteristics. For instance, some midwater fishes undertake diurnal vertical feeding migrations of 1000 m or more (e.g., ref. 96). Other species migrate vertically through ontogeny, with different life history stages developing at different depths in the water column (e.g., refs. 51, 52, 55, 65, 90, 91). Other species migrate for reproduction (e.g., ref. 28). Thus individuals may experience a change of tens to hundreds of atmospheres pressure either daily, or through their life history. In addition to the differences in pressure regime experienced by vertically migrating fishes, depending on surface temperatures, there may be a temperature change of more than 10°C in going from cold deep waters to warm surface waters.

The topics to be addressed in this chapter concern the potential of hydrostatic pressure to disrupt the biochemical and physiological processes of deep-sea fishes. What are the bases of the effects of hydrostatic pressure? What biochemical processes are affected by increased hydrostatic pressure? Are the pressures experienced in the marine environment sufficient to elicit adaptation, and how much evolution is required?

II. Bases of pressure effects

Pressure acts on the volume changes associated with a process. There are a number of reviews detailing the effects of hydrostatic pressure and the theoretical framework for these effects (e.g., refs. 34, 37–39, 43, 53, 95).

The pressure sensitivity of a chemical reaction is dependent upon the sign and magnitude of the volume change associated with the equilibrium constant (K) or rate constant (k), according to the following:

$$\frac{\delta \ln K}{\delta P} = -\frac{\Delta V}{RT}$$

$$\frac{\delta \ln k}{\delta P} = -\frac{\Delta V^{\ddagger}}{RT}$$

where P is pressure, R is the gas constant, T is the absolute temperature and ΔV is the difference in volume between the final and initial states of the system at equilibrium and $\Delta V^{\ddagger}$ is the difference in volume of the system between the transition and ground states. These volume changes are determined by the changes in the volume of the entire solute-solvent system. The response of a reaction or a step in a reaction to an increase in pressure is governed by LeChatelier's principle. If a reaction step proceeds with an increase in volume, increased pressure will inhibit that step. Processes which proceed with a decrease in volume will be enhanced by an increase in pressure. Only if there is no net volume change will a process be independent of pressure changes. Table 1 illustrates the magnitude of the effects of pressure for several combinations of P and $\Delta V^{\ddagger}$.

TABLE 1

The effects of various pressure – volume change combinations on the ratio of the rate constants k_p/k_o^a

$\Delta V^{\ddagger}$ (cm³/mol)	k_p/k_o			
	50 atm	100 atm	400 atm	1000 atm
+10	0.98	0.96	0.84	0.65
+20	0.96	0.92	0.70	0.42
+100	0.80	0.64	0.17	0.01
+200	0.64	0.42	0.03	0.0002
−10	1.02	1.04	1.19	1.55
−20	1.04	1.09	1.42	2.40
−100	1.24	1.55	5.77	80
−200	1.55	2.40	33	6412

[a] Rate constants k_p/k_o were calculated at 278°K according to the following equation: $k_p = k_o \exp(-P\Delta V^{\ddagger}/RT)$, where k_p and k_o are the rate constants at elevated and atmospheric pressure, respectively, and P is pressure, $\Delta V^{\ddagger}$ is the activation volume change, R is the gas constant, and T is absolute temperature.

Changes in the system volume may result from changes in the volumes of the macromolecular components and from changes in the density and organization of water. The volume of proteins may change during catalysis due to changes in the packing efficiency of the amino acids or exposure of various amino acid side chains. The organization of lipid membranes may change on increase of hydrostatic pressure because of changes in the packing of acyl chains.

Volume changes in the solvent may result from exposure of charges on macromolecules, increasing the electrostriction of water. The hydrophobic inter-subunit contact regions of multimeric proteins may be imperfectly packed, leading to their exposure and interaction with water upon application of pressure[95]. This would result in a denser organization of the water molecules around non-polar residues, a decrease of the system volume, and the depolymerization of the multi-subunit protein. Examples of volume changes are tabulated in Table 2. Formation or exposure of charge groups and the exposure of hydrophobic groups result in changes in the hydration of the exposed groups, causing a decrease in the volume of the solute – solvent system.

III. Perturbation by pressure

Based on studies of the effects of hydrostatic pressure on model compounds and other preparations (e.g., Table 2 and refs. 14, 34, 37, 95), it is apparent that fundamental biochemical processes are potentially susceptible to perturbation by hydrostatic pressure. Because the sign and magnitude of volume changes associated with different processes differ, not only might individual biochemical reactions be disrupted, but changing pressure could perturb the entire regulatory organization and balance of metabolism. Only a few proteins and membrane

TABLE 2

Volume changes of chemical processes [a]

Reaction	ΔV(cm^3/mol)	Reference
Neutralization/ion-pair formation		
$H^+ + OH^- > H_2O$	21.3	8
Imidazole-H^+ > imidazole + H^+	1.1	41, 64
$HPO_4^{2-} + H^+ > H_2PO_4^-$	24	56
Protein-$COO^- + H^+$ > protein-COOH	10	41, 64
Protein-$NH_3^+ + OH^-$ > protein-$NH_2 + H_2O$	20	41, 64
Tris-H^+ > Tris + H^+	1	56
Hydrogen bond formation		
Polylysine helix formation	−1.0	58
Poly-(adenylic and uridylic acid) helix formation	1.0	92
Hydrophobic hydration		
$C_6H_6 > (C_6H_6)_{water}$	−6.2	40
$(CH_4)_{hexane} > (CH_4)_{water}$	−22.7	40
Polar group hydration		
n-Propanol > (n-propanol)$_{water}$	−4.5	24
Protein subunit dissociation		
Actin	−9 to −139	93
Lactate dehydrogenase, tetramer > 4 monomers	−170 to −220	42
Solubilized sarcoplasmic reticulum		
Ca-ATPase dimer > monomer	−167	94
70S ribosomal subunit > 30S + 50S	−242	60
Tubulin	−90	69
Hexokinase dimer > monomer	−115 to −160	68
β_2 dimer of tryptophan synthetase > monomer	−162	85
Activation volumes		
Acetylcholinesterase	28 to 21	36
Alkaline phosphatase	−25 to −2	39
Lactate dehydrogenase	0 to 13	81
Na^+/K^+-ATPase	24 to 105	15, 25
Pyruvate carboxylase	7	63

[a] More extensive compilations are provided in refs. 34, 37 and 53. The measurement conditions of these reactions may influence the apparent volume changes observed.

systems of marine fishes have been studied with regard to the effects of hydrostatic pressure. It is likely however that the differences in the pressure responses observed in these comparisons of deep- and shallow-occurring species may reflect general trends for other biochemical systems.

IV. Proteins: subunit assembly and stability

As noted by Ruan and Weber[68], the association volumes of dimers and tetramers are in the range of 100 to 200 cm^3/mol (see Table 2). These large volume changes upon association of subunits of enzymes from terrestrial warm-adapted organisms, suggest that the compressibilities of multisubunit complexes may be an important

TABLE 3

The pressures at which M_4-lactate dehydrogenase homologs of six macrourid species are half inactivated ($P_{1/2}$) [a]

Species	Depth of maximal abundance (m)	$P_{1/2}$ (atm)
Nezumia bairdii	600	840 ± 80
Coryphaenoides rupestris	1000	565 ± 100
Coryphaenoides acrolepis	1200	770 ± 40
Coryphaenoides carapinus	2000	1265 ± 90
Coryphaenoides armatus	2900	1715 ± 95
Coryphaenoides leptolepis	3500	1570 ± 20

[a] The C. rupestris data are at 3°C; all other data are at 4°C. From Hennessey and Siebenaller[32].

locus of pressure perturbation (e.g., ref. 63). The dead space or free volume between associated subunits is a source of the of the positive volume changes associated with subunit aggregation[95].

For the muscle-type (M_4) lactate dehydrogenase (LDH, EC 1.1.1.27, NAD^+: L-lactate oxidoreductase) of six species from the deep-sea fish family Macrouridae, the pressures that dissociate the enzyme homologs are correlated with the depth of occurrence of the species[32] (Table 3). Although there is a correlation, the pressures that inactivate the tetrameric enzymes are much greater than the pressures which the enzymes are exposed to *in situ*. This is analogous to studies of thermal denaturation of proteins, which demonstrate a correlation of denaturation temperature with cell temperature, but find that the denaturation temperatures are much greater than the cell temperatures[35]. Because thermal stability and resistance to proteolysis are often correlated, the thermal denaturation temperature may reflect the resistance of enzymes to proteolysis *in vivo*[20,27,49].

The resistance of LDH to pressure denaturation correlates with resistance to inactivation by proteolysis[33] (Table 4). The LDH homologs of 5 shallow-occurring cold-adapted fishes are more susceptible to trypsinolysis at both atmospheric pressure and 1000 atm than are the homologs of 6 deep-living species. At 200 atm pressure, a pressure typical of the bathyal environment, tryptic inactivation of the LDH from the shallow-living scorpaenid, *Sebastes melanops*, is increased 14% over the rate at atmospheric pressure. In contrast, 200 atm pressure did not increase the rate of tryptic inactivation of the homologs of two bathyal macrourid species[33] (Fig. 1). Minor conformational perturbations of proteins by pressure may increase their susceptibility to proteolysis. Increased structural stability of the proteins of deep-sea species may prevent too rapid turnover of proteins, which would be energetically wasteful in the food-poor deep sea.

The muscle protein actin from deep-occurring fishes also displays increased structural stability. The depolymerization of filamentous actin from the macrourid *Coryphaenoides armatus*, which commonly occurs down to 5000 m, has a smaller ΔV of assembly than other marine fishes[93] (Table 5).

TABLE 4

Inactivation of M_4-lactate dehydrogenase homologs by hydrostatic pressure and proteases [a]

Species	Depths of abundance (m)	$P_{1/2}$ (atm)	Rate of proteolytic inactivation (h^{-1})	
			Trypsin	Subtilisin
Shallow-living				
Sebastes melanops	0– 100	1085 ± 60	0.744 ± 0.047	0.342 ± 0.005
Sebastes pinniger	0– 300	1085 ± 60	0.238 ± 0.012	0.798 ± 0.046
Platichthys stellatus	0– 300	800 ± 45	0.230 ± 0.012	0.540 ± 0.022
Parophrys vetulus	20– 330	1350 ± 70	0.395 ± 0.017	0.524 ± 0.024
Microstomus pacificus	18– 914	935 ± 55	0.708 ± 0.032	0.351 ± 0.008
Mean ± S.E.		1051 ± 205	0.463 ± 0.249	0.504 ± 0.185
Deep-living				
Anoplopoma fimbria	200–1670	1750	0.099 ± 0.008	0.081 ± 0.005
Antimora microlepis	400–3300	1330 ± 65	0.122 ± 0.010	0.177 ± 0.015
Antimora rostrata	825–2500	1620 ± 80	0.076 ± 0.010	0.165 ± 0.004
Coryphaenoides rupestris	550–1960	565 ± 100	0.088 ± 0.007	0.210 ± 0.011
Coryphaenoides acrolepis	475–2825	770 ± 40	0.124 ± 0.011	0.111 ± 0.009
Coryphaenoides leptolepis	1188–4693	1570 ± 20	0.139 ± 0.012	0.194 ± 0.005
Mean ± S.E.		1171 ± 478	0.108 ± 0.024	0.157 ± 0.050

[a] The pressures ($P_{1/2}$) at which the teleost enzymes are half inactivated were determined at 4°C. Proteolytic inactivation was determined at 10°C and atmospheric pressure with 0.5 mg/ml trypsin and 0.05 mg/ml subtilisin. From ref. 33.

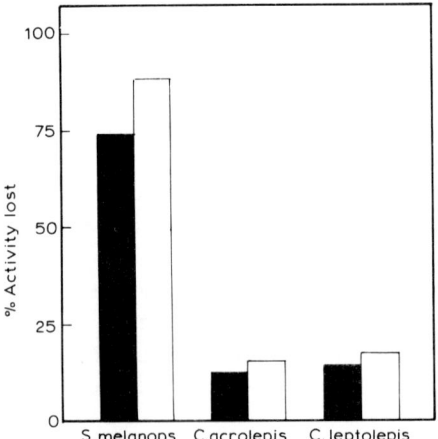

Fig. 1. Inactivation of M_4-lactate dehydrogenase homologs of *Sebastes melanops*, *Coryphaenoides acrolepis* and *C. leptolepis* by trypsin at atmospheric pressure and 200 atm pressure. The enzyme homologs were incubated for 60 min at 10°C with 0.5 mg/ml trypsin at atmospheric pressure (filled bars) and with 0.5 mg/ml trypsin at 200 atm pressure (open bars). There was no loss of activity during incubations of the LDH in the presence of 0.5 mg/ml albumin at 200 atm pressure. From ref. 33.

TABLE 5

The volume change of assembly of F-actin from skeletal muscle of cold-adapted teleost fishes [a]

Species	Depths of maximal abundance (m)	ΔV of assembly (cm^3/mol)
Sebastolobus alascanus	250	56
Coryphaenoides acrolepis	1200	63
Halosauropsis macrochir	2300	58
Coryphaenoides armatus	2900	9

[a] The ΔV of self-assembly is extrapolated to atmospheric pressure. Adapted from ref. 93.

V. Proteins: catalytic rate

The apparent activation volume changes ($\Delta V^{\ddagger}$) of enzyme reactions tend to be on the order of $\pm$ tens of cm^3/mol (Table 2; see also ref. 53). At saturating substrate concentrations, the presssure dependence of the reaction depends solely on the $\Delta V^{\ddagger}$. Although these small volume changes result in only modest inhibition of catalysis, the LDH homologs of deep-living fishes tend to have smaller apparent volume changes than do the homologs of shallow species[81] (Table 6).

VI. Proteins: ligand binding

The specific recognition and binding of ligands is a critical step in enzymatic catalysis, respiratory pigment function, and activation of receptors. In evolutionary temperature adaptation, the binding characteristics of substrate (pyruvate) and cofactor (NADH) are strongly conserved for homologs of the enzyme M_4-LDH[29,35,98,99] (Fig. 2). For instance, despite a 6-fold variation in K_m of NADH

TABLE 6

Apparent volume changes calculated from the pressure inhibition of optimal velocities of shallow- and deep-living fish species M_4-lactate dehydrogenase [a]

Species	$\Delta V \pm$ S.E. (cm^3/mol)	% Inhibition of 1 atm rate at		
		68 atm	204 atm	340 atm
Shallow-living				
Sebastolobus alascanus	12.8 ± 0.53	4	11	17
Scorpaena guttata	12.8 ± 0.94	4	11	17
Pagothenia borchgrevinki	5.2 ± 1.35	2	5	7
Deep-living				
Sebastolobus altivelis	8.1 ± 0.22	2	7	11
Antimora rostrata	0.3 ± 1.45	0	0	0
Coryphaenoides acrolepis	2.6 ± 0.83	1	2	4
Halosauropsis macrochir	4.2 ± 0.40	1	3	5

[a] Reactions measured at 10°C. From ref. 81.

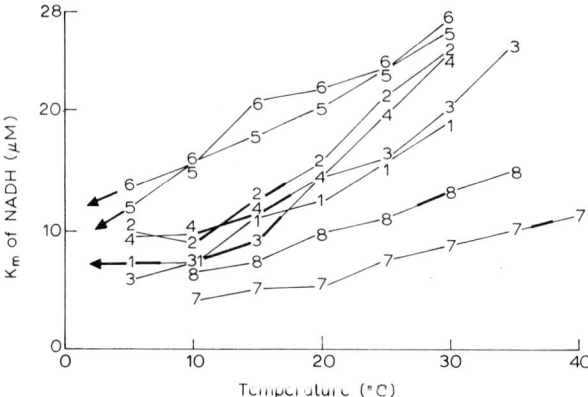

Fig. 2. Effects of temperature on the apparent K_m of coenzyme, NADH, for purified M_4-lactate dehydrogenase homologs of 7 teleost fishes and a mammal: (1) *Sebastolobus alascanus*, (2) *Sebastes pinniger*, (3) *Parophrys volitans*, (4) *Platichthys stellatus*, (5) *Antimora rostrata*, (6) *Coryphaenoides acrolepis*, (7) pig, (8) *Pterois vetulus*. The body temperatures of the species are indicated with a heavy solid line. Assays were conducted at atmospheric pressure in 100 mM potassium phosphate buffer, pH 7.5. From ref. 98.

values with measurement temperature, at the cell temperatures at which the enzyme homologs function the K_m are within the range 7.4 to 13.8 μM[98] (Fig. 2). *In vivo*, dehydrogenases are thought to be saturated with coenzyme due to the high affinities of these enzymes for coenzyme. The direction of dehydrogenase function will be determined by the redox balance of the coenzyme pool[5].

The K_m values of substrate approximate *in vivo* substrate concentrations, which tend to be similar among species[35]. By maintaining K_m values within a narrow range during evolutionary temperature adaptation, the relationship of K_m and *in vivo* substrate concentration is maintained, and the catalytic reserve capacity and regulatory characteristics of the enzyme are preserved (see ref. 35).

The effects of hydrostatic pressure on the K_m of NADH for dehydrogenase homologs of deep- and shallow-occurring cold-adapted demersal and benthopelagic species indicate two distinct responses[73,75,78,80,81] (Fig. 3–6). At atmospheric pressure and 5°C, the K_m values of coenzyme for all of the enzyme homologs are identical. With increasing measurement pressures, the K_m values increase markedly for species occurring shallower than approximately 550 m. The K_m values increase by a factor of 2 or more with the application of 68 atm pressure for these species. K_m values tend to increase with increased application of pressure. The K_m values for the homologs of species occurring deeper than approximately 600 m are relatively insensitive to pressure increases. K_m values are either unaffected by pressure increases, or increased less than 25% (Figs. 3–6).

For M_4-LDH homologs of demersal and benthopelagic species, the K_m of pyruvate follows a similar pattern. Species occurring shallower than approximately 550 m have K_m of pyruvate values that are increased approximately 33% by 68 atm of pressure. In contrast, for species occurring deeper than 600 m, K_m of pyruvate

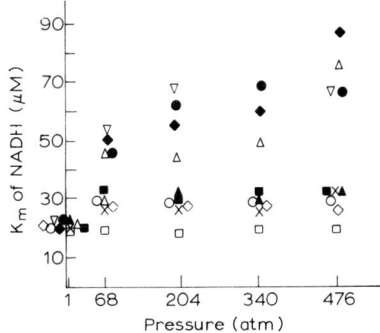

Fig. 3. Effects of hydrostatic pressure on the apparent K_m of coenzyme, NADH, for M_4-lactate dehydrogenase homologs of shallow- and deep-living fishes. Assays were at 5°C in 80 mM Tris-HCl, pH 7.5. Symbols: deep-living fishes — *Coryphaenoides armatus* (□), *C. carapinus* (⊗), *C. acrolepis* (▲), *C. leptolepis* (▼), *Sebastolobus altivelis* (■), *Nezumia bairdii* (×), *Antimora microlepis* (○), *Halosauropsis macrochir* (◇). Shallow-living fishes – *Sebastolobus alascanus* (●), *Pagothenia borchgrevinki* (△), *Sebastes melanops* (▽), *Scorpaena guttata* (♦). From refs. 78, 80, 81.

values are unaffected by pressures as high as 476 atm, the highest pressure tested[80,81].

A number of conclusions may be drawn from these studies:

1. Hydrostatic pressures typical of the bathyal environment (tens of atmospheres) are sufficient to perturb the function of dehydrogenases.

2. Adaptation to these moderate, bathyal pressures is important. This is evidenced by the remarkable convergent evolution of the LDH homologs of species belonging to four different fish families. Each of these homologs has independently evolved resistance to perturbation by hydrostatic pressure (Fig. 3).

3. Shallow-living species adapted to cold temperatures do not necessarily have enzymes that are pre-adapted for function in the deep-sea environment.

4. Coenzyme binding seems particularly susceptible to pressure perturbation. The coenzyme binding domain of dehydrogenases may contain one-half of the

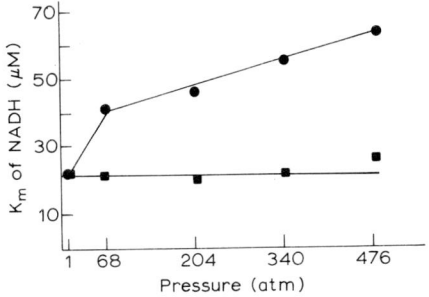

Fig. 4. Effects of hydrostatic pressure on the apparent K_m of NADH for cytoplasmic malate dehydrogenase-1 isozymes (EC 1.1.1.37, NAD^+ :L-malate oxidoreductase) of shallow- and deep-living fishes. The enzymes were purified to homogeneity and the assays were conducted at 5°C in 80 mM Tris-HCl, pH 7.5, 100 mM KCl and 6 mM oxaloacetate. Symbols as in Fig. 3. From ref. 75.

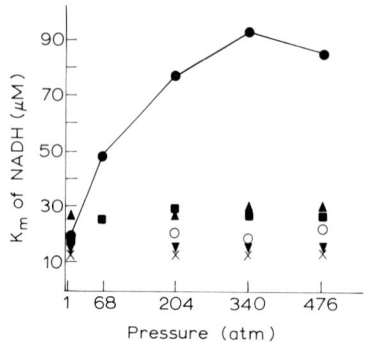

Fig. 5. Effects of hydrostatic pressure on the apparent K_m of NADH for cytoplasmic malate dehydrogenase-2 isozymes (EC 1.1.1.37, NAD^+ : L-malate oxidoreductase) of shallow- and deep-living fishes. Symbols as in Fig. 3. The enzymes were purified to homogeneity and the assays were conducted at 5°C in 80 mM Tris-HCl, pH 7.5, 100 mM KCl and 0.3 mM oxaloacetate. From refs. 75, 78.

molecule[66]. Even minor pressure-induced changes in protein conformation might significantly perturb coenzyme binding.

5. Enzymes which have evolved K_m values which are not perturbed by moderate pressures seem to be able to resist perturbation by higher pressures.

These observations suggest that for a species colonizing the deep-water environment, the adjustment of enzymes to maintain appropriate K_m values may be an important evolutionary requirement for success. A structural comparison of the M_4-LDH homologs of two scorpaenid species suggests that acquisition of pressure tolerance may require only minimal changes in the primary structure of a dehydrogenase[76]. Demersal adults of *Sebastolobus alascanus* are common between 180 and 330 m; *S. altivelis* adults are common between 550 and 1300 m. The M_4-LDH homologs of these two species differ markedly in their susceptibility to perturbation by pressure[80] (Fig. 3). Because these species are genetically close[73] and may

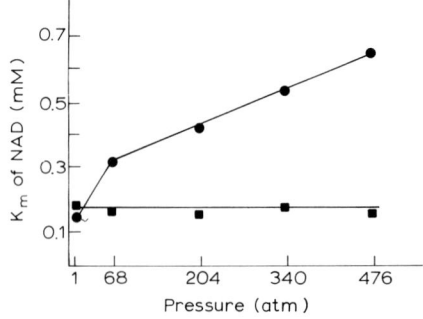

Fig. 6. Effects of hydrostatic pressure on the apparent K_m of NAD for glyceraldehyde 3-phosphate dehydrogenase (EC 1.2.1.12, NAD^+ : D-glyceraldehyde 3-phosphate oxidoreductase [phosphorylating]) of shallow- and deep-living fishes. The enzymes were purified to homogeneity and the assays were conducted at 5°C in 80 mM Tris-HCl, pH 8.5, 100 mM KCl, and 1.5 mM D-glyceraldehyde 3-phosphate. Symbols as in Fig. 3. From ref. 75.

not have accumulated many random amino acid replacements, the tryptic peptides of these LDH homologs were mapped in order to quantify the amount of change in the primary sequence required to evolve adaptation to hydrostatic pressure[76]. There was only one clear amino acid replacement between the two species. The pressure-sensitive enzyme of *S. alascanus* has a histidine residue at position 115. The pressure-resistant LDH of *S. altivelis* has an asparagine. Ionization of the imidazole of histidine is pressure-sensitive (see Table 2), although the volume change associated with the ionization is not large. However, histidines at critical locations in a protein may give rise to anomalously large volume changes[41,64]. Histidine 115 is adjacent to a region of the protein which undergoes a large conformational change upon coenzyme binding[1]. This information, and the pH-dependence of the effects of pressure on coenzyme binding[74] suggest that a histidine residue(s) may be involved in the pressure sensitivity of the *S. alascanus* enzyme. Ionization of the imidazole of histidine may influence the conformational changes associated with coenzyme and substrate binding.

The peptide mapping protocol used may not have identified all of the amino acid replacements between the two *Sebastolobus* LDH homologs. Clearly, however, only minor changes in the amino acid sequence are required to acquire pressure tolerance. Thus, although the need for acquiring pressure-tolerance may be widespread for a class of enzymes such as dehydrogenases, pressure-resistance might be achieved with only minor changes in the proteins.

A number of mesopelagic (e.g., Myctophidae) and bottom-associated bathyal and abyssal species (e.g., Macrouridae and Moridae) maintain gas-filled swim bladders[50]. The primary gas found in the swimbladder is oxygen[23,71]. At depth, oxygen must be secreted against a partial pressure gradient into an oxygen-filled space in which the swimbladder gas pressure is approximately equal to the ambient hydrostatic pressure. The concentrating ability of the rete mirabile countercurrent mechanism is effected by salting-out and a reduced affinity of hemoglobin for oxygen due to the lowered blood pH. Noble *et al*.[57] examined the hemoglobins of six demersal deep-living fishes which maintain gas-filled swimbladders and two species without swimbladders. All of the hemolysates exhibited the Root effect. At lowered pH in the presence of organic phosphate modulators, the hemoglobins of species with swimbladders were functionally heterogeneous, having two distinct ligand binding affinities. The hemoglobins from the deeper dwelling fishes with gas-filled swimbladders had ligand binding affinities that are among the lowest reported for any vertebrate. Species with deeper depth distributions have hemoglobins with lower affinities as would be expected if such low affinities were to facilitate oxygen pumping into an oxygen gas phase at high pressure.

Alexander[2,3] and Mcdonald[44] discuss the problems posed by pressure for a vertically migrating fish secreting and resorbing gas to maintain neutral buoyancy.

VII. Membranes: homeoviscous adaptation

Membranes are extremely sensitive to changes in temperature and hydrostatic pressure. Pressure effects on cell membranes may be the most extensive problem faced by a species colonizing deep water (e.g., refs. 16, 17, 45, 46, 54). The proper physiological function of membranes presumably requires the maintenance of precise levels of microviscosity[72], and perturbation of the degree of ordering of membrane lipid acyl chains by environmental factors will influence the function of membrane associated proteins (e.g., ref. 15). Increased pressure and decreased temperature increase the degree of order of lipid membranes by increasing the organization of the acyl chains of lipids. The relationship of pressure and temperature effects on membrane phase transitions are described by the Clayperon relationship:

$$\frac{dT}{dP} = \frac{\Delta V T}{\Delta H}$$

where dT/dP is the change in phase temperature with application of pressure, ΔV is the volume change, T temperature and ΔH the enthalpy change.

A pressure increase of 1000 atm orders membranes by an amount equivalent to a 15 to 25°C temperature decrease (e.g., refs. 14, 18, 45, 46). Thus, the apparent temperature experienced by the membrane of an organism living at the average depth of the ocean, where temperatures are typically 1 to 2°C, would be reduced to -5 to -12°C. The resultant increase in membrane viscosity may adversely affect the function of enzymes dependent on membrane fluidity (e.g., ref. 15). Compensatory adjustments in the composition of membranes to maintain an optimal state of 'fluidity' has been termed homeoviscous adaptation[19,86]. These compositional adjustments may include changes in cholesterol content, changes in the degree of lipid saturation and changes in the length of the acyl chain[19].

Cossins and McDonald[16] used the fluorescence polarization of the fluorophore 1,6-diphenyl-1,3,5-hexatriene to estimate and compare the degree of membrane order in fish brain myelin membrane fractions. Fluorescence polarization correlated significantly with the depth of capture of the fish (200 to 4000 m). This represented partial adaptation under the predictions of homeoviscous theory.

In liver mitochondria from deeper occurring fishes, the unsaturated fatty acid fraction of phospholipids showed relative increases in phosphatidylcholine and phosphatidylethanolamine[17]. Consistent with homeoviscous theory, the deep-living species had a higher proportion of unsaturated fatty acids than did shallow-living species.

A study of the composition of brain ganglioside fatty acids from 39 fish species with different body temperatures and habitat depths also supports homeoviscous theory[6]. Cold-adapted fishes had a higher degree of unsaturation of brain gangliosides. The gangliosides of the deep-living *Antimora rostrata* (2000 m) had the lowest relative content of saturated fatty acids and stearic acids, the highest monoenoic acid content and a high amount of polyenoic acids. These data suggest that the effects of pressure and low temperature on ganglioside composition are indeed synergistic.

In a barophilic marine bacterium the relative amounts of polyunsaturated fatty acids varied as a function of pressure at which the cultures were grown, consistent with the predictions of homeoviscous theory[22]. The adjustment in fatty acid composition occurred on a short, acclimatory time scale.

VIII. Membrane-associated processes

The theory of homeoviscous adaptation is based on the observation that the viscosity of the lipid bilayer is adjusted to offset environmentally induced changes[19]. By maintaining an appropriate state, membrane function is conserved. Among the functions which may be sensitive to the fluidity state of the membrane are ion transport, signal transduction, and the role of the membrane as a permeability barrier (e.g., refs. 45–47).

A variety of biochemical and physiological studies indicate the importance of pressure effects on membranes for aquatic organisms. One example of a pressure-induced disruption of physiological function is the high pressure neurological (or nervous) syndrome (HPNS) which affects both terrestrial and aquatic organisms. The manifestations of this syndrome include tremors and convulsions[9]. Onset of the symptoms is dependent on the rate of pressurization and the pressures achieved. Shallow-living animals are affected by pressures in the range of 20 to 100 atm[9,30,48]. The origins and mechanisms of HPNS are not well established, but the syndrome is most likely a complex set of phenomena at the level of excitable tissues (e.g., refs. 4, 67). Disruption of excitable tissue function may result from, among other things, the effects of pressure on ion channels (see refs. 31, 59) or disruption of signal transduction (e.g., refs. 61, 79). Membrane function might also be impaired through loss of membrane components 'squeezed' from membranes rigidified by increased pressure and decreased temperature (see ref. 21).

The effects of pressure on transmembrane signaling have been studied in a deep-sea fish. The A_1 adenosine receptor and its associated effector elements, a guanine nucleotide binding regulatory protein and adenylyl cyclase, were studied in brain tissue of the deep-living morid, *A. rostrata*[79]. Adenosine agonists acting at the A_1 adenosine receptor negatively modulate adenylyl cyclase[89]. In the tissues of *A. rostrata* the functioning of this transmembrane signaling complex was not impaired by 272 atm, a pressure similar to the pressures experienced by this bathyal fish (Fig. 7). Pressure does not inhibit basal adenylyl cyclase activity, and the efficacy of adenosine agonists in inhibiting the adenylyl cyclase activity is not decreased. In contrast, both basal adenylyl cyclase activity, and modulation of the cyclase activity adenosine agonists in the brain tissue of shallower-occurring species were significantly affected by pressure. Basal adenylyl cyclase activity in these species was inhibited 11 to 25% by 136 atm pressure.

Another membrane-associated system from marine teleosts has been studied. The integral membrane protein Na,K-ATPase, prepared from fish gills, shows a depth-related pattern of pressure adaptation[25]. The enzymes from deep-living and cold-adapted species are less inhibited by hydrostatic pressure than are the enzymes of shallow-living and warm-adapted species. The ordering of the lipids in the gill membranes, as measured by the steady-state anisotropy of a hydrophobic

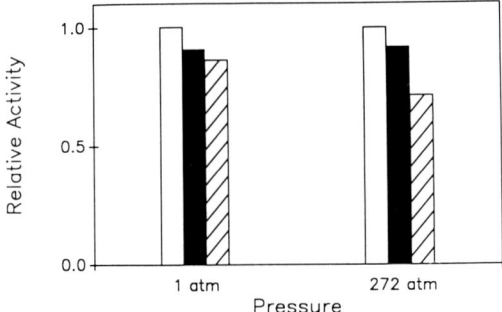

Fig. 7. The effects of hydrostatic pressure on *Antimora rostrata* basal adenylyl cyclase activity (open bar) and inhibition of basal adenylyl cyclase activity by the adenosine analogs N^6-cyclopentyladenosine (100 μM; filled bar) and 5'-N-ethylcarboxamidoadenosine (100 μM; hatched bar). Membranes were incubated at atmospheric pressure or 272 atm for 2 h at 5°C. All values are standardized to the atmospheric pressure basal adenylyl cyclase activity. The 1 atm and 272 atm basal activities were 3.3 pmol/min/mg protein. From ref. 79.

fluorescence probe, 1,6-diphenyl-1,3,5-hexatriene, increased with decreased temperature[26] (Fig. 8). The activity of the gill Na,K-ATPases decreased with increased pressure and decreased temperature. The Na,K-ATPase activity depends on membrane fluidity; equivalent fluorescence polarization values, induced by different combinations of temperature and pressure changes, caused equivalent changes in the enzyme activity (Fig. 9).

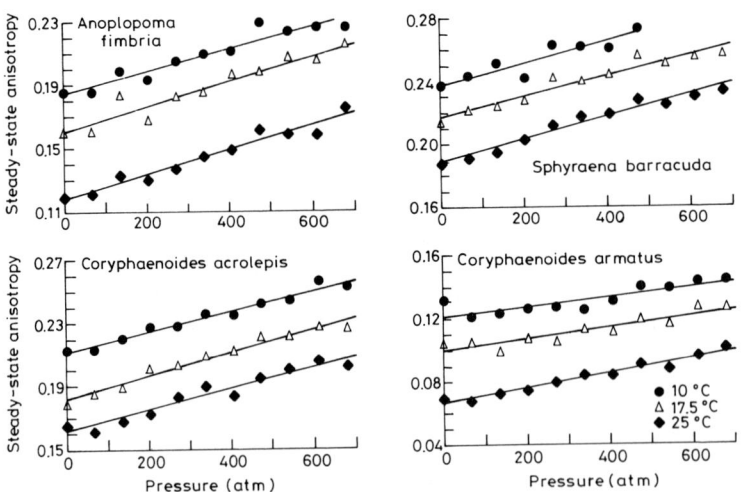

Fig. 8. The effects of temperature and hydrostatic pressure on the steady-state anisotropy of the fluorescent probe, 1,6-diphenyl-1,3,5-hexatriene in teleost gill membrane preparations. The steady state anisotropy was used to estimate the degree of ordering of the membrane lipids. ♦, 25°C; △, 17.5°C; ●, 10°C. *Sphyraena barracuda* is a warm-adapted, shallow-living species. The other three species are cold-adapted. Their depth distributions are given in Tables 3 and 4. From ref. 26.

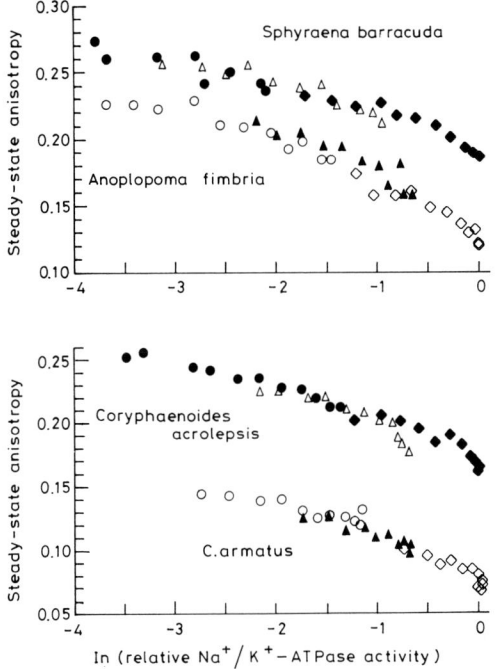

Fig. 9. The relationship of the steady-state anisotropy of the fluorescent probe, 1,6-diphenyl-1,3,5-hexatriene, and relative gill Na,K-ATPase activity in four teleost fishes. Measurements of steady-state anisotropy, as an estimate of membrane fluidity, and Na,K-ATPase activity were made at different pressures and temperatures. Symbols and species as in Fig. 8. Increased pressure increased the degree of lipid order (Fig. 8) and decreased the Na,K-ATPase activity in all the preparations. The enzyme activities are expressed relative to the 25°C, atmospheric pressure values. From ref. 26.

IX. Biosynthesis and gene regulation

Pressure may influence processes in the cell cycle. Meiosis, mitosis and gene transcription are known to be affected by pressure increases (see references in ref. 82). Pressure has been observed to block both nucleic acid synthesis and protein synthesis[100]. However, no data are available on marine teleosts.

Bartlett et al.[7] have reported that pressure modulates the expression of a gene of a deep-sea barophilic bacterium. Environmental pressure may play a role in the control of the genetic or developmental program of an organism. This observation of control of gene expression and the sensitivity of mitosis and macromolecular synthesis[100] to pressure raise intriguing questions about the adaptations of these processes in fishes which display ontogenetic migrations. Does pressure pose an adaptational problem for development or are developmental events triggered by pressure? Currently, there are no data to address this.

X. Coping with pressure variation

The magnitudes of the pressure perturbation which elicit an adaptive response differ for different biochemical systems. For example, coenzyme binding to dehydrogenases from cold-adapted, shallow-living fishes is susceptible to perturbation by tens of atmospheres of pressure. That this perturbation is clearly important is seen in the fact that deep-living species have independently converged on the solution of pressure-resistant dehydrogenases[75,81] (Figs. 3–6). In contrast, adaptation of the membrane systems of cold-adapted species may require higher pressures to evoke an adaptive response. Based on their study of teleost gill Na,K-ATPase, Gibbs and Somero[25] hypothesize that adaptive responses in membrane-based systems require pressures of 200 atm or more. The data available are too limited to evaluate this hypothesis. However, the study of the A_1 receptor – G protein – adenylyl cyclase complex from the deep-living *Antimora rostrata*[79] is in accord with the hypothesis.

Vertically migrating or broadly distributed fishes may experience pressure changes of tens to hundreds of atmospheres. The deep-living scorpaenid fish, *Sebastolobus altivelis* is demersal as an adult. The reported depth range of the adults is 200 to 1550 m. An electrophoretic survey of populations of the demersal adults from over 70% of the known depth range of the species found no variation in the isozyme composition of 20 proteins related to depth of occurence[73]. In *Sebastolobus* gelatinous egg masses are produced which float to the surface[62], and the larval stages and part of the juvenile period are spent in the water column. Approximately 20 months are spent in the water column before metamorphosis into a demersal juvenile occurs[55]. The depth range traversed during development is up to 1550 m.

Pressure increases of tens of atmospheres are sufficient to disrupt the function of dehydrogenases from the shallower-living congener of *S. altivelis*, *S. alascanus*[75,81] (Figs. 3–6). A species which occurs at different depths during ontogeny might rely on depth-specific isozymes appropriate for the pressure-temperature regime experienced, or the species might rely on a single form of a protein throughout development. Based on an electrophoretic study of 22 proteins, *S. altivelis* does not rely on different, depth-specific isozymes[77]. The pelagic and demersal life history stages expressed the same protein forms. The LDH and malate dehydrogenase forms expressed by the pelagic stages were identical in functional characteristics and pressure-insensitivity of K_m values to the forms of the demersal adults[77] (Figs. 3 and 4).

Under changing pressure regimes, the pressure-insensitive dehydrogenases of *S. altivelis* will always have the appropriate K_m value. By using a single-pressure insensitive form of an enzyme throughout its life history, an individual insures that the enzyme will always have appropriate functional characteristics. For a species which experiences a broad ontogenetic migration and commonly maintains populations over an approximately 800 m depth gradient, this may be an optimal solution. The varying metabolic needs of the life history stages appear to be met by

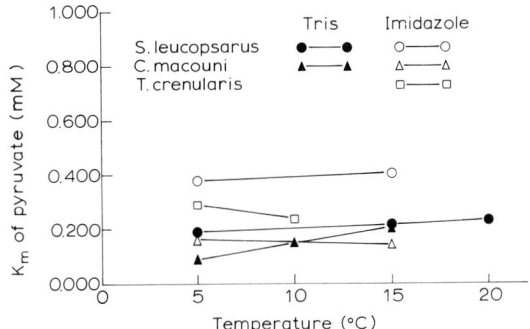

Fig. 10. The effect of temperature on the K_m of pyruvate of the purified muscle-type (M_4)-lactate dehydrogenase homologs of three mesopelagic fishes. *Stenobrachius leucopsarus* (0 to 600 m) and *Tarletonbeania crenularis* (0 to 400 m) are diurnally migrating myctophids. *Chauliodus macouni* (Chauliodontidae) is common from 400 to 500 m. The K_m of pyruvate was determined at the indicated temperature in either 80 mM Tris-HCl, pH 7.5 at the assay temperature, or in 80 mM imidazole-HCl, pH 6.98 at 20°C, in which the buffer pH was allowed to vary with temperature (see ref, 99). Data from ref. 97 and P.H. Yancey, unpublished data.

alteration of the quantities of enzymes maintained in muscle tissue rather than developmental changes in isozymes[77].

For a vertically migrating mesopelagic species, the pressure and temperature regime will vary over the course of a day. Yancey et al.[97] have studied the effects of temperature and hydrostatic pressure on the apparent K_m of pyruvate for LDH homologs from three midwater fishes, two vertically migrating myctophids and a viper-fish (Figs. 10 and 11). For all three species, the K_m of pyruvate varies little over a temperature change of 10 to 15°C (Fig. 10). Pressure has virtually no effect on the K_m of pyruvate of these species (Fig. 11). This contrasts with the marked pressure sensitivity of the K_m of pyruvate of LDH homologs of cold-adapted species which occur in shallow water (e.g., ref. 81). Although based on single study, it appears that the enzymes of midwater species have K_m values with reduced

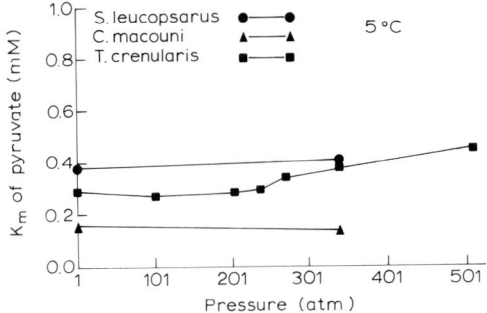

Fig. 11. The effect of hydrostatic pressure on the apparent K_m of pyruvate of the purified muscle-type (M_4)-lactate dehydrogenase homologs of 3 mesopelagic fishes. The K_m of pyruvate was determined at 5°C in the imidazole-HCl buffer described in Fig 8. Data are from ref. 97 and P.H. Yancey, unpublished data.

sensitivity to temperature and pressure changes. These migrating species do, however, maintain relatively high levels of enzymes for locomotion relative to non-migrating species[13,84].

Although based on a small sampling of total protein complement of fishes, it appears, at least for enzymes such as the dehydrogenases, that fishes colonizing the deep water habitat cope with high hydrostatic pressure by evolving enzymes which are resistant to perturbation. This evolutionary response of insentitivity to hydrostatic pressure additionally solves the problem of the variations in pressure regimes encountered by conspecific populations and individuals. Whether other proteins and membrane-associated systems show other patterns remains to be explored.

Acknowledgements. Portions of this work and the preparation of this manuscript were supported by National Science Foundation Grant DCB-8710155 and Office of Naval Research grant N00014-89-J-1865.

XI. References

1. C. Abad-Zapatero, J. Griffith, J. Sussman and M. Rossmann. Refined crystal structure of dogfish M_4 apo-lactate dehydrogenase. *J. Mol. Biol.* 198: 445–467, 1987.
2. Alexander, R.McN. Physical aspects of swimbladder function. *Biol. Rev.* 41: 141–176, 1966.
3. Alexander, R.McN. The energetics of vertical migration by fishes. *Symp. Soc. Exp. Biol.* 26: 273–294, 1972.
4. A. Angel and M.J. Halsey. The effect of increased ambient pressure on lumbosacral motoneurone excitability in the rat. In: *Current Perspectives in High Pressure Biology*, edited H.W. Jannasch, R.E. Marquis and A.M. Zimmerman. London: Academic Press, 1987, pp. 149–158.
5. Atkinson, D.E. *Cellular Energy Metabolism and its Regulation*. New York: Academic Press, 1977.
6. Avrova, N.F. The effect of natural adaptations of fishes to environmental temperature on brain ganglioside fatty acid and long chain base composition. *Comp. Biochem. Physiol.* 78B: 903–909, 1984.
7. Barlett,D., M. Wright, A.A. Yayanos and M. Silverman. Isolation of a gene regulated by hydrostatic pressure in a deep-sea bacterium. *Nature* 342: 572–574, 1989.
8. Bodansky, A. and W. Kauzmann. The apparent molar volume of sodium hydroxide at infinite dilution and the volume change accompanying the ionization of water. *J. Phys. Chem.* 66: 177–179, 1962.
9. Brauer, R.W. Hydrostatic pressure effects on the central nervous system: perspectives and outlook. *Phil. Trans. R. Soc. Lond. B* 304: 17–30, 1984.
10. Brauer, R.W., M.R. Jordan, C.G. Miller, E.D. Johnson, J.A. Dutcher and M.E. Sheehan. Interaction of temperature and pressure in intact animals. In: *High Pressure Effects on Selected Biological Systems*, edited by A.J.R. Pequeux and R. Gilles. New York: Springer-Verlag, 1985, pp.3–28.
11. Brauer, R.W. Introductory considerations regarding high pressure physiology in intact animals. In: *Current Perspectives in High Pressure Biology*, edited by Jannasch, H.W., R.E. Marquis and A.M. Zimmerman. London: Academic Press, 1987, pp. 129–135.
12. Carney, R.S., R.L. Haedrich and G.T. Rowe. Zonation of fauna in the deep sea. In: *The Sea, Vol. 8*, edited by G.T. Rowe. New York: John Wiley, 1983, pp. 371–398.
13. Childress, J.J. and G.N. Somero. Depth-related enzymic activities in muscle, brain and heart of deep-living pelagic marine teleosts. *Mar. Biol.* 52: 273–283, 1979.
14. Chong, P.L.-G., A.R. Cossins and G. Weber. A differential polarized phase fluorometric study of the effects of high hydrostatic pressure upon the fluidity of cellular membranes. *Biochemistry* 22: 409–415, 1983.
15. Chong, P.L.-G., P.A.G. Fortes and D.M. Jameson. Mechanisms of inhibition of (Na,K)-ATPase by hydrostatic pressure studied with fluorescent probes. *J. Biol. Chem.* 260: 14484–14490, 1985.

16. Cossins, A.R. and A.G. Macdonald. Homeoviscous theory under pressure. II. The molecular order of membranes from deep-sea fish. *Biochim. Biophys. Acta* 776: 144–150, 1984.
17. Cossins, A.R. and A.G. Macdonald. Homeoviscous adaptation under pressure. III. The fatty acid composition of liver mitochondrial phospholipids of deep-sea fish. *Biochim. Biophys. Acta* 860: 325–335, 1986.
18. Cossins A.R. and A.G. Macdonald. The adaptations of biological membranes to temperature and pressure: fish from the deep and cold. *J. Bioenerg. Biomem.* 21: 115–135, 1989.
19. Cossins, A.r. and M. Sinensky. Adaptation of membranes to temperature, pressure and exogenous lipids. In: *Physiology of membrane fluidity, Vol. II*, edited by M. Shinitzky. Boca Raton, FL: CRC Press, 1984, pp. 1–20.
20. Daniel, R.M., D.A. Cowan, H.W. Morgan and M.P. Curran. A correlation between protein thermostability and resistance to proteolysis. *Biochem. J.* 207: 641–644, 1982.
21. Deckmann, M., R. Haimovitz and M. Shinitzky. Selective release of integral proteins from human erythrocyte membranes by hydrostatic pressure. *Biochim. Biophys. Acta* 821: 334–340, 1985.
22. DeLong, E.F. and A.A. Yayanos. Adaptation of the membrane lipids of a deep-sea bacterium to changes in hydrostatic pressure. *Science* 228: 1101–1103, 1985.
23. Enns, T., E. Douglas and P.F. Scholander. Role of the swimbladder rete of fish in secretion of inert gas and oxygen. *Adv. Biol. Med. Phys.* 11: 231–244, 1967.
24. Friedman, M.E. and H.A. Scheraga. Volume changes in hydrocarbon-water systems. Partial molar volumes of alcohol-water solutions. *J. Phys. Chem.* 69: 3795–3800, 1965.
25. Gibbs, A. and G.N. Somero. Pressure adaptation of Na^+/K^+-ATPase in gills of marine teleosts. *J. Exp. Biol.* 143: 475–492, 1989.
26. Gibbs, A. and G.N. Somero. Pressure adaptation of teleost gill Na^+/K^+-adenosine triphosphatase: role of the lipid and protein moieties. *J. Comp. Physiol. B.* 160: 431–439, 1990.
27. Goldberg, A.L. and J.F. Dice. Intracellular protein degradation in mammalian and bacterial cells. *Annu. Rev. Biochem.* 43: 835–869, 1974.
28. Gordon, J.D.M. Lifestyle and phenology in deep-sea anacanthine teleosts. *Symp. Zool. Soc. Lond.* 44: 327–359, 1979.
29. Graves, J.E. and G.N. Somero. Electrophoretic and functional enzymic evolution in four species of eastern Pacific barracudas from different thermal environments. *Evolution* 36: 97–106, 1982.
30. Harper, A.A., A.G. Macdonald, C.S. Wardle and J.-P. Pennec. The pressure tolerance of deep sea fish axons: results of Challenger cruise 6B/85. *Comp. Biochem. Physiol.* 88A: 647–653, 1987.
31. Heinemann, S.H., F. Conti, W. Stuhmer and E. Neher. Effects of hydrostatic pressure on membrane processes. *J. Gen. Physiol.* 90: 765–778, 1987.
32. Hennessey, J.P., Jr. and J.F. Siebenaller. Pressure inactivation of tetrameric lactate dehydrogenase homologues of confamilial deep-living fishes. *J. Comp. Physiol.* B155: 647–652, 1985.
33. Hennessey, J.P., Jr. and J.F. Siebenaller. Pressure-adaptive differences in proteolytic inactivation of M_4-lactate dehydrogenase homologues from marine fishes. *J. Exp. Biol.* 241: 9–15, 1987.
34. Heremans, K. High pressure effects on proteins and other biomolecules. *Annu. Rev. Biophys. Bioeng.* 11: 1–21, 1982.
35. Hochachka, P.W. and G.N. Somero. *Biochemical Adaptation*. Princeton, NJ: Princeton University Press, 1984.
36. Hochachka, P.W., K.B. Storey and J. Baldwin. Design of acetylcholinesterase for its physical environment. *Comp. Biochem. Physiol.* 52B: 13–18, 1975.
37. Jaenicke, R. Enzymes under extremes of physical conditions. *Annu. Rev. Biophys. Bioeng.* 10: 1–67, 1981.
38. Jannasch, H.W., R.E. Marquis and A.M. Zimmerman (eds.). *Current Perspectives in High Pressure Biology*. London: Academic Press, 1987.
39. Johnson, F.H., H. Eyring and B.J. Stover. *The Theory of Rates Processes in Biology and Medicine*. New York: John Wiley, 1974.
40. Kauzmann, W. Factors in interpretation of protein denaturation. *Adv. Protein Chem.* 14: 1–63, 1959.
41. Kauzmann, W., A. Bodanszky and J. Rasper. Volume changes in protein reactions. II. Comparison of ionization reactions in proteins and small molecules. *J. Am. Chem. Soc.* 84: 1777–1788, 1962.
42. King, L. and G. Weber. Conformational drift in dissociated lactate dehydrogenases. *Biochemistry* 25: 3626–3637, 1986.
43. Laidler, K.J.. The influence of pressure on the rates of biological reactions. *Arch. Biochem.* 30: 226–236, 1951.
44. Macdonald, A.G. *Physiological Aspects of Deep Sea Biology*. London: Cambridge University Press, 1975.

45. Macdonald, A.G. The effects of pressure on the molecular structure and physiological functions of cell membranes. *Phil. Trans. R. Soc. Lond. B* 304: 47–68, 1984.
46. Macdonald, A.G. The role of membrane fluidity in complex processes under high pressure. In: *Current Perspectives in High Pressure Biology*, edited by H.W. Jannasch, R.E. Marquis and A.M. Zimmerman. London: Academic Press, 1987, pp. 207–223.
47. Macdonald, A.G. Application of the theory of homeoviscous adaptation to excitable membranes: pre-synaptic processes. *Biochem. J.* 256: 313–327, 1988.
48. Macdonald, A.G., I. Gilchrist and C.S. Wardle. Effects of hydrostatic pressure on the motor activity of fish from shallow water and 900 m depths; some results of Challenger cruise 6B/85. *Comp. Biochem. Physiol.* 88A: 543–547, 1987.
49. McLendon, G. and E. Radany. Is protein turnover thermodynamically controlled? *J. Biol. Chem.* 253: 6335–6337, 1987.
50. Marshall, N.B. *Deep-Sea Biology*. New York: Garland STPM Press, 1979.
51. Mead, G.W., E. Bertelsen and D.M. Cohen. Reproduction among deep-sea fishes. *Deep-Sea Res.* 11: 569–596, 1964.
52. Merret, N.R. On the identity and pelagic occurrence of larval and juvenile stages of rattail fishes (Family Macrouridae from 60°N, 20°W and 53°N, 20°W. *Deep-Sea Res.* 25: 147–160, 1978.
53. Morild. E. The theory of pressure effects on enzymes. *Adv. Protein Chem.* 34: 93–166, 1981.
54. Morita, R.Y. Microbial life in the deep sea. *Can. J. Microbiol.* 26: 1375–1385, 1980.
55. Moser, H.G. Development and distribution of juveniles of *Sebastolobus* (Pisces; Family Scorpaenidae). *U.S. Nat. Mar. Fish. Serv. Fish. Bull.* 72: 865–884, 1974.
56. Neuman, R.C., Jr., W. Kauzmann and A. Zipp. Pressure dependence of weak acid ionization in aqueous buffers. *J. Phys. Chem.* 77: 2687–2691, 1973.
57. Noble, R.W., L.D. Kwiatkowski, A. DeYoung, B.J. Davis, R.L. Haedrich, L.-T. Tam and A.F. Riggs. Functional properties of hemoglobins from deep-sea fish: correlations with depth distribution and presence of a swimbladder. *Biochim. Biophys. Acta* 870: 522–563, 1986.
58. Noguchi, H. Studies of the helix-coil transition in poly-L-lysine in film and solution. *Biopolymers* 4: 1105–1113, 1966.
59. Otter, T. and E.D. Salmon. Pressure-induced changes in Ca^{2+}-channel excitability in *Paramecium*. *J. Exp. Biol.* 117: 29–43, 1985.
60. Pande, C. and A. Wishnia. Pressure dependence of equilibria and kinetics of *Escherichia coli* ribososmal subunit association. *J. Biol. Chem.* 261: 6272–6278, 1986.
61. Parmentier, J.L., B.B. Shrivastav and P.B. Bennet. Hydrostatic pressure reduces synaptic efficiency by inhibiting transmitter release. *Undersea Biomed. Res.* 8: 175–183, 1981.
62. Pearcy, W.G. Egg masses and early developmental stages of the scorpaenid fish *Sebastolobus*. *J. Fish. Res. Bd. Canada* 19: 1169–1173, 1962.
63. Penniston, J.T. High hydrostatic pressure and enzymic activity: Inhibition of multimeric enzymes by dissociation. *Arch. Biochem. Biophys.* 142: 322–332, 1971.
64. Rasper, J. and W. Kauzmann. Volume changes in protein reactions. I. Ionization of proteins. *J. Am. Chem. Soc.* 84: 1771–1777, 1962.
65. Robison, B.H. and T.M. Lancraft. An upward transport mechanism from the benthos. *Naturwissenschaften* 71: 322–323, 1984.
66. Rossmann, M.G. and P. Argos. The taxonomy of binding sites in proteins. *Mol. Cell. Biochem.* 21: 161–182, 1978.
67. Rostain, J.-C. The high pressure nervous syndrome at the central nervous system level. In: *Current Perspectives in High Pressure Biology*, edited by H.W. Jannasch, R.E. Marquis and A.M. Zimmerman. London: Academic Press, 1987, pp. 137–148.
68. Ruan, K. and G. Weber. Dissociation of yeast hexokinase by hydrostatic pressure. *Biochemistry* 27: 3295–3301, 1988.
69. Salmon, E.D. Pressure-induced depolymerization of brain microtubules *in vitro*. *Science* 189: 884–886, 1975.
70. Saunders, P.M. and N.P. Fofonoff. Conversion of pressure to depth in the ocean. *Deep-Sea Res.* 23: 109–111, 1976.
71. Scholander, P.F., L. van Dam. Composition of the swimbladder gas in deep sea fishes. *Biol. Bull.* 104: 75–86, 1953.
72. Shinitzky, M. (ed.) *Physiology of Membrane Fluidity, Vols. I and II*. Boca Raton, FL: CRC Press, 1984.
73. Siebenaller, J.F. Genetic variability in deep-sea fishes of the genus *Sebastolobus* (Scorpaenidae). In: *Marine Organisms: Genetics Ecology and Evolution*, edited by B. Battaglia and J. Beardmore. New York: Plenum, 1978, pp. 95–122.

74. Siebenaller, J.F. The pH-dependence of the effects of hydrostatic pressure on the M_4-lactate dehydrogenase homologs of scorpaenid fishes. *Mar. Biol. Lett.* 4: 233–243, 1983.
75. Siebenaller, J.F. Pressure-adaptive differences in NAD-dependent dehydrogenases of congeneric marine fishes living at different depths. *J. Comp. Physiol.* 154: 443–448, 1984.
76. Siebenaller, J.F. Structural comparison of lactate dehydrogenase homologs differing in sensitivity to hydrostatic pressure. *Biochim. Biophys. Acta* 786: 161–169, 1984.
77. Siebenaller, J.F. Analysis of the biochemical consequences of ontogenetic vertical migration in a deep-living teleost fish. *Physiol. Zool.* 57: 598–608, 1984.
78. Siebenaller, J.F. Biochemical adaptation in deep-sea animals. In: *Current Perspectives in High Pressure Biology*, edited by H.W. Jannasch, A.M. Zimmerman and R.E. Marquis. London: Academic Press, 1987, pp. 33–48.
79. Siebenaller, J.F. and T.F. Murray. A_1 adenosine receptor modulation of adenylyl cyclase of a deep-living teleost fish, *Antimora rostrata*. *Biol. Bull.* 178: 65–73, 1990.
80. Siebenaller, J.F. and G.N. Somero. Pressure adaptive differences in lactate dehydrogenases of congeneric fishes living at different depths. *Science* 201: 255–257, 1978.
81. Siebenaller, J.F. and G.N. Somero. Pressure-adaptive differences in the binding and catalytic properties of muscle-type (M_4 lactate dehydrogenases of shallow- and deep-living marine fishes. *J. Comp. Physiol.* 129: 295–300, 1979.
82. Siebenaller, J.F. and G.N. Somero. Biochemical adaptation to the deep sea. *Rev. Aquat. Sci.* 1: 1–25, 1989.
83. Siebenaller, J.F., G.N. Somero and R.L. Haedrich. Biochemical characteristics of macrourid fishes differing in their depths of distribution. *Biol. Bull.* 163: 240–249, 1982.
84. Siebenaller, J.F. and P.H. Yancey. The protein composition of white skeletal muscle from mesopelagic fishes having different water and protein contents. *Mar. Biol.* 78: 129–137, 1984.
85. Silva, J.L., E.W. Miles and G. Weber. Pressure dissociation and conformational drift of the β-dimer of tryptophan synthase. *Biochemistry* 25: 5780–5786, 1986.
86. Sinensky, M. Homeoviscous adaptation – a homeostatic process that regulates the viscosity of membrane lipids in *E. coli*. *Proc. Natl. Acad. Sci. U.S.A.* 71: 522–525, 1974.
87. Somero, G.N. Life at low volume change: hydrostatic pressure as a selective factor in the aquatic environment. *Am. Zool.* 30: 123–135, 1990.
88. Somero, G.N., J.F. Siebenaller and P.W. Hochachka. Biochemical and physiological adaptations of deep-sea animals. In: *The Sea, Vol. 8*, edited by G.T. Rowe. New York: John Wiley, 1983, pp. 261–330.
89. Snyder, S.H. Adenosine as a neuromodulator. *Annu. Rev. Neurosci.* 8: 103–124, 1985.
90. Stein, D.L. Description and occurrence of macrourid larvae and juveniles in the Northeast Pacific Ocean off Oregon, U.S.A. *Deep-Sea Res.* 27A: 889–900, 1980.
91. Stein, D.L. and W.G. Pearcy. Aspects of reproduction, early life history and biology of macrourid fishes off Oregon, U.S.A. *Deep-Sea Res.* 29: 1313–1329, 1982.
92. Suzuki, K. and Y. Taniguchi. Effect of pressure on biopolymers and model systems. In: *The Effects of Pressure on Living Systems*. edited by M.A. Sleigh and A.G. Macdonald. New York: Academic Press, 1972, pp. 103–124.
93. Swezey, R.R. and G.N. Somero. Pressure effects on actin self-assembly: interspecific differences in the equilibrium and kinetics of the G to F transformation. *Biochemistry* 24: 852–860, 1985.
94. Verjovski-Almeida, S., E. Kurtenbach, A.F. Amorim and G. Weber. Pressure-induced dissociation of solubilized sarcoplasmic reticulum ATPase. *J. Biol. Chem.* 261: 9872–9878, 1986.
95. Weber, G. and H.G. Drickamer. The effect of high pressure upon proteins and other biomolecules. *Quart. Rev. Biophys.* 16: 89–112, 1983.
96. Willis, J.M. and W.G. Pearcy. Vertical distribution and migration of fishes of the lower mesopelagic zone off Oregon. *Mar. Biol.* 70: 87–98, 1982.
97. Yancey, P.H., W. Bement and M. Maier. Temperature, pH and pressure effects on lactate dehydrogenases of vertically migrating midwater fishes. *Am. Zool.* 25: 19A, 1985.
98. Yancey, P.H. and J.F. Siebenaller. Coenzyme binding ability of homologs of M_4-lactate dehydrogenase in temperature adaptation. *Biochim. Biophys. Acta* 924: 483–491, 1987.
99. Yancey, P.H. and G.N. Somero. Temperature dependence of intracellular pH: its role in the conservation of pyruvate apparent K_m values of vertebrate lactate dehydrogenases. *J. Comp. Physiol.* 125: 129–134, 1978.
100. Zimmerman, A.M., S. Tahir and S. Zimmerman. Macromeolecular synthesis under hydrostatic pressure. In: *Current Perspectives in High Pressure Biology*, edited by H.W. Jannasch, A.M. Zimmerman and R.E. Marquis. London: Academic Press, 1987, pp. 49–63.

Species index

Acanthocybium solandri, 272, 301 f.
Acanthocybium sp., 272
Acanthonus armatus, 236
Acanthurus bahianus, 222
Acanthurus chirugus, 221
Acipenser transmontanus, 39
Aethotoxis mitopteryx, 229
Allocytus verrucosus, 230
Allothynnus sp., 272
Amia calva, 55, 143, 149
Ammodytes lancea, 234
Amphipnous cuchia, 141
Anabas scandens, 57
Anabas testudineus, 141
Anguilla anguilla, 55
Anguilla japonica, 228, 231
Anisotremus davidsoni, 216
Anomalops katoptron, 95
Anoplopoma fimbria, 220–222, 232, 234, 327, 336 f.
Antimora microlepis, 327, 331 f.
Antimora rostrata, 212, 214 f., 220, 232, 327, 329 f., 333–336, 338
Apogon ellioti, 98
Arapaima gigas, 213, 232
Arctocephalus townsendi, 219
Argyropelecus affinis, 81
Argyropelecus hemigymnus, 98
Aristomstomias scintillans, 91
Artemia, 15
Astyanax fasciatus, 87
Auxis sp., 272

Bairdiella icistia, 234
Balistes capriscus, 228
Barathrodemus iris, 214 f.
Bassogigas profundissimus, 211
Bassozetes sp., 214 f.
Bathygobius soporator, 110
Bathylagus bercoides, 89, 91
Bathylagus longirostris, 89, 91
Bathylagus milleri, 236
Bathylagus pacificus, 236
Bathylagus stilbius, 89

Bathylagus wesethi, 89, 91
Bathysaurus sp., 227
Benthosema glaciale, 229
Berryteuthis magister, 226
Blennius pholis, 57
Boleophthalmus pectinirostris, 57
Brachydanio rerio, 112–114
Brochis splendens, 231

Calanus plumchrus, 223
Callorhinus ursinus, 219
Callorhynchus milii, 57
Carassius auratus, 36, 55, 57, 111 f., 216
Caulolatilus princeps, 232
Centrolophus sp., 228, 230
Centrophorus granulosa, 228
Centrophorus spp., 225
Centroscymnus sp., 230, 233
Cetorhinus maximus, 230
Chaetodon ornatissimus, 232
Chaetodon trifascialis, 232
Channa gachua, 57
Chauliodus macouni, 236, 339
Cheilinus rhodochrous, 220
Chimaera barbouri, 226
Chimaera ogilbyi, 226
Chlamydomonas reinhardtii, 63
Cirrhitus pinnulatus, 220
Cirrhitus rivulatus, 221
Clarias batrachus, 141
Clarias gariepinus, 111–114, 124, 129
Clupea harengus, 55, 221, 235
Clupea harengus pallasi, 122
Cocculinella minutissima, 225
Coregonus clupeaformis, 231
Coryphaena hippurus, 235
Coryphaenoides abyssorum, 214
Coryphaenoides acrolepis, 214–216, 218, 220, 232, 327–332, 336 f.
Coryphaenoides armatus, 215, 218, 227, 327, 329, 331 f., 336 f.
Coryphaenoides carapinus, 327, 331 f.
Coryphaenoides fernandezianus, 215
Coryphaenoides leptolepis, 327 f., 331 f.

Coryphaenoides rupestris, 233, 327 f.
Cottocomephorus grewingki, 111 f., 124
Crithidia fasciculata, 63
Cubiceps gracilis, 229 f.
Cybiosarda sp., 272
Cyclopterus lumpus, 219, 230, 236
Cynoscion nebulosus, 235
Cypridina, 98 f.
Cyprinus carpio, 55, 57, 143, 231, 254 f., 261

Dalatias licha, 226, 228, 235
Dasyatis americana, 56
Dasyatis sabina, 47, 50–53, 56, 64
Deania sp., 230, 233
Delphinus delphis, 219
Dermochelys coriacea, 224
Dexea solandri, 230
Diaphus coeruleus, 98
Diaphus elucens, 98
Dicrolene intronigra, 214
Dissostichus mawsoni, 220, 228 f., 231
Drosophila melanogaster, 30, 39, 84

Electrona antarctica, 229
Electrophorus, 231
Engraulis mordax, 57
Engraulis ringens, 57
Enhydra lutris, 219
Eptatretus stouti, 143
Erignathus barbatus, 219
Erilepis zonifer, 230
Erythrinus erythrinus, 232
Escherichia coli, 46, 62 f.
Esox lucius, 55
Eumetopias jubata, 219
Euthynnus sp., 272
Euthynnus alletteratus, 235
Exallias brevis, 221

Fundulus heteroclitus, 35, 172

Gadus callarias, 55
Gadus macrocephalus, 55
Gadus morhua, 212, 234 f., 251–253, 258 f., 264
Galeocerdo cuvieri, 226
Gambusia affinis, 28
Gasterochisma melampus, 271–4, 277, 281, 292 f., 304–306
Genypterus maculatus, 67
Gobius jozo, 110–113, 123 f., 126, 128 f.
Gonatopsis makko, 226
Grammatorcynus sp., 272
Gymnosarda sp., 272
Gymnoscopelus nicholsi, 229

Halichoerus grypus, 219
Halosauropsis macrochir, 329, 331 f.
Hemiscyllium plagiosum, 57
Heteropneustes fossilis, 47, 49, 55, 141, 154
Hippoglossus hippoglossus, 234 f.
Hoplerythrinus unitaeniatus, 232
Hoplias malabaricus, 232
Hoplosternum thoracatum, 231
Hoplostethus atlanticus, 214, 221, 223 f., 229 f., 235
Hoplostethus gilchristi, 230
Hoplostethus islandicus, 230
Howella sherborni, 91
Hydrolagus colliei, 47, 50 f., 56–58, 143, 220, 226
Hydrolagus novaezealandiae, 226
Hyphessobrycon pulchripinnus, 128

Ictalurus natalis, 55
Ictalurus punctatus, 47, 55, 57, 228, 232
Istiophorus platypterus, 220, 272, 282
Istiophorus sp., 290
Isurus oxyrinchus, 228

Katsuwonus sp., 272
Kryptophanaron alfredi, 95

Laemonema (Podonema) longipes, 228
Laemonema morosum, 228, 230
Lamna ditropis, 305
Lampanyctus leucopsarus, 213, 230
Lampanyctus ritteri, 230
Latimeria chalumnae, 56, 64, 141, 143, 214, 219, 230 f.
Leiognathus equulus, 212
Leiostomus xanthurus, 212
Leishmania donovanii, 63
Lepidocybium flavobrunneum, 230
Lepidopsetta bilineata, 55
Lepidosiren paradoxa, 142, 213
Lepisosteus platostomus, 55, 143
Lepomis cyanellus, 229
Lepomis macrochirus, 47, 55
Leptocottus armatus, 57
Leptonychotes weddelli, 219
Leucicthys reighardi, 231
Leuresthes tenuis, 221
Limanda ferruginea, 35
Lophius setigerus, 228
Lotella phycis, 228
Lumpaneus maculatus, 232
Lutjanus campechanus, 235

Macruronus novaezelandiae, 235
Makaira indica, 272, 302

Makaira nigricans, 272, 275, 277, 280, 282, 297, 302
Makaira sp., 272, 290, 292
Malacocephalus laevis, 91
Malacosteus niger, 91
Mallotus villosus, 230, 234
Maurolicus muelleri, 229
Melanogrammus aeglefinus, 234
Melanonous zugmayeri, 221
Merlangius merlangus, 55, 234
Merluccius capensis, 235
Micropterus salmoides, 55, 58, 60, 142, 143
Microstomus achne, 228
Microstomus pacificus, 328
Mirounga angustirostris, 219
Mordacia mordax, 233
Morone saxatilis, 263
Mugil cephalus, 28, 235
Mugil japonica, 235
Mullus barbatus, 55
Myoxocephalus scorpius, 254, 256 f.
Mystus vittatus, 57

Negaprion brevirostris, 226
Neoceratodus forsteri, 142
Neoscopelus microchir, 98
Neurospora crassa, 63, 66
Nezumia bairdii, 327, 331 f.
Notothenia neglecta, 263

Odobenus rosmarus, 219
Odontaspis taurus, 231
Oncorhynchus gorbusha, 222, 227, 232
Oncorhynchus keta, 220, 222, 232
Oncorhynchus kisutch, 222
Oncorhynchus mykiss, 27, 32, 37, 140, 143, 231
  see also *Salmo gairdneri*
Oncorhynchus nerka, 28, 57, 222
Oncorhynchus tshawytscha, 47
Ophioblennius atlanticus, 227
Opsanus beta, 47, 55, 141, 143, 151 f., 156 f., 193, 195
Opsanus tau, 47, 53, 55, 141, 143, 151, 213, 262
Orcyopsis sp., 272
Oreochromis alcalicus grahami, 55, 57, 141, 154
Oreochromis nilotica, 55, 57
Orthopristis forbesi, 216, 217, 227
Osteapelta mirabilis, 225
Oxyjulus californica, 222

Pachystomias microdon, 91
Pagellus erythrinus, 28
Pagothenia borchgrevinki, 220, 229, 231, 233, 329, 331 f.

Parabassogigas sp., 215
Paralabrax clathratus, 232
Paralabrax nebulifer, 232
Paralepsis rissoi, 230
Parapriacanthus ransonnetti, 98
Parophrys vetulus, 328, 330
Peprilus simillinus, 220 f.
Perca fluviatilis, 55
Periophthalmus cantonensis, 57
Periophthalmus sobrinus, 57
Phoca vitulina, 219
Photobacterium leiognathi, 92–95, 97
Photobacterium phosphoreum, 92–95
Physeter catadon, 229
Pimelodus claria, 231
Pimelometopon pulcher, 221 f.
Platichthys stellatus, 47, 55, 57, 328, 330
Pleurogramma antarcticum, 220, 224, 229, 231
Pleuronectes platessa, 221, 234
Podonema longipes, 228
Poecilia reticulata, 57, 110, 140
Pollachius virens, 254
Polyodon spathula, 55, 143
Polypterus sp., 143
Pomoxis nigromaculatus, 47, 55
Porichthys notatus, 55, 58, 81, 98 f., 143
Potamotrygon hystrix, 57
Potamotrygon sp., 56 f., 146
Priacanthus arenatus, 222
Prionace glauca, 233, 297
Pristopomoides macrophthalmus, 221
Protomyctophum bolini, 229
Protopterus aethiopicus, 26, 142 f.
Protopterus annectens, 142 f.
Pseudocyttus maculatus, 221, 224
Pseudopleuronectes americanus, 34
Pterois radiata, 89
Pterois vetulus, 330

Rachycentron canadum, 235
Raja binoculata, 47, 56
Raja circularis, 56
Raja erinacea, 57, 141, 143, 147
Rastrelliger sp., 272
Rhinochimaera pacifica, 226
Rhizobium, 96
Rhomboplites aurorubens, 228
Rutilus rutilus, 55, 260
Ruvettus pretiosus, 221 f., 229 f., 272

Saccharomyces cerevisiae, 49, 62 f.
Salmo gairdneri, 55, 57, 175, 231 f., 261 see also *Oncorhynchus mykiss*
Salmo salar, 55

Salmo trutta, 33, 214 f.
Salvelinus namaycush, 149, 231
Sardina pilchardus, 234
Sardinops caerula, 233
Scaphirhynchus platorhynchus, 55, 143
Scardinus erithrophthalmus, 55
Schedophilis medusophagus, 220
Sciaenops ocellata, 235
Scoliodon terrae-novae, 56
Scomber scombrus, 249 f., 272
Scomberomorus cavalla, 272
Scomberomorus maculatus, 220, 272
Scomberomorus sp., 272
Scombrolabrax sp., 272
Scorpaena guttata, 329, 331 f.
Scyliorhinus canicula, 56, 264
Scyliorhinus torazame, 226
Sebastes caurinus, 47, 55
Sebastes melanops, 328, 331 f.
Sebastes pinniger, 328, 330
Sebastes ruberrimus, 220 f.
Sebastolobus alascanus, 329–333, 338
Sebastolobus altivelis, 89, 329, 331–333, 338
Sebastolobus sp., 338
Semicossyphus pulcher, 221, 232
Seriola mazatlana, 216, 227
Sicyases sanguineus, 57
Siluris glanis, 55
Solea vulgaris, 55
Sphoeroides annulatus, 216, 227
Sphyraena barracuda, 336 f.
Sphyrna tiburo, 56
Sphyrna tudes, 56
Sphyrna zygaena, 56
Sprattus sprattus, 234
Squalus acanthias, 47 f., 50–53, 56–59, 64, 67, 143, 147, 157, 194, 226 f., 230, 233, 235 f.

Stenella caeruleo-alba, 230
Stenobrachius leucopsarus, 230, 339
Stromateus maculatus, 228, 230

Tactostoma macropus, 236
Taeniura lymma, 56
Tarletonbeania crenularis, 339
Tetrapturus albidus, 220, 272, 275
Tetrapturus angustirostris, 272, 275
Tetrapturus audax, 272
Tetrapturus sp., 290, 292, 294
Thaleichthyes pacificus, 225
Thunnus sp., 272
Thunnus albacares, 272
Thunnus thynnus, 272, 306
Tinca vulgaris, 55
Trachurus symmetricus, 57
Trematomus bernacchi, 231, 233
Trematomus sp., 220
Trichiurus lepturus, 272
Trichogaster cosby, 235
Triphoturus mexicanus, 230

Urolophus jamaicensis, 56

Vargula, 98 f.
Vibrio fischeri, 92–97
Vibrio harveyi, 95 f.

Xenopus, 11
Xiphias gladius, 220, 270, 272, 275–277, 281, 292, 294, 296
Xiphophorus maculatus, 30

Yarella illustris, 98

Zalophus californianus, 219

Subject index

11-cis-3-dehydroretinal
  structure, 85
11-cis-retinal
  structure, 85
15-keto-prostaglandin $F_{2\alpha}$
  as olfactory stimulant, 117
  receptor, 120
$17\alpha,\beta$-dihydroxy-4-pregnen-3-one
  glucuronidation, 117
  in blood, 122
  milt production,122
  milt volume, 115
  oocyte maturation, 115
  plasma clearance, 128
  receptor, 120f., 126
  release of gonadotropin, 116
  release, 122
  source, 122
  synergistic actions with prostaglandins, 119
  testis, 115
17α-hydroxyprogesterone, 117
2,3-bisphosphoglycerate
  and Root effect, 212
2-oxoglutarate dehydrogenase
  substrate and product specificity, 186
3-ketoacyl-ACP synthase
  substrate and product specificity, 186
6-phosphogluconate dehydrogenase
  substrate and product specificity, 186
A-band, 260
A_1 adenosine receptor, 335
acceleration, 264
acetazolamide, 197
  carbonic anhydrase, 189
acetoacetate decarboxylase
  substrate and product specificity, 186
acetylcholinesterase, 318
activation volume, 326
acetyl CoA carboxylase, 188
  substrate and product specificity, 186
acid secretion, 177
acid–base regulation, 192
acidosis, 49
acrocentric chromosomes, 27

actin filaments, 254, 260
actin structure
  stability, 327
action potential, 254
active space, 129
active transport (see also individual ions)
  Cl^-, 167
  Na^+, 167
adenylate kinase
  in muscle, 263
adenylyl cyclase
adenosine receptor, 335, 338
adrenergic receptor, 86
aerobic capacity
  fast muscle, 262
  slow muscle, 262
  swimming, 249
aerobic metabolism, 264
  muscle, 262
air-breathing, 49, 53, 139, 141, 185, 197
air-exposure, 195
alanine
  in blood, 201
  in liver, 201
  in muscle, 200
alanine aminotransferase, 44
alanine transport, 201
alanine uptake, 201
alanine–glucose cycle, 138
alkaline lake, 141
alkaline phosphatase
  activation volume, 326
alkyldiacylglycerol
  density, 210
  for buoyancy, 209, 226
  in muscle, 230
  in skin, 228
  turnover, 210
  upthrust, 210
allantoic acid, 139
allantoicase, 44, 140
allozymes, 43
amiloride, 175
ammonia, 44, 188

acid–base, 193
anoxia, 193
catabolic product, 183
detoxification, 153 f.
dogfish pup, 145
efflux, 66, 200
excretion rates, 57
excretion, 49, 52, 137, 139, 141, 149, 182, 200
exercise, 193
formation, 66
from AMP, 150
metabolic substrate, 183
metabolism, 193
muscle load, 200
non ionic diffusion, 201
permeability in fish, 201
permeability, 66
pK', 181
proton flux, 66
reactions producing, 187, 193
release from muscle, 61
solubility in water, 181
tolerance, 195
toxicity, 153, 193
transport, 65, 182, 200
urea synthesis, 194
waste product, 184
ammonia detoxification, 48, 65
　birds, 65
　crocodiles, 65
ammonia transfer
　in blood, 201
　in liver, 201
ammonium, see ammonia buoyancy in squid, 237
AMP
　in muscle, 263 f.
AMP-deaminase, 194
　in muscle, 263
　substrate and product specificity, 187
AMP deamination, 200
anaerobic heterotrophs, 183
anaerobic metabolism, 263
　swimming, 249
androstenedione
　as olfactory stimulant, 123
　inhibitory function, 123
androstenedione, 111
anoxia, 200
antifreeze genes, 35
antifreeze protein, 34
antiport process
　definition, 167
aphotic zone, 79, 80

apolipoproteins, 233
aquatic turtles, 319
arginase, 59, 61, 138 f., 146, 148, 157
　compartmentation, 59, 66, 142–144
　gene, 66
　isozymes, 66
　Mn^{2+} dependence mRNA, 67
　subunits, 66
　synthesis, 67
arginine
　biosynthesis, 54
　essential amino acid, 140
　oxidation, 148
　substrate for arginase, 67
　synthesis, 140
　uptake, 148
asparagine deaminase
　substrate and product specificity, 187
aspartate, 152, 153, 156
　in muscle, 200
　transport, 151
aspartate aminotransferase, 46, 156
aspartate transcarbamylase, 49, 60, 63
　in invertebrates, 63
　in protozoons, 63
　shark spleen, 63
assembly
　volume exchange, 329
assembly–disassembly processes, 18
association–induction hypothesis (AIH), 12
atmosphere, 183
ATP
　and Root effect, 212
ATPase, 261
　anion-dependent, 167
autogenic, 97
autoinducer, 96
autosomes, 26

backflux, 168
bacteria, 91
bacterial bioluminesence (see also luminescence), 92 ff.
　characteristics, 93
　mixed function oxidase, 95
　occurrences, 92
　symbionts, 92
　topology, 92, 94
bacterial luciferase
　coenzyme, 95
　gene sequence, 95
　substrates, 95
bacterial symbionts, 92
bacteriorhodopsins, 87

behavioural thermoregulation, 296 ff.
β-glucuronidase, 113, 115
11β-hydroxytestosterone
   pheromone, 111
bicarbonate (see also CO_2 and HCO_3^-), 118 ff.
   as metabolic substrate, 183
   availability, 189
   consumption, 195
   cytoplasmic, 176
   dehydration, 198
   diffusion, 197
   during exercise, 189
   efflux, 196
   membrane permeability, 191, 196
   permeability, 182, 188
   plasma, 176
   transport, 197
biochemical inefficiencies, 321
bioluminescence (see also bacterial bioluminescence), 91
   evolution in fishes, 78
   gene, 78
   intensity, 81
   phylogeny, 78
   symbiotic, 78
blood
   lipids, 233
   solutes, 233
Bohr effect
   in swimbladder, 211
   pressure effects, 217
bone
   oil content, 220
   porosity, 220
   wax esters, 222
bone lipid
   synthesis, 222
   turnover, 222
   utilization, 222
bone ossification, 219 f.
'bound' ions, 16
boundary layer, 188, 197, 198
brain glutaminase, 154
brain heaters, 304
brain myelin
   membrane order, 334
brain, 154
branchial epithelia, 173
   cytoplasmic pH, 175
   Na accumulation, 176
buffering capacity
   fish muscle, 194
bulk phase, 14
bumetinide, 170

buoyancy, 209 ff.
   of blood, 233
   of eggs, 233

C-banding, 30
Ca-ATPase, 167
   density in heater organ, 285
   in heater organ, 283 ff.
calcium cycling
   in heater organ, 285–289
calcium channel regulation, 318
calcium diffusion, 318
calcium pump
   in heater organ, 283 ff.
calcium uptake, 259
   in heater organ, 284 ff.
camouflage, 82
carbamoylphosphate synthetase (see also CPS), 46, 188, 196
   ammonia as substrate, 58
   cDNA, 61
   compartmentation, 59, 145
   evolution, 46, 61 f., 144, 149
   function, 142
   glutamine dependence, 46, 58, 142
   in fishes, 55 f., 142
   in chloroplasts, 63
   in invertebrates, 58
   properties, 58
   regulation, 61, 142
   reptilians, 65
   substrate and product
   subunits, 62
   specificity, 62, 151, 186
   tortoise, 65
carbon dioxide (CO_2), 181 ff.
   and urea synthesis, 194
   as catabolic product, 183
   as metabolic substrate, 183
   as waste product, 184
   diffusion, 188, 196 f.
   efflux, 197
   erythrocyte, 188
   excretion, 182, 184, 197 f.
   generation, 196
   gradient, 188
   permeability, 183, 188
   permeability constant, 188
   pK', 181
   solubility in water, 181
   transport, 182 f., 188, 199
carbonic anhydrase, 145, 153, 184, 191
   compartmentation, 193
   fish liver, 189

in CO_2 excretion, 184
in cyanobacteria, 184, 188
in red blood cell, 197
pH dependence, 193
substrate and product specificity, 186f.
urea synthesis, 193
cardiolipin, 45
cartilage
  calcification, 219
  in teleosts, 219f.
catalase, 44, 145
catalytic efficiency, 43
caudal peduncle muscle fibre, 255
cell regulation, 16
cell volume, 16
cell water, 16
cell water rotation, 16
cell water translation, 16
cellular transport, 166
centromere, 27
cGMP, 85
channelling, 44, 147–149, 157, 195
chaperonin, 45
chloride (see also Cl^-)
chloride/bicarbonate exchange, 176, 197–199
chloride cells, 170, 173f.
chloride channel, 174
chloride efflux, 177
chloride transport, 198
cholesterol
  in fish liver, 228
  in plasma, 233
  in swimbladder, 214–216
cholesterol biosynthesis
  oxygen dependence, 217
choroid, 80
chromatophores, 100
chromophore
  isomerization, 85
chromosomal morphology, 25
chromosomal rearrangements, 26, 30
chromosome replication, 31
chrysopsins, 89
circannual cycle, 141
cistrons, 34
citrate synthase, 144
citrulline, 43ff.
  synthesis, 61, 147f., 157
  transport, 146
Cl^- channel, 177
  apical, 172
Cl^- diffusion potential, 170
Cl^- electrochemical gradient, 171
Cl^- flux, 171, 178

Cl-specific channel, 170
cluster model, 14
coevolution of luminescence and vision, 82
colligative properties of cytoplasm, 16
colloidal systems, 10
communication
  definition, 125
communication vs. spying, 127
compartmental isozymes, 45f., 52
compartmentation, 44, 59, 138, 147, 158, 191
cone pigments, 83
cones
  in retina, 82
consensus sequence, 45
continuum model (water), 3
contraction speed, 255, 258
contrast hypothesis
  in fish vision, 84
cooperativity of water molecules, 5
cornea, 80
  bioluminescent fishes, 81
  ontogeny, 81
cost of contraction, 264
counter-current
  heat exchange system, 270
  in swimbladder, 212
countershading, 82
covalent modifications, 192
CPS, 189, 194
  ammonia as substrate, 58
  cDNA, 61
  compartmentation, 59
  evolution, 62
  glutamine as substrate, 58
  immunological cross reactivity, 63
  in chloroplasts, 63
  in fish, 55f.
  in invertebrates, 58
  molecular evolution, 61
  properties, 58
  regulation, 61
  reptilians, 65
  subunits, 62
  tortoise, 65
creatine phosphokinase
  in muscle, 263
creatine, 139, 264
creatine phosphate (see also phosphocreatine), 264
crustecdysone
  behaviour regulation, 110
cyanobacteria, 183f., 188
cysts, 14
cytoskeleton, 6

cytomatrix, 17, 18
cytoplasmic water, 6

D-loop, 39
demersal fish, 323
density (water), 3
detoxification, 44
dexamethasone, 153, 156
diacylglyceryl ethers
  metabolism, 227
dielectric properties, 18
differential Q_{10} values, 318
dihydroorotase, 54, 60, 63
dipole–dipole interactions, 3
disomic segregation, 32
dispensable sequences, 38
diurnal fluctuations
  carbon dioxide, 182
  pCO_2, 182
  pH, 182
diurnal migrations, 275, 324
DNA content, 31
duplex retinas, 82
dysphotic zone, 79 f.

ectothermy, 269
  vs. endothermy, 320 f.
efficiency
  of muscle, 264
egg
  buoyancy, 234
  phospholipids, 234
ejection of 2-oxo-glutarate, 196
ejection of citrate, 196
electro-olfactogram (EOG), 113 f., 117
electrochemical gradient, 13, 166 f., 170
electrolyte flux, 166
electromyography, 258
electrostatic interactions, 18
EMG, 256
end plate receptor channel
  function, 318
  temperature effects, 319
end plate receptor, 318
  temperature effects, 318
endothermy, 269, 270
  and metabolic rate, 302 f.
  and muscle fibre, 304 ff.
  convergence, 299
  evolution, 271 f., 305
  in turtle, 224
  regional, 270, 295, 304
  whole body, 295
endplate, 254

endurance, 250
entropy of liquid water, 4
enzyme-complexes, 196
enzyme–modulator interactions
  pressure 329 f.
  Q_{10} effects, 316 f.
  temperature effects, 316
enzyme–substrate affinities
  pressure effects, 329 f.
  Q_{10} effects, 316
  temperature effects, 316
enzymes
  consuming ammonia 185, 187
  consuming CO_2, 185 f.
  producing ammonia 185, 187
  producing CO_2, 185 f.
epinephrine, 156
equilibrium distribution ratio, 168
estradiol
  as pheromone, 115
etiocholanolone glucuronide, 110, 123
euphotic zone, 79
evolution of metabolism, 44
excretion
  ammonia, 49, 52, 57, 137, 149, 182, 200
  proton, 175
  urea 52 f., 57
exons, 37
extraocular muscle
  heater organ, 273
  innervation, 273
eye lid, 82

fast twitch muscle, 252
fibre types, 252
fingerprints, 37
fish bioluminescence, 77
fish hemoglobin, 192
fish pigments
  selection pressure, 84
fish vision, 77
flanking region, 37
force–velocity curve, 259
force–velocity relationship, 254
Fourier law of conduction, 279
'free' ions, 16
frog skin, 176
  transport, 175
fructose bisphosphatase, 316
fuel
  in heater organ, 292
furosemide, 170
futile cycling, 149

G-protein, 85f.
  adenosine receptor, 335, 338
gangliosides, 334
gas bladder (see also swimbladder)
  as reflector, 100
gas gland
  cholesterol synthesis, 215
gene arrangements, 29
gene content, 26
gene duplication, 25, 31, 36, 46, 49
gene maps, 28f.
gene silencing, 33
gene structure, 37
genome
  evolution, 25
  organization, 25
gill lipids, 232
gill–water interface, 197
glucagon, 153, 155f.
  induction of urea cycle, 155
  regulation of CPS, 155
glucagon-like peptide, 156
gluconeogenesis, 44, 152, 189–191
glucose turnover, 303
glucose-6-phosphate
  in swimbladder, 212
glucose 6-phosphate dehydrogenase, 153
glucose–alanine cycle, 152
glutamate deamination, 66
glutamate dehydrogenase, 144, 149, 158, 199
  substrate and product specificity, 187
glutamate oxidation, 66
glutaminase, 139, 148f.,154f.,194
glutamine, 49
  analogs, 59
  oxidation, 148f.
glutamine deaminase
  substrate and product specificity, 187
glutamine shuttle, 154
glutamine synthesis, 154
  liver, 149
  muscle, 149
glutamine synthetase (GS), 44, 46, 49, 59, 61, 149–151, 153f., 156, 194f.
  ammonia binding, 48
  amphibians, 65
  brain, 47, 154
  compartmentation, 59, 158
  gene duplication, 49
  isozymes, 49
  kidney, 47
  K_m for ammonia, 61
  liver, 47
  mRNA, 50

  prokaryotes, 48
  sequence conservation, 50
  species-specific cleavage, 51
  subunit mass, 49
  tissue-specific expression, 50
  tortoise, 65
  zonation, 154
γ-glutamylhydroxamate, 47
glyceryl ethers
  and buoyancy, 226
glycine synthase
  substrate and product specificity, 186f.
glycogen
  synthesis in muscle, 189
glycogen particles, 9
glycogen phosphorylase, 317
glycogenolysis
  in muscle, 263
glycolysis, 183, 264
  in swimbladder, 211
gonadotropin (GtH), 115ff.
growth hormone (GH), 37
  cDNA, 38
  evolution, 35
  gene, 37
  mRNA, 38
  sequence conservation, 38
guanine
  in retina, 212
  in swimbladder, 212
guanine deaminase
  substrate and product specificity, 187
gynogenesis, 33
gynogenetic diploids, 29

H^+ gradient, 177
H-ATPase, 167
  electrogenic, 177
H-pumps, 177
hagfish erythrocytes, 198
HCO_3^- (see also bicarbonate)
  cytoplasmic, 177
  excretion, 195
  gradient, 177
  retention, 195
heat exchanger, 270
heat generation mechanisms, 321
heat loss, 269
heat shock proteins, 45
heater organ, 273
  adipose tissue, 276
  blood supply, 277f.
  eye muscle, 276f., 279
  heat exchanger, 278f.

heat generation, 279
mitochondrial density, 280f.
myofibrillar lattice, 277
oxidative enzymes, 281
phenotype, 290
scaling, 277
structure, 276
hemoglobin (Hb), 333
  Root effect, 212
heterochromatin, 28
heteromorphic sex chromosome, 27, 30
hexoseaminidase, 30
hexose monophosphate shunt, 183, 215
  in swimbladder, 211
high pressure neurological syndrome, 335
histidase
  substrate and product specificity, 187
homeobox complexes, 39
homeoviscous adaptation, 238, 318, 333f.
hormonal pheromones, 109
hormones
  as pheromones, 128
hybrid dysgenesis, 31
hybridogenesis, 33
hydration forces, 10
hydrodynamic constraints
  swimming, 249
hydrogen bond formation
  volume exchange, 326
hydrogen-bonding, 3, 16
hydrophobic effects, 10
hydrophobic hydration
  volume exchange, 326
hydrophobic interactions, 18
hydrostatic pressure (see also pressure), 323
hydroxylamine, 47
hypercapnia, 192
hypotonicity
  in buoyancy, 235ff.
hypoxanthine
  in swimbladder, 212
hypoxia, 200

IMP
  in muscle, 200, 263
innervation
  dual, 254
  focal, 254
  muscle fibres, 254
  mutiple, 254
  polyneural, 254, 256
inoculation
  with luminescent bacteria, 93f.
internal light organs, 100

intracellular architecture, 6
intracellular microviscosity, 15
intracellular water, 17
intraspecific communication
  evolution, 82
intravening sequences, 37
introns, 37f.
ion binding, 16
ion homeostasis, 192
ion selectivity, 12
ion transport, 13
ion-pair formation
  volume exchange, 326
ion-selective channels, 166
isocitrate dehydrogenase
  substrate and product specificity, 186
isolated fibre, 256
isomyosins, 261
Ito cells, 228

junction potential, 254

K^+ flux, 172
K^+ gradient, 178
  membrane potential, 178
K-channel
  basolateral, 172
K/Cl-cotransporter, 177
karyotype, 28, 31
karyotypic evolution, 26

L-amino acid oxidase
  substrate and product specificity, 187
lactate
  in muscle, 263
  in swim bladder, 211f.
lactate dehydrogenase (LDH, Ldh), 35
  activation volume, 326
  coenzyme binding, 329ff., 338
  evolution, 35
  homologs, 330
  in heater organ, 281
  isozymes, 35
  locus, 35
  pressure inactivation, 327–329, 338
  proteolytic inactivation, 327f.
  sequence conservation, 38
  substrate binding, 329ff.
  subunit dissociation, 326f.
  temperature effects, 339
lactate oxidation, 191
lactate turnover, 303
Lake Baikal, 91
lamprey erythrocytes, 198f.

lanosterol
　in cholesterol synthesis, 217
length–force curve, 259
length–tension relationship, 254
lens, 80, 83
　bioluminescent fishes, 81
　ontogeny, 81
Leydig cells, 110
light
　absorption, 79
　attenuation, 79
　downwelling, 79
　scattering, 80
light-chain, (see also myosin), 261
light emission, 82
light organ, 100
　location, 92, 94
light-organ symbioses, 92
linkage, 28 f.
linkage distances
　conservation, 30
lipid bilayer, 166
liquid structure, 3
liquid water, 3
liver
　buoyancy, 227
　heterogeneity, 154
locomotion, 250
luciferase (see also bacterial luciferase), 95, 99 f.
　evolution, 97
　occurrence, 98
　operon, 96
　origin, 98 f.
　substrates, 98
luciferin
　absorption, 99
　acquisition, 99
　biosynthesis, 99
　convergence, 99
　Cypridina-type, 98
　evolution, 99
　occurrence, 98
　recycling, 99
　transport, 99
　Vargula-type, 98, 99
luminescence
　bacterial, 100
　selection pressure, 79
　coevolution with vision, 80
luminescent display, 100
luminous bacteria
　inoculation of fish, 93 f.
lungfish, 26, 65, 144, 146, 155
lux operon, 96

malate dehydrogenase (MDH), 31, 46
　pressure effects, 331
malate–aspartate shuttle, 151, 191
male pheromone, 110
malic enzyme, 152 f., 156
　substrate and product specificity, 186
mate fidelity
　in goldfish, 125
melanin
　in pigmented layers, 80
melanin shutters, 100
membrane fluidity, 336
mesorchial gland, 110
metabolism
　swimming, 249
metacentric, 27
methioninesulfoximine, 151
micellar aggregation, 10
micelles, 9
microenvironment, 17
microtrabecular lattice (MTL), 6
microviscosity, 16
midwater fish, 324
migration
　diurnal 209 ff., 324 ff.
　ontogeny, 209 ff., 324 ff.
　reproduction, 324
mitochondria
　permeability, 44
　protein import, 45
mitochondrial density
　fast muscle, 262
　slow muscle, 262
mitochondrial DNA (mtDNA), 26, 34, 38, 273
mitochondrial fractional volume, 263
mitochondrial targetting, 63
mixture models, 3
mobile genetic elements, 30
molecular evolution, 431
monophyletic lineage
　of scombroid fishes, 273
motional properties, 12
mRNA transcript, 38
muscarinic receptor, 86
muscle biochemistry, 250
muscle fibre types, 252
muscle strain, 259 f.
musculature
　reduction for buoyancy, 230
myofibril diameter, 259
myofibrillar packing, 260
myofibrillar protein
　genes, 260
　isoforms, 260

myofibrils, 260
myoglobin
  muscle, 281
  heater organ, 281, 290
myosepta, 250f.
myosin filaments, 254, 260
myosin heavy chain, 260
myosin light chain, 260
  kinase, 261
  phosphatase, 261
  temperature adaptation, 262
myotomes, 250, 251

N-acetylglutamate, 54, 139, 142, 149
N-acetylglutamate synthase, 148, 157
Na absorption, 177
Na channel
  electrogeneic, 177
Na electrochemical gradient, 171
Na flux, 172
Na^+/K^+-ATPase, 167, 170, 173, 175
  activation volume, 326
  gill, 337
  pressure effects, 335, 337
  temperature effects, 337
Na^+/H^+ exchange, 200
Na^+/NH_4^+ exchange, 201
Na,K,2Cl-cotransport, 171, 174
NaCl
  branchial absorption, 175, 177
  bicarbonate, 175
NaCl cotransporter, 171
Nernst equation, 168
neutralization
  volume exchange, 326
NH_3 (see also ammonia and ammonium)
  in muscle, 263
nitrogen transport, 141, 150, 158
NMR titration model of water, 11
non-coding DNA, 39
non-coding regions, 37
nuclear DNA, 26, 31
nuclear genome, 26
nuclear targeting, 63
nucleolus organizer regions, 28
nucleoplasm, 6
null alleles, 32, 33

oculomotor nerve, 286
opsin, 84, 88
  phosphorylation, 85
opsin-shift, 86
organization of enzymes, 17
ornithine, 147
  channelling, 67
  oxidation, 148
  transport, 146
  transporter, 147
  uptake, 67
ornithine aminotransferase, 139, 147
ornithine decarboxylase, 148
ornithine transaminase, 44, 67
ornithine transcarbamylase, 46, 54, 59, 63, 67
  compartmentation, 59
  evolution, 63
  isozymes, 64
ornithine–urea cycle
  compartmentation, 137, 147
  evolution, 143
  genes in fish, 145
  in elasmobranchs, 150
  in fish ontogeny, 145
  metabolic cost, 145
  monophyletic trait, 59, 143
  stoichiometry, 150, 151
ornithine/citrulline antiport, 146
osmoregulation, 142
osmotic equilibrium, 17
osmotic pressure gradients, 13
ovulation, 113, 115, 117, 234
oxaloacetate
  transport, 191
oxidative muscle
  topology, 271
oxygen
  clathrate hydrate, 218
  in swimbladder, 211ff., 333
  role in luminescence symbiosis, 96
  solubility in swimbladder lipids, 218
  solubility in water, 182

P-element, 30, 31
parvalbumin, 259
passive diffusion, 166
Pasteur effect
  in swimbladder, 211
pCO_2
  fish blood, 182
pentose phosphate shunt (see also hexose monophosphate shunt)
  CO_2 generation, 197
PEP-carboxykinase (PEPCK), 44, 46, 190
  compartmentation, 191
  fish muscle, 189
  substrate and product specificity, 186
peptide assembly, 45
peroxisomes, 44
$PGF_{2\alpha}$

as olfactory stimulant, 117
  metabolites, 117
  receptor, 120
pH regulation, 192
pH_i
  muscle, 189
phase transition, 318
  lipids, 318
  proteins, 318
phenotype, 43
pheromone, 109
  definition, 125
  evolution, 126, 129 ff.
  preadaptation, 126
  release, 111
phosphate flux, 196
phosphate influx, 196
phosphate/hydroxyl transporter, 196
phosphocreatine (see also creatine phosphate), 264
  in muscle, 263
phosphofructokinase (PFK-1), 194, 316
phosphoglucomutase-1
  expression, 39
phospholipids
  in swimbladder, 214
phosphorylase b kinase, 317
phosphorylation-dephosphorylation reactions, 317
photobleaching, 6
photogenic tissue, 100
  hormonal control, 100
  nervous control, 100
photons
  processing, 80
photophores, 99
photoreceptor, 80
photosynthesis, 79, 184
photosynthetic bacteria, 183
pigment
  absorbance maxima, 83
  lens, 82
pigment patterns, 30
placental lactogen
  evolution, 35
pNH_3
  fish blood, 182
polar group hydration
  volume exchange, 326
polarization of water, 12
polyamine synthesis, 148
polyenic alcohol
  pheromone, 111
polymorphisms, 37

polyploidy, 31
pore water, 13
porphyropsin, 86
  absorption spectrum 85
postovulatory pheromone, 117
potential difference, 170
power output, 264
  fast fibre, 258
  isolated fibre, 256
  slow fibre, 258
pre-biotic, 183
predator–prey interactions
  evolution, 82
pressure effects
  A_1-adenosine receptor, 335 f
  catalytic rate, 329
  Na,K-ATPase 335–337
  ontogeny, 338
  substrate binding, 329 f.
  subunit assembly, 326 f.
  trans-membrane signalling, 335
  volume, 324
pristane, 210
prolactin
  evolution, 35
promoters, 37
propionyl CoA carboxylase, 188
  substrate and product specificity, 186
prostaglandin $F_{2\alpha}$
  release, 115
  synthesis, 115
protamine gene, 31, 38
protein
  subunit assembly, 326
  subunit dissociation, 326
  subunit stability, 326
  targetting, 44
  volume exchange, 326
proton
  ejection from mitochondrion, 191
  excretion, 175
  generation in mitochondrion, 191
pseudogenes, 35 f.
purine biosynthesis, 44
purine degradation, 139, 193
purine platelets, 100
  in reflectors, 80
purines
  as reflector, 100
pyrimidine biosynthesis, 54, 60
pyruvate
  excretion, 97
pyruvate carboxylase (PC), 185, 188 f., 191, 196
  activation volume, 326

fish muscle, 189
  substrate and product specificity, 186
pyruvate dehydrogenase (PDH), 192
  substrate and product specificity, 186
pyruvate kinase, 145, 189

Q-banding, 30
Q_{10}
  additive effect, 318
  differential effect, 318
  evolved property, 319f.
  for biochemical processes, 320
  for chemical processes, 320
  for physical processes, 320
  matching values, 320

rectal gland, 170, 172
red muscle, 252, 270
  distribution, 271, 299
red muscle fibre
  distribution, 295
  myoglobin, 295
  oxidative capacity, 295
reflective tissue, 100
reflector, 80, 100
regional duplication, 31, 34f.
regional endothermy, 270, 295, 304
repeated DNA, 36
repetitive DNA, 36
reproductive isolation, 31
reproductive success, 127
  and prostaglandin release, 128
respiratory acidosis, 195
rete mirabile, 213, 333
  brain, 271
  heat exchange, 271
  in red muscle, 296
  in swimbladder, 212
  muscle, 271
  viscera, 271
retina, 80f.
rhodopsin, 84, 86
  absorbance maxima, 90
  absorption structure, 85
  blue shift, 89
  evolution, 89
  molecular biology, 78
  phylogeny, 89
  rate of evolution, 87
  retinal binding, 87
  sequence conservation, 87
  size, 87
  spectral modulation, 89
RNA polymerase, 36

Robertsonian rearrangements, 28
rod pigments, 83
  absorption maxima, 84
rods
  in retina, 82
Root effect, 212
  pressure effect, 217, 333
rRNA genes, 39

salt balance, 166f.
sarcomere length, 260
sarcomere structure, 259f.
sarcoplasmatic reticulum, 259
  in heater organ, 282–284
scaling effects
  on muscle enzymes, 263
  on swimming speed, 250
secretory cycle, 173
selfish DNA, 26
seminal vesicle
  production of male pheromone, 111
seminal vesicle fluid, 114
semistability
  definition, 317
  Q_{10} effects, 317
sensitivity hypothesis
  in fish vision, 84
serine dehydratase
  substrate and product specificity, 187
sex chromosome, 30
sex linkage, 30
sex pheromone, 110
shutters
  in luminescent organs, 81
signal peptide, 45
signal peptidase, 45
signal specialization, 124
  selection pressure, 127
site-directed mutagenesis
  in rhodopsin, 86, 88
  in CPS, 62
solute
  lipid solubility, 166
  diffusion characteristics, 166
solute gradients, 174
solvent properties, 13
spawning behaviour, 111
species specificity
  of hormonal pheromones, 129
spectral quality, 83
  transmitted light, 82
sphingomyelin
  in swimbladder, 215
spying vs. communication, 126

squalene
    and buoyancy, 209, 225
    cyclization, 216
    density, 210, 216
    in shark liver, 225
    in teleosts, 225
    metabolism, 225
    turnover, 210
    upthrust, 210
steady-state swimming, 254
steroid glucuronides
    as pheromone, 113
    induction of ovulation, 113
steroidal pheromones, 113
stop codon, 36
streamlining, 301
sub-carangiform swimming, 254
subcellular localization (see compartmentation), 44
succinate
    in muscle, 200
superoxide dismutase
    in swimbladder, 214
surface tension, 8
surface to volume ratio, 2
swimbladder, 333
    cholesterol, 217
    fat-filled, 211, 214–218
    fat-invested, 211, 213 f.
    fatty acids, 217
    gas diffusion, 212
    gases, 211–213
    guanine, 212
    inflation, 213
    membrane lipids, 217
    muscle, 262
    oxygen secretion, 211–213
    oxygen tension, 211
    physoclist, 211
    physostome, 211
    sound production, 262
    surface tension, 213
    surfactants, 213
    tunica externa, 212
    volume, 213
swimming speed
    blue marlin, 302
    bluefin tuna, 302
    control, 250
symbionts, 94
symbioses, 95 f.
symbiotic, 91
symport process
    definition, 167

T tubule
    in heater organ, 282
tapetum, 80
targeting signals, 44 f.
TATA box, 38
telocentrics, 27
temperature (see also Q_{10})
    and power output, 300 ff.
    swordfish muscle, 296
temperature acclimation
    myosin light chain composition, 262
    myosin heavy chain expression, 262
    mitochondrial density, 263
temperature regulation
    in brain, 274
    in eyes, 274
testicular mesorchial gland, 124
testosterone glucuronides
    as pheromone, 115
testosterone, 111, 123
tetradecanal
    synthesis, 96
tetraploid, 31 f.
tetrasomic segregation, 32 f.
thermal anomalies, 8
thermal diffusion, 269
thermal hysteresis, 298
thermal stability, 43
thermodynamic constraints
    on transport, 167
thermogenic organ, 270, 271
threonine dehydratase
    substrate and product specificity, 187
thyroid hormone, 321
tight junction, 169
tonic muscle fibre, 252
trans-apical Cl^- flux, 174
transaminases, 193
transdeamination, 44, 199
transepithelial absorption
    Na^+, 170
    Cl^-, 170
transepithelial secretion
    Na^+, 169
    Cl^-, 169
translocation, 34, 44
    compartmentation, 44
    enzyme translocation, 44
transmembrane segments, 87
    of rhodopsin, 86
transmission genetics, 26
transport, 146
transposable elements, 38
transposition, 31

transposons, 30
triacylclycerol
  density, 210
  for buoyancy, 209
  in bone, 220
  in eggs, 235
  in muscle, 230
  turnover, 210
  upthrust, 210
trimethylamine oxide (TMAO), 233
  osmolyte, 236
troponin, 259f.
twitch duration, 258, 259
twitch kinetics, 254

uncoupling protein
  brown fat, 282
urate, 65
urea
  absorption, 53
  biosynthesis, 46
  excretion rates, 57
  excretion, 52, 53
  osmolyte, 46, 62, 141, 144, 150, 236
  permeability, 46, 61
  reabsorption, 46
  turnover, 67
urea cycle enzymes
  expression in teleosts, 53
  in fish, 55f., 141
  monophyletic trait, 59
  urea excretion, 137, 153
  ovarian fluid, 145
  permeability, 140, 142
urea synthesis, 44, 137, 142, 194
  acid–base regulation, 195
  bicarbonate, 195
  carbonic anhydrase, 193, 195
  embryo, 61
  osmolyte, 62
  teleosts, 141
ureagenesis
  carbonic anhydrase, 189
urease (microbial)
  substrate and product specificity, 186
ureosmotic fish, 46
uric acid (see also urate), 193
  excretion, 65, 139
  degradation, 44, 139
  synthesis, 48, 59
uricase
  substrate and product specificity, 186

vasoactive peptides, 156
vesicle stability, 10
vicinal water, 10, 18
  structure, 9
viscosity
  in fish blood, 233
vision, 77ff.
  coevolution with luminescence, 80
visual pigments, 83
visual systems
  adaptation, 79
  evolution, 78
vitamin A, 85
  in fish liver, 227
vitamin D
  in fish liver, 227
vitellogenesis, 234

water, 3ff.
  as solvent, 2
  intracelluar, 2
  light environment, 78
  pore, 13
  properties, 2
  pure, 2, 12
  solvent properties, 17
  structure, 4,5
  viscous resistance, 264
water–enzyme interactions, 17
wax esters
  density, 210, 223
  distribution, 223
  for buoyancy, 209
  in diet, 225
  in eggs, 235
  in muscle, 230
  in serum, 225
  in skin, 229
  in spermaceti oil, 224
  in swimbladder, 214
  in teleost liver, 228
  metabolism, 224
  phase transition, 223
  pressure effects, 223
  turnover, 210
  upthrust, 210, 225
weak bonds
  Q_{10} effects, 316
white fibre, 252

zonation
  liver, 154
zonula occludens, 169

WITHDRAWN